Chemistry of Biomolecules

Second Edition

Chemistry of Biomolecules

Second Edition

S. P. Bhutani

Formerly Associate Professor
Department of Chemistry, Rajdhani College
University of Delhi, India

CRC Press
Taylor & Francis Group
Boca Raton London New York

CRC Press is an imprint of the
Taylor & Francis Group, an **informa** business

CRC Press
Taylor & Francis Group
6000 Broken Sound Parkway NW, Suite 300
Boca Raton, FL 33487-2742

First issued in paperback 2022

ISBN-13: 978-0-367-20855-4 (hbk)
ISBN-13: 978-1-03-233795-1 (pbk)
DOI: 10.1201/9780429266423

Library of Congress Cataloging-in-Publication Data

Names: Bhutani, S. P., author.
Title: Chemistry of biomolecules / S.P. Bhutani, Formerly Associate Professor, Department of Chemistry, Rajdhani College, University of Delhi, India.
Description: Second edition. | Boca Raton : CRC Press, 2019. | Includes bibliographical references and index. | Summary: "Biomolecules are molecules that are involved in the maintenance and metabolic processes of all living organisms. This fully revised second edition offers extensive coverage of important biomolecules from an organic chemistry point of view. The author discusses carbohydrates, amino acids, peptides, proteins, enzymes, pyrimidines, purines, nucleic acids, terpenoids, and lipids. The various topics are described in simple, lucid language and explain the mechanisms of the reactions wherever required. Ideal for upper level undergraduates, graduates and researchers"-- Provided by publisher.
Identifiers: LCCN 2019030756 | ISBN 9780367208554 (hardback) | ISBN 9780429266423 (ebook)
Subjects: LCSH: Biomolecules. | Biochemistry.
Classification: LCC QD415 .B44 2019 | DDC 572--dc23
LC record available at https://lccn.loc.gov/2019030756

Visit the Taylor & Francis Web site at
http://www.taylorandfrancis.com

and the CRC Press Web site at
http://www.crcpress.com

Dedicated to my
Granddaughter
Little Princess 'Aarna'
with
Love and Blessings

Contents

4. Pyrimidines, Purines, and Nucleic Acids   **247-311**

Preface

The subject of organic chemistry is quite fascinating. The study of biomolecules becomes much more interesting because biomolecules constitute a large number of organic compounds with varying functional groups. We know biomolecules are involved in the maintenance and regulation of various metabolic processes of living systems. The study of these molecules becomes crucial because they are essential for sustaining life processes.

My aim in writing the second edition of *"Chemistry of Biomolecules"* has been the same as with the first edition, to make it an effective and up-to-date textbook to meet the needs of graduate students of chemistry, biochemistry and biology. In addition, the book will be an ideal reference book for the teachers of the same subjects. My experience of teaching these topics to undergraduate students at Delhi University for more than 37 years has prompted me to write this text.

The book retains many features of the first edition. In the first edition there were six chapters on biomolecules. However, in this second edition one more chapter on Akaloids has been added. Moreover, all the six chapters have been updated. Additions have been made in each chapter so as to make the book more comprehensive.

The **1st Chapter** on **Carbohydrates**, deals with the structures of glucose, fructose, disaccharides-maltose, lactose, cellobiose, and sucrose and polysaccharides – cellulose and starch. The new additions include epimers, reactions of monosaccharides, the polysaccharides – glycogen, chitin and heparin and biosynthesis of carbohydrates. At the end of the chapter some natural products derived from carbohydrates like ascorbic acid, anthocyanins, anthraquinone glycosides, cardiac glycosides, carbohydrate antibiotics *etc.* have been described.

The **2nd Chapter** is on **Amino acids, Peptides, and Proteins**. It describes the usual chemistry of these topics. In this book, the new additions are biosynthesis of amino acids; resolution of racemic mixture and stereoselective synthesis of amino acids, introduction to chromatographic methods of separation of amino acids, the Fmoc method of solid phase peptide synthesis, some interesting peptides like aspartame, glutathione, bradykinin, oxytocin and insulin. Lastly, molecular shape of proteins, factors that influence molecular shape, some features of fibrous and globular proteins are the new topics, which have been added in this 2nd chapter.

Chapter 3 is on "**Enzymes**. The new additions include cofactors-coenzymes; the structures of important coenzymes like NAD^+, $NADP^+$, FMN and FAD, TPP, pyridoxal-5-phosphate, and Coenzyme A. Also introduction to Biocatalysis and Green Chemistry is given in this chapter.

The **4th Chapter** deals with the study of **Pyrimidines, Purines, and Nucleic Acids**–DNA, RNA. In addition to a detailed chemistry of purine and pyrimidine bases, an introduction to the genetic code, transcription, translation and gene therapy is given here.

The additions in the **5th Chapter** on **Lipids-Oils and Fats** include an introduction to waxes and glycolipids, the biological role of fatty acids; the structure and importance of cholesterol and triglycerides, lipid membranes, and liposomes.

The **6th Chapter** is on **Terpenoids**, which deals with the chemistry of some monoterpenoids like citral, dipentene, geraniol, α-terpenol, menthol, α-pinene and camphor. Now in this book the chemistry of selected diterpenoids, triterpenoids and carotenoids has been discussed. In this chapter I have added the chemistry of phytol, squalene, β-carotene, α-carotene, γ-carotene and lycopene.

The last **Chapter 7** on '**Alkaloids**' has been added in this edition. Alkaloids are important physiologically active biomolecules occurring in plants. The chemistry of some typical alkaloids namely Coniine, Nicotine, Hygrine, Atropine, Cocaine, Quinine and Morphine have been described.

I have made an attempt to present the topics in simple and lucid language. Mechanisms of the reactions have explained wherever required. A classroom touch has been given in developing each topic. I hope the book will fulfill the requirements of undergraduate/graduate students of almost all global universities and colleges.

I express my sincere gratitude to my wife Komal, son Nikhil and daughter-in-law Pallavi for their moral support, patience and understanding.

I wish to express my thanks to Lafoe Hilary, Senior Acquisitions Editor and Jessica Poile, Editorial Assistant, Chemistry for the help and cooperation extended to me for the preparation of this book. I am especially grateful to Lafoe Hilary for getting the book reviewed by the concerned professors and for the useful suggestions made by them.

I am also indebted to Professor J.M. Khurana of Department of Chemistry, University of Delhi for giving useful inputs for the improvement of the book.

I invite constructive comments and valued suggestions from the students and learned teachers for improving the book in subsequent editions.

S.P. Bhutani
New Delhi, India

Carbohydrates

Learning Objectives

In this chapter we will study

- Carbohydrates – Occurrence in Nature and Definitions
- Formation of Carbohydrates in Plants
- Classification of Carbohydrates-Monosaccharides, Oligosaccharides and Polysaccharides
- Configurational Relationship of Monosaccharides—Aldotetroses, Aldopentoses and Aldohexoses
- Epimers
- Structure and Properties of D-Glucose—Open-Chain Structure, Cyclic Structure, Configuration of D-Glucose, Haworth-Projections, Conformations of D-Glucose
- Determination of Ring size of D-Glucose
- Mutarotation and its Mechanism
- Osazone Formation and its Mechanism
- Ketohexoses-Structure of D-Fructose
- Reactions of Monosaccharides
- Lengthening and Shortening of Carbon Chain of Aldoses
- Interconversions
- Structure of Various Disaccharides—Maltose, Lactose, Cellobiose and Sucrose
- Aldopentoses—The sugars of Nucleic Acids
- Polysaccharides-Starch, Cellulose, Glycogen, Chitin and Heparin
- Biosynthesis of Carbohydrates
- Some Natural Products derived from Carbohydrates including Ascobic Acid, Anthocyanisns, Cardiac Glycosides *etc.*

1.1   INTRODUCTION

Carbohydrates are one of the essential food ingredients which we all require. Every one of us is quite familiar with the words glucose and cane sugar. They are simple carbohydrates. We consume carbohydrates in one form or the other every time we take our meals, whether it is breakfast, lunch or dinner. They are the primary source of energy *i.e* in our body and constitute an important part of any well-balanced diet. We encounter carbohydrates at every turn of our lives. The paper we use for writing, the cotton clothes we wear and the wooden furniture around us are all made of cellulose. The bread we eat, rice, potatoes, peas, etc., all contain starch. The sweetening agents in fruits are nothing but simple carbohydrates.

1.2   OCCURRENCE AND BIOLOGICAL IMPORTANCE

Carbohydrates are the most widespread organic compounds occurring in nature. They are present in almost all the plants and comprise about 80 per cent of the dry weight. The most important are cellulose—the chief structural material of plants, starches and the sugars—sucrose and glucose. In higher animals and human beings, glucose is an essential constituent of blood and takes part in various metabolic reactions. Carbohydrates occur in a bound form in biologically important compounds like adenosine triphosphate (ATP), which is a key material in biological energy storage and transport system. ATP is called the *"energy currency"* of the cell. Carbohydrates are also present in nucleic acids, which control the production of enzymes and the transfer of genetic information. We shall discuss nucleic acids in Unit 4 of this book.

Carbohydrates are the major source of energy. We usually think of carbohydrates as "quick energy". Our body can mobilise carbohydrates more easily than fat, even though fats contain more energy on a gram-for-gram basis. Free glucose in the blood stream is quickly depleted, but we have stored glucose in readily accessible form as the polymer-glycogen. When we require energy glycogen breaks down to glucose in a quick response to the need for energy. Oligosaccharides play a key role in processes that take part on the surfaces of the cells. Polysaccharides such as cellulose are essential components of grass and trees and other polysaccharides are major components of bacterial cell walls. Starch is the prinicipal food reserve of plants.

1.3   DEFINITIONS

Originally, the name carbohydrate was given to all such compounds having the general formula $C_x(H_2O)_y$, *i.e.*, they were considered to be hydrates of carbon. Now, we know that carbohydrates are not simply hydrates of carbon but have a variety of other structural features. They are usually defined as polyhydroxy aldehydes and ketones or substances that hydrolyse to yield polyhydroxy aldehydes and ketones. We shall see later that this definition is not entirely satisfactory as they exist primarily as hemiacetals and acetals or as hemiketals and ketals.

Polyhydroxy aldehyes are also called aldoses and polyhydroxy ketones are called ketoses. Here *ald-* is taken for aldehydes and *ket-* for ketones and *-ose* is the common suffix used for these.

Low molecular weight carbohydrates are also called as sugars or saccharides, *e.g.*, glucose and sucrose are simple sugars. More common sugars like glucose and fructose, and some less common sugars also occur in the combined state with various hydroxy compounds. Such derivatives are called glycosides and the non-sugar component is called 'aglycone'. When we have glucose as the sugar component, the compound is called glucoside, if fructose is present, it is a fructoside and so on.

Glycosides are widely distributed in plants and animals. Structurally, these compounds are related to simple methyl glucosides. When the sugar moiety forms an ether linkage with the aglycone, it is called *O*-glycoside and when there is a formation of a C—C bond, it is called *C*-glycoside.

1.4 FORMATION OF CARBOHYDRATES IN PLANTS

Carbohydrates are produced in green plants as a result of photosynthesis. It involves chemical combination or fixation of carbon dioxide and water by utilisation of light energy. In this process, water is oxidised to oxygen and carbon dioxide is reduced. The overall equation for photosynthesis can be written as follows:

$$x\, CO_2 + y\, H_2O + \text{solar energy} \rightarrow C_x(H_2O)_y + x\, O_2$$
$$\text{Carbohydrate}$$

Photosynthesis involves many individual enzyme-catalysed reactions, but all are not fully understood.

However, we know that photosynthesis begins with the absorption of light by the green plant pigment, chlorophyll. The chlorophyll has the ability to absorb sunlight in the visible region. This is due to the extended conjugated system present in chlorophyll. As photons of sunlight are trapped by the chlorophyll, energy becomes available to the plant in a chemical form that can be used to carry out series of reactions that reduce carbon dioxide to carbohydrates and oxidise water to oxygen. The first reduction product of carbon dioxide is not known definitely but seems to be closely related to D-glyceric acid (1)

$$\begin{array}{c} COOH \\ | \\ H{-}C{-}OH \\ | \\ CH_2{-}OH \end{array}$$
(1)
D-Glyceric acid

The plant carries out a series of enzyme-catalysed reactions which result in the synthesis of simple sugars like glucose and more complex ones like starch and cellulose. Glucose is the most imporant key product in photosynthesis. It is usually stored in the plants in the form of starch to serve as food energy or to support the framework of the plant and is present as cellulose (Fig. 1.1)

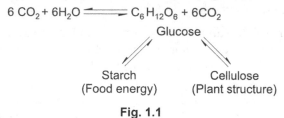

$$6\ CO_2 + 6H_2O \rightleftharpoons C_6H_{12}O_6 + 6CO_2$$

Fig. 1.1

Thus, in glucose nature stores the energy utilised in the photosynthetic process. This serves as the main source of energy in the living organisms. When we consume starch it is converted into glucose in our body and provides us energy. The remaining glucose in the body gets converted into animal starch, known as glycogen, which is stored in the muscle tissues. When we require energy at times when we are on fast, the glycogen is broken down once again into glucose molecules.

Besides providing energy, a major metabolic product of glucose is acetyl Coenzyme A, which acts as a starting point for the synthesis of fats, fatty acids, amino acids and a large number of other important compounds required by our body (Fig. 1.2)

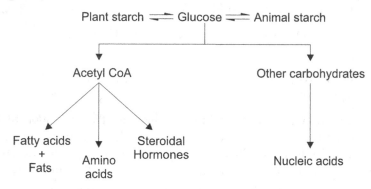

Fig. 1.2

1.5 CLASSIFICATION OF CARBOHYDRATES

Carbohydrates are classified on the basis of the number of aldose and/or ketose units produced upon hydrolysis (summarised in Table 1.1).

A. Monosaccharides

The carbohydrates, which cannot be further hydrolysed into smaller simpler units, are called monosaccharides. Examples of aldose monosaccharides are arabinose, glucose and mannose. The most common ketose is fructose.

Monosaccharides are further characterised in terms of the number of carbon atoms present in an aldose or a ketose. Thus, pentoses and hexoses contain five and six carbons respectively. For example, glucose is a hexose but it is completely characterised as an aldohexose, which means it is a six carbon sugar containing an aldehyde group. Similarly, fructose is a ketohexose and arabinose an aldopentose. In aldoses, the aldehyde group is at the top of the chain and the aldehyde carbon is considered as carbon number 1.

All naturally occurring ketoses have the keto group on the second carbon in the chain.

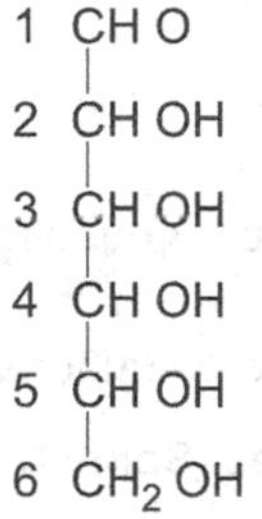

$$
\begin{array}{ll}
1 & CHO \\
2 & CH\ OH \\
3 & CH\ OH \\
4 & CH\ OH \\
5 & CH\ OH \\
6 & CH_2\ OH
\end{array}
\qquad
\begin{array}{ll}
1 & CH_2\ OH \\
2 & C{=}O \\
3 & CH\ OH \\
4 & CH\ OH \\
5 & CH\ OH \\
6 & CH_2\ OH
\end{array}
$$

An aldohexose 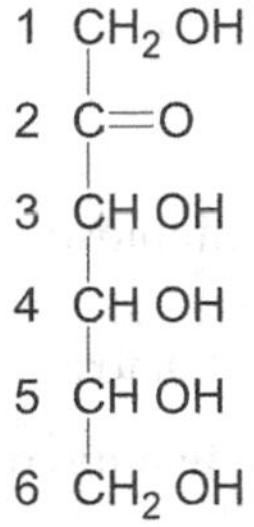   A ketohexose

The simplest sugars have three carbon atoms and are called trioses. The simplest aldose is glyceraldehyde, an aldotriose, and the simplest ketose is dihydroxy acetone, a ketotriose.

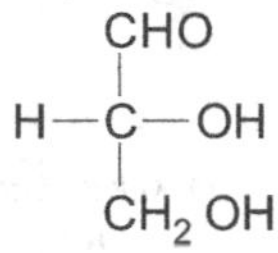

$$
\begin{array}{l}
CHO \\
H{-}C{-}OH \\
CH_2\ OH
\end{array}
\qquad
\begin{array}{l}
CH_2\ OH \\
C{=}O \\
CH_2\ OH
\end{array}
$$

Glyceraldehyde 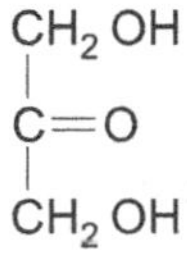   Dihydroxy acetone

Addition of an extra carbon atom to a triose gives successively a tetrose, a pentose, a hexose and so on. Most of the naturally occurring monosaccharides are either pentoses or hexoses. Trioses and tetroses do not occur in nature.

The most abundant pentoses are L-arabinose, D-ribose, 2-deoxy-D-ribose, and D-xylose, which are all aldopentoses (Table 1.2). The common hexoses are D-glucose, D-fructose, D-mannose and D-galactose. D-Fructose is a ketohexose whereas all others are aldohexoses (Table 1.3).

B. Oligosaccharides

These are low molecular weight condensation polymers containing 2-9 monosaccharide units. More often the monosaccharide units are hexoses.

Oligosaccharides can be further classified as disaccharides, trisaccharides, tetrasaccharides, etc., depending upon the number of monosaccharide units obtained on hydrolysis.

i. Disaccharides

A sugar that yields two monosaccharide units on hydrolysis is known as a disaccharide.

For example, sucrose, the common table sugar is a disaccharide because on hydrolysis it produces one mole of glucose and one mole of fructose. The disaccharides maltose and cellobiose both produce two moles of glucose on hydrolysis. They differ in the linkage between the two units. Lactose has glucose and galactose as the two monosaccharide units.

$$\text{Sucrose} \xrightarrow{\text{H}_2\text{O}} \text{Glucose} + \text{Fructose}$$

$$\text{Maltose} \xrightarrow{\text{H}_2\text{O}} \text{Glucose} + \text{Glucose}$$

$$\text{Cellobiose} \xrightarrow{\text{H}_2\text{O}} \text{Glucose} + \text{Glucose}$$

$$\text{Lactose} \xrightarrow{\text{H}_2\text{O}} \text{Glucose} + \text{Galactose}$$

TABLE 1.1. Classification of Carbohydrates (Some Common Examples)

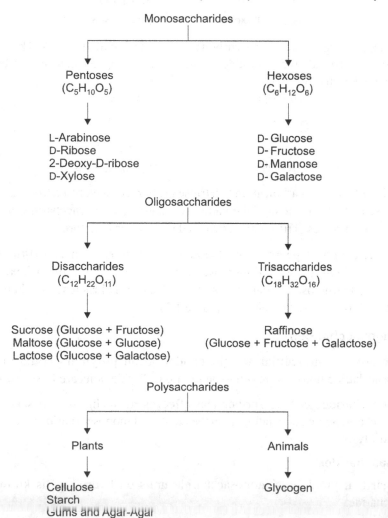

TABLE 1.2   Some Common Aldopentoses, their Occurrence and Structure

Occurrence of Pentose	Open Chain Structure	Cyclic Structure
L-ARABINOSE It is of wide occurrence in plants. It is usually found combined with other monosaccharides in gums and as glycosides.	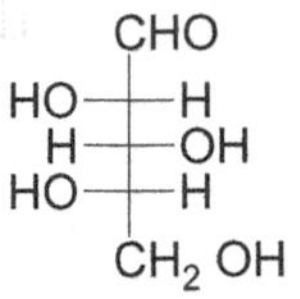	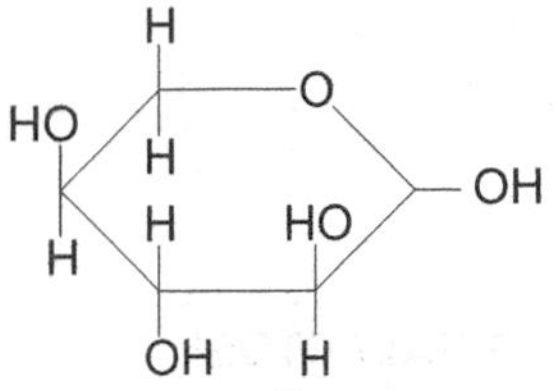
D-RIBOSE It is a component of the nucleic acid (RNA) and several vitamins and coenzymes. It is also a component of antibiotics Neomycin A and B.	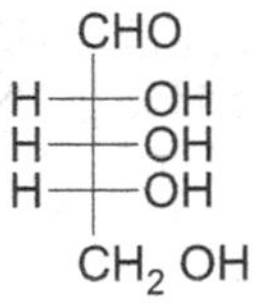	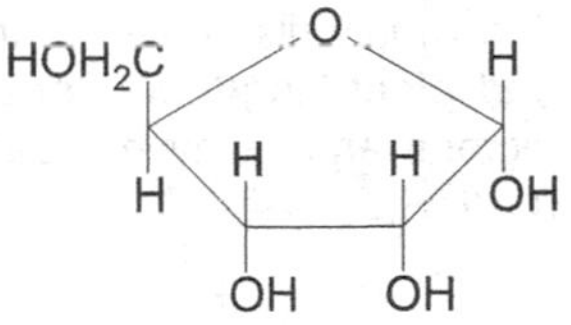 α-anomer
2-DEOXY-D-RIBOSE It occurs in cell nuclei as a component of deoxyribonucleic acid (DNA).	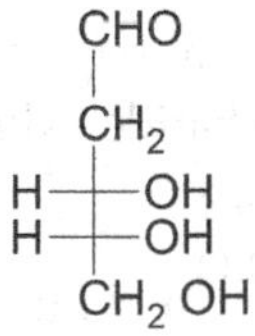	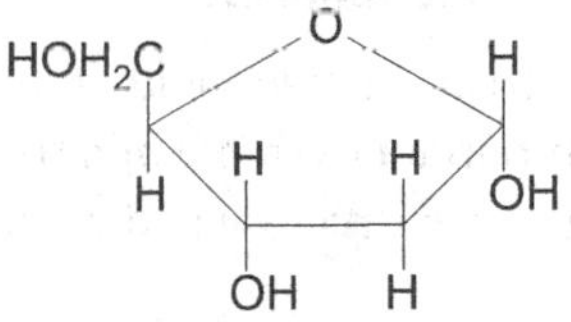 α-anomer
D-XYLOSE It is a component of the polysaccharides present in woody materials and of plant gums.	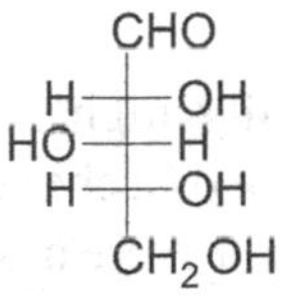	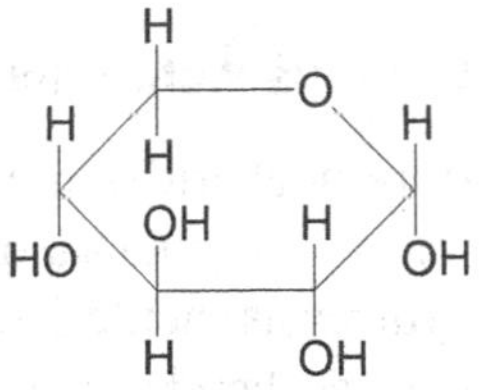 α-anomer

TABLE 1.3 Some Common Hexoses

Hexose	Open Chain Structure	Cyclic Structure
D-GLUCOSE It is the most abundant of all the hexoses. It occurs in the free state in many fruits, honey and plants, also in the combined state as glycosides. It is the basic unit of many polysaccharides like starch and cellulose. The most important disaccharide sucrose contains glucose as one unit.	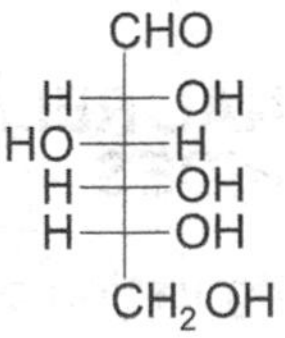	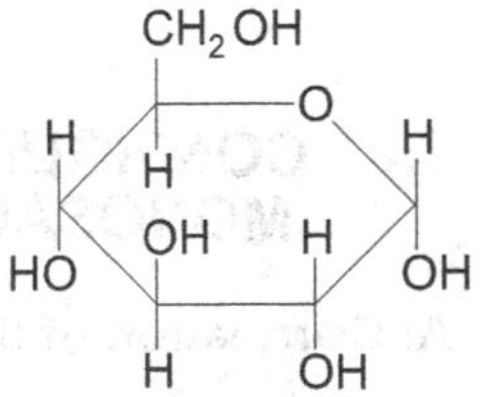 α-anomer
D-FRUCTOSE It occurs free in fruit juices and honey and in the combined state in sucrose and polysaccharides.	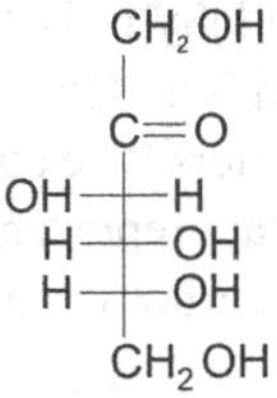	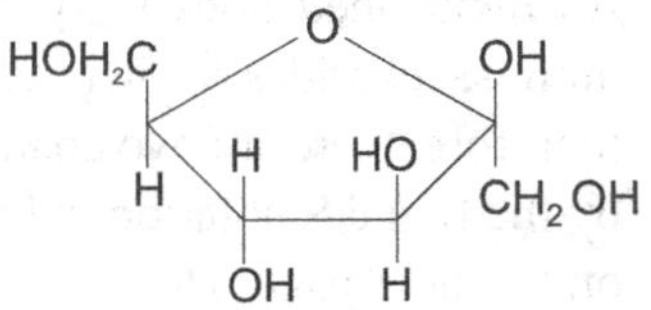 β-anomer

D-MANNOSE

It is reported to occur free in the peel of oranges and is a constituent of many polysaccharides.

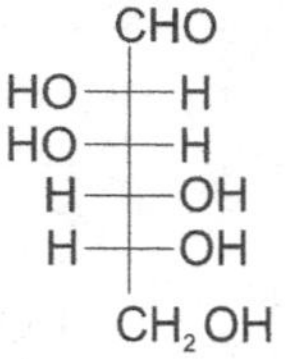

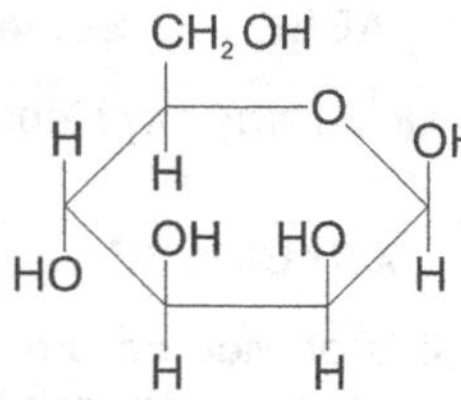

CHO
HO——H
HO——H
H——OH
H——OH
CH$_2$OH

CH$_2$OH

β-anomer

D-GALACTOSE

It is widely distributed in the animal and plant kingdom. Constituent of lactose and many polysaccharides (GUMS).

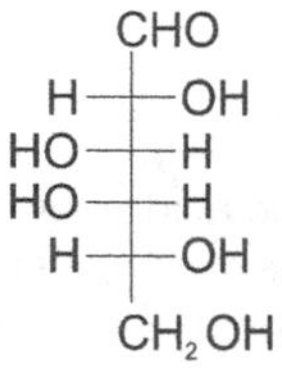

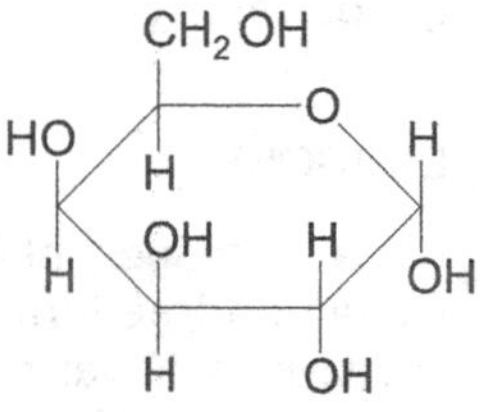

CHO
H——OH
HO——H
HO——H
H——OH
CH$_2$OH

CH$_2$OH

α-anomer

ii. Trisaccharides

A trisaccharide yields three monosaccharide units on hydrolysis. Raffinose is an important example of a trisaccharide, which on hydrolysis gives one mole each of glucose, fructose and galactose.

$$\text{Raffinose} \xrightarrow{H_2O} \text{Glucose} + \text{Fructose} + \text{Galactose}$$

C. Polysaccharides

Polysaccharides are generally high molecular weight polymers built up by repeated condensation of monosaccharide units. A polysaccharide on complete hydrolysis should furnish ten or more monosaccharide units. However, the natural polysaccharides consist of 100-3000 sub-units.

Polysaccharides are present both in plants and animals. Common examples of plant polysaccharides are cellulose, starches and the various gums. Glycogen is a polysaccharide present in animal tissues.

1.6 CONFIGURATIONAL RELATIONSHIP OF MONOSACCHARIDES

A. Correlation of the Configuration of D-Glucose and

D-Glyceraldehyde

As mentioned above glyceraldehyde is the parent compound of the class of monosaccharides known as aldoses. Glyceraldehyde contains one chiral centre, therefore, there are two enantiomers. The configuration of these has been described by the R and S notation, which are represented by the Fischer projections as given on the next page. However, in carbohydrate chemistry, it is customary to use the older system of representation, *i.e.*, D and L. According to that the Fischer projection is oriented in such a manner so that the most oxidised carbon is on the top. Now if the group of highest priority is on the right it is called D and when this group is on the left, it is called L.

$$
\begin{array}{ccc}
& \text{CHO} & \\
\text{H}- & \text{C} & -\text{OH} \\
& \text{CH}_2\text{OH} &
\end{array}
\qquad\qquad
\begin{array}{ccc}
& \text{CHO} & \\
\text{HO}- & \text{C} & -\text{H} \\
& \text{CH}_2\text{OH} &
\end{array}
$$

D(+)-Glyceraldehyde   L(−)-Glyceraldehyde
or   or
R(+)-Glyceraldehyde   S(−)-Glyceraldehyde

So under this convention all monosaccharides with the same configuration at the last chiral carbon atom as D-glyceraldehyde are called D-sugars. Those with opposite stereochemistry at the last chiral centre are called L-sugars.

$$
\begin{array}{ccc}
& \text{CHO} & \\
\text{H}- & \text{C} & -\text{OH} \\
& \text{CH}_2\text{OH} &
\end{array}
\qquad\qquad
\begin{array}{ccc}
& \text{CHO} & \\
\text{HO}- & \text{C} & -\text{H} \\
& \text{CH}_2\text{OH} &
\end{array}
$$

D-Sugars   Last   L-Sugars
chiral centre

D-Glyceraldehyde was chosen as an arbitrary standard by **Emil Fischer** for correlating the configuration of higher sugars. But his guess was later shown to be correct by an X-ray diffraction analysis of the tartarate salts.

We can extend the aldose family starting from glyceraldehyde. As naturally occurring sugars generally belong to the D-family, so we will consider here only those belonging to the D-series, although there are an equal number of enantiomers for each class belonging to the L-series (Fig. 1.5).

B. Aldotetroses

D-Glyceraldehyde may be stepped up to give two isomeric tetroses (2 and 3) called D-erythrose and D-threose. They have the same configuration about the last chiral carbon as does D-glyceraldehyde. The question remains- How to decide the configuration of the generated carbon atom or which is which?

D-Glyceraldehyde

$$
\begin{array}{cc}
\text{CHO} & \\
\text{H} \!-\!\!-\! \text{OH} & \\
\text{H} \!-\!\!-\! \text{OH} & \\
\text{CH}_2\text{OH} &
\end{array}
\qquad\qquad
\begin{array}{cc}
\text{CHO} & \\
\text{HO} \!-\!\!-\! \text{H} & \\
\text{H} \!-\!\!-\! \text{OH} & \\
\text{CH}_2\text{OH} &
\end{array}
$$

(2)   (3)

This is decided by oxidising the two tetroses to the corresponding dicarboxylic acids and determining the optical rotation of these acids. Thus, on oxidation D-erythrose forms meso-tartaric acid; so its configuration must be (2). On the other hand, D-threose (3) gives optically active dicarboxylic acid.

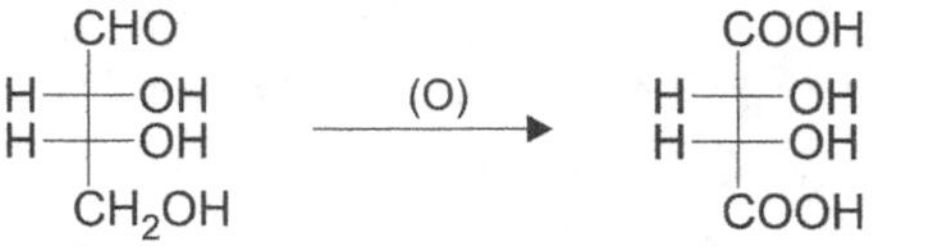

D-Erythrose   (O)   Mesotartaric acid

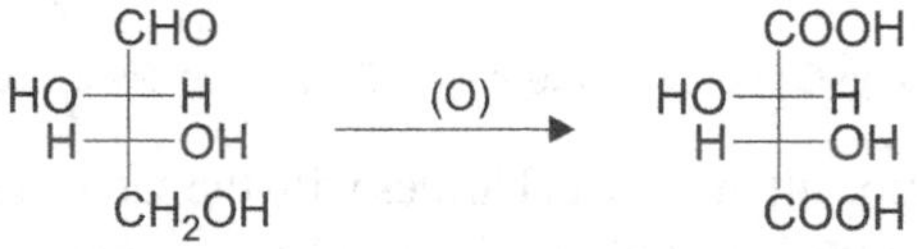

D-Threose   (O)   Optically active tartaric acid

C. Erythro and Threo Diastereomers

We know stereoisomers that are not mirror images are called diastereomers. They are either geometrical isomers or compounds containing two or more chiral centres. For example, 2-bromo-3- chlorobutane has two chiral carbon atoms. It can have 2 pairs of enantiomers as given below:

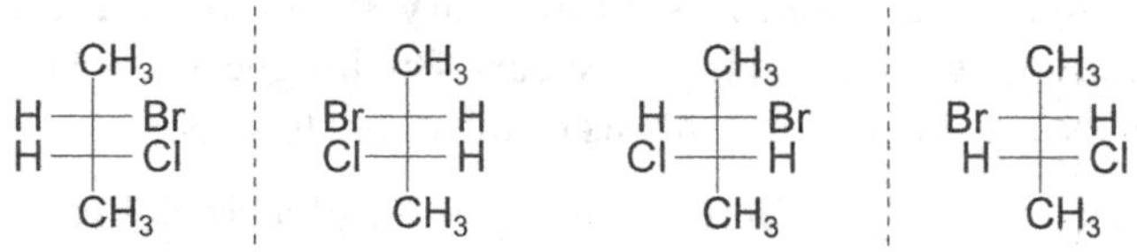

I   II   III   IV

Enantiomers   Enantiomers

Structures II and III are not the same. They are not enatiomers because they are not mirror images of each other. Since these compounds are stereoisomers, but not enantiomers, they are called diastereomers. Thus, we have two pairs of enantiomers. Either member of one pair of enantiomers is a diasteromer of either member of the other pair. We have seen earlier erythrose is the aldotetrose with the —OH groups of its two chiral centres situated on the same side of the Fischer projection and threose is the diastereomer with —OH groups on the opposite sites of the Fischer projection. These names have evolved into a method of naming diastereomers with two adjacent chiral carbon atoms. A diastereomer is called *erythro* if its Fischer projection shows similar groups on the same side of the molecule. It is called *threo* if similar groups are on the opposite sides of the Fischer projection. For example, hydroxylation of *trans*-crotonic acid gives two enantiomers of the threo-2, 3-dihydroxybutanoic acid whereas the same reaction with *cis*-crotonic acid gives the erythro enantiomers.

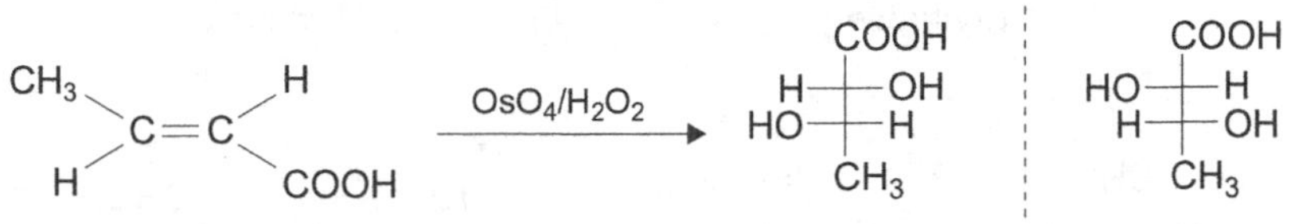

trans-Crotonic acid

Threo-2, 3-dihydroxy
butanoic acid

Each pair of *threo* is a diastereomer of each pair of *erythro* isomer. Similarly,

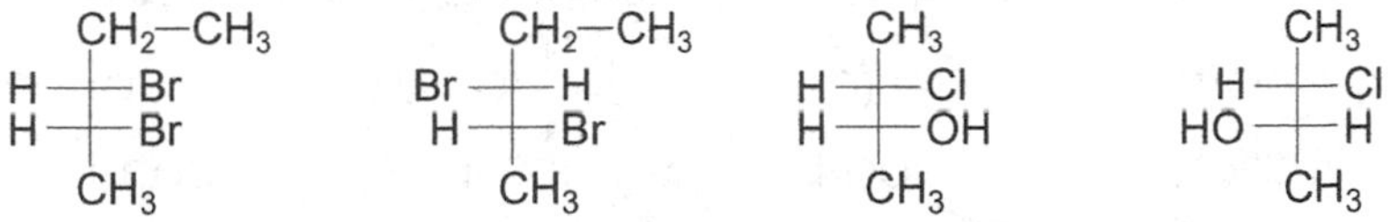

Erythro   Threo   Erythro   Threo

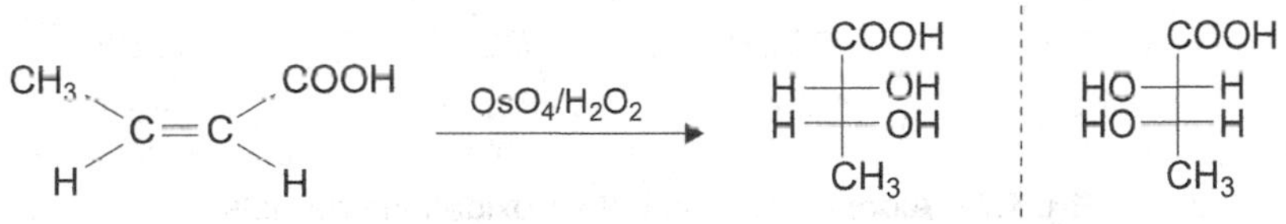

Erythro-2, 3-dihydroxy
butanoic acid

The terms erythro and threo are generally used only with molecules that do not have symmetric ends. In symmetric molecules such as 2, 3-dibromobutane and tartaric acid, the terms meso and (d, l) are preferred. The following examples show the use of these terms:

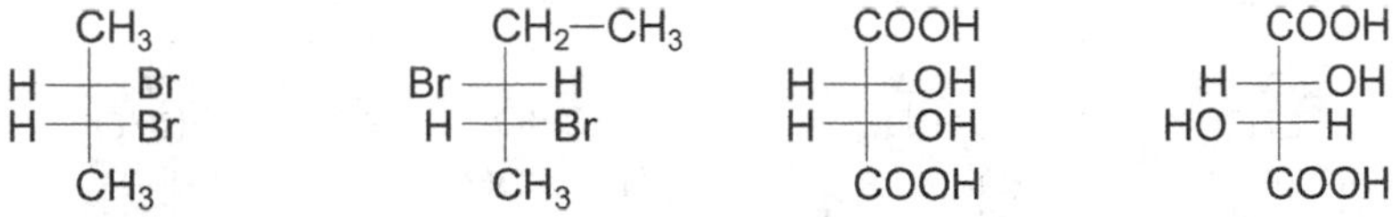

(meso)   (d, l)   (meso)   (d, l)

2, 3-Dibromobutane   Tartaric acid

D. Aldopentoses

There are eight optical isomers, *i.e.*, four pairs of enantiomers. Four belong to D-series and four to L-series. All are known. By Kiliani-Fischer synthesis (page 47) D-erythrose can be converted to D-ribose and D-arabinose whereas D-xylose and D-lyxose are obtained from D-threose by the same reaction.

On oxidation with nitric acid D-arabinose gives an optically active dicarboxylic acid whereas D-ribose gives an optically inactive dicarboxylic acid. Similarly, D-xylose on oxidation gives an optically inactive acid and the last D-lyxose affords an optically active dicarboxylic acid. Thus, the configuration of all aldopentoses belonging to D-family can be established (Fig. 1.3).

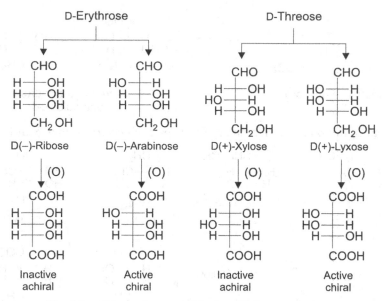

Fig. 1.3 Aldopentoses and their oxidation products.

E. Aldohexoses

Of the sixteen aldohexose stereoisomers, eight belong to the D-series. The configuration of these sugars has been worked out by Fischer. They are represented in Fig. 1.4.

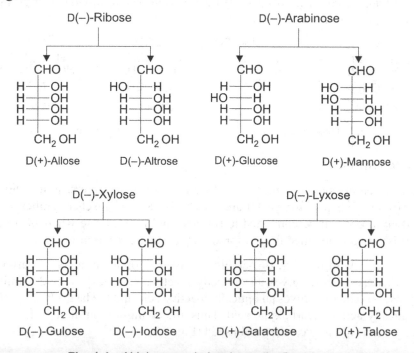

Fig 1 4   Aldohexoses belonging to the D-series

All these are diastereoisomeric. D-Arabinose gives D-glucose and D-mannose, which are epimers. Diastereoisomers which differ in configuration at one of the chiral centres are called epimers. Thus, D-glucose and D-galactose are also epimers.

Fig. 1.5 gives the correlation of D-glyceraldehyde with aldohexoses belonging to the D-Series.

1.7   EPIMERS

We have seen in the previous section that the pair of aldoses obtained from a lower sugar differ only in configuration about C-2. All these are diastereoisomers. Diastereoisomeric aldoses that differ only in configuration at C-2 are called **epimers**. For example, D-glucose and D-mannose are epimers because the configuration in these differs only about C-2, the first chiral centre.

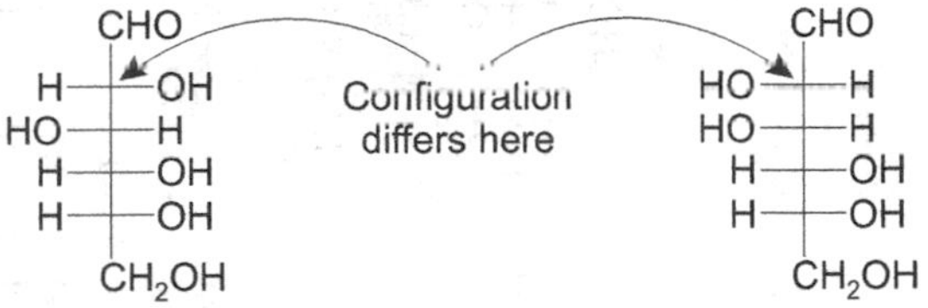

D-Glucose   Configuration differs here   D-Mannose

We will see that a pair of aldoses can be recognised as epimers not only by their conversion into the same osazone (Section 1.11) but also by their formation in the same **Kiliani-Fischer Synthesis** (Section 1.15.).

Many other common sugars are also closely related. They differ only by the stereochemistry at a single chiral centre. The term epimer now is used with a broader meaning. All such diastereoisomers, which differ by the stereochemistry at a single chiral centre, are called epimers. Thus, D-glucose and D-galactose are also epimers because they differ in configuration at C-4. The chiral centre where they differ is generally stated. If the number of chiral centre is not specified, it is assumed to be C-2. Therefore the C-4 epimer of D-glucose is D-galactose and the C-2 epimer of D-erythrose is D-threose.

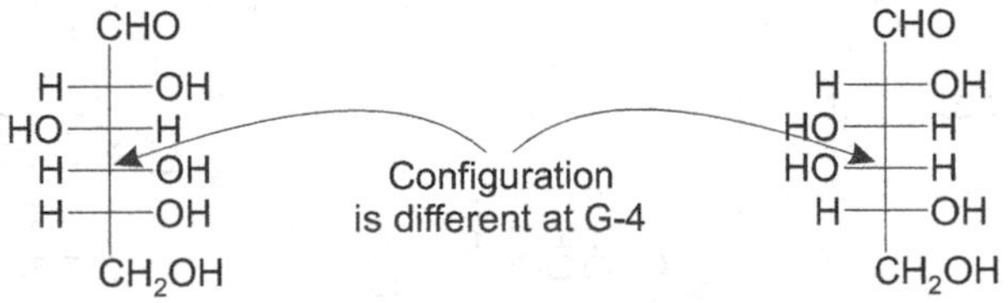

G-Glucose   Configuration is different at G-4   D-Galactose

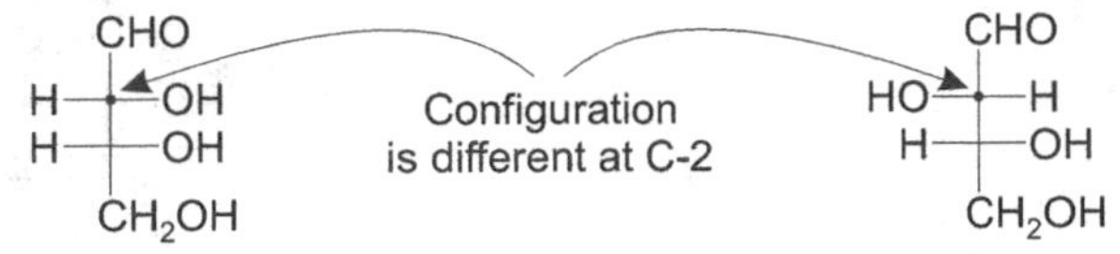

D-Erythrose   Configuration is different at C-2   D-Threose

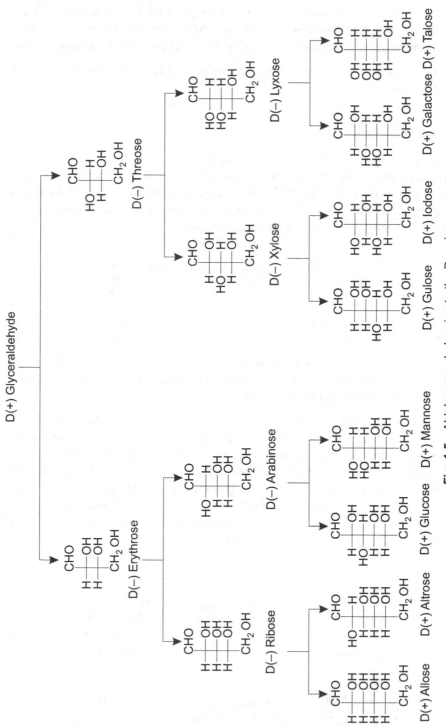

Fig. 1.5 Aldehexoses belonging to the D-series.

1.8   STRUCTURE AND PROPERTIES OF D-GLUCOSE

Glucose is the most impotant monosaccharide. It is by far the most abundant monosaccharide which occurs in nature. There are more (+)–glucose units in nature than any other organic molecule.

(+)–Glucose is typical of all monosaccharides. The chemistry of all other carbohydrates revolves around the glucose molecule. By learning about its structure and properties we can learn about the structure and properties of all other members of this family.

A. Open-Chain Structure of D-Glucose

The open-chain structure is derived from the following facts:

1. Glucose has the molecular formula, $C_6H_{12}O_6$ as shown by elemental analysis and molecular weight determination.

2. *Presence of an aldehyde group.*

   That the glucose molecule contains the carbonyl group is shown by the formation of phenylhydrazone and oxime.

   Mild oxidation of glucose with Br_2 water gives a monocarboxylic acid, known as gluconic acid.

3. Glucose is a reducing sugar which is shown by the fact that it reduces Tollen's reagent and Fehling solution. Both these tests are given by aldehydes and not by ketones.

4. On reduction with sodium-alcohol glucose gives sorbitol, a hexa-hydric alcohol.

5. *Oxidation with nitric acid:*

   Dilute nitric acid oxidises both the —CHO group and the terminal —CH$_2$OH group of an aldose to —COOH groups. These dicarboxylic acids are known as aldaric acids.

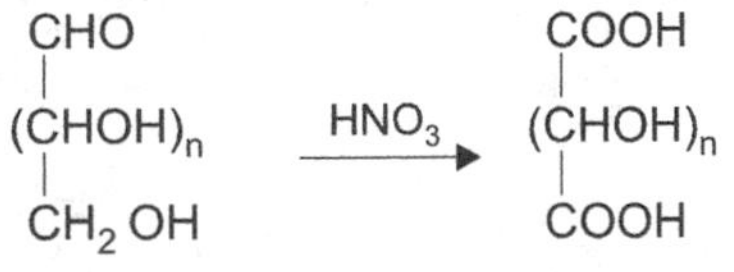

$$
\begin{array}{c}
\text{CHO} \\
| \\
\text{(CHOH)}_n \\
| \\
\text{CH}_2\text{OH} \\
\text{Aldose}
\end{array}
\quad \xrightarrow{\text{HNO}_3} \quad
\begin{array}{c}
\text{COOH} \\
| \\
\text{(CHOH)}_n \\
| \\
\text{COOH} \\
\text{Aldaric acid}
\end{array}
$$

   Oxidation of glucose with dilute nitric acid gives a dicarboxylic acid known as D-glucaric acid or saccharic acid.

6. *Reaction with HI/P:*

   Reduction of glucose with hydriodic acid and red phosphorus gives *n*-hexane.

The cyanohydrin, obtained from glucose, on hydrolysis followed by treatment with HI/P gives *n*-heptanoic acid.

All the above facts suggest that glucose is a six-carbon, straight chain pentahydroxy aldehyde *i.e.*, glucose is an aldohexose. The structure of glucose is thus written as (4).

CH O
|
CH OH
|
CH OH
|
CH OH
|
CH OH
|
CH$_2$ OH

(4)

All the above reactions of glucose are summarised in Fig. 1.6.

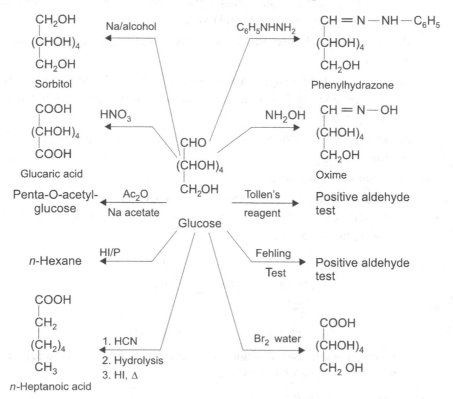

Fig. 1.6.   Reactions of glucose.

B. Configuration of D (+)-Glucose

The above structure (4) of the (+)-glucose shows that there are four chiral centres in the molecule and hence there should be $2^4 = 16$ stereoisomers, *i.e.*, eight pairs of enantiomers. All these 16 possible stereoisomers are now known. Some have been isolated from natural sources and others have been synthesised in the laboratory.

The question now is "which structure represents glucose out of these 16 stereoisomers?" The problem was solved by **Emil Fischer**. He began his work on the stereochemistry of (+)-glucose in 1888. He gave the configuration of (+)-glucose and was also able to synthesise and characterise all the other isomers. For this research work he was awarded the **Nobel Prize** in 1902.

Fischer had recognised that 16 stereoisomers consisted of two enantiomeric families. He arbitrarily guessed that (+)-glucose has the same configuration at C-5 as D(+) glyceraldehyde. The configuration of D(+)-glyceraldehyde was arbitrarily assigned as (5).

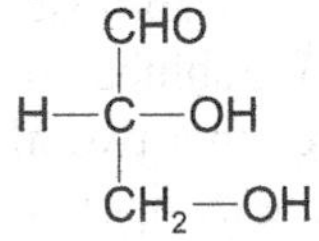

$$
\begin{array}{c}
CHO \\
| \\
H-C-OH \\
| \\
CH_2-OH
\end{array}
$$

(5)

So all sugars having —OH at C-5 on the right were designated as D-sugars and those with —OH at C-5 on the left were designated as L-sugars. Fischer, therefore, rejected eight enantiomers retaining only those in which —OH at C-5 were on the right. It was not until 1957, when **Bijuoet** determined the absolute configuration of D(+)-tartaric acid (and, hence of D(+)-glyceraldehyde) that Fischer's arbitrary assignment of (+)-glucose to the D-family was known to be correct.

Fischer's assignment of structure (6) to (+)-glucose was based on the following arguments:

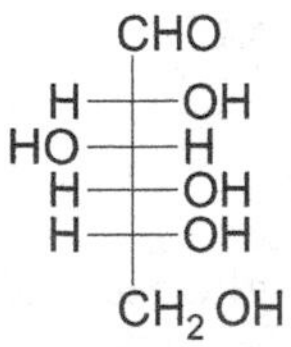

$$
\begin{array}{c}
CHO \\
H \!-\!\!\!-\!\!\!-\! OH \\
HO \!-\!\!\!-\!\!\!-\! H \\
H \!-\!\!\!-\!\!\!-\! OH \\
H \!-\!\!\!-\!\!\!-\! OH \\
CH_2 OH
\end{array}
$$

(6)

1. Out of the four possible aldopentoses belonging to the D-series, only D-arabinose could be converted to a mixture of D(+)-glucose and D(+)-mannose by Kiliani-Fischer Synthesis (*see* page 47). Here again Fischer had retained only those enantiomers in which the bottom chiral carbon atom

carried OH on the right. This means that at C-3, C-4 and C-5 (+)-glucose has the same configuration as that of arabinose.

2. Since (+)-glucose and (+)-mannose give the same osazone derivative, these sugars must have identical configurations at all chiral centres except C-2.

Sugars which have different configurations at one of the chiral centres are called epimers. Thus, Fischer realised that (+)-glucose and (+)-mannose were epimeric at C-2 (Fig. 1.7).

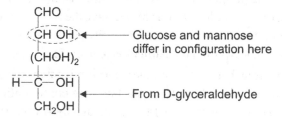

CHO

CH OH ←———— Glucose and mannose
differ in configuration here
(CHOH)₂

H—C—OH ←———— From D-glyceraldehyde
CH₂OH

Fig. 1.7. Configurations of (+) glucose and (+) mannose.

3. Nitric acid oxidation of (–)-arabinose gives an optically active aldaric acid. This indicates that (–)-arabinose must have either of the two possible configurations (7 or 8). The other two aldopentoses (9 and 10) would have afforded an optically inactive compound, *i.e.*, meso aldaric acids.

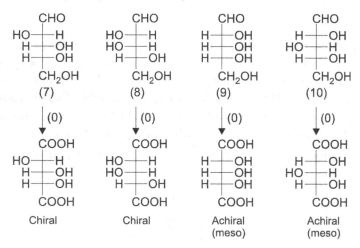

4. Nitric acid oxidation of (+)-glucose and (+)-mannose gives in each case an optically active aldaric acid. This means that —OH at C-4 in both these molecules is on the right. If it were on the left then at least one aldaric acid would have been achiral (meso). So, arabinose must have configuration (7) and not (8). This is illustrated by the followin scheme (Fig. 1.8).

Once the configuration of (–)-arabinose is decided as (7), (+)-glucose and (+)-mannose should have structures (11) and (12). The question remains, however, which one is which.

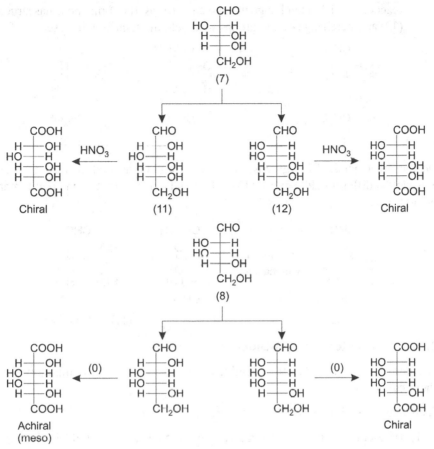

Fig. 1.8.

5. Another aldohexose (+)-gulose on oxidation with nitric acid gives the same aldaric acid as (+)-glucose. Only acid (14) can be derived from two different hexoses. This means that (+)-glucose must have configuration (11) and (+)-gulose the inverted configuration (13). (+)-Mannose, therefore, has configuration (12).

6. Fischer had already developed a method for effectively interchanging the two end groups —CHO and —CH₂OH of an aldose chain and with brilliant logic he arrived at the structure of (+)-glucose as (11). So the problem can be viewed by considering the overall effect on glucose and mannose of a sequence of

reactions which interchange the two end groups. If (+)-mannose has structure (12) an interchange of end groups will yield the same aldohexose.

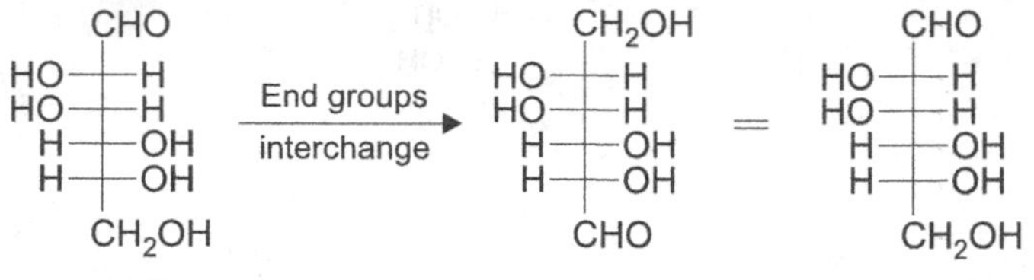

```
     CHO                          CH₂OH                      CHO
HO──┼──H                      HO──┼──H                  HO──┼──H
HO──┼──H     End groups       HO──┼──H                  HO──┼──H
 H──┼──OH   ───────────▶       H──┼──OH        =         H──┼──OH
 H──┼──OH    interchange        H──┼──OH                  H──┼──OH
    CH₂OH                         CHO                       CH₂OH

     (12)                                                    (12)
```

On the other hand, if (+)-glucose has structure (11), an end group interchange will yield a different aldohexose (13), which is L-sugar and it is an enantiomer of D(–)-gulose.

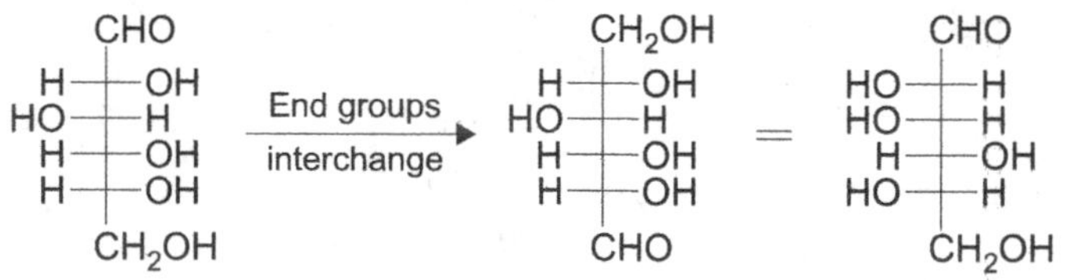

```
     CHO                          CH₂OH                      CHO
 H──┼──OH                      H──┼──OH                  HO──┼──H
HO──┼──H      End groups      HO──┼──H                   HO──┼──H
 H──┼──OH   ───────────▶       H──┼──OH        =          H──┼──OH
 H──┼──OH    interchange        H──┼──OH                  HO──┼──H
    CH₂OH                         CHO                       CH₂OH

     (11)                                              (13), L(+) Gulose
```

C. Cyclic Structure of D(+)-Glucose

The above structure (11) for D(+)-glucose could not account completely for its chemical behaviour.

Facts that are still to be accounted for are the following:

1. D(+)-Glucose fails to undergo certain reactions typical of aldehydes, *e.g.*, it does not form a bisulphite addition product and gives a negative test with Schiff's reagent, although it is readily oxidised.

2. The penta-acetate and the pentamethyl ether derivatives of glucose are not oxidised by either Tollen's reagent or Fehling solution or also do not form phenylhydrazone or bisulphite addition product.

   One of the ether groupings in pentamethyl derivative is highly reactive. It undergoes hydrolysis in dilute acids to a tetramethyl derivative. The tetramethyl derivative reduces Tollen's reagent as effectively as glucose itself.

3. D(+)-Glucose exists in two isomeric forms, which undergo mutarotation.

   When crystals of ordinary glucose, m.p. 146°C, are dissolved in water, its specific rotation is +112°. On keeping this solution the value drops to +52.5°.

   When glucose is obtained from fresh hot concentrated solution, its specific rotation changes from an initial value of +19° to an equilibrium value of +52.5°. This form of D(+)-glucose is termed as β-D(+)-glucose (m.p. 150°C). The α-isomer is the previously used isomer. The change in rotation of these to the equilibrium value is called mutarotation.

$$\alpha\text{-D(+)-Glucose} \longrightarrow +52.5° \longleftarrow \beta\text{-D(+)-Glucose}$$
$$\alpha\text{-isomer} \qquad\qquad\qquad\qquad \beta\text{-isomer}$$
$$[\alpha]_D = +112° \longrightarrow +52.5° \longleftarrow [\alpha]_D = +19°$$

4. D(+)-Glucose forms two isomeric methyl-D-glucosides; methyl α-D-glucoside and methyl-β-D-glucoside under conditions which normally convert an aldehyde to a hemiacetal or acetal derivative.

$$C_6H_{12}O_6 + CH_3\,OH \xrightarrow{\text{HCl}} (C_6H_{12}O_5) - OCH_3$$
Methyl-α- and β-
D-glucoside

We know aldehydes react with alcohols in the presence of anhydrous hydrochloric acid to form hemiacetal or acetal.

Hemiacetal   Acetal

These glucosides do not reduce Tollen's reagent and Fehling solution and do not undergo mutarotation. That means they are comparatively quite stable under mild acidic or mild basic conditions. Furthermore, they have different m.p. and specific rotation. The isomer of higher positive rotation is called methyl-α-D-glucoside (m.p. 165°C, $[\alpha]_D = +158°$) and the other methyl-β-D-glucoside has m.p. 107°C, $[\alpha]_D = -33°$. (*See* Table 1.4.)

TABLE 1.4   Properties of D(+)-Glucose and its Methyl Glucosides

Compound	m.p.	$[\alpha]_D$
α-D-(+) Glucose	146°C	+ 112°
β-D(+)-Glucose	150°C	+ 19°
Methyl-α-D-glucoside	165°C	+ 158°
Methyl-β-D-glucoside	107°C	- 33°

The above observations cannot be explained on the basis of the open-chain structure for the glucose molecule. It is obvious that the aldehyde group in the glucose molecule is not free but is tied up with one of the hydroxyl groups, which turns out to be at C-5 to form a hemiacetal structure as explained above. Thus, we get a cyclic structure for the glucose molecule.

A new chiral centre is created at C-1 by hemiacetal formation. There are, therefore, two stereoisomeric forms of D(+)-glucose. They are α-D(+)-glucose and β-D(+)-glucose, represented by structures (15) and (16). These are diastereoisomers differing in configuration at C-1. Such pairs of diastereoisomers are called **anomers**.

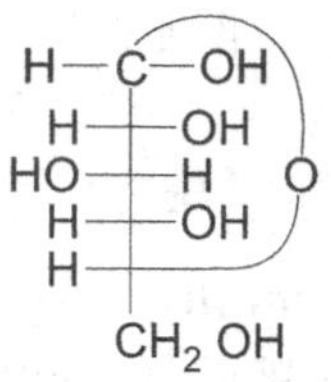

(15) α-D(+)-Glucose 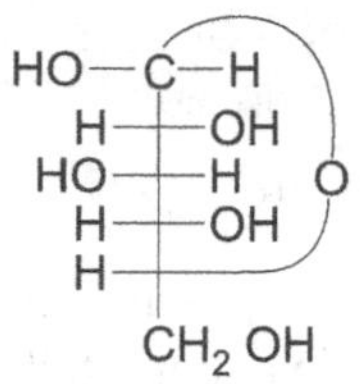   (16) β-D(+)-Glucose

X-ray analysis of D(+)-glucose indicates that in the α-anomer, the hydroxyl groups at C-1 and C-2 are *cis* to one another and *trans* in the β-anomer. The above representations are according to Fischer Projections. Hydroxyl groups on the same side in the Fischer Projection of the hemiacetal are *cis* to each other and those on the opposite side are *trans*. Thus, C-1 hydroxyl group is on the right in the Fischer projection of α-D(+)-glucose and on the left in β-D(+)-glucose.

i. Haworth Projections

Structures (15) and (16) drawn for the isomers of D(+)-glucose are not consistent with their molecular shapes. Haworth projections were given to overcome this difficulty. In this, hemiacetal structure is represented by a planar ring containing the oxygen atom.

To write the Haworth projection for a Fischer projection, we rotate the open-chain form in such a manner so that –OH at C-5 and the carbonyl group forming the hemiacetal are in close proximity. This is possible when the C_4–C_5 bond is rotated. After rotating, hemiacetal formation results in giving the Haworth structure. For example, we can write Haworth structure for the glucose molecule as given below.

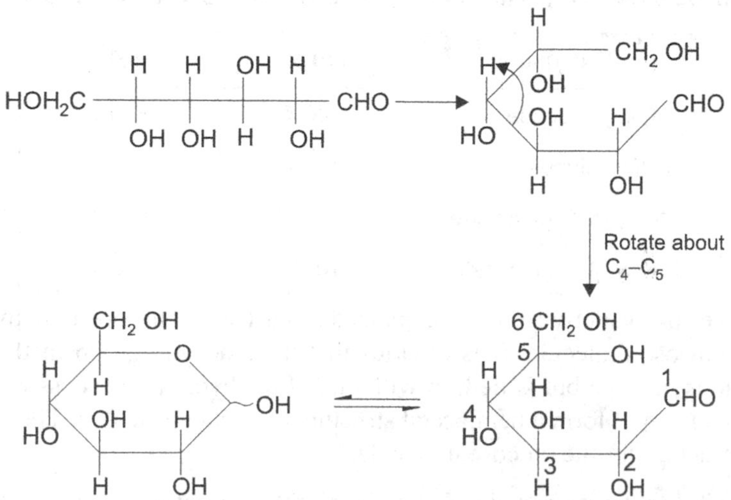

Rotate about
C_4–C_5

Haworth structure
of D-glucose

Thus, the Haworth structures for the α-D(+)-glucose and β-D(+)-glucose are represented by the structures (17) and (18) respectively. As they are derived from the parent heterocycle pyran, they are named α-D(+)-glucopyranose and β-D(+) glucopyranose. The corresponding five-membered cyclic hemiacetals are referred to as furanoses, named after furan.

In these projections, the —CH_2OH group is written as up for D-sugars and down for L-sugars. For D-sugars the —OH at C-1 will be down in the α-form and up for the β-form.

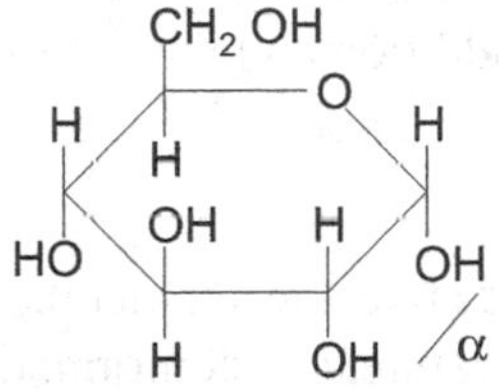

(17) α-D(+)-Glucopyranose 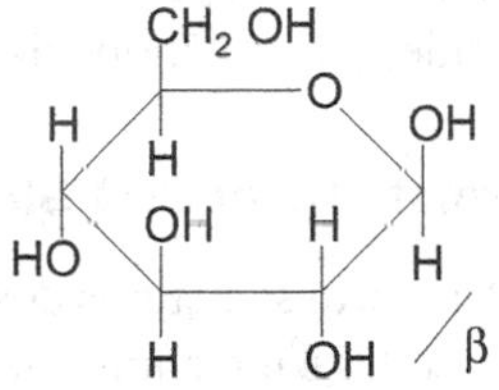   (18) β-D(+)-Glucopyranose

ii. Correlation between Open-Chain and Ring Structures

α- and β-D(+)-Glucose are readily hydrolysed by water. In aqueous solution either anomer is converted *via* the open chain form into an equilibrium mixture containing both the cyclic isomers. Thus, mutarotation results from the ready opening and closing of the hemiacetal ring.

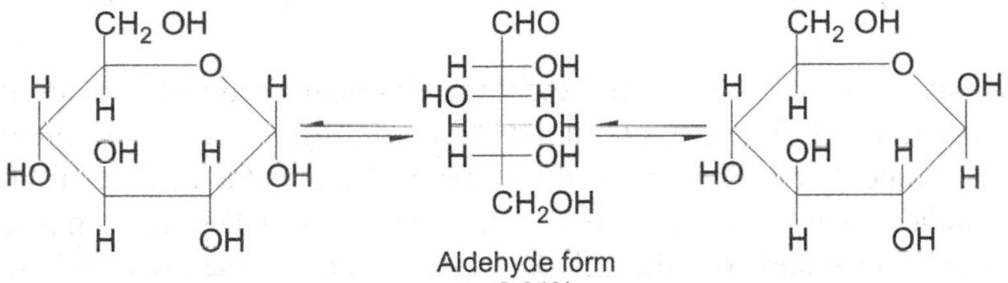

α-anomer
(37%)

Aldehyde form
~0.01%

β-anomer
(63%)

- No bisulphite

- No colour reaction with Schiff's reagent

- Oxidation by Tollen's reagent or Fehling solution

- Formation of oxime or osazone

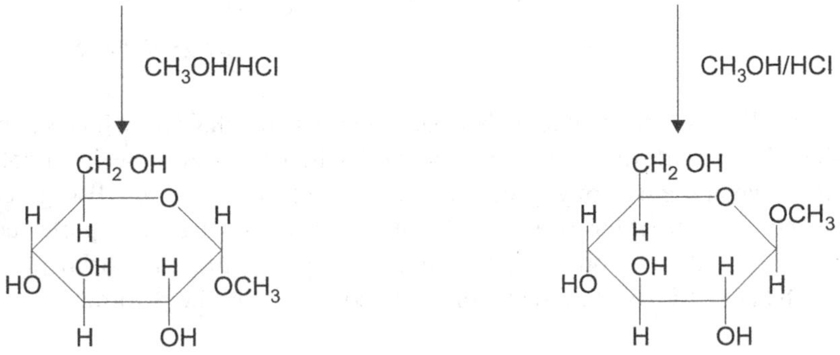

CH_3OH/HCl CH_3OH/HCl

Methyl-α-D(+)-glucoside   Methyl-β-D(+)-glucoside

The typical aldehyde reactions of D(+)-glucose, for example, the oxidation of glucose to gluconic acid (reduction of Tollen's reagent and Fehling solution), the formation of oxime or osazone and the conversion of glucose to a methyl glucoside are presumably due to the open-chain structure of the glucose. This is present in very low concentratins at equilibrium (~ 0.01%). As the aldehyde is consumed by the chemical reaction, it is rapidly replaced by opening of the α- or β-anomers. The concentration of this open-chain structure, is, however, too low for certain easily reversible aldehyde reactions like bisulphite addition and the Schiff's test.

D. Conformations of D(+) Glucose

Although Haworth structures describe sugars as five and six membered rings, they do not represent the most important structural feature, *i.e.*, conformations. We know that cyclohexane is not planar but has chair conformations. There is least strain and minimum repulsion in these conformations.

By analogy with cyclohexane we expect the six-membered cyclic oxide structure of glucose to exist preferentially in a chair conformation. To avoid interactions, bulky substituents prefer to be in the equatorial rather than the axial positions. X-ray studies have in fact established that crystalline α-D-glucose has the chair conformation (19) and crystalline β-D(+)-glucose has a structure (20) in which all the substituents can be seen to occupy equatorial positions.

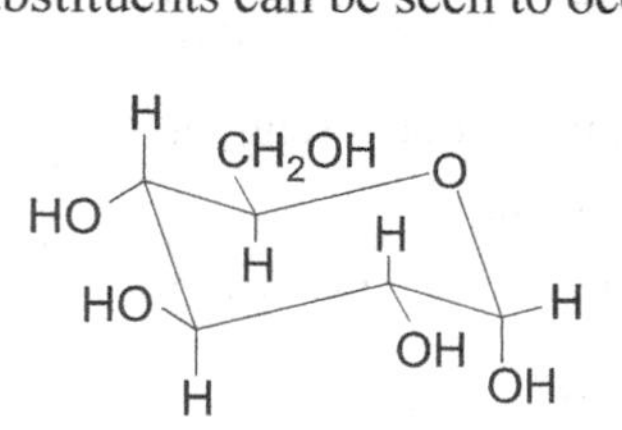

(19) α-D(+)-Glucose

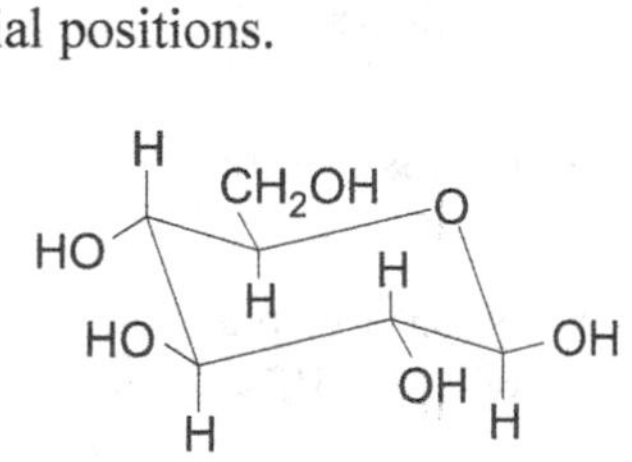

(20) β-D(+)-Glucose

Due to small difference in energy between the boat and chair forms, it is uncertain whether these conformations are preserved when D(+)-glucose is in solution. However, when the hydroxyl groups are acetylated to give α- and β-D-glucose-2, 3, 4, 6-tetra-acetate, it has been found that a chair conformation is preferred with the largest number of bulky substituents occupying the equatorial positions. The —OH at C-1 is axial in the α-form and equatorial in the β-form.

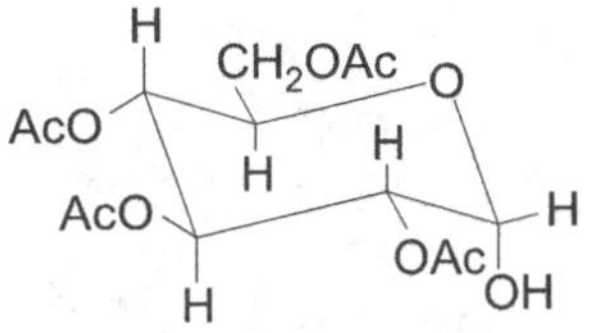

α-Isomer

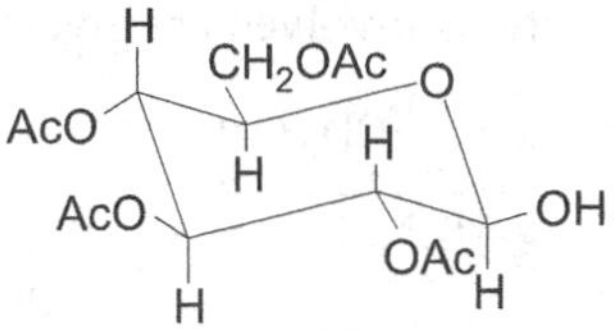

Ac = – COCH₃

β-Isomer

As it is clear that in β-D(+)-glucose the —OH at C-1 occupies equatorial position, the β-isomer is more stable than the α-isomer. It is, therefore, understandable that β-D(+)-glucose is the most predominant structure occurring in nature.

1.9   DETERMINATION OF THE RING SIZE OF GLUCOSE

We have already derived the structure of glucose as a six-membered cyclic oxide that is known as pyranose. While discussing this structure it was presumed that hemiacetal formation takes place between the aldehyde group and —OH at C-5. This pyranose structure for the glucose molecule has been proved by a number of methods including X-ray analysis and periodate oxidations. Here we give the standard chemical method *i.e.*, by methylation studies which is used for determining the ring size of any sugar.

A. Evidence for the Pyranose Structure of D(+)-Glucose

The method was developed by **Haworth** and **Hirst** in 1927.

D(+)-Glucose is refluxed with methanol in the presence of a small amount of hydrochloric acid. The methyl glucoside so obtained is methylated with an excess of dimethyl sulphate in the presence of sodium hydroxide to give methyl-tetra-*O*-methyl-D-glucoside. This on hydrolysis with dilute hydrochloric acid gives tetra-*O*-methyl-D-glucose. A lactone is isolated when the tetra-*O*-methyl-D-glucose is oxidised with bromine at 90°C. Finally, the lactone on further oxidation with HNO_3 affords trimethoxy glutaric acid.

One —COOH group of this acid comes from the lactone group that is combined with the hydroxyl in the formation of the lactone ring in gluconolactone. The second carboxyl is derived from the non-methylated carbon *i.e.*, from the —CHOH group that is involved in the ring formation of the sugar. So, there are three methoxyl groups in the lactone ring and thus C-5 hydroxyl must be involved in the ring formation. Therefore, the lactone is 2,3,4,6-tetra-*O*-methyl gluconolactone. Working backwards in this way D(+)-glucose must have structure as D(+)-glucopyranose. This is true with both α- and β-isomers.

The reactions involved are given below:

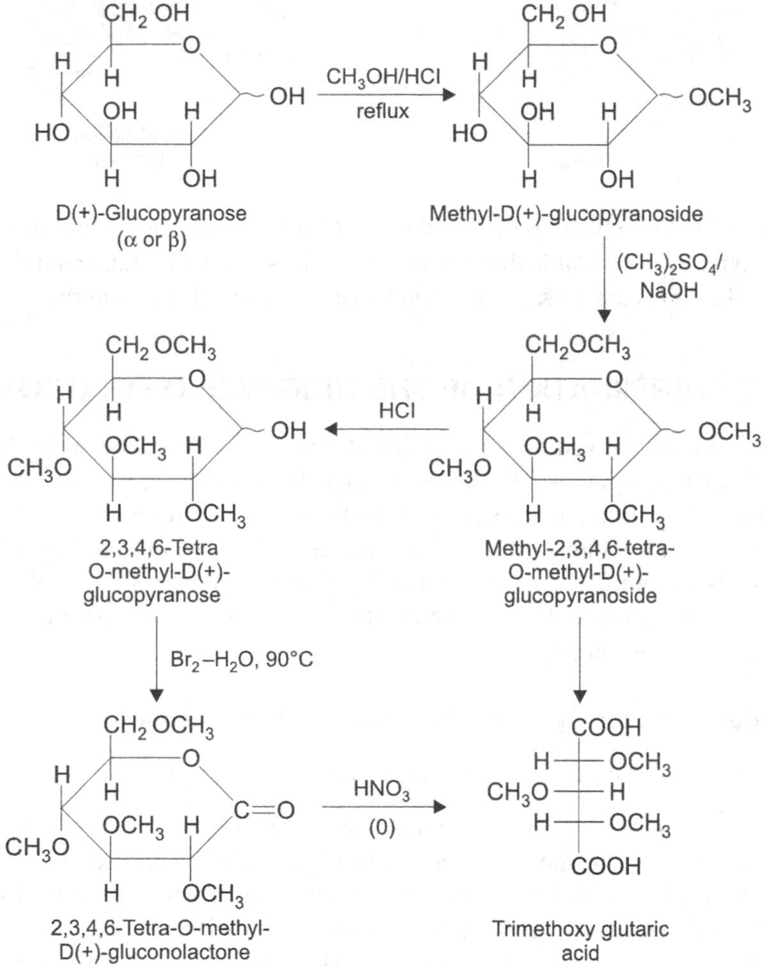

D(+)-Glucopyranose
(α or β)

CH$_3$OH/HCl
reflux

Methyl-D(+)-glucopyranoside

(CH$_3$)$_2$SO$_4$/
NaOH

2,3,4,6-Tetra
O-methyl-D(+)-
glucopyranose

HCl

Methyl-2,3,4,6-tetra-
O-methyl-D(+)-
glucopyranoside

Br$_2$–H$_2$O, 90°C

2,3,4,6-Tetra-O-methyl-
D(+)-gluconolactone

HNO$_3$
(O)

COOH
H——OCH$_3$
CH$_3$O——H
H——OCH$_3$
COOH

Trimethoxy glutaric
acid

B. Evidence for the Furanose Structure of D(+)-Glucose

Fischer in 1914 prepared the methyl glucoside in a different way. He dissolved glucose in methanol and added to that 1%-hydrochloric acid. The mixture was allowed to stand at 0°C. He obtained a syrup instead of a crystalline product as obtained in the earlier case. This syrup was subsequently shown to be a mixture of α- and β-glucofuranosides. The procedure adopted here was just the same as given above. The end product in this case comes out to be dimethyl tartaric acid.

The formation of dimethyl-D-tartaric acid indicates that there are two methoxy groups within the ring; *i.e.*, lactone is a γ-lactone. Therefore, its structure is given as 2,3,5,6-tetra-*O*-methyl gluconolactone and hence the glucose has got a furanose structure named D-glucofuranose. It must be noted here that D-glucofuranose has not been isolated. The structure is deduced on the basis of isolation of methyl-D-glucofuranoside (Fig. 1.9).

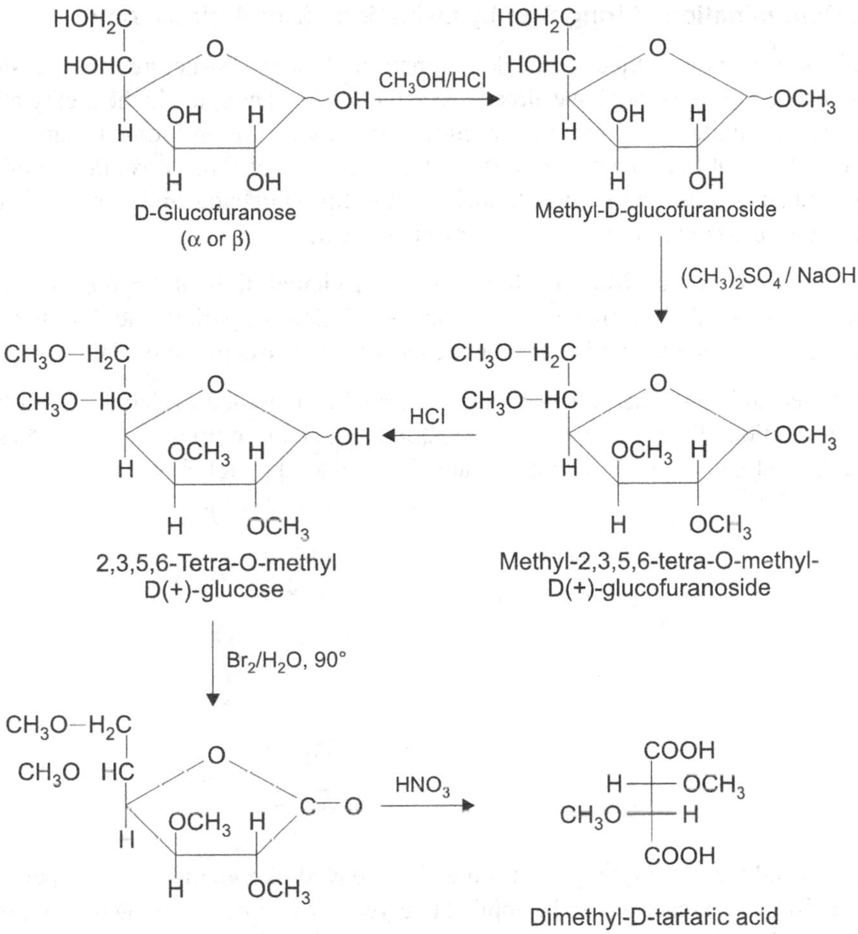

D-Glucofuranose
(α or β)

Methyl-D-glucofuranoside

2,3,5,6-Tetra-O-methyl
D(+)-glucose

Methyl-2,3,5,6-tetra-O-methyl-
D(+)-glucofuranoside

Dimethyl-D-tartaric acid

Fig. 1.9.

Both the above schemes favour different ring sizes for the glucose molecule. So, it is possible to decide the actual ring size on the basis of methylation studies.

Methylation at different temperatures gives different methyl glucosides. At refluxing temperatures a pyranoside is formed whereas at 0°C the formation of a furanoside is favoured. The important question now is: What is the actual size of the ring in the original sugar? Oxidation of an aldose with hypobromite produces δ-lactone. This is the first product of oxidation.

Further investigations have proved that δ-lactone obtained by the oxidation of glucose is quite unstable. That slowly gets converted into more stable γ-lactone. It, therefore, follows that the size of the ring in normal sugars is pyranose.

McDonald *et al.* examined α-D-glucose by X-ray analysis and confirmed the presence of a six-membered ring.

C. Determination of Ring Size by Oxidations with Periodic Acid

This is an elegant method for the determination of the ring structures of glycosides and hence of sugars. We have already seen the methods adopted in the methylation studies are quite laborious and structures are deduced in an indirect manner. In contrast to that here the procedures are facile and direct. The consumption of the oxidant and liberation of products such as CO_2, formaldehyde, and formic acid can be determined quantitatively on a semi-micro scale.

It was discovered by **Malaprade** in 1928 that vicinal diols undergo cleavage of their C—C bond when treated with aqueous solutions of periodic acid or its salts. This cleavage is quantitative and can be used for the structure elucidations.

The mechanism of the periodic oxidation has been studied in detail. It involves the formation of a cyclic diester of the periodic acid. Decomposition of the diester yields the two carbonyl fragments and iodic acid as shown below:

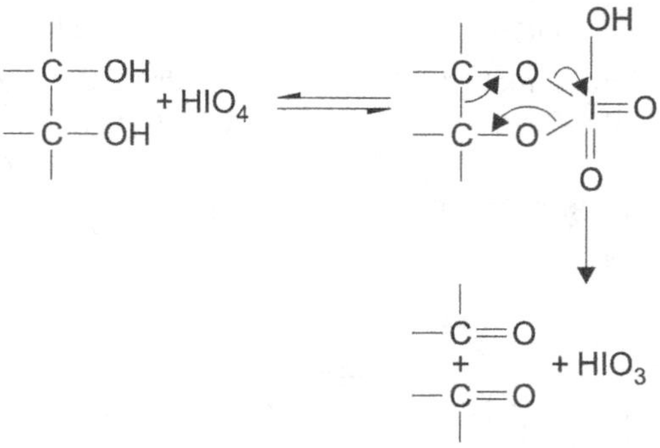

Sugars and their methyl glycosides are also cleaved like vicinol diols by periodic acid. The above reaction can be applied to glycosides for determining the ring size, *i.e.*, whether they have pyranose or furanose structure. Under mild conditions (0.01 M to 0.05 M of HIO_4 at 15°C for 23 hours) the reaction can be controlled and the amount of periodic acid used is determined quantitatively.

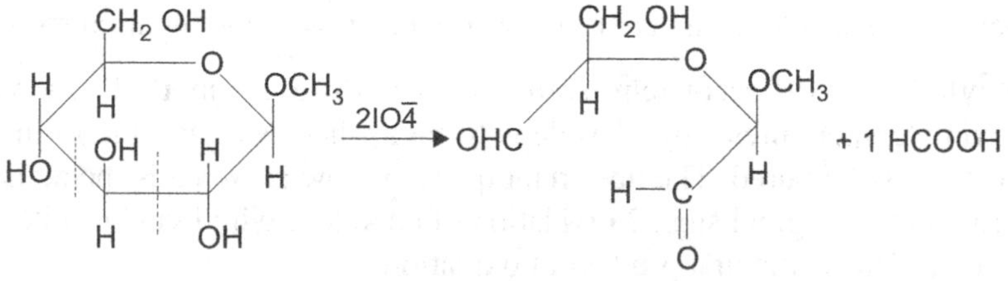

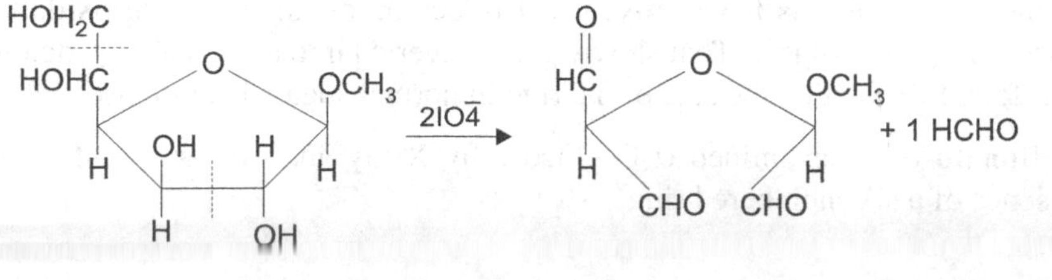

Thus, methyl-α- and β-D-glucopyranoside each reacts with only two moles of periodate and give one mole of formic acid and no formaldehyde. This proves that the pyranoside structure assigned from methylation studies were correct. Furanoside, however, reacts with two equivalents of the reagent but yields only one equivalent of formaldehyde and no formic acid.

Glucose itself under strong conditions (1% aqueous $NaIO_4$ in 1 N H_2SO_4) consumes five equivalents of HIO_4 and yields five equivalents of formic acid and one equivalent of formaldehyde. It appears that with free sugars, it is the open-chain form that is oxidised and complete oxidative cleavage takes place.

$$
\begin{array}{c}
\text{CHO} \\
\text{H} \!-\!\!-\! \text{OH} \\
\text{HO} \!-\!\!-\! \text{H} \\
\text{H} \!-\!\!-\! \text{OH} \\
\text{H} \!-\!\!-\! \text{OH} \\
\text{CH}_2\,\text{OH}
\end{array}
\quad + 5HIO_4 \longrightarrow 5HCOOH + 1HCHO
$$

1.10   MUTAROTATION

We have already seen that glucose can be isolated in two different crystalline forms. These forms are α-D(+)-glucose and β-D(+)-glucose. They have different melting points and different specific rotations. The α-form is soluble in water to the extent of 82 g/100 ml at 25°C and the β-isomer is soluble in water to a greater extent, 154 g/100 ml at 15°C. Both these crystalline forms of α- and β-D(+)-glucose are quite stable. In solution (water) each form slowly changes into an equilibrium mixture of both. The process can be readily observed as a decrease in the specific rotation of the α-anomer (+112°) or an increase for the β-anomer (+19°) to an equilibrium value of +52.5°. This phenomenon is known as mutarotation and is commonly observed for reducing sugars. Aldopentoses, aldohexoses and ketohexoses all undergo mutarotation.

We have seen that mutarotation takes place in water solution. However, mutarotation can take place in any solvent, which is amphiprotic in nature, *i.e.*, a solvent which can function both as an acid as well as a base. It has been observed that mutarotation does not take place when pyridine or cresol are used as solvents separately but it takes place in a mixture of the two. That is, mutarotation is particularly effective if both an acid and a base are present in the same solution. We know that water acts both as an acid as well as a base. Therefore, mutarotation takes place readily in water.

Mechanism of Mutarotation

The phenomenon of mutarotation indicates that the two forms of D(+)-glucose are interconvertible and they have a common intermediate. This change presumably takes place *via* the open-chain structure of glucose. There is sufficient evidence for the existence of this open-chain form. The generally accepted mechanism is virtually the same as described for acid and base-catalysed hemiacetal and hemiketal formation of aldehydes and ketones.

Acid Catalysed Hemiacetal Formation:

$$CH_3{-}(H){>}C{=}O + H^+ \rightleftharpoons CH_3{-}(H){>}C{=}\overset{\oplus}{O}H$$

$$CH_3{-}(H){>}C{=}\overset{\oplus}{O}H + \overset{..}{O}(H)CH_3 \rightleftharpoons CH_3{-}\underset{OH}{\overset{H}{C}}{-}\overset{\oplus}{O}(H){-}CH_3$$

$$CH_3{-}\underset{OH}{\overset{H}{C}}{-}O{-}CH_3$$

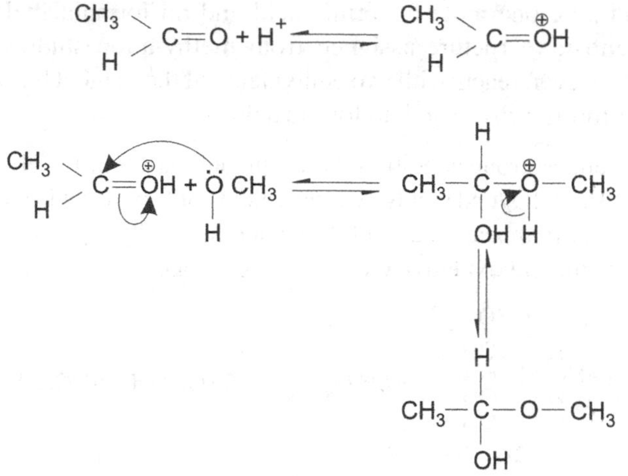

Hemiacetal

Base Catalysed Hemiacetal Formation

$$CH_3{-}OH + \bar{O}H \rightleftharpoons CH_3\bar{O} + H_2O$$

$$CH_3{-}\underset{H}{C}{+}\overset{O}{}\overline{O}CH_3 \rightleftharpoons CH_3{-}\underset{H}{\overset{\bar{O}}{C}}{-}OCH_3$$

$$CH_3{-}\underset{H}{\overset{\bar{O}}{C}}{-}OCH_3 + H_2O \rightleftharpoons CH_3{-}\underset{OH}{CH}{-}OCH_3$$

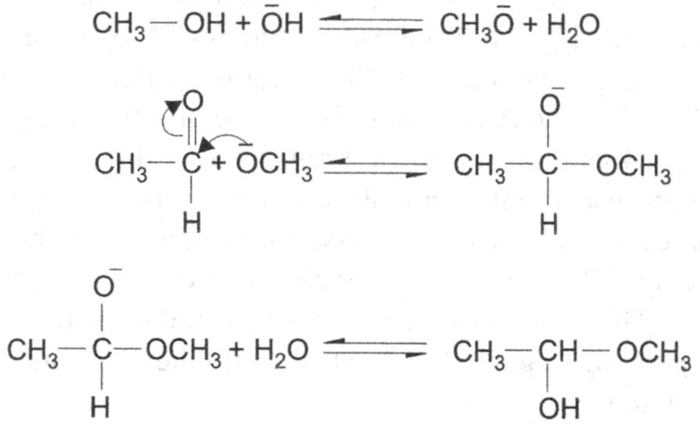

Hemiacetal

When mutarotation takes place, the ring of one form opens up and then recloses in the inverted positon to give the 2nd form.

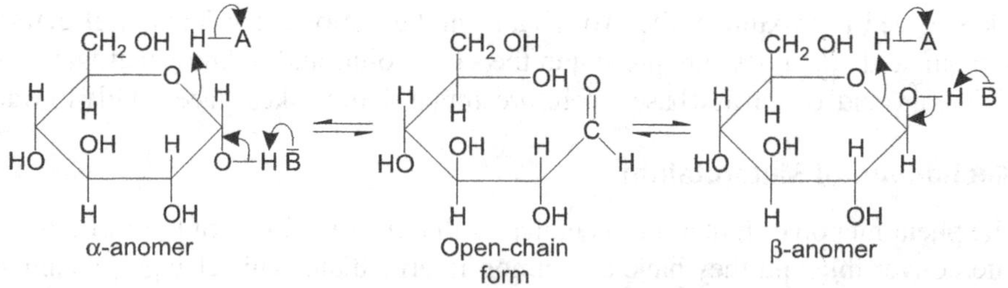

α-anomer   Open-chain form   β-anomer

As already mentioned at equilibrium about 63% of the β-anomer and 37% of the α-anomer are present. The amount of free aldehyde form is very small (~0.01%).

1.11 OSAZONE FORMATION

The reaction of glucose with phenylhydrazine is quite interesting. With excess of phenylhydrazine glucose forms a crystalline derivative known as glucose-phenyl osazone or glucosazone. The characteristic feature is that all aldoses form these derivatives, called osazones. Osazone formation requires the oxidation of the carbon group adjacent to the carbonyl carbon. Under normal conditions of the reaction phenylhydrazone is formed first which then reacts with excess of phenylhydrazine to form the osazone.

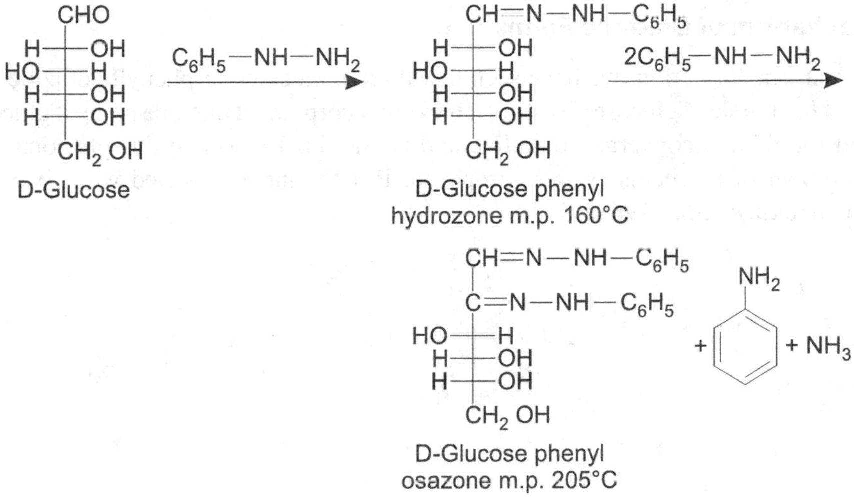

D-Glucose

$C_6H_5-NH-NH_2$ →

D-Glucose phenyl
hydrozone m.p. 160°C

$2C_6H_5-NH-NH_2$ →

D-Glucose phenyl
osazone m.p. 205°C

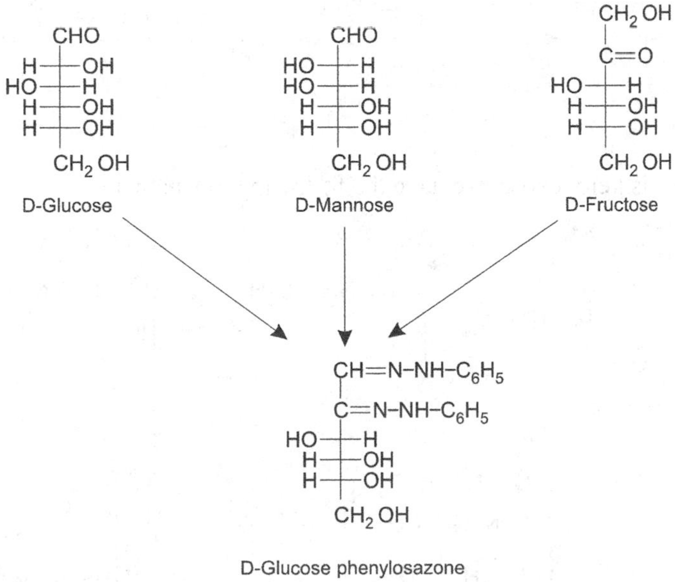

$+ NH_3$

D-Glucose

D-Mannose

D-Fructose

D-Glucose phenylosazone

Osazones are crystalline derivatives which can be handled very easily. Hence, they are used for characterising the sugars through their characteristic melting points.

Osazones are also useful for establishing the stereochemical relationships in isomeric sugars. Osazone formation destroys the chirality at C-2 position of an aldose. Therefore, epimeric sugars give the same osazone, *e.g.*, D-glucose and D-mannose form the same osazone. D-Fructose also gives an osazone identical with the one obtained from D-glucose. This indicates that two sugars, D-glucose and D-fructose have the same configuration at C-3, C-4 and C-5.

Mechanism of Osazone Formation

In osazone formation, we have seen that three molecules of phenylhydrazine are used for 1 mole of glucose. Two molecules are incorporated into one mole of glucose and the third is converted to aniline and ammonia. To explain this phenomenon a number of mechanisms were proposed. But the most expected was given by Shemyakin *et al.* in 1965.

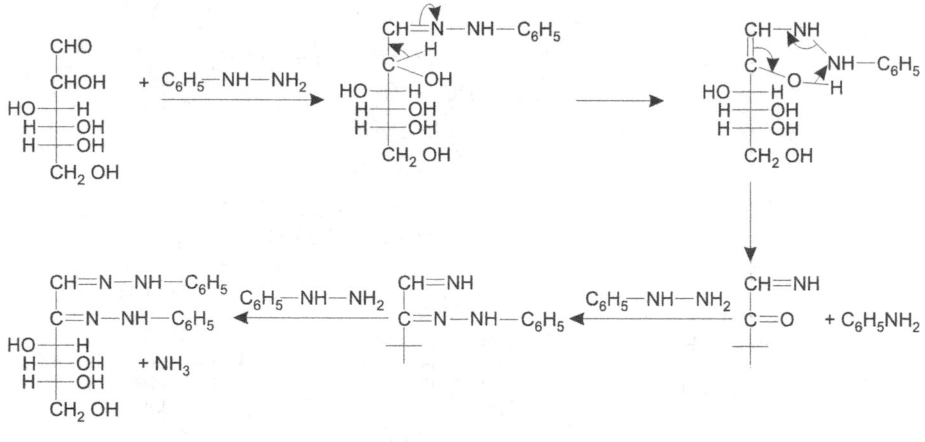

When there is ketohexose, we have the following mechanism:

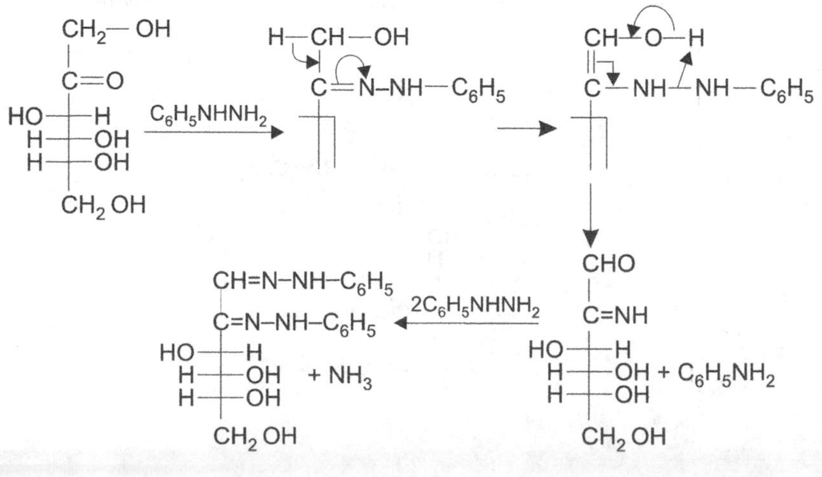

1.12 KETOHEXOSES

All the ketohexoses that occur in nature have the keto group adjacent to a terminal —CH_2OH group. Thus, there are three chiral carbon atoms present in the structural formula of a ketohexose. There are eight stereoisomeric forms possible, *i.e.*, four pairs of enantiomers. Six are known. These are D-(–) and L(+)-fructose, D(+) and L(–)-sorbose, D(+)-tagatose and L(–)-psicose. Out of these only D(–)-fructose, L(–)-sorbose and D(+)-tagatose occur in nature.

CH_2OH	CH_2OH	CH_2OH	CH_2OH
C=O	C=O	C=O	C=O
HO——H	H——OH	HO——H	HO——H
H——OH	HO——H	HO——H	HO——H
H——OH	HO——H	H——OH	HO——H
CH_2OH	CH_2OH	CH_2OH	CH_2OH
D(–)–Fructose	L(–)–Sorbose	D(+)–Tagatose	L( ) Psicose

1.13 STRUCTURE OF D(–)-FRUCTOSE

D(–)-Fructose is present in the free state in some fruits and honey and in the combined form with glucose in the most important disaccharide, sucrose. Fructose is also quite sweet. Naturally occurring fructose is laevorotatory and is, therefore, called as laevulose.

Open-chain structure of D(–)-fructose is derived on the basis of the following observations:

1. D(–)-Fructose has the molecular formula $C_6H_{12}O_6$ as deduced from elemental analysis and molecular weight determination.
2. Fructose forms the oxime when treated with hydroxylamine and a cyanohydrin is formed with hydrogen cyanide. These reactions indicate the presence of a carbonyl group.
3. Fructose is a reducing sugar. It reduces Tollen's reagent and Fehling solution.
4. Fructose is reduced with sodium-amalgam and water to a mixture of hexahydric alcohols, sorbitol and mannitol. The alcohols on treatment with hydriodic acid and red phosphorus at 100°C give a mixture of *n*-hexane and 2-iodohexane.
5. Fructose forms the penta-acetate when treated with acetic anhydride and anhydrous sodium acetate indicating the presence of five hydroxyl groups.
6. Fructose is not oxidised by bromine water. However, oxidation with nitric acid affords a mixture of trihydroxy glutaric acid, tartaric acid and glycollic acid. Since all these acids contain fewer carbon atoms than the fructose molecule, the carbonyl group in the fructose must be present as a keto group.

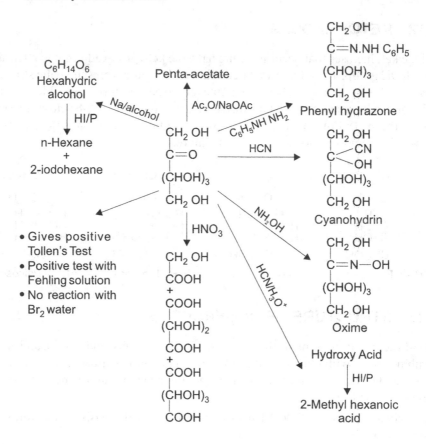

Fig. 1.10.   Reactions of fructose.

The above reactions (given in Fig. 1.10) show that fructose has a six carbon straight chain structure as given above:

$$
\begin{array}{l}
CH_2\,OH \\
C{=}O \\
CH\,OH \\
CH\,OH \\
CH\,OH \\
CH_2\,OH
\end{array}
$$

Fructose

A. Configuration of D-Fructose

We have already seen that D-fructose gives the same osazone as is obtained from D-glucose. This means that the configuration at C-3, C-4 and C-5 is the same in the two sugars, D-glucose and D-fructose. Thus, the configuration of D-fructose is established.

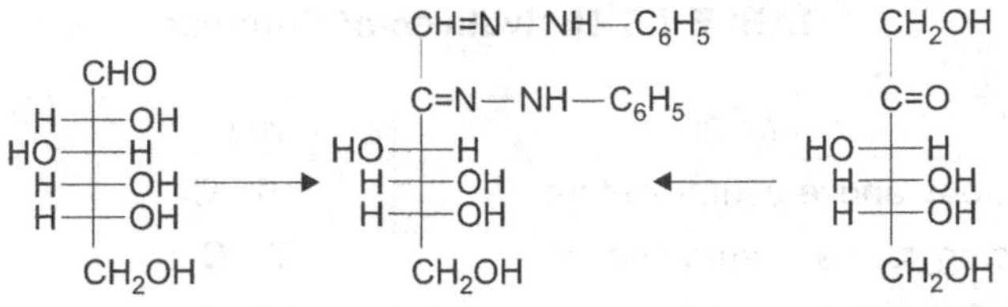

D(+)-Glucose   Osazone   D(−)-Fructose

B. Cyclic Structure of D-Fructose

The above open chain structure of fructose does not explain all the observations/ reactions given by the fructose molecule.

The following observations cannot be explained by the open-chain structure of D-fructose:

1. Fructose does not form a bisulphite addition product.
2. Fructose undergoes mutarotation in aqueous solutions, *e.g.*, $[\alpha]_D = 133°$ changes to −92°.
3. As has been shown in the case of glucose, fructose also gives two methyl glycosides, known as methyl fructosides when treated with methyl alcohol in the presence of hydrochloric acid.

To account for the above facts a cyclic hemiketal structure for fructose has been proposed. We get two forms of this cyclic structure with the generation of an anomeric carbon. Thus, we can have two structures for the fructose designated as α-D-fructopyranose and β-D-fructopyranose represented by the following Haworth projections:

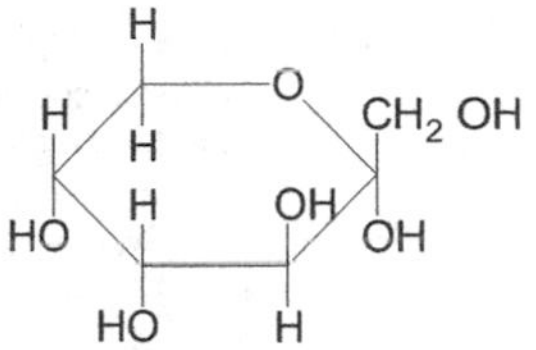

α-D–Fructopyranose 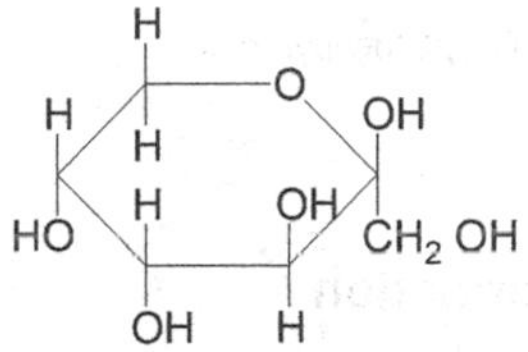   β-D–Fructopyranose

In the crystalline form fructose probably exists in the pyranose form with β-isomer predominating (α-form has not been isolated).

However, fructose always occurs in the furanose form when combined in natural products. There is good evidence that a considerable amount of furanose form is present in a solution of fructose.

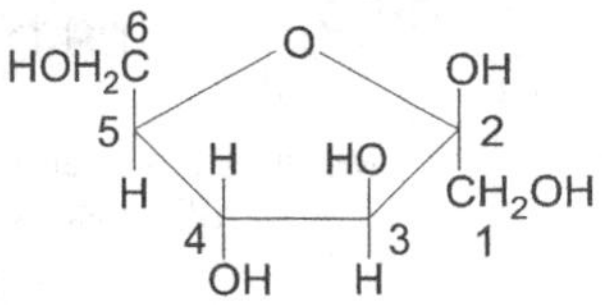

β–D-Fructofuranose 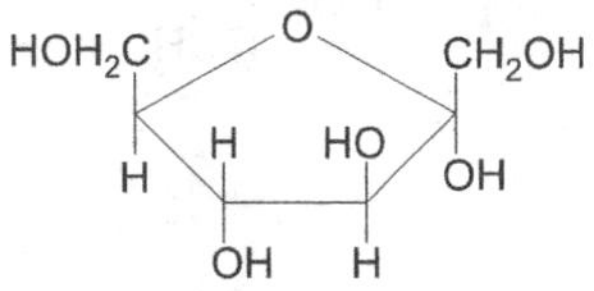   α–D-Fructofuranose

TABLE 1.5 Derivatives of Fructose

Compounds	m.p.	$[\alpha]_D$
β-D-Fructopyranose penta-acetate	109°C	− 121°
α-D-Fructopyranose penta-acetate	70°C	+ 35°
Methyl β-D-Fructopyranoside	120°C	− 172°
Methyl-α-D-Fructopyranoside	102°C	+ 90°.

1.14   REACTIONS OF MONOSACCHARIDES

A. Esterification

We know that alcohols can be esterified to give esters. Carbohydrates contain many hydroxyl groups and thus react with acetic anhydride to give esters. When glucose is treated with acetic anhydride, it forms a penta-acetate indicating that glucose contains five hydroxyl groups. This also explains the open-chain structure of glucose.

However, the penta-acetate has been assigned the cyclic structure with an acetyl group at carbon 1 and not at carbon 5.

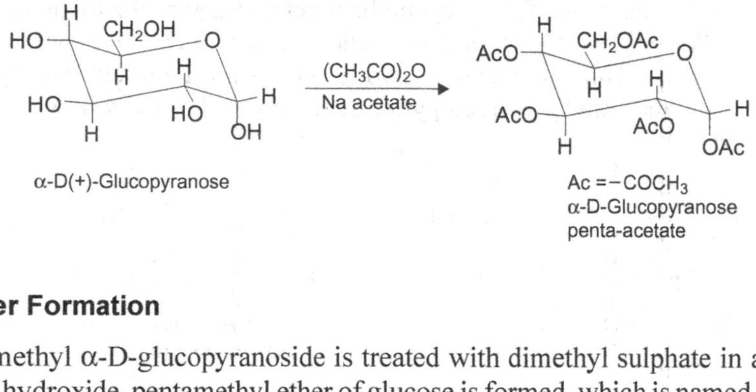

α-D(+)-Glucopyranose

Ac =−COCH₃
α-D-Glucopyranose
penta-acetate

B. Ether Formation

When methyl α-D-glucopyranoside is treated with dimethyl sulphate in aqueous sodium hydroxide, pentamethyl ether of glucose is formed, which is named methyl-2,3,4,6-tetra-O-methyl-α-D-glucopyranoside.

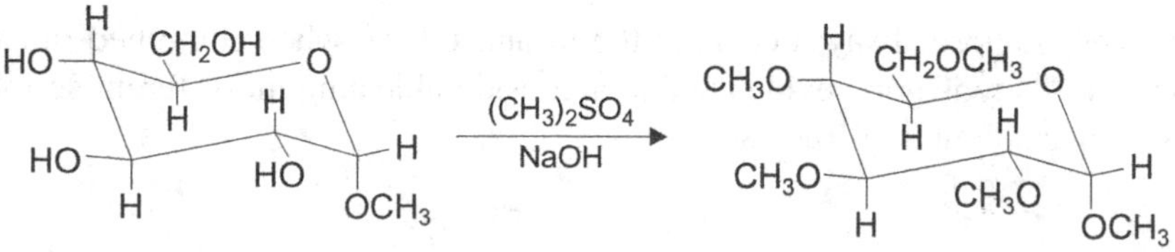

Methyl 2,3,4,6-tetra-O-methyl-α-D-glucopyranoside

The methyl ethers formed from carbohydrates are stable to bases and dilute acids and serve to protect the hydroxyl groups.

The methoxyl group at the anomeric carbon is different in its reactivity from the other groups in the molecule. When treated with dilute acids, this methyl group is removed without affecting other methoxyl groups and we get the formation of tetra-O-methyl glucose.

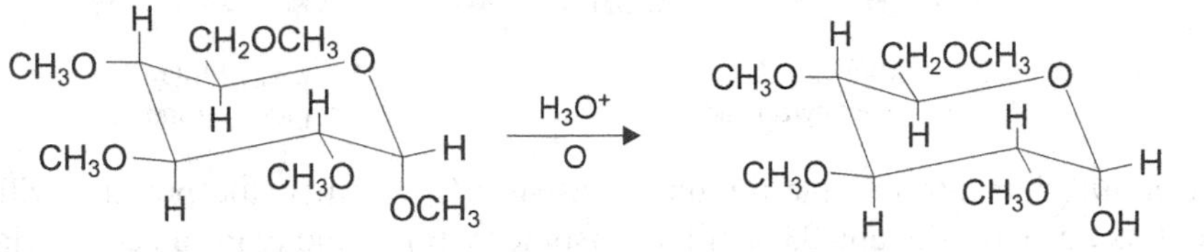

Methyl 2,3,4,6-tetra-O-methyl-
α-D-glucopyranoside

2,3,4,6-Tetra-O-methyl-
α-D-glucopyranose

C. Formation of Glycosides

When gaseous hydrogen chloride is passed into a solution of D(+)-glucose in methanol, methylation of the anomeric hydroxyl takes place giving rise to the formation of acetals.

Carbohydrate acetals are generally called glycosides and acetal of glucose is called a glucoside. The structures of the acetals obtained depend upon reaction conditions. At low temperature the principal products are the five-membered ring methyl glucofuranosides. Thus, from glucose we get a mixture of methy-α-D-glucofuranoside and methyl-β-D-glucofuranoside.

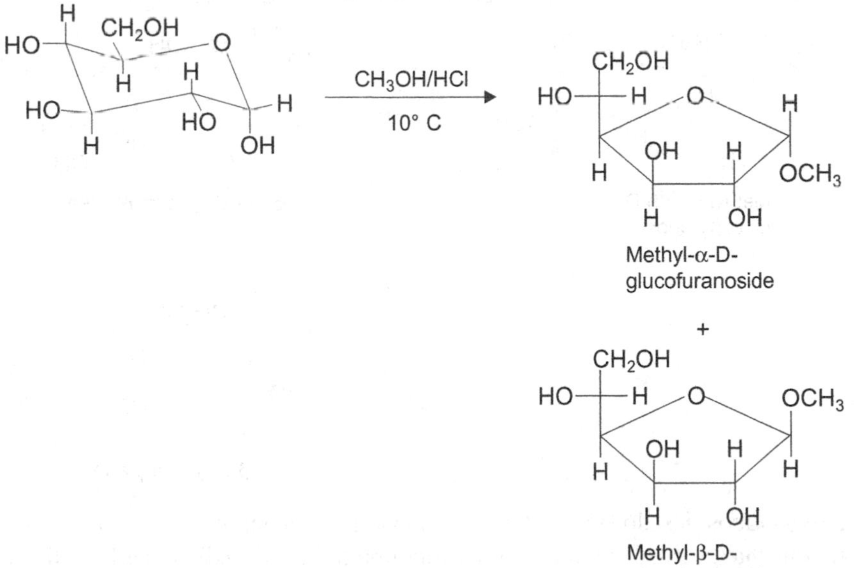

Methyl-α-D-
glucofuranoside

+

Methyl-β-D-
glucofuranoside

When glucose is treated with methanol and hydrogen chloride at higher temperature, we get glucopyranosides

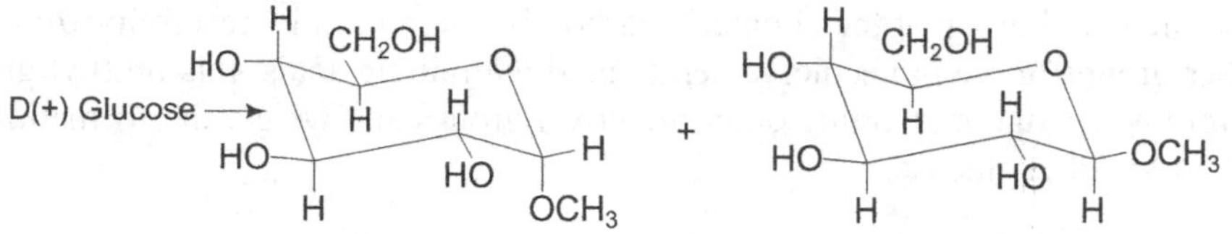

Methyl-α-D-
glucopyranoside

Methyl-β-D-
glucopyranoside

At equilibrium 66% of the mixture consists of α-isomer, the product with the methoxy group axial and 33% of the β-isomer with methoxy group equatorial.

The methyl-D-glucofuranosides formed at lower temperatures are kinetically favoured products as five-membered rings are formed faster than six-membered rings. However, at higher temperatures rings can open and equilibrium is established among all the different forms.

As mentioned earlier, the mixed acetal of an aldose or ketose is called glycoside. They are stable in basic aqueous solutions. In acidic solutions glycosides undergo hydrolysis to produce a sugar and an alcohol. The alcohol produced from a glycoside is known as *aglycone*. Enzymes can hydrolyse glycosides selectively. The enzyme maltase will cleave only α-glucopyranosides and the enzyme emulsin cleaves only β-glucopyranosides.

Under acidic conditions glucopyranosides (α or β) are hydrolysed to give an equilibrium mixture of α-D-glucopyranose and β-D- glucopyranose.

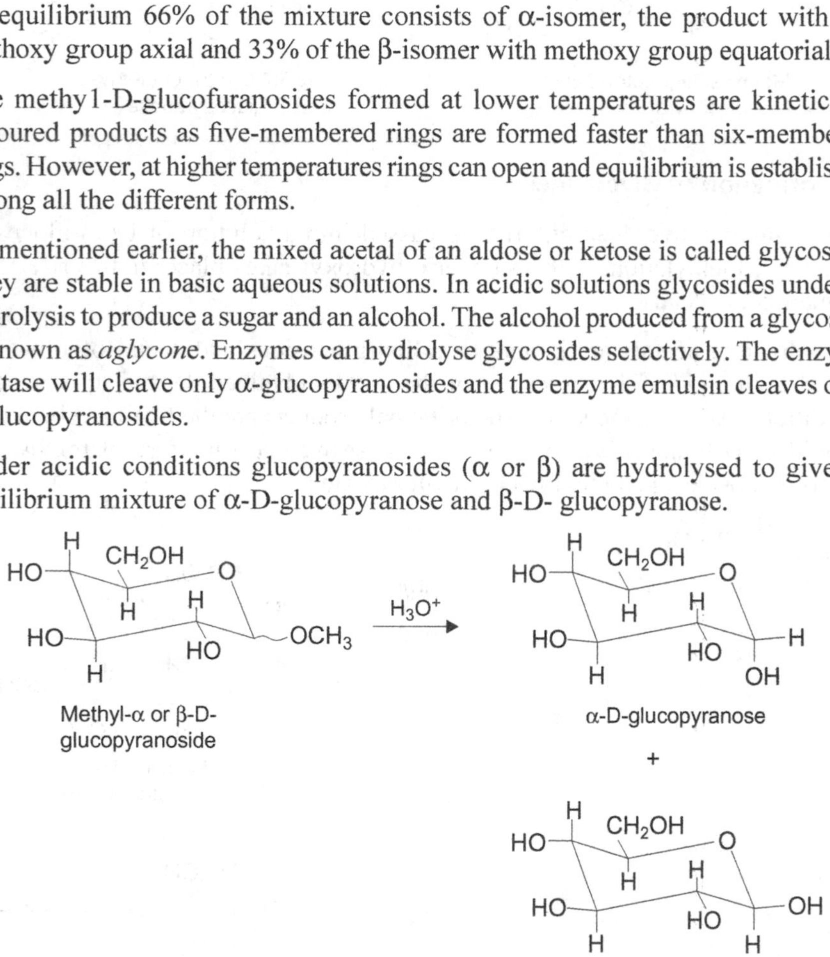

Methyl-α or β-D-
glucopyranoside

H_3O^+

α-D-glucopyranose

+

β-D-glucopyranose

Methyl-glucosides do not undergo mutarotation or show any of the aldehyde reactions that glucose exhibits. They cannot be oxidised easily to carboxylic acids.

D. Esters of Phosphoric Acid

In biological systems, the most important esters of sugars are the ones formed with phosphoric acid. Phosphoric acid forms ester linkages with the —OH groups of the sugars. Sugar phosphates are intermediates in many metabolic processes such as the degradation of glycogen to lactic acid, the fermentation of sugar to alcohol and the biosynthesis of cabohydrates in plants by the process known as photosynthesis.

Sugar phosphates, which are involved in the biosynthesis and degradation of glycogen and starch, are α-D-glucopyranosyl phosphate and D-glucose-6-phosphate.

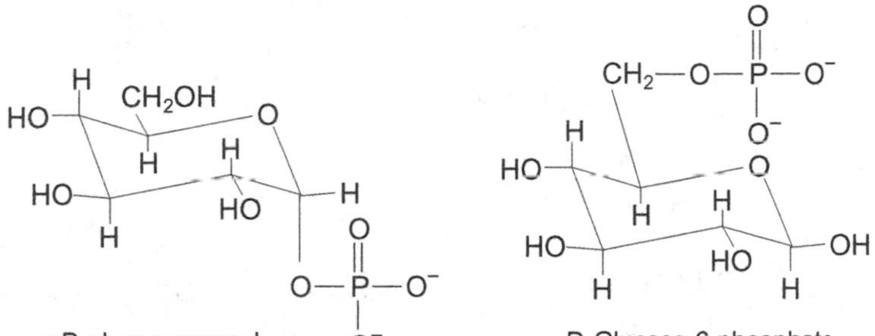

α-D-glucopyranosyl phosphate

D-Glucose-6-phosphate

The polysaccharides are synthesised in organisms by an enzyme- catalysed process. The glucose units are added in a stepwise fashion on to the growing polysaccharide chain. The glucose units are in the form of α-D-glucosylphosphate. In this form, the anomeric carbon is activated toward nucleophilic substitution. Thus, nucleophilic substitution takes place and the polysaccharide chain is built up. The process is reversible. The reversible process is the method by which degradation of the polysaccharide takes place to generate glucose molecules.

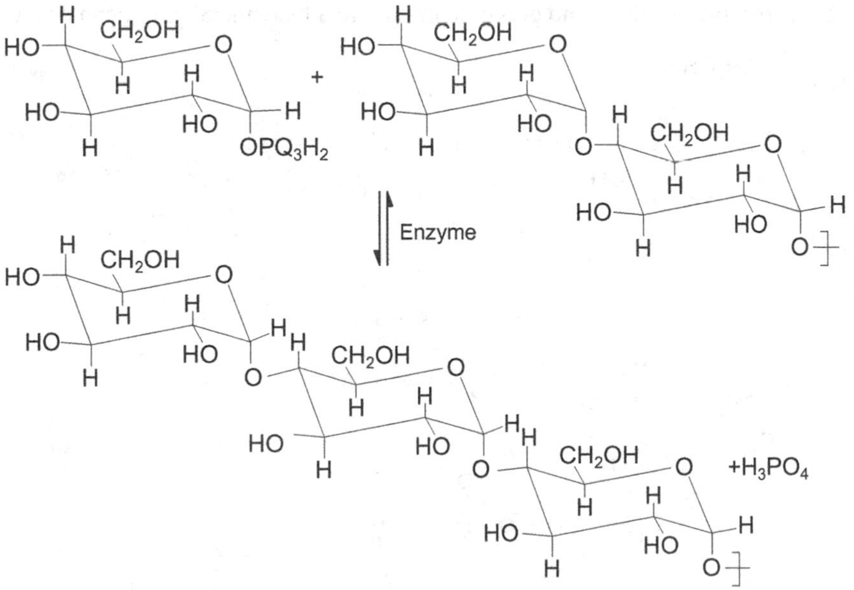

Similarly, enzymes catalyse the formation and cleavage of the $1 \rightarrow 6$ glycosidic bond by way of D-glucose-6-phosphate. α-D- Glucopyranosyl phosphate is also involved in the formation and degradation of sucrose.

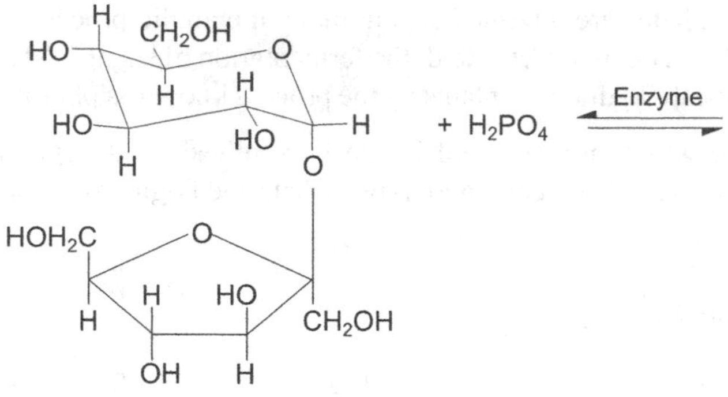

Sucrose

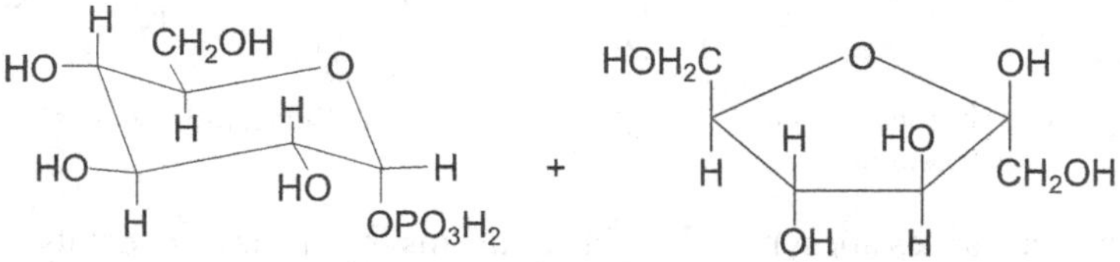

α-D-Glucopyranosyl phosphate   D-Fructose

In nature phosphoric acid units are found in nucleotides, which are the phosphoric acid esters of nucleosides. Adenosine 5'-triphosphate (ATP) is found extensively in living systems. Adenosine triphosphate stores chemical energy and makes it available to specific cell processes. It gives up energy by transfering a phosphate group to another molecule. In glycolysis glucose takes up one unit of phosphate to give glucose-6-phosphate and gets converted into adenosine 5'-diphosphate (ADP).

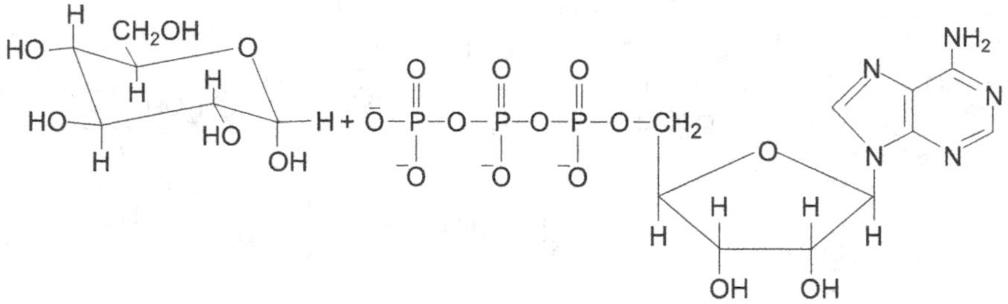

Adenosine 5'-triphosphate (ATP)

Hexokinase Mg^{2+}

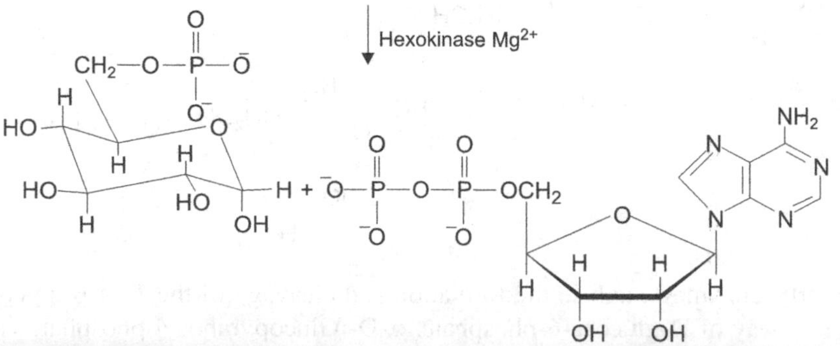

Glucose 6 phosphate   Adenosine 5'-diphosphate (ADP)

The reaction of glucose with adenosine triphosphate is a nucleophilic substitution reaction. The reaction is catalysed by an enzyme, hexokinase, and magnesium ion is required as a cofactor. The magnesium ion forms a complex with the two terminal phosphate groups of adenosine triphosphate, reducing the charge on the ion and making it easier for the substitution to take place.

E. Oxidation of Monosaccharides

We know sugars such as glucose and fructose are reducing agents. That means they can be easily oxidised even by mild oxidising agents. Sugars in which the carbonyl group is tied up in the form of an acetal linkage are non-reducing sugars. Thus, methyl glucosides are nonreducing in nature.

i. Oxidation with Tollen's Reagent

Aldehydes react with Tollen's reagent to give carboxylate ions and metallic silver. This is obtained in the formation of silver mirror. That's why the test is known as the silver mirror test. Aldehydes give this test whereas ketones do not.

Tollen's reagent is ammoniacal silver nitrate. It is obtained by adding ammonia solution dropwise to a solution of silver nitrate till the precipitate obtained just dissolves. Aldehydes give a positive test with Tollen's reagent whereas ketones do not.

$$R-\overset{\overset{O}{\|}}{C}-H + Ag(NH_3)_2 \overset{+}{} \overset{-}{OH} + OH^-$$

$$\longrightarrow R-\overset{\overset{O}{\|}}{C}-\overset{-}{O} + 2Ag^+\downarrow + NH_3 + H_2O$$

An aldose has an aldehyde group in its open form. It reacts with Tollen's reagent to give an aldonic acid and a silver mirror.

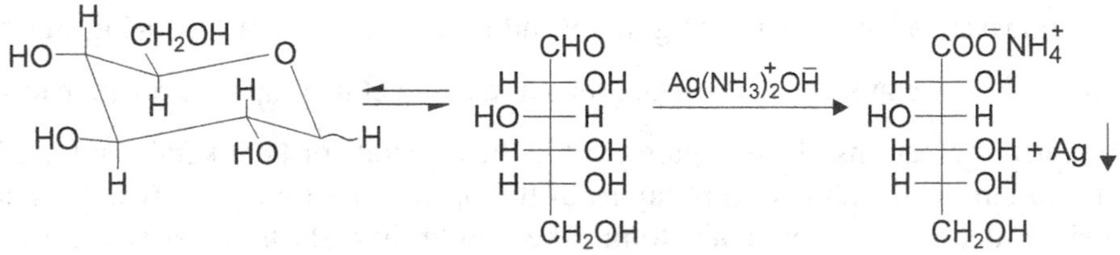

| α-or β-D-glucose | Open-chain form | D-gluconic acid salt |

Ketoses also give a positive Tollen's test because under basic conditions, the open-chain form of a ketose can isomerise to an aldose, which reacts to give a positive Tollen's test.

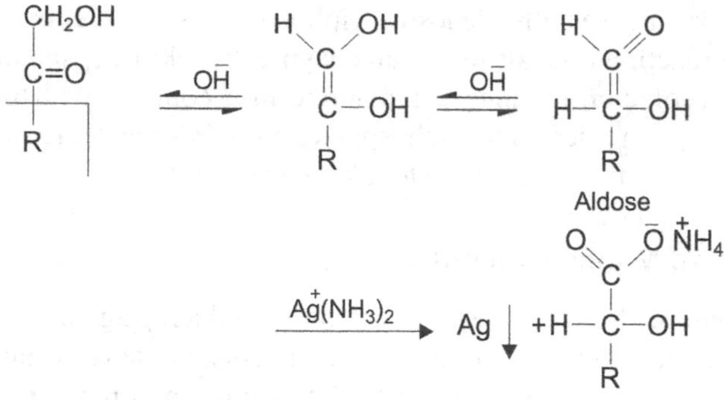

Thus, D-fructose gives a positive test with Tollen's reagent.

ii. Oxidation with Benedict's Reagent and Fehling Solution

Benedict's reagent is an alkaline solution of copper sulphate and sodium citrate and Fehling solution contains copper sulphate and sodium potassium tartarate. Both these solutions (cupric citrate complex and cupric tartarate complex) oxidise aldoses and ketoses to aldonic acids and deposit a brick-red precipitate of cuprous oxide. Since both these solutions are blue in colour the appearance of a brick red precipitate is a positive test.

$$\underset{\text{(Complex)}}{Cu^{2+}} + \text{CHO}-(\text{CHOH})_4-\text{CH}_2\text{OH} \longrightarrow \underset{\text{Brick red}}{Cu_2O\downarrow} + \text{Oxidation products}$$

Benedict's reagent
or
Fehling Solution

The tests are positive with both D-glucose and D-fructose. We know α-hydroxyketones in general are oxidised very easily to the diketones.

Sugars that give positive tests with Tollen's or Benedict's solution or Fehling solution are known as reducing. Thus, D-glucose and D-fructose are both reducing sugars.

All sugars that contain a hemiacetal group or a hemiketal group give positive tests.

In aqueous solutions, these sugars exist as hemiacetals or hemiketals and are in equilibrium with small concentrations of the open chain form. The reactions are positive due to the open chain form. The equilibrium shifts towards it till one reactant is completely exhausted.

iii. Oxidation with Bromine Water

Bromine water is a general reagent that oxidises the aldehyde group of an aldose to a carboxylic acid, *i.e.*, it converts an aldose to an aldonic acid.

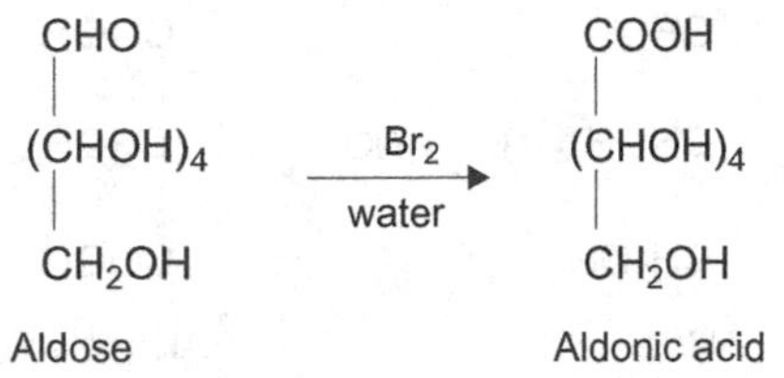

CHO

(CHOH)₄ $\xrightarrow[\text{water}]{\text{Br}_2}$ COOH

CH₂OH

(Aldose)   (Aldonic acid)

Bromine water is a mild oxidising agent that does not oxidise the alcoholic groups of the sugar and ketoses are untouched. Bromine water is acidic in nature and does not cause epimerisation or rearrangement of the carbonyl group; for example, bromine water oxidises D-glucose to D-gluconic acid.

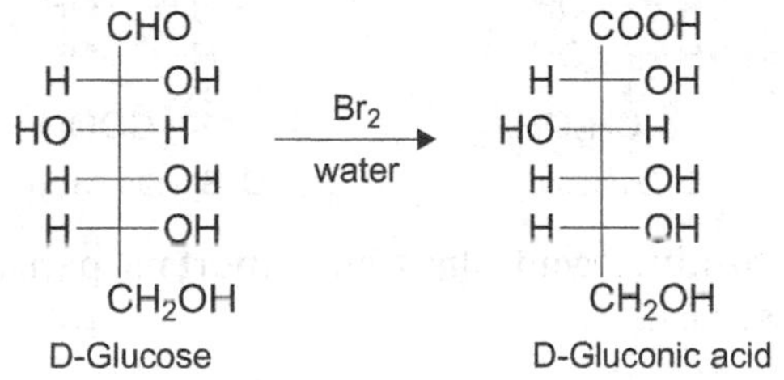

D-Glucose $\xrightarrow[\text{water}]{\text{Br}_2}$ D-Gluconic acid

Thus, mannose is converted to the corresponding mannonic acid

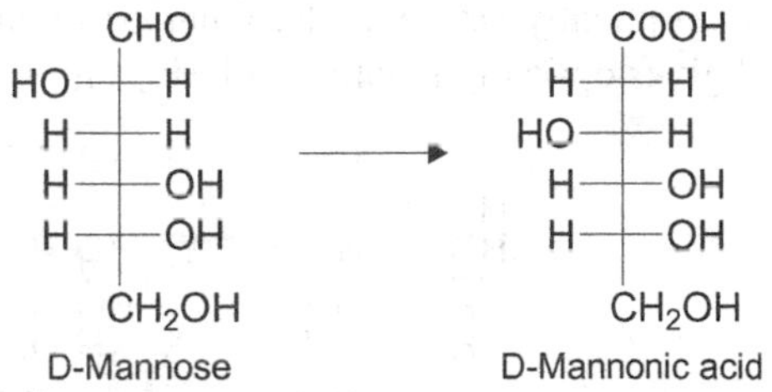

D-Mannose $\longrightarrow$ D-Mannonic acid

The initial product is a δ-lactone which hydrolyses to an aldonic acid.

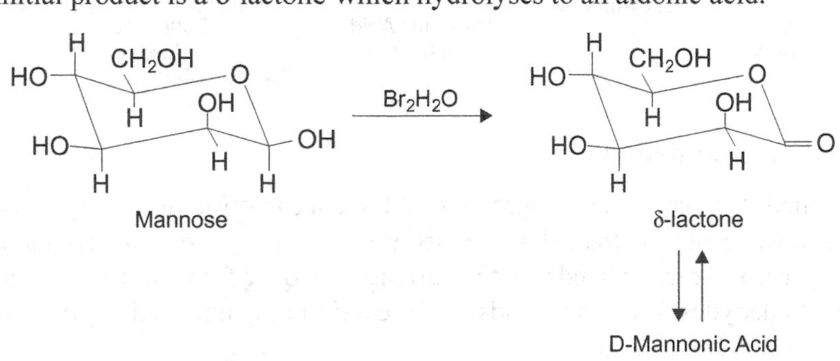

Mannose $\xrightarrow{\text{Br}_2\text{H}_2\text{O}}$ δ-lactone

D-Mannonic Acid

iv. Oxidation with Nitric Acid

When oxidation of an aldose is carried with dilute nitric acid, it oxidises both the —CHO group and the terminal —CH₂OH group to —COOH groups. The resulting carboxylic acid is called an aldaric acid.

$$
\begin{array}{c}
\text{CHO} \\
| \\
\text{(CHOH)}_4 \\
| \\
\text{CH}_2\text{OH}
\end{array}
\quad + \quad \text{HNO}_3 \quad \longrightarrow \quad
\begin{array}{c}
\text{COOH} \\
| \\
\text{(CHOH)}_4 \\
| \\
\text{COOH}
\end{array}
$$

Aldose   Aldaric acid

For example, nitric acid oxidises D-glucose to D-glucaric acid

```
      CHO                          COOH
  H——OH                        H——OH
 HO——H                        HO——H
  H——OH           ⟶           H——OH
  H——OH                        H——OH
     CH₂OH                        COOH
```

D-Glucose   D-Glucaric acid

This oxidation with nitric acid plays an important part in determining the stereochemistry of sugars.

In this reaction, the carbon 1 and 6 are converted into the same functional group. If there is any chirality present in the molecule it becomes clear. For example, glucose and mannose both give optically active acids, whereas D-galactose on oxidation with dilute nitric acid gives optically inactive aldaric acid.

```
      COOH                 COOH                 COOH
  H——OH               HO——H                 H——OH
 HO——H                HO——H                HO——H
  H——OH                H——OH               HO——H      Plane
  H——OH                H——OH                H——OH      of symmetry
      COOH                 COOH                 COOH
```
          No plane
          of symmetry

Glucaric   Mannaric Acid   Galactaric
Acid   Chiral   Acid
Chiral   Achiral

v. Periodate Oxidations

Compounds that have hydroxyl groups on adjacent carbon atoms undergo oxidative cleavage when they are treated with aqueous solution of periodic acid or its sodium salt. The carbon-carbon bond is broken giving rise to the formation of two carbonyl groups (aldehydes, ketones or acids). The reaction is symbolised as given below:

$$
\begin{array}{c}
| \\
-\text{C}-\text{OH} \\
| \\
-\text{C}-\text{OH} \\
|
\end{array}
\quad + \text{HIO}_4 \quad \longrightarrow \quad
\begin{array}{c}
| \\
-\text{C}=\text{O} \\
+ \\
-\text{C}=\text{O} \\
|
\end{array}
\quad + \text{HIO}_3
$$

As the reaction is quantitative, valuable information can be deduced by estimating the moles of HIO_4 used and by identifying the carbonyl compounds.

Periodate oxidations take place, through formation of a cyclic intermediate already given on page 28.

Let us study the periodate oxidations of various polyhydroxy compounds:

1. Ethylene glycol on periodate oxidation gives two moles of formaldehyde and consumes only one mole of periodate.

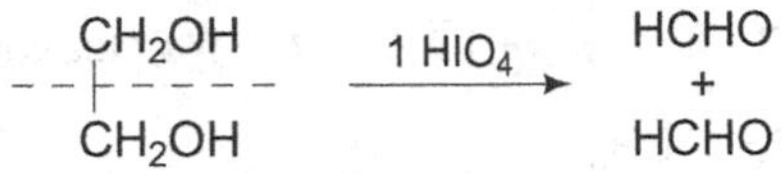

$$
\begin{array}{c}
CH_2OH \\
\text{- -} | \text{- - - -} \\
CH_2OH
\end{array}
\quad \xrightarrow{1 \; HIO_4} \quad
\begin{array}{c}
HCHO \\
+ \\
HCHO
\end{array}
$$

2. Periodate oxidation of glycerol consumes two moles of periodate and gives two molar equivalents of formaldehyde and one molar equivalent of formic acid.

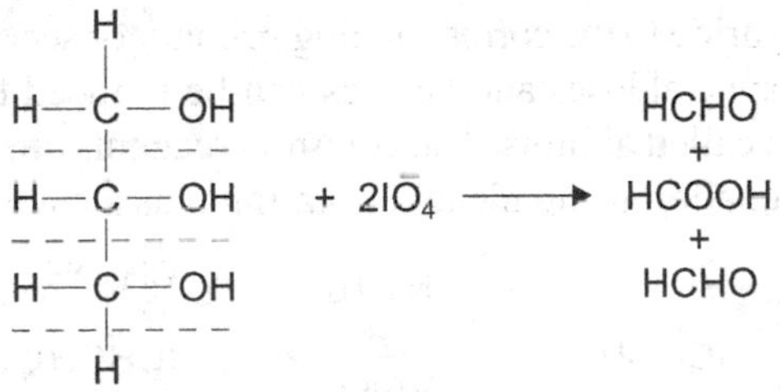

$$
\begin{array}{c}
H \\
| \\
H-C-OH \\
\text{- - -} | \text{- - - -} \\
H-C-OH \\
\text{- - -} | \text{- - - -} \\
H-C-OH \\
\text{- - -} | \text{- - - -} \\
H
\end{array}
\quad + \; 2I\bar{O}_4 \quad \longrightarrow \quad
\begin{array}{c}
HCHO \\
+ \\
HCOOH \\
+ \\
HCHO
\end{array}
$$

3. Oxidative cleavage also takes place when an –OH group is adjacent to the carbonyl group of an aldehyde or ketone. For example, glyceraldehyde yields two molar equivalents of formic acid and one molar equivalent of formaldehyde while dihydroxy ketone gives two molar equivalents of formaldehyde and one molar equivalent of carbon dioxide.

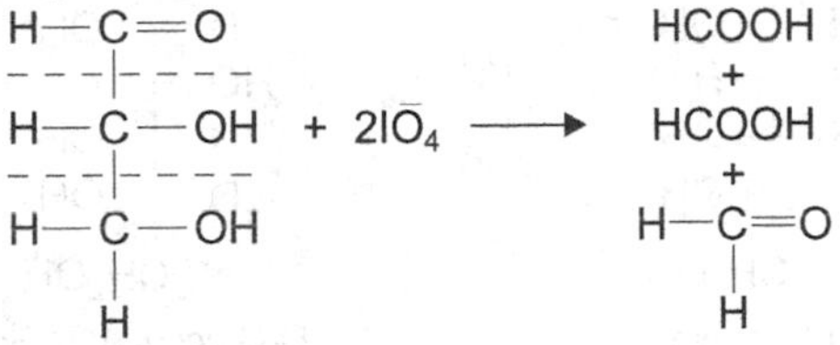

$$
\begin{array}{c}
H-C=O \\
\text{- - -} | \text{- - - -} \\
H-C-OH \\
\text{- - -} | \text{- - - -} \\
H-C-OH \\
| \\
H
\end{array}
\quad + \; 2I\bar{O}_4 \quad \longrightarrow \quad
\begin{array}{c}
HCOOH \\
+ \\
HCOOH \\
+ \\
H-C=O \\
| \\
H
\end{array}
$$

Glyceraldehyde

$$
\begin{array}{c}
H \\
| \\
H-C-OH \\
\text{- - -} | \text{- - - -} \\
C=O \\
\text{- - -} | \text{- - - -} \\
H-C-OH \\
| \\
H
\end{array}
\quad + \; 2I\bar{O}_4 \quad \longrightarrow \quad
\begin{array}{c}
HCHO \\
+ \\
CO_2 \\
+ \\
HCHO
\end{array}
$$

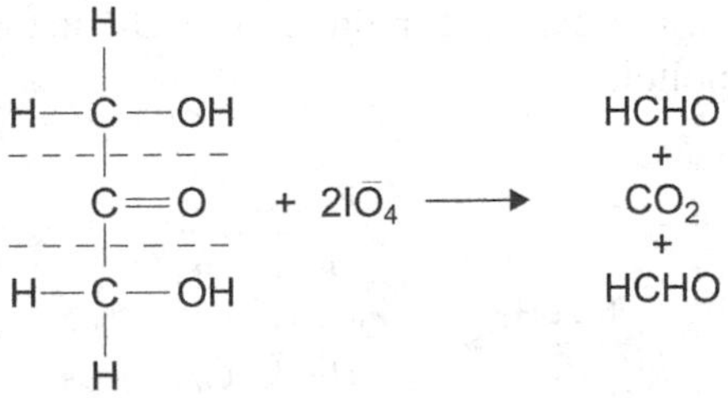

Dihydroxy acetone

4, Periodic acid does not cleave compounds in which the hydroxyl groups are separated by a —CH_2-group nor those in which a hydroxyl group is adjacent to an ether or acetal function

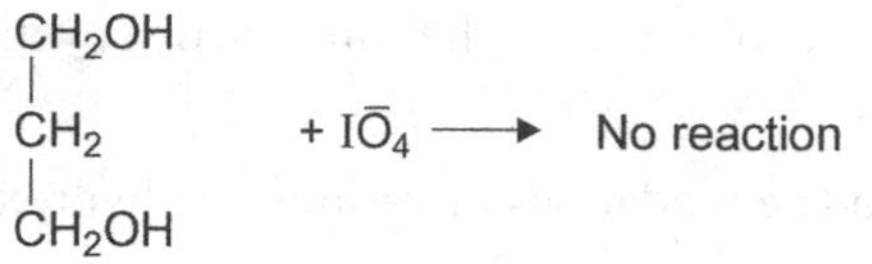

$$
\begin{array}{l}
CH_2OH \\
| \\
CH_2 \\
| \\
CH_2OH
\end{array}
\quad + \ I\bar{O}_4 \longrightarrow \quad \text{No reaction}
$$

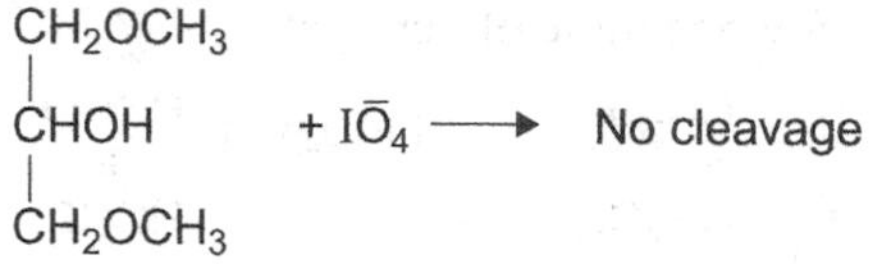

$$
\begin{array}{l}
CH_2OCH_3 \\
| \\
CHOH \\
| \\
CH_2OCH_3
\end{array}
\quad + \ I\bar{O}_4 \longrightarrow \quad \text{No cleavage}
$$

F. Reduction of Monosaccharides

We know reduction of aldehydes and ketones can be carried out catalytically or with sodium borohydride to the corresponding primary or secondary alcohols. Like aldehydes and ketones, aldoses and ketoses can be reduced to the corresponding polyhydric alcohols called alditols. The common reagents are sodium borohydride or catalytic hydrogenation using nickel or platinum as a catalyst.

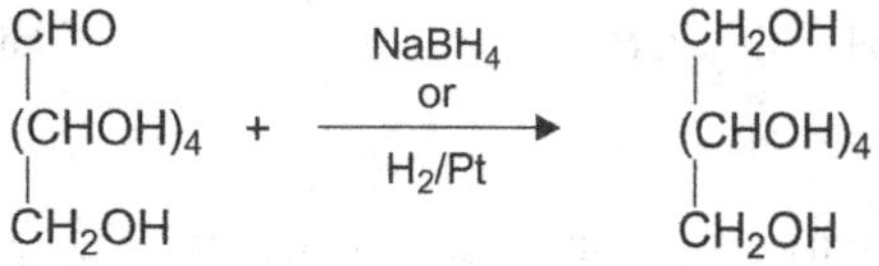

$$
\begin{array}{l}
CHO \\
| \\
(CHOH)_4 \\
| \\
CH_2OH
\end{array}
\quad + \quad
\xrightarrow[\;\;H_2/Pt\;\;]{\substack{NaBH_4 \\ or}}
\quad
\begin{array}{l}
CH_2OH \\
| \\
(CHOH)_4 \\
| \\
CH_2OH
\end{array}
$$

Aldose   Alditol

For example, reduction of D-glucose yields D-glucitol or D-sorbitol.

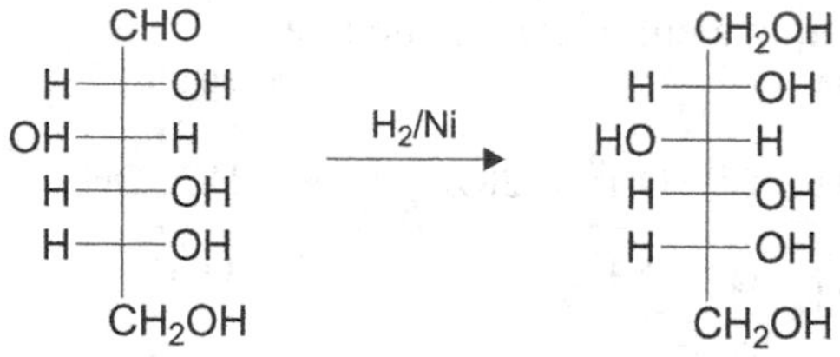

D-Glucose   D-Glucitol or D-Sorbitol

Reduction of a ketose creates a new chiral centre resulting in the formation of two epimeric alcohols. For example, reduction of D-fructose gives a mixture of D-sorbitol and D-mannitol.

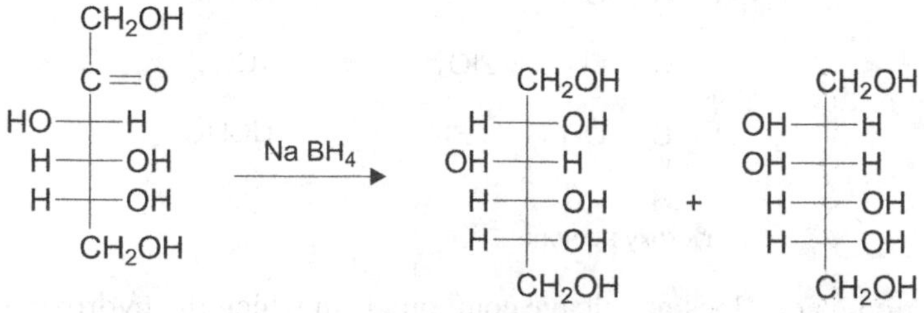

D-Fructose   D-Sorbitol   D-Mannitol

Sugar alcohols are used in industry as food additives and sugar substitutes. For example, sorbitol is used as a sugar substitute, a moistening agent and as a starting material for synthesising vitamin C. Mannitol is obtained commercially from seaweed or it can be made by catalytic hydrogenation of mannose. Similarly, galacitol can be made by catalytic hydrogenation of galactose. It can also be obtained from many plants.

1.15 LENGTHENING THE CARBON CHAIN OF ALDOSES THE KILIANI-FISCHER SYNTHESIS

Aldose chains may be lengthened by one carbon through the cyanohydrin formation. The reaction sequence is known as **Kiliani-Fischer** synthesis. A new chiral centre is created at the cyanohydrin stage.

Thus, we get a pair of epimeric stereoisomers at this site. A pair of aldoses can be recognised as epimers not only by their conversion into the same osazone, but also by their formation in the same Kiliani-Fischer synthesis.

Heinrich Kiliani at the **Technische Hochschule** in **Munich**, showed that an aldose can be converted into the glyconic acids of the higher sugars by addition of HCN and hydrolysis of the resulting cyanohydrins.

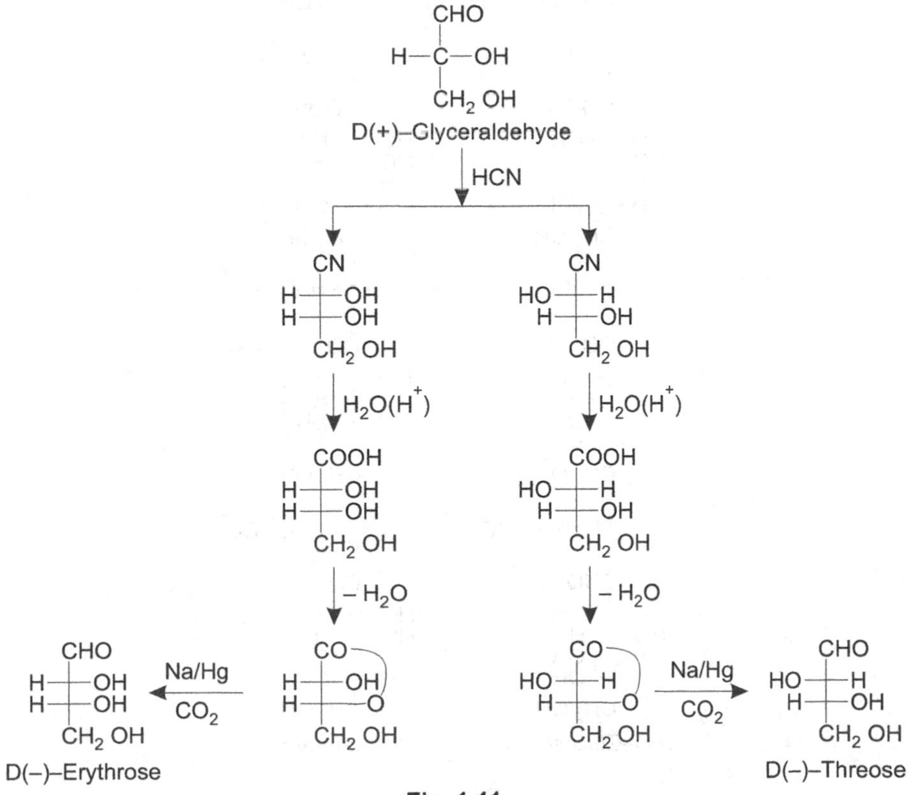

Fig. 1.11.

At the same time **Fischer** reported that reduction of a glyconic acid, in the form of its lactone, can be controlled to yield the corresponding aldose.

Thus, a lower sugar can be converted to two epimeric higher sugars and the total reaction is called Kiliani-Fischer synthesis.

If we subject D(+)-glyceraldehyde to the Kiliani-Fischer synthesis, two stereoisomeric tetroses, erythrose and threose, are formed. The reaction sequence is shown in Fig. 1.11.

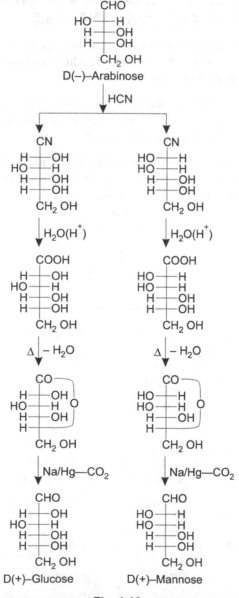

Fig. 1.12.

Similarly, D(–)-arabinose, an aldopentose, can be converted into two epimeric aldohexoses, D(+)-glucose and D(+) mannose (Fig. 1.12). The separation can be achieved at the diastereoisomeric glyconic acids. We would expect these acids to form lactones under acidic conditions. These lactones are finally reduced to the corresponding aldoses. Reduction can also be carried out selectively by using NaBH$_4$.

1.16 SHORTENING THE CARBON CHAIN OF ALDOSES

Aldose chains may be selectively shortened from the aldehyde end by one carbon atom by a number of methods.

A. The Wohl Degradation

This is the reverse of Kiliani-Fischer synthesis. The aldose is converted to its oxime, which undergoes simultaneous acetylation and dehydration to the poly-O-acetyl nitrile when treated with acetic anhydride in the presence of anhydrous sodium acetate. Loss of HCN together with deacetylation occurs when the nitrile is treated with ammoniacal silver oxide to give the lower sugar.

The reaction sequence can be illustrated by taking D-glucose as an example. D-Arabinose is the product (Fig. 1.13).

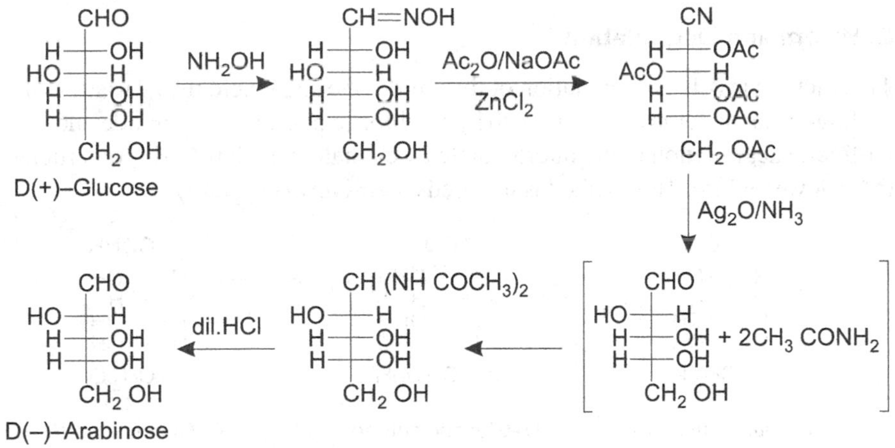

Fig. 1.13. Wohl Degradation.

B. The Ruff Degradation

This is the most widely used method. In this, a soluble salt of an aldonic acid undergoes oxidative decarboxylation with H$_2$O$_2$ in the presence of Fe^{3+} ions to give the lower aldose sugar. It is assumed that the intermediate product is an α-keto acid which undergoes decarboxylation.

D(+)-Glucose is converted into D(+)-arabinose in the following manner (Fig. 1.14):

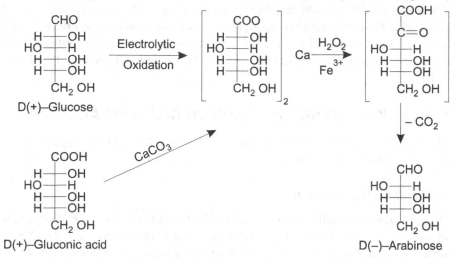

Fig. 1.14.   Ruff Degradation.

Roy L. Whistler improved this method by oxidising glucose with 3 moles of hypochlorite at pH 11 to gluconic acid and then oxidising this with 1.4 moles of the reagent at pH 4.5–5.0.

C. Weermann Degradation

This method involves degradation of the amide of a sugar acid by a variation of the Hofmann reaction. On treatment with hydrochlorite or bromite the amide undergoes Hofmann degradation to the intermediate isocyanate, which then disproportionates to the lower aldose. This method is not used extensively (Fig. 1.15).

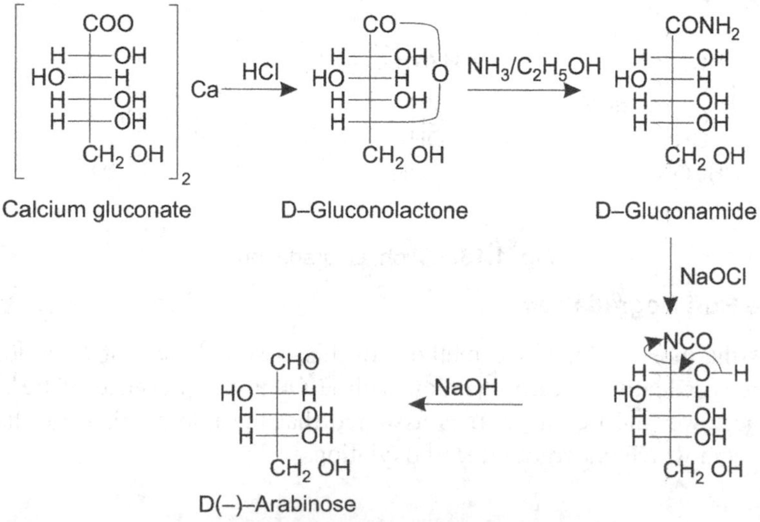

Fig 1.15   Weermann Degradation

1.17 INTERCONVERSIONS

A. The Interconversion of Glucose, Mannose and Fructose

Glucose, mannose and fructose are interconvertible in solution in the presence of a base. This transformation is called **Lobry de Bruyn-Alberda van Ekenstein** rearrangement. If D-glucose is put into a solution of dilute sodium hydroxide and allowed to stand for some days, some of it gets converted into D-fructose and D-mannose. Thus, we get a mixture of the three.

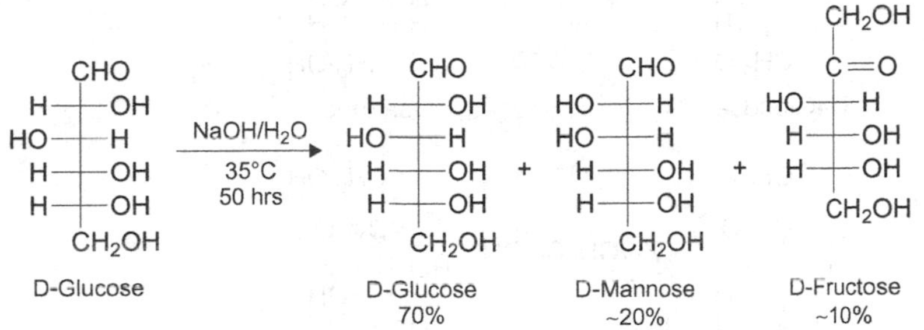

| | D-Glucose | NaOH/H₂O
35°C
50 hrs → | D-Glucose
70% | + | D-Mannose
~20% | + | D-Fructose
~10% |

$$
\begin{array}{c}
\text{CHO} \\
\text{H}-\text{OH} \\
\text{HO}-\text{H} \\
\text{H}-\text{OH} \\
\text{H}-\text{OH} \\
\text{CH}_2\text{OH}
\end{array}
\xrightarrow[\substack{35°C\\50\ hrs}]{\text{NaOH/H}_2\text{O}}
\begin{array}{c}
\text{CHO} \\
\text{H}-\text{OH} \\
\text{HO}-\text{H} \\
\text{H}-\text{OH} \\
\text{H}-\text{OH} \\
\text{CH}_2\text{OH}
\end{array}
+
\begin{array}{c}
\text{CHO} \\
\text{HO}-\text{H} \\
\text{HO}-\text{H} \\
\text{H}-\text{OH} \\
\text{H}-\text{OH} \\
\text{CH}_2\text{OH}
\end{array}
+
\begin{array}{c}
\text{CH}_2\text{OH} \\
\text{C}=\text{O} \\
\text{HO}-\text{H} \\
\text{H}-\text{OH} \\
\text{H}-\text{OH} \\
\text{CH}_2\text{OH}
\end{array}
$$

D-Glucose ⟶ D-Glucose 70% + D-Mannose ~20% + D-Fructose ~10%

If the reaction mixture is kept for long period, more fructose and mannose are formed.

The conversion of D-glucose to D-mannose takes place *via* the enolate ion under basic conditions. The enolate ion is converted to enediol which finally gives D-fructose.

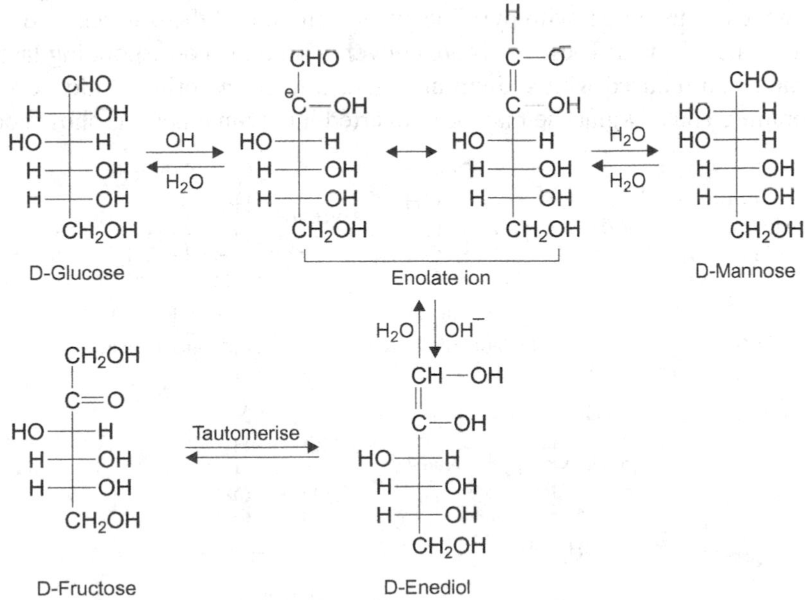

D-Glucose ⇌ Enolate ion ⇌ D-Mannose

D-Fructose ⇌ (Tautomerise) ⇌ D-Enediol

B. Conversion of an Aldose to the Corresponding Ketose

An aldose is converted to the corresponding ketose *via* the osazone. We know that an aldose and its corresponding ketose give the same osazone. So, an aldose is first converted to an osazone using an excess of phenylhydrazine. The osazone is hydrolysed with dilute hydrochloric acid to give osone, which on reduction with zinc and glacial acetic acid gives the corresponding ketose. This is due to the fact that zinc reduces the aldehyde group in preference to the ketonic group. For example, D-glucose may be converted into D-fructose in the following manner:

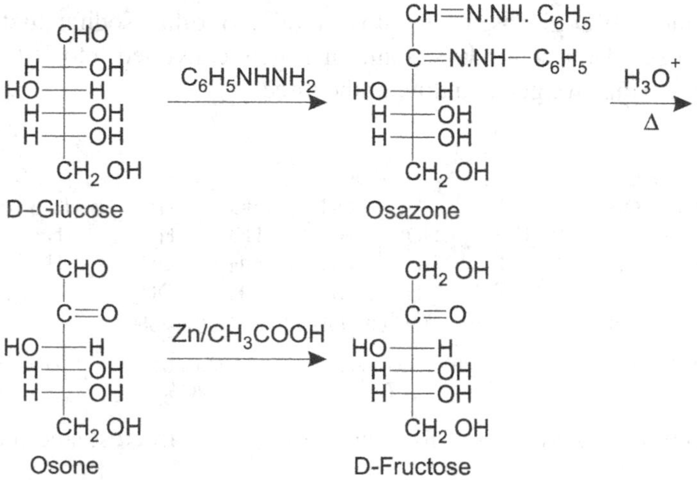

```
        CHO                                    CH=N.NH. C6H5
    H ──── OH                                  C=N.NH─C6H5
   HO ──── H      C6H5NHNH2          HO ──── H              H3O+
    H ──── OH     ───────────→        H ──── OH            ─────→
    H ──── OH                          H ──── OH              Δ
        CH2 OH                              CH2 OH
      D–Glucose                             Osazone
```

```
        CHO                                    CH2 OH
        C=O                                    C=O
   HO ──── H     Zn/CH3COOH         HO ──── H
    H ──── OH    ───────────→        H ──── OH
    H ──── OH                         H ──── OH
        CH2 OH                            CH2 OH
        Osone                           D-Fructose
```

C. Conversion of an Aldose to its Epimer

The aldose is oxidised with bromine water to give the corresponding aldonic acid, which on treatment with pyridine gives a mixture of the original acid and its epimeric aldonic acid. These acids are converted into the corresponding lactones, separated and reduced with sodium amalgam to give the original aldose and its C-2 epimer. Thus, D-glucose may be converted into D-mannose as shown below:

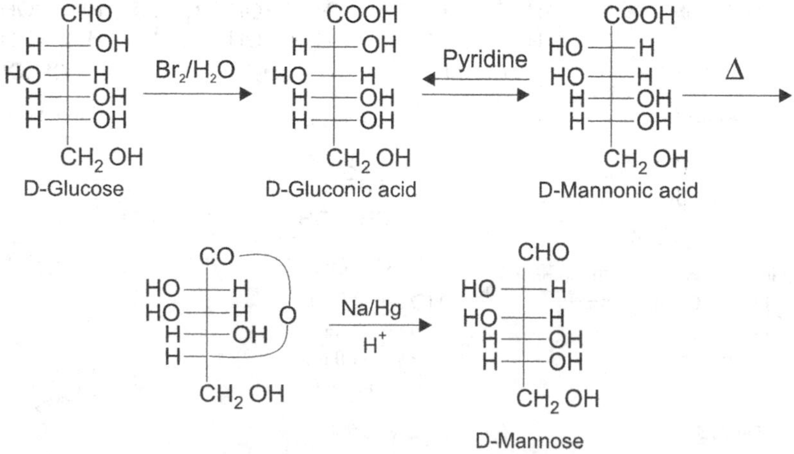

```
     CHO                  COOH                      COOH
  H ── OH              H ── OH       Pyridine    HO ── H
 HO ── H    Br2/H2O   HO ── H      ←─────────   HO ── H       Δ
  H ── OH   ───────→   H ── OH                    H ── OH    ─────→
  H ── OH              H ── OH                     H ── OH
     CH2 OH               CH2 OH                      CH2 OH
  D-Glucose          D-Gluconic acid           D-Mannonic acid
```

```
            CO ─┐                       CHO
        HO ── H │                   HO ── H
        HO ── H │ O    Na/Hg        HO ── H
         H ── OH┘      ─────→        H ── OH
         H ──             H+         H ── OH
            CH2 OH                      CH2 OH
                                     D-Mannose
```

D. Conversion of a Ketose into the Corresponding Aldose

The ketose is reduced catalytically (H_2/Ni) to the corresponding polyhydric alcohol. As a new chiral centre is created, this process results in the formation of two different polyhydric alcohols. These are oxidised together with nitric acid to give the corresponding aldonic acids. These acids on heating give a mixture of γ-lactones, which are separated by fractional crystallisation. The lactones are then reduced with sodium amalgam in a weakly acidic solution to give aldoses, which are isomeric with the original ketose. Thus, D-fructose may be converted into D-glucose and D-mannose as shown below.

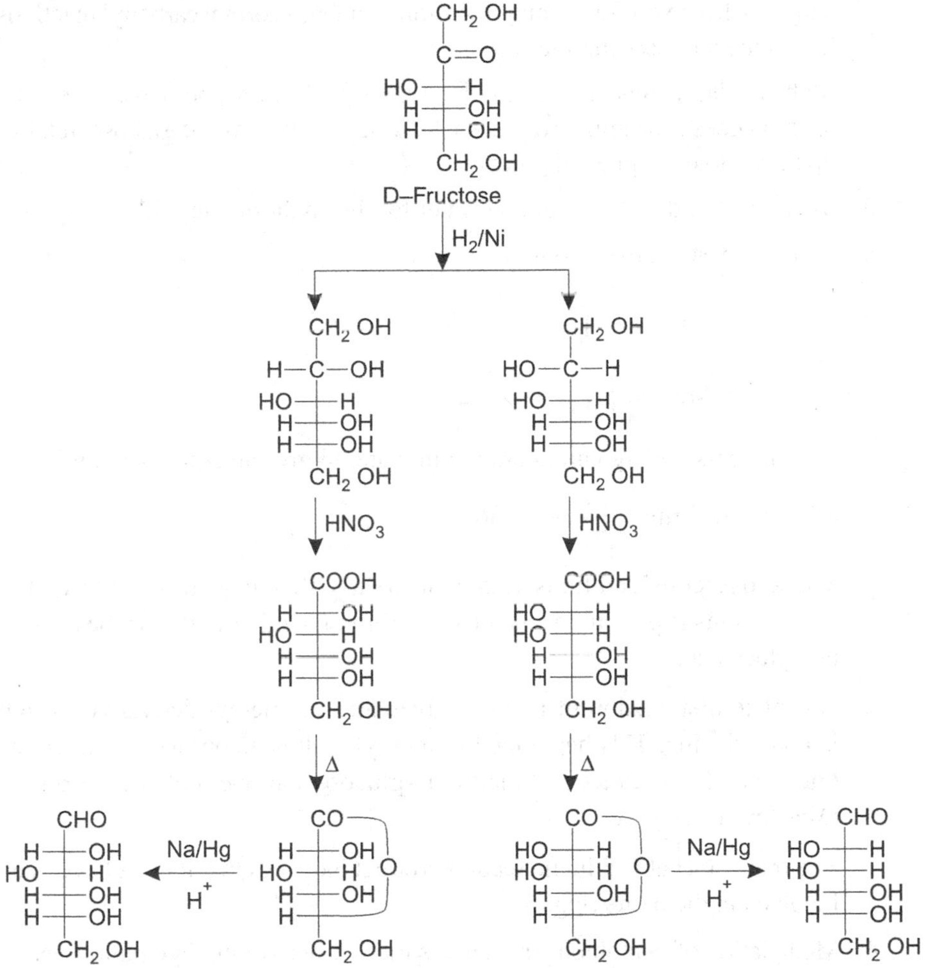

1.18   MALTOSE

Maltose is a simple disaccharide, which is produced by the enzymatic hydrolysis of starch. Diastase (α- and β-amylases) acts on starch to give maltose in about 80% yield. Maltose is also found free in small quantities in some plants, *e.g.*, barley grains.

Maltose is readily fermented by yeast. Under ordinary conditions, it is first converted to glucose by the enzyme, maltase, but at pH 4.5 (conditions which inhibit the action of maltase) direct fermentation occurs.

Structure of Maltose

The structure of maltose is established on the basis of the following observations:

1. Maltose is hydrolysed by dilute acids to yield two moles of D(+)-glucose.

2. Maltose is a reducing sugar, *i.e.*, it gives positive tests with Fehling, Benedict and Tollen's reagents. That means in maltose one of the glucose units has its anomeric hydroxyl free. It may, therefore, undergo normal carbonyl reactions just as the monosaccharides do.

   Maltose also reacts with phenylhydrazine to form a phenylosazone. The reaction goes in a similar way as we have seen in the case of glucose. It takes up three moles of phenylhydrazine.

3. Maltose is oxidised by bromine water to give maltobionic acid.

4. Maltose exists in two anomeric forms:

$$\alpha\text{-Maltose } [\alpha]_D^{25} = + 168°$$

$$\beta\text{-Maltose } [\alpha]_D^{25} = + 112°$$

   The anomers undergo mutatoration in water to give an equilibrium mixture of α- and β-forms $[\alpha]_D^{25} = + 136°$.

   The above considerations establish clearly that in maltose one of the glucose units is present in a hemiacetal form; the other unit must be present as a glucoside.

5. Complete methylation of maltose gives an octa-methyl derivative, which is non-reducing. This hepta methyl-methyl maltoside on hydrolysis yields one mole of 2,3,4,6-tetra-O-methyl-D-glucose and one mole of 2,3,6-trio-O-methyl-D-glucose.

   The free —OH at C-5 in the second product indicates that it must have been involved in the oxide ring.

6. Methylation of maltobionic acid gives methyl-octa-O-methyl maltobionate, which is hydrolysed under acidic conditions to give 2,3,4,6-tetra-O-methyl-D-glucose and 2,3,5,6-tetra-O-methyl-D-gluconic acid (Fig. 1.16).

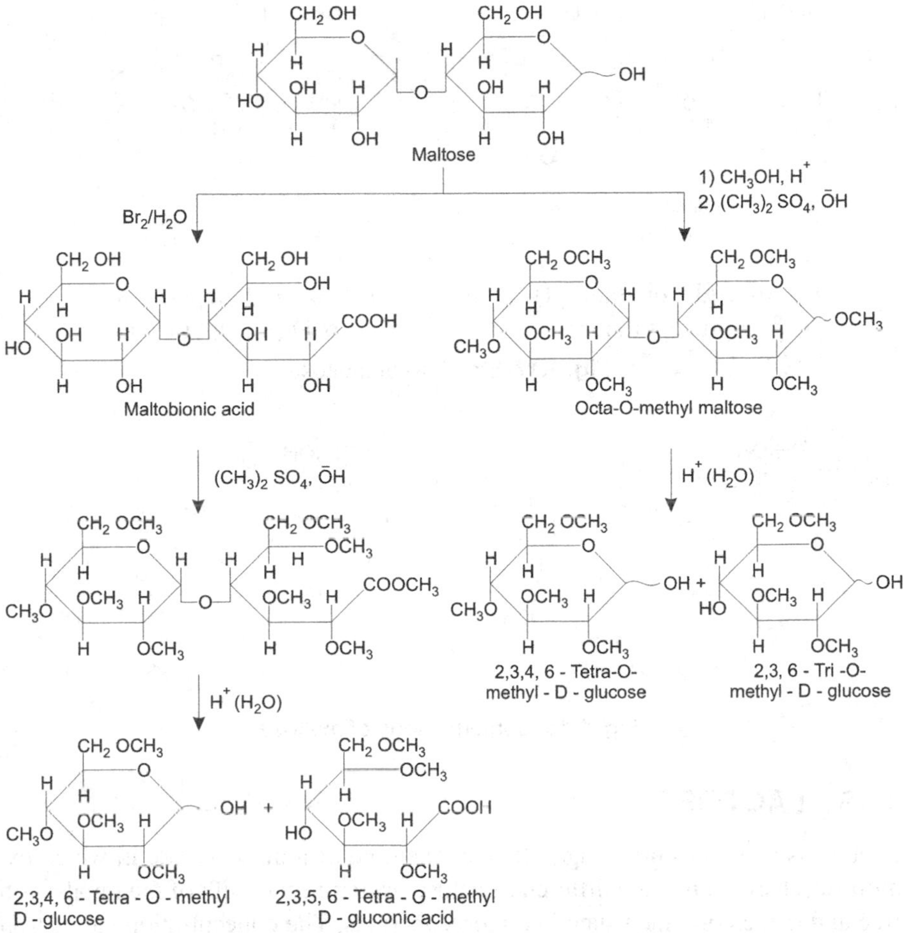

Fig. 1.16

The second product has free hydroxyl at C-4, which must have been involved in the glycosidic linkage.

These facts show that both the glucose units in maltose are in the pyranose form and C-4 hydroxyl of the reducing sugar is involved in a glycosidic linkage with C-1 hydroxyl of the non-reducing glucose.

7. The only remaining question is the stereochemistry of the glycosidic linkage in maltose. This is established by the fact that maltose is hydrolysed by maltase, an enzyme specific for α-glucosides whereas β-glucosidase does not affect the disaccharide. Thus, the glycosidic linkage in maltose is α. The high specific rotation of maltose and its derivatives is consistent with the above facts.

Maltose is, therefore, designated as 4-0-(α-D-glucopyranosyl)-D-glucopyranose. Thus, we have two anomers of maltose as shown below (Fig. 1.17 and Fig. 1.18)

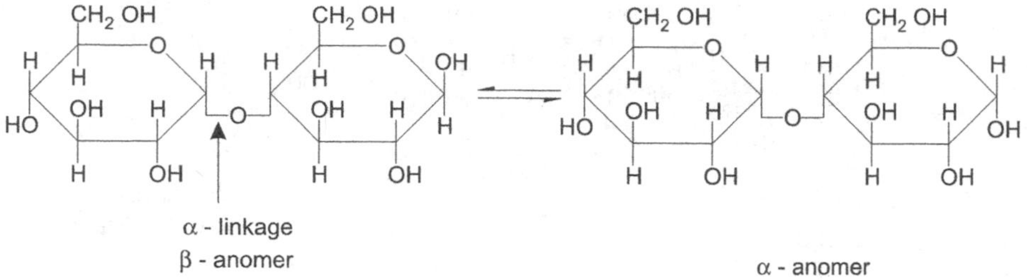

α - linkage
β - anomer
                                                        α - anomer

4-0-(α-D-glucopyranosyl)-   4-0-(α-D-glucopyranosyl)-
    β-D-glucopyranose   α-D-glucopyranose

Fig. 1.17 Structure of maltose.

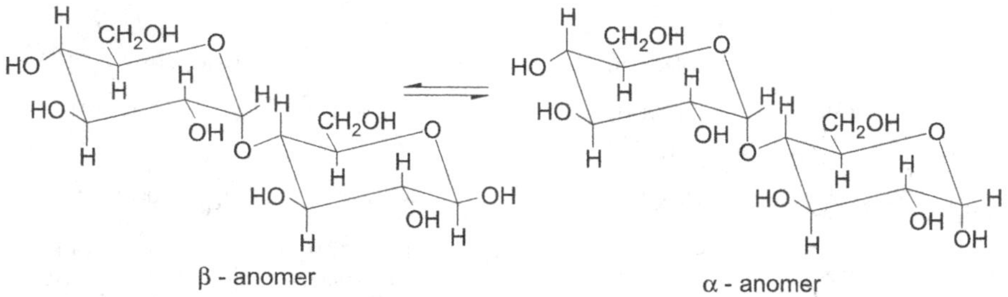

β - anomer   α - anomer

Fig. 1.18 Conformations of maltose.

1.19 LACTOSE

Lactose is known as milk sugar. It is an example of a disaccharide in which two monosaccharide units are different. Lactose occurs in the milk of mammals, both free and in the combined state in oligosaccharides. The concentration varies from 2% to 9%, the average being about 5% by weight of mammalian milk. It also occurs in a number of fruits. Crystals of lactose are gritty to the tongue and only faintly sweet. The disaccharide is readily converted to lactic acid by the lactic bacillus. This process is responsible for the souring of milk.

Lactose is produced commercially from whey, which is obtained as a byproduct in the manufacture of cheese. Evaporation of whey at temperatures below 95°C causes the less soluble α-anomer to precipitate out. Large amounts of lactose are being used for the preparation of homoeopathic medicines.

Structure of Lactose

The following facts establish the structure of lactose as 4-0-(β-D-galactopyranosyl)-D-glucopyranose:

1. Hydrolysis of lactose affords one mole of galactose and one mole of glucose.

2. Lactose is a reducing disaccharide. It responds to various tests of reducing sugars as given by reducing monosaccharides. It also forms an osazone.

3. Lactose exists in α- and β-forms, which undergo mutarotation.

   The α-isomer is more stable, has m.p. 252°C, $[\alpha]_D = +90°$ changes to +55.4° (water), β-D(+)-lactose has $[\alpha]_D = +35°$.

4. That the galactose moiety is in the non-reducing part of the molecule is clearly shown by the hydrolysis of lactobionic acid to galactose and gluconic acid (Fig. 1.19).

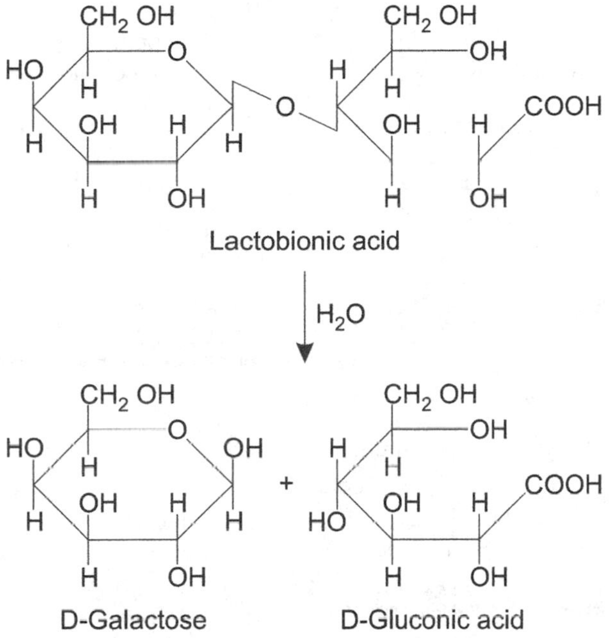

Lactobionic acid

H_2O

D-Galactose   D-Gluconic acid

Fig 1.19

5. The glycosidic linkage between galactose and glucose in the lactose molecule is shown by application of the methylation technique.

   Hydrolysis of octa-O-methyl lactose affords 2,3,4,6-tetra-O-methyl-D-galactopyranose and 2,3,6-tri-O-methyl-D-glucopyranose. This means that C-4 hydroxyl of the glucose moiety is bound in the glucosidic linkage (Fig. 1.20).

6. Further complete methylation of lactobionic acid gives methyl-octa-O-methyl lactobionate, which on acid hydrolysis yields 2,3,4,6-tetra-O-methyl-D-galactopyranose and 2,3,5,6,-tetra-O-methyl-D-gluconic acid. (Fig. 1.20).

   The above facts show that anomeric hydroxyl (C-1) of the galactose is bound in the form of glycosidic linkage to C-4 of the glucose moiety.

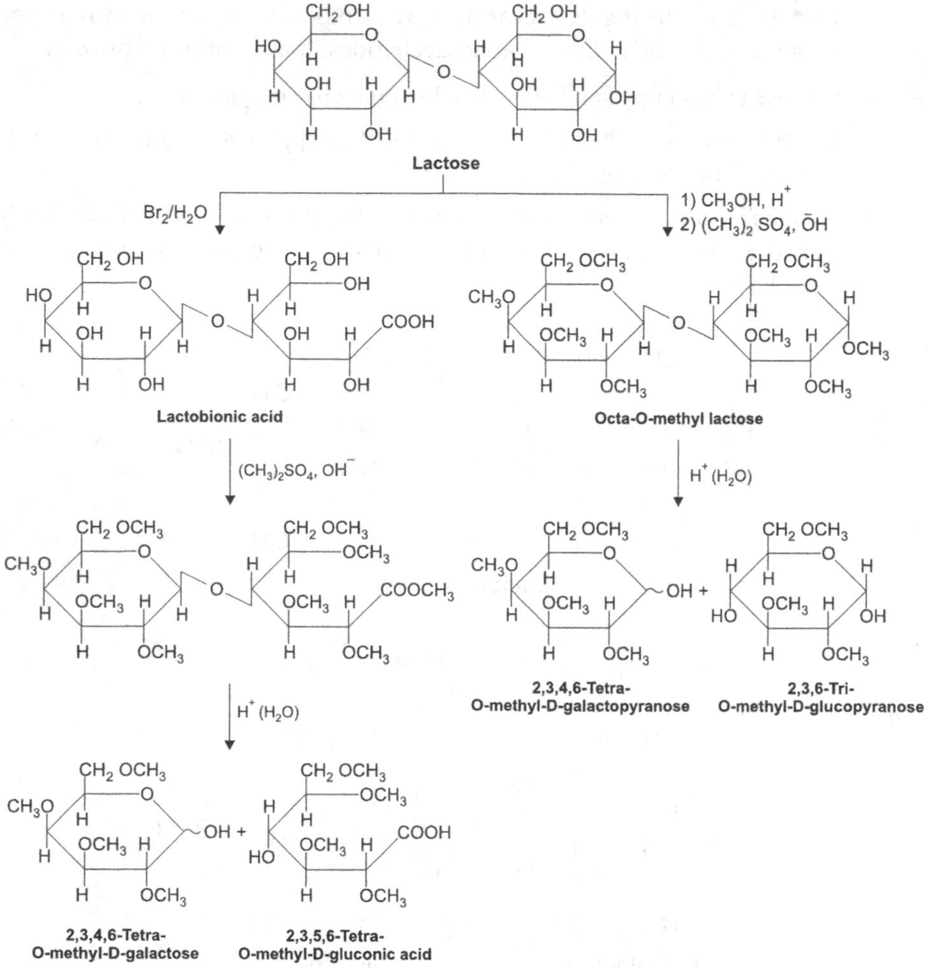

Fig. 1.20

7. Lastly, the hydrolysis of lactose is catalysed by an enzyme called β-galactosidase which is specific for β-galactoside linkages.

Therefore, the structure of lactose is given as in Fig. 1.21:

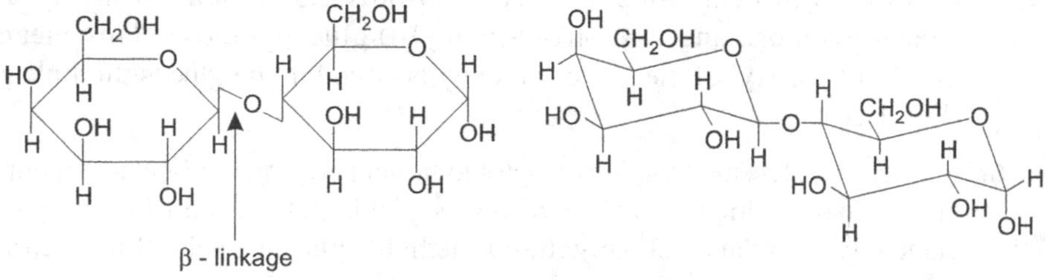

Fig. 1.21 Structure of lactose.

1.20 CELLOBIOSE

We have already mentioned that cellobiose is obtained as a result of partial hydrolysis of cellulose. It is generally isolated as the octa-acetate when cellulose is treated with acetic anhydride and sulphuric acid. Acetylation and hydrolysis take place simultaneously in this process. Alkaline hydrolysis of the octa-acetate gives cellobiose.

Cellobiose with molecular formula $C_{12}H_{22}O_{11}$ resembles maltose in every respect except one and that is the configuration of its glycosidic linkage.

Structure of Cellobiose

The structure of cellobiose is deduced on the basis of the following observations:

1. Like maltose cellobiose also yields two moles of D-glucose on acid hydrolysis.
2. It is a reducing sugar and forms an osazone. On oxidation with bromine water cellobiose gives cellobionic acid.
3. Cellobiose exists in α- and β-forms, which undergo mutarotation. As we have seen in other sugars the β-form of cellobiose is more stable.

    It has m.p. 225°C, $[\alpha]_D = +14°$ changes to +35° in water.

    α-D(+)-Cellobiose, $[\alpha]_D = +72°$

    β-D(+)-Cellobiose, $[\alpha]_D = +14°$

4. Methylation of cellobionic acid gives octa-O-methyl derivative, which on hydrolysis affords the same products, *i.e.*, 2,3,4,6-tetra-O-methyl-D-glucose and 2,3,5,6-tetra-O-methyl-D-gluconic acid as hydrolysis of octa-O-methyl-maltobionic acid.

    That means cellobiose is isomeric with maltose.

5. The glycosidic linkage in cellobiose is, however, decided on the basis of enzymatic hydrolysis. The hydrolysis of cellobiose is catalysed by β-glucosidase and not by α-glucosidase.

Therefore, the structure of β-D-cellobiose is given as 4-0-(β-D-glucopyranosyl)-β-D-glucopyranose (Fig. 1.22).

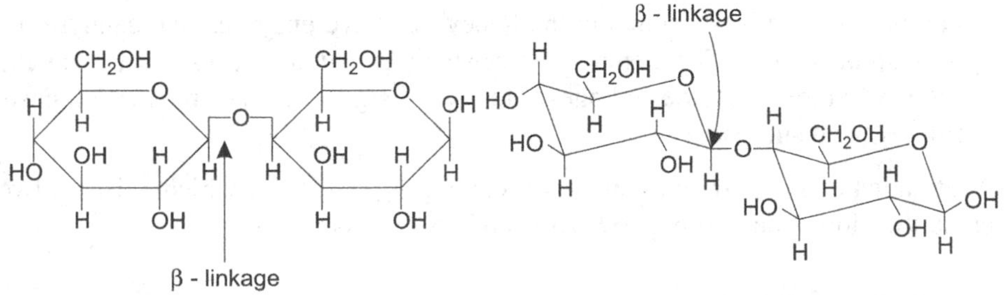

Fig. 1.22 Structure of cellobiose.

1.21 SUCROSE

We are quite familiar with ordinary table sugar, which is a disaccharide called sucrose. Large amounts of it are being consumed daily the world over.

Sucrose is one of the most widespread sugars in nature and is one of the few pure organic compounds produced industrially in very large amounts. It occurs almost universally in the plant kingdom and is found in all parts of the plant. The richest sources are sugar cane (*Saccharum officinarum*) and the sugar beet (*Beta vulgaris*). In sugar cane, it occurs to the extent of about 20% of the juice. Sucrose is commercially obtained from these sources.

Sucrose exists in two crystalline forms, sucrose A, m.p. 184°–185°C, which is the stable form obtained on recrystallisation from most solvents and sucrose B, m.p. 169°–170°C, obtained by recrystallisation from methanol. Molten sugar on cooling forms a glass which slowly becomes crystalline and loses its transparency. That is how we get various preparations of sucrose. At 195°–200°C sucrose changes into a brown uncrystallisable product.

Sucrose is dextrorotatory having $[\alpha]_D = +66°$.

D-Glucose is also dextrorotatory. The equilibrium mixture of α- and β-anomers of glucose has $[\alpha]_D = +52.5°$. But D-fructose is laevorotatory, the equilibrium mixture having $[\alpha]_D = -92.4°$.

Sucrose is hydrolysed by invertase or on warming with dilute acids to give an equimolar mixture of D-glucose and D-fructose. In the process of hydrolysis the dextrorotatory sucrose solution becomes laevorotatory because an equilibrium mixture of D-glucose and D-fructose has $[\alpha]_D = -20°$. The commonly encountered mixture is called 'invert sugar'. The name is given from the process of inversion in the sign of rotation that occurs during its formation.

$$\text{Sucrose} \xrightarrow[\substack{\text{or} \\ \text{Invertase}}]{H^+ (H_2O)} \alpha\text{- and }\beta\text{-D-glucose} + \alpha\text{- and }\beta\text{-D-fructose}$$

$[\alpha] = +66°$

$[\alpha]_D = +52.5°$ $[\alpha]_D = -92.4°$

$[\alpha]_D = -20°$

A number of organisms including the honey bee have enzymes that catalyse the hydrolysis of sucrose. These enzymes are usually called invertases and are specific for the β-fructofuranoside linkages. Honey is largely a mixture of D-glucose, D-fructose, and sucrose.

Acetylation of sucrose gives an octa-acetate, $[\alpha]_D = +59.6°$, which exists in two crystalline forms having m.p. 69–70°C and 75°C respectively.

Structure of Sucrose

The structure of sucrose is based on the following observations:

1. Acid hydrolysis of one mole of sucrose yields one mole of D-glucose and one mole of D-fructose.

2. Sucrose is a non-reducing sugar. It does not reduce Tollen's reagent and gives a negative test with Benedict's solution.

3. Sucrose does not form an osazone. However, on treatment with dilute hydrochloric acid inversion takes place. The mixture then deposits osazone on treatment with phenylhydrazine. It results in inversion of the sign of rotation. The resulting mixture is known as invert sugar.

4. Sucrose does not undergo mutarotation.

   These facts suggest that there is no hemiacetal or hemiketal group present in the glucose or fructose portion of sucrose. The two units are linked together by a glycosidic linkage between the two anomeric carbons. Thus, C-1 of D-glucose is linked to C-2 of D-fructose.

5. The ring size of D-glucose and D-fructose in sucrose is established on the basis of usual methylation studies. Thus, methylation of sucrose gives an octamethyl derivative which on acid hydrolysis gives 2,3,4,6-tetra-*O*-methyl glucose and 1,3,4,6-tetra-*O*-methyl fructose. The formation of these products indicates that in sucrose, glucose is in the pyranose form and fructose is in the furanose form (Fig. 1.23).

6. The configuration at the glycosidic linkage follows from studies with enzymes.

Sucrose is hydrolysed by an α-glucosidase obtained from yeast and not by β-glucosidase. This shows α-configuration at the glucoside portion. It is unaffected by emulsion (β-glucosidase).

Sucrose is also hydrolysed by sucrase, an enzyme specific for β-fructofuranosides. This indicates that fructose has β-configuration.

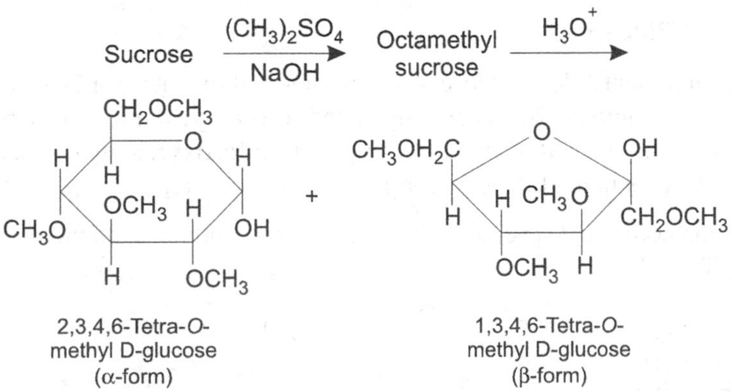

Fig. 1.23.

Thus, the structure of sucrose is α-D-glucopyranosyl-β-D-fructofuranoside (Fig. 1.24).

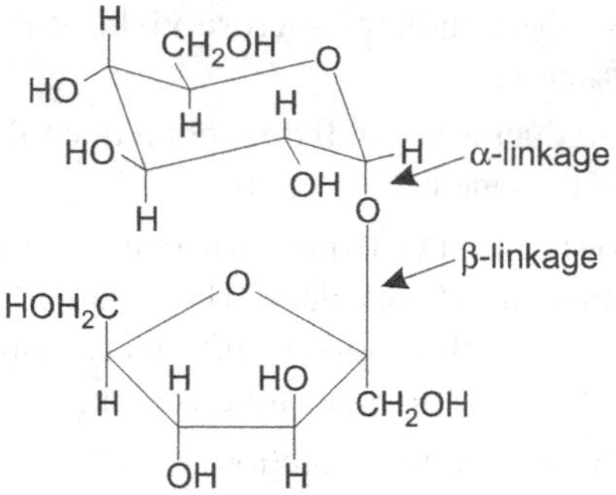

Fig. 1.24 Structure of sucrose.

1.22 ALDOPENTOSES—THE SUGARS OF NUCLEIC ACIDS

A. Deoxy Sugars

Deoxy sugars are those compunds in which one or more hydroxyl groups have been replaced by hydrogens. Mono, di- and trihydroxy sugars occur naturally. The 6-deoxyaldohexoses occur widely in nature, *e.g.*, 6-deoxy-l-mannose, known as l-rhamnose and 6-deoxy- L-galactose (L-fucose) are quite common.

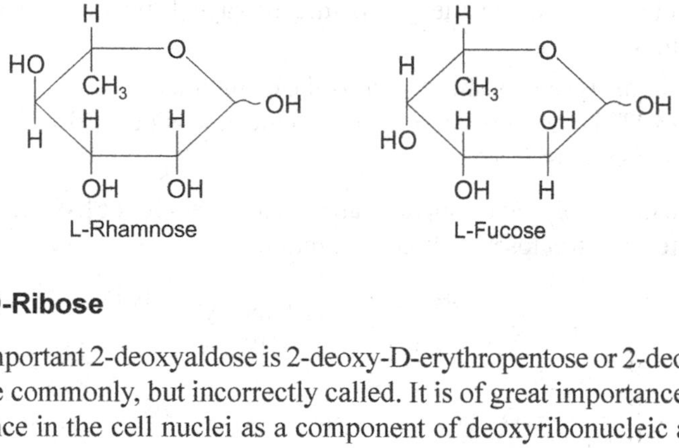

L-Rhamnose   L-Fucose

2-Deoxy-D-Ribose

The most important 2-deoxyaldose is 2-deoxy-D-erythropentose or 2-deoxy D ribose as it is more commonly, but incorrectly called. It is of great importance because of its occurrence in the cell nuclei as a component of deoxyribonucleic acid (DNA) in which it exists in the furanosyl form.

The melting points and specific rotations of the two anomers and their acetates are given in Table 1.5

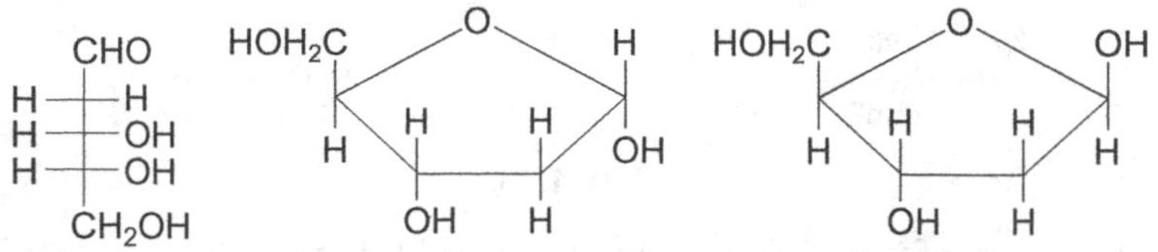

Open-chain   α-anomer   β-anomer

TABLE 1.6

2-Deoxy-D-ribose	m.p	$[\alpha]_D$
α-anomer (furanose)	78-82°C	− 55°
β-anomer (furanose)	96-98°C	−91 → − 58°C
		(mutarotates)
Tri-O-actate ((β-furanose)	98-99°C	− 139°
Tri-O-acetate (β-pyranose)	90-91°C	− 169°

Crystalline 2-deoxy-β-D-erythropentose has been shown by X-ray studies to exist in the seemingly unfavourable pyranose conformation.

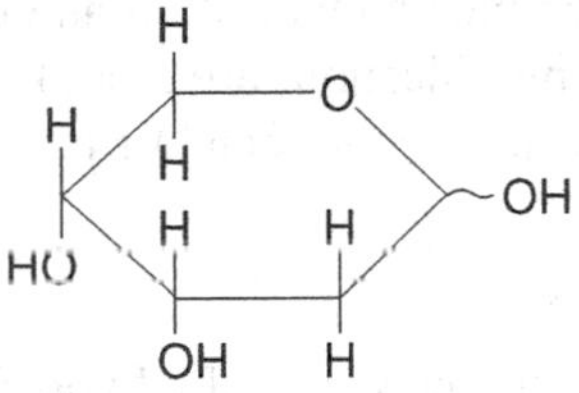

Haworth structure

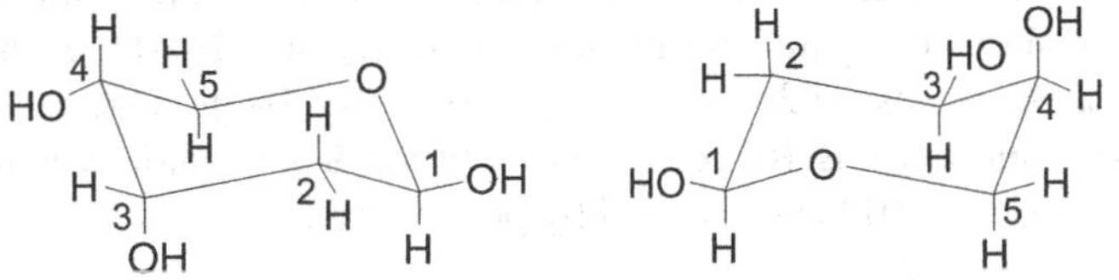

Synthesis of 2-Deoxy-D-ribose

2-Deoxy-D-ribose has been synthesised by the nitromethane condensation method as given below:

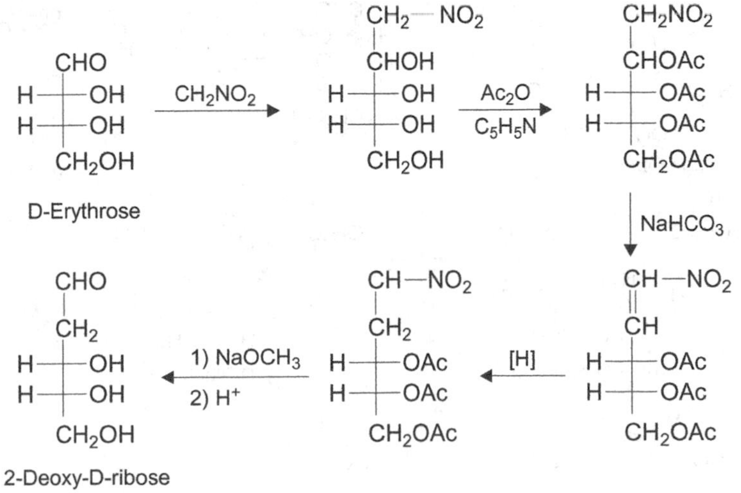

B. D-Ribose

D-Ribose, m.p. 87°C, $[\alpha]_D = -23.1° \rightarrow -23.7°$, is a component of the nucleic acid, RNA and several vitamins and coenzymes. Therefore, it is present in the cells of all animal and plant material. It is also a component of the antibiotics Neomycins A and B and Paromomycin. D-Ribose arises in nature from D-glucose as the 6-phosphate on oxidative decarboxylation. The pentose may be prepared by the partial hydrolysis of yeast nucleic acids either with magnesium oxide or enzymes to nucleosides, which are then cleaved by acid hydrolysis to give D-ribose after the removal of the aglycone.

The sugar occurs in natural products in the furanosyl form. In aqueous solution, the free sugar exhibits a complex mutarotation and considerable quantities of the furanose and straight chain modifications are thought to be present. D-Ribose is not fermentable by ordinary yeasts. Commercial D-ribose is best purified *via* its anilide.

Mutarotation of D-Ribose

Evidence for the pyranose, furanose and aldehydo forms of D-ribose in aqueous solution has been obtained by NMR studies in D_2O, but none of the other aldopentoses showed significant amounts of furanoses. It was estimated that at equilibrium at 35°, ribose yields 6% of α-furanose, 18% β-furanose, 20% α-pyranose and 50% β-pyranose forms. Ribose forms about 8.5% of the aldehydo form whereas glucose forms only 0.01% as revealed by polarography.

The conformation of the furanose ring has an important bearing upon the equilibrium composition. Both pyranose and furanose forms of D-ribose contain unfavourable non-bonded interactions which are probably responsible for the abnormal concentration of the aldehydo form.

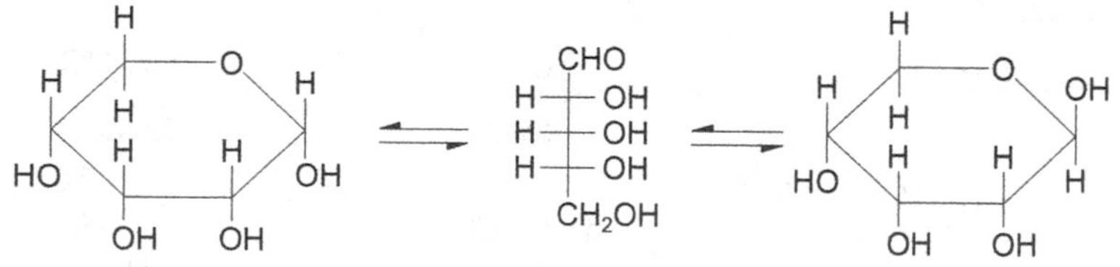

α-D-Ribopyranose   β-D-Ribopyranose

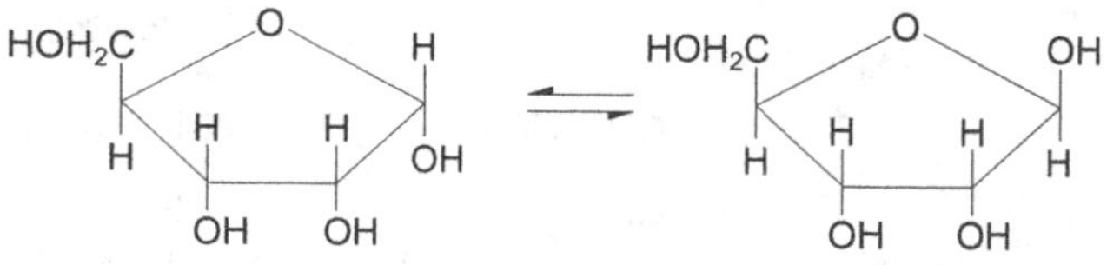

α-D-Ribofuranose   β-D-Ribofuranose

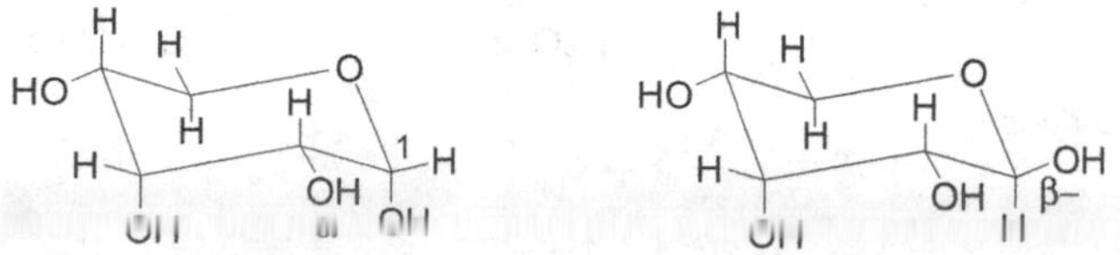

The concentration of aldehydo and furanose forms in solution tends to increase with conformational instability of the pyranose form. As axial-hydroxyl on the stable chair conformations as in D-galactose and D-arabinose results in greater amounts of aldehydo forms in solution than are produced by all equatorial D-glucose and D-xylose, which exist as expected predominantly in the favourable pyranose chair conformations.

It has been shown by NMR that there is a conformational equilibrium between the chair forms of α-D-ribopyranose tetra-acetate.

1.23 POLYSACCHARIDES

A polysaccharide is a naturally occurring sugar polymer. As the name implies the polymer contains many sugar units; the actual number runs into thousands. Molecular weights have been reported as high as 1,000,000. In spite of the large number of —OH groups, only the smaller polymers exhibit water solubility. The bulk of the polysaccharides are insoluble in water. They may be broken into soluble derivatives only by enzymes or by acid or base catalysed hydrolysis.

Starch and cellulose are the most important of the polysaccharides. We have already mentioned that starch is the reserve food supply for plants while cellulose is the structural unit which provides both rigidity and form. We use starch and cellulose in the same way as nature does, starch as food and cellulose as basic structural material.

A. Cellulose

As mentioned earlier, cellulose is the main constituent of cell walls of plants to support the framework. Thus, cellulose is the chief component of wood and plant fibres and is very widely distributed in nature. Cotton, for example, is almost 100% pure cellulose. The cell walls in wood are only 50% cellulose. The chief impurities which are associated with raw cotton are waxes and fats. These impurities are removed by washing the raw cotton with alcohol. It is then treated with hot caustic soda solution, washed with water and dried to give pure amorphous cellulose. It is insoluble in water and is tasteless. We use cellulose for its structural properties. We use wood for houses and furniture, as cotton for clothing, as paper for communications and packaging.

Structure of Cellulose

1. Cellulose is a polysaccharide with molecular formula $(C_6H_{10}O_5)_n$.
2. It is composed entirely of D-glucose units as complete hydrolysis by acid yields D(+)-glucose as the only monosaccharide. That means, D(+)-glucose is the monomer of cellulose.
3. Hydrolysis of completely methylated cellulose gives a high yield of 2,3,6-tri-O-methyl-D-glucose. Therefore, cellulose is made up of chains of D-glucose units, each unit joined by a glycoside linkage to C-4 of the next.

4. On treatment with acetic anhydride and sulphuric acid, cellulose yields octa-O-acetyl-cellobiose. Therefore, cellulose may be considered as a polymer of cellobiose.

5. Further, cellulose on hydrolysis with enzymes yields cellobiose. We have already seen that cellobiose has β-glycosidic linkage. That means all glycosidic linkages in cellulose are β-linkages.

Therefore, the structure of cellulose is represented as in Fig. 1.25.

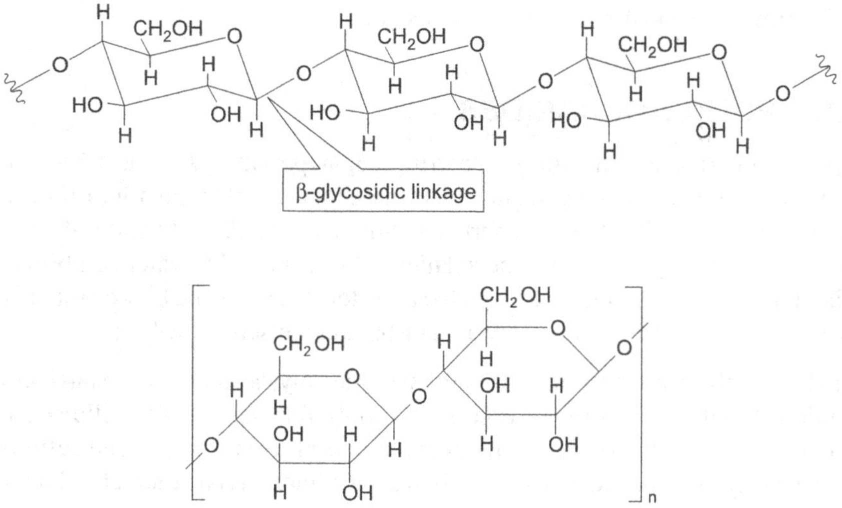

Fig. 1.25 Structure of cellulose.

The above structure of cellulose indicates that it is made up of a long chain of cellobiose units in which glucose is linked alternately through the carbon atoms in positions 1 and 4. The linkage is β and conformation is completely equatorial. Each glucose unit contains three free hydroxyl groups except at the ends of the chain. One end contains a free or potential aldehyde group and the other end has an extra hydroxyl group at position 4.

The cellulose molecule has been shown to have molecular weights ranging from 250,000 to 1,000,000. That means there are at least 1500 glucose units present in cellulose. These long chains lie side by side in bundles, which are held together by hydrogen bonds between the neighbouring —OH groups. These bundles are twisted together to form rope-like structures which are grouped to form the fibres. In wood, these cellulose ropes are embedded in lignin to give the compact structure.

This compact structure of cellulose gives another interesting and important fact about cellulose. We, human beings, cannot digest cellulose. The digestive enzyme of humans cannot attack its β, 1:4 linkages. Hence, cellulose cannot serve as a food source for us as can starch. However, cows and other animals can use the cellulose of grass as a food source because their digestive juices contain β-glucosidases.

B. Starch

We have already mentioned above that starch makes up the reserve food supply of plants and occurs chiefly in seeds. It is more water soluble than cellulose, more easily hydrolysed and hence more readily digested. We use starch as a food. The chief sources are potatoes, corn, wheat, rice, barley *etc*. Starch occurs as granules in the roots, tubers and seeds of plants which vary in size and shape depending upon their plant source.

Starch is not a pure compound. It is a mixture of two polysaccharides – amylose and amylopectin. Amylose is a water-soluble fraction and gives a deep blue colour with iodine. In general, starch contains 20% of amylose and 80% of amylopectin, which is water-insoluble and gives no colour with iodine.

Both these forms of starch correspond to different carbohydrates of high molecular weight having the general formula $(C_6H_{10}O_5)_n$. When starch is treated with dilute acid or appropriate enzymes, it is slowly hydrolysed to maltose and then to D-glucose. That means both amylose and amylopectin are made up of D(+)-glucose units, but differ in molecular size and shape.

i. Structure of Amylose

1. Hydrolysis of amylose gives maltose as the only disaccharide and D(+)-glucose as the only monosaccharide. Therefore, amylose is made up of chains of many D(+)-glucose units. As maltose is also the product of hydrolysis of amylose, that means each glucose unit is joined by a-glycosidic linkage to C-4 of the next one.

2. When amylose is methylated and hydrolysed, we get mainly 2,3,6-trio-O-methyl-D-glucose. That is expected when each D(+)-glucose unit in amylose is joined to two others; it contains only three free —OH groups which are methylated to give the tri-methyl product (Fig. 1.26).

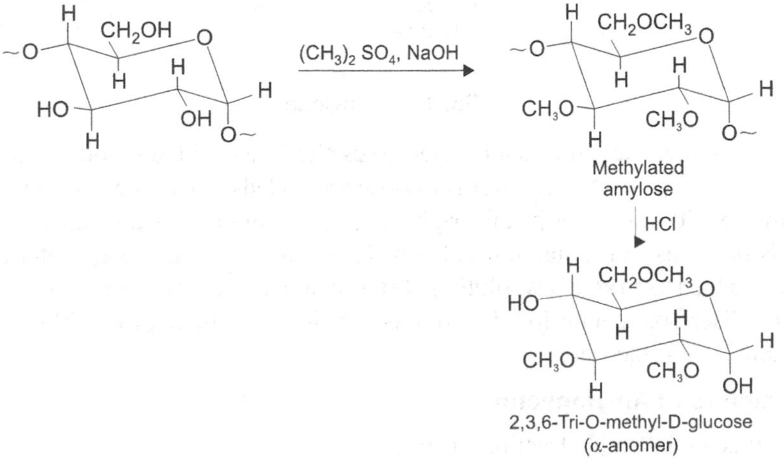

Methylated amylose

2,3,6-Tri-O-methyl-D-glucose
(α-anomer)

Fig. 1.26.

However, we also get a small amount of 2,3,4,6-tetra-O-methyl-D-glucose (Fig. 1.27), which is about 0.2 – 0.4% of the total product.

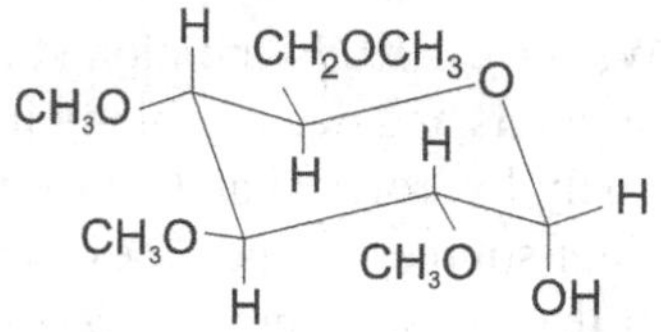

Therefore, we conclude that each D-glucose unit in amylose is attached to two other D-glucose units, one through C-1 and the other through C4. The formation of the above products indicates that amylose is a straight chain polymer with one end having a free aldehyde group. The other end has a D-glucose unit that has a free —OH at C4.

2,3,4,6-Tetra-O-methyl-D-glucose
(α-anomer)

Fig. 1.27.

Thus, each molecule of completely methylated amylose on hydrolysis should give one molecule of 2,3,4,6-tetra-O-methyl-D-glucose along with a number of tri-O-methyl-D-glucose molecules. The formation of 0.2% of tetra-O-methyl derivative indicates the chain length of about 500 units. But physical methods suggest that the chains are longer, 1000 to 4000 glucose units per molecule of amylose. That means some degradation results during methylation and hydrolysis of the amylose molecule.

Therefore, amylose is a linear polymer of D-glucose units joined together by α-linkages as in maltose. There is no branching of the chain (Fig. 1.28). Amylose differs from cellulose only in the stereochemistry of the glycosidic linkage.

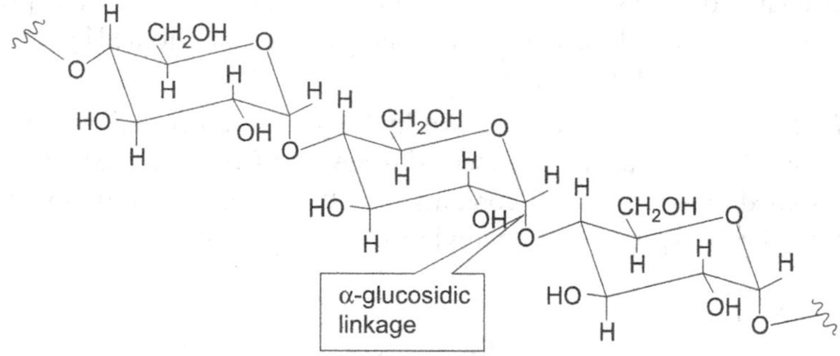

Fig. 1.28 Amylose.

Amylose is the fraction of starch that gives the intense blue colour with iodine. X-ray analysis shows that chains of amylose are coiled to give a helix-like structure. The inside of the helix has just the right space to accommodate an iodine molecule. This is the basis of the starch iodide test for oxidising agents. The material to be tested is added to an aqueous solution of starch and potassium iodide. If the material is an oxidiser, some of the iodide is oxidised to iodine, which forms a blue complex with amylose of the starch.

ii. Structure of Amylopectin

1. It is the insoluble fraction of starch.

    Amylopectin is also hydrolysed to the single disaccharide, maltose.

2. On methylation and hydrolysis of amylopectin, we get mainly 2,3,6-tri-*O*-methyl-D-glucose. Therefore, amylopectin is made up of chains of D-glucose with each unit joined by the α-glycoside linkage of C-4 of the next unit.

3. Hydrolysis of methylated amylopectin also gives about 5% of 2,3,4,6-tetra-*O*-methyl-D-glucose along with an equal amount of 2,3-di-O-methyl-D-glucose (Fig. 1.29).

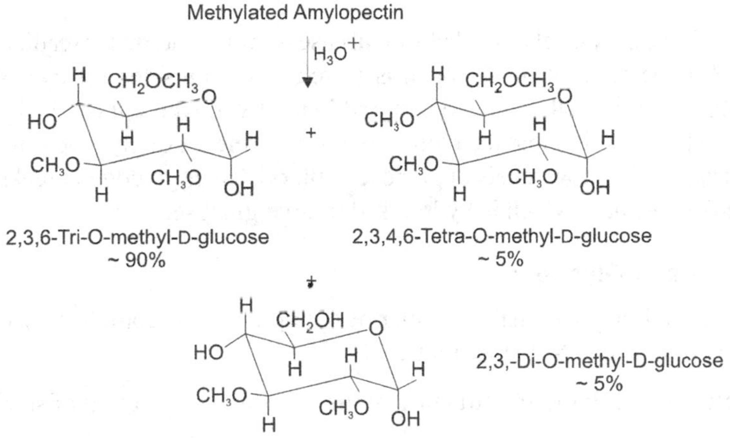

Fig. 1.29.

Formation of 5% of tetra-*O*-methyl-glucose indicates that there are 20 end D-glucose units per chain and the formation of dimethyl-glucose suggests that amylopectin has a highly branched structure consisting of several hundred short chains of about 20-25 D-glucose units. One end of each of these chains is joined through C-1 to C-6 on the next chain.

Therefore, amylopectin is a branched α-1, 4′ polymer of glucose with α-1,6′ linkage that provides the attachment point for another chain (Fig. 1.30).

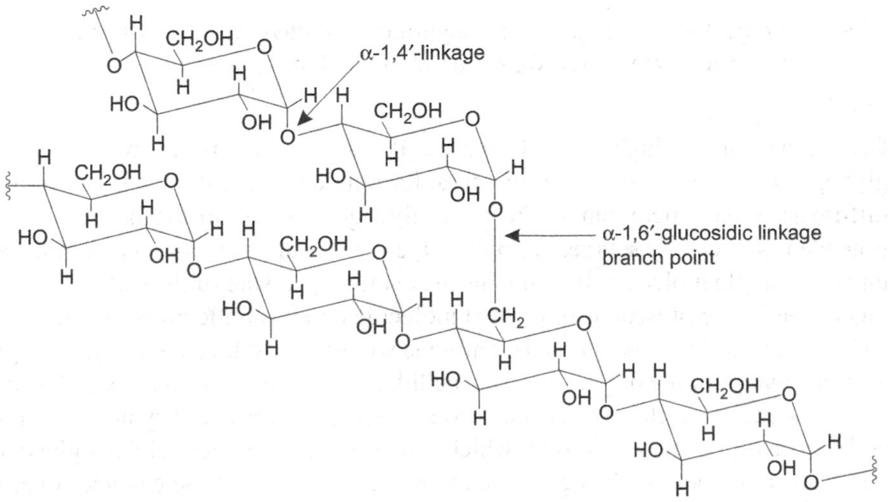

Fig. 1.30 Structure of amylopectin.

C. Glycogen

i. Introduction

Glycogen is found is animal cells in granules similar to the starch granules in plant cells. Glycogen granules are observed in well-fed liver and muscle cells, but they are not seen in some other cell types such as brain and heart cells under normal conditions.

Glycogen is the polysaccharide that animals use to store glucose for readily available energy. A large amount of glycogen is stored in the museles for hydrolysis and metabolism. Additional glycogen is stored in the liver, where it can be hydrolysed to glucose for secretion into the blood stream. The release of glycogen stored in the liver is triggered by low levels of glucose in blood. Liver glycogen breaks down to glucose-6-phosphate which is hydrolysed to give glucose.

ii. Structure of Glycogen

Glycogen is a branched-chain polymer of α-D-glucose as complete hydrolysis of glycogen gives only D-glucose units.

Its structure is similar to the structure of the amylopectin fraction of starch.

Like amylopectin, glycogen consists of a chain of α-(1-4′) linkages with α-(1-6′) linkages at the branch points.

The main difference between glycogen and amylopectin is that glycogen is more highly branched. Branch points occur about every 10 residues in glycogen and to about every 25 residues in amylopectin.

In glycogen, the average chain length is 13 glucose residues and there are 12 layers of branching.

iii. Glycogen as a Reserve Source of Energy

When we digest a meal high in carbohydrates, we have a supply of glucose that exceeds our immediate needs. We store the extra glucose as a polymer in the form of glycogen.

Glycogen has a very high molecular weight. Its size and structure are responsible for glycogen to act as a reserve carbohydrate for animals. First, its size is too large to diffuse across cell membranes. Therefore, glycogen remains inside the cell where it is needed as an energy source. Secondly, glycogen has tens of thousands of glucose units in a single molecule. It maintains the osmotic pressure of the cell. If so many glucose units are present as individual molecules, the osmotic pressure within the cell would be so large that the cell membrane would almost break. Finally, glycogen acts as a ready source of glucose when cellular glucose concentrations are low and even when cellular glucose concentrations are high. There are enzymes within the cell that catalyse the reactions by which glucose units are released from glycogen and converted back to glycogen when they are in excess. These enzymes operate at end points by hydrolysing or forming α-(1 . 4′) glycosidic linkages

However, the overall concentration of glycogen is quite low because of its enormous molecular weight.

Glycogen acts as a reserve source of energy because glucose released is highly water soluble. As a result, it diffuses rapidly through the aqueous medium of the cell and serves as an ideal source of ready energy. In the degradation of glycogen, several glucose residues can be released simultaneously, one from each end of a branch chain. This feature is useful to an organism in meeting short term demands for energy by increasing the glucose supply as quickly as possible.

D. Chitin

Chitin is a polysaccharide structurally similar to cellulose. It is the major component of shells of lobsters, crabs and shrimps and the exoskeletons of insects. It also occurs in cell walls of algae, fungi and yeasts.

Chitin is a homopolysaccharide with all the residues linked in β(1-4′) glycosidic bonds. Chitin differs from cellulose in the nature of the monosaccharide unit. In cellulose, the monomer is β-D-glucose; in chitin, the monomer is N-acetyl glucosamine. In N-acetylglucosamine the hydroxyl group on C-2 of glucose is replaced by an amino group and that amino group is acetylated as given below:

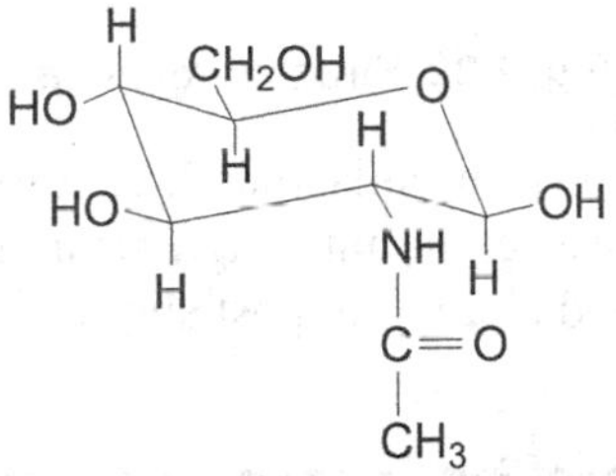

N-Acetylglucosamine

Chitin is bonded like the cellulose using N-acetyl glucosamine instead of glucose. Like cellulose, chitin plays a structural role and has a fair amount of mechanical strength because the individual strands are held together by hydrogen bonds. The structure of chitin is given below (Fig. 1.31):

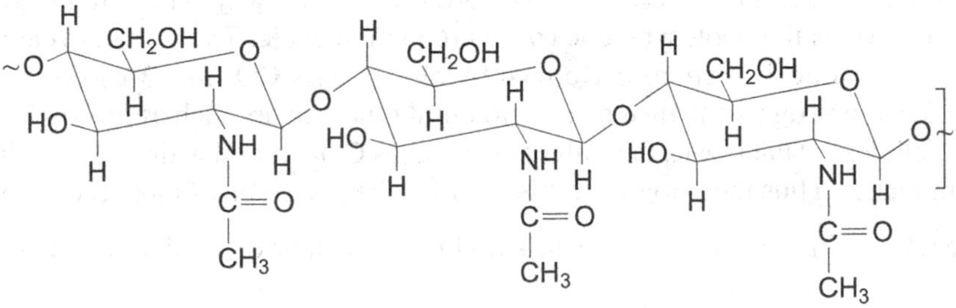

Fig. 1.31 Structure of chitin.

E. Heparin

Heparin is an anticoagulant that is released to prevent excessive blood clot formation when an injury occurs. It is a polysaccharide made up of glucosamine and glucuronic acid subunits.

The C-6 OH groups of the glucosamine subunits and C-2 OH groups of the glucuronic acid are sulphonated. Some of the amino groups are sulphonated and some are acetylated.

A partial structure of heparin is given below (Fig. 1.32):

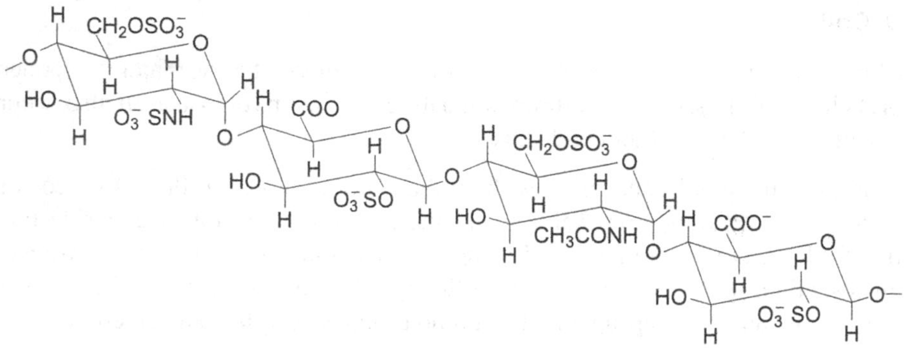

Fig. 1.32 Structure of heparin.

Heparin is found principally in cells that line arterial walls. The longer the polysaccharide chain, the greater the anticoagulant activity. Heparin is widely used in medicine to prevent blood clotting in post surgical patients.

1.24 BIOSYNTHESIS OF CARBOHYDRATES—THE CALVIN CYCLE

We have already seen in Section 1.4 that carbohydrates are obtained in plants by photosynthesis, which takes place in the presence of light energy and carbon dioxide. The pathway for the fixation of CO_2 in photosynthesis was worked out by **Calvin** *et al*. He used labelled CO_2 tracer to work out the pathway. The Calvin cycle is also known as the photosynthetic carbon reduction cycle. The Calvin cycle is the cycle of chemical reactions performed by plants to fix CO_2 into 3-carbon sugars. The 3 carbon sugars can then be used to build other sugars such as glucose, starch and cellulose. That means the Calvin cycle takes CO_2 from the air and turns it into plant matter. Thus the Calvin cycle is vital for the existence of most ecosystems.

The CO_2 assimlation process *via* the Calvin cycle can be divided into three stages:

• Stage I-Fixation

CO_2 condenses with a five carbon acceptor, ribulose-1,5-bisphosphate to form 3-phosphoglycerate.

• Stage II- Reduction

3-Phosphoglycerate is reduced to form glyceraldehyde-3-phosphate.

• Stage III-Regeneration

Ribulose-1,5 bisphosphate is regenerated using glyceraldehyde-3-phosphate. We give below the above stages in detail.

1. Stage I

D-Ribulose-1,5-bisphosphate takes up one molecule of carbon dioxide and the product so obtained is split into two molecules of 3-phosphoglycerate. The reaction is catalysed by the enzyme ribulose-1, 5-bisphosphate carboxylase oxygenase.

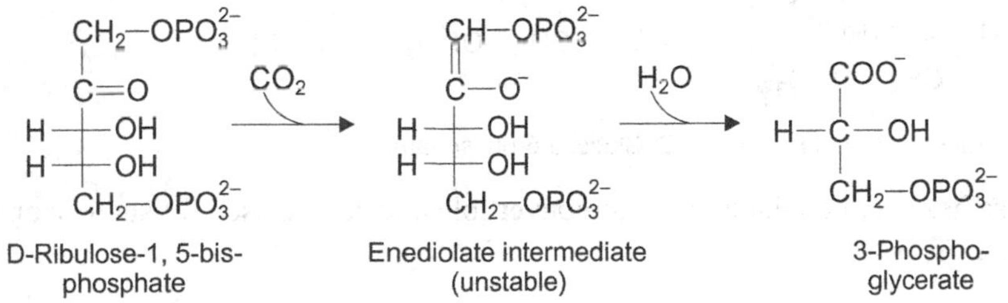

$CH_2-OPO_3^{2-}$
$C=O$
$H-OH$
$H-OH$
$CH_2-OPO_3^{2-}$

D-Ribulose-1, 5-bis-
phosphate

CO_2 →

$CH-OPO_3^{2-}$
$C-O^-$
$H-OH$
$H-OH$
$CH_2-OPO_3^{2-}$

Enediolate intermediate
(unstable)

H_2O →

COO^-
$H-C-OH$
$CH_2-OPO_3^{2-}$

3-Phospho-
glycerate

2. Stage II

In the second stage of the Calvin cycle 3-phosphoglycerate is converted to 1, 3-bisphosphoglycerate, which is then reduced to glyceraldehyde-3-phosphate. The process is catalysed by 3-phosphoglycerate kinase and uses ATP and NADPH.

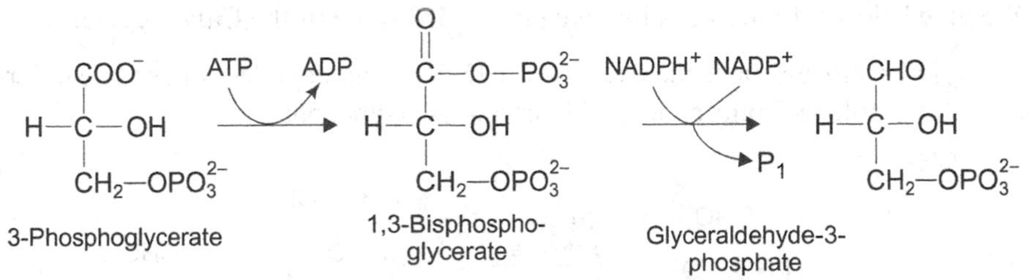

COO^-
$H-C-OH$
$CH_2-OPO_3^{2-}$

3-Phosphoglycerate

ATP   ADP

O
$\|$
$C-O-PO_3^{2-}$
$H-C-OH$
$CH_2-OPO_3^{2-}$

1,3-Bisphospho-
glycerate

$NADPH^+$ $NADP^+$

P_1

CHO
$H-C-OH$
$CH_2-OPO_3^{2-}$

Glyceraldehyde-3-
phosphate

The second step in the above reaction is catalysed by glyceraldehyde-3-phosphate dehydrogenase.

Now D-glyceraldehyde-3-phosphate is converted into D-glucose-6-phosphate *via* dihydroxyacetone phosphate, D-fructose-1,6-bisphosphate and fructose-6-phosphate.

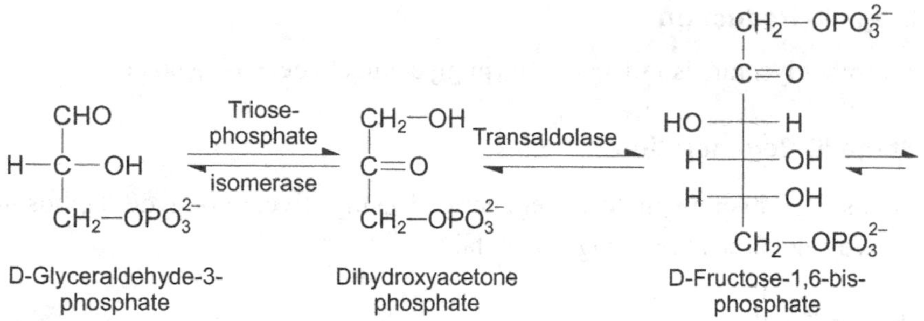

D-Glyceraldehyde-3-phosphate   Dihydroxyacetone phosphate   D-Fructose-1,6-bis-phosphate

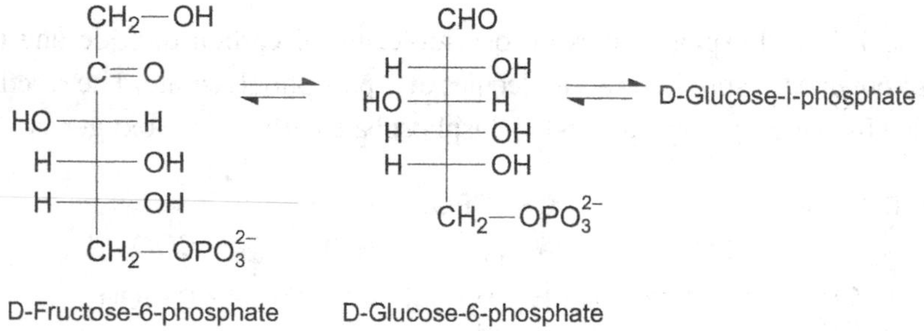

D-Fructose-6-phosphate   D-Glucose-6-phosphate

These hexose phosphates are interconvertible and can be used for sucrose or starch synthesis.

3. Stage III

Some glyceraldehyde-3-phasphate molecules go to make glucose while others are recycled to generate ribulose-1, 5-bisphosphate that is used to accept new CO_2 molecules. The regeneration process requires ATP.

It is a complex process inviting many steps.

We give below various steps involved in the 3rd stage of the Calvin cycle.

1. D-Fructose-6-phosphate reacts with D-glyceraldehyede-3 phosphate to form D-xylulose-5 phosphate and D-erythrose 4-phosphate

D-Fructose-6-phosphate   D-Glyceraldehyde-3-phosphate   D-Xylulose-5-phosphate   D-Erythrose-4-phosphate

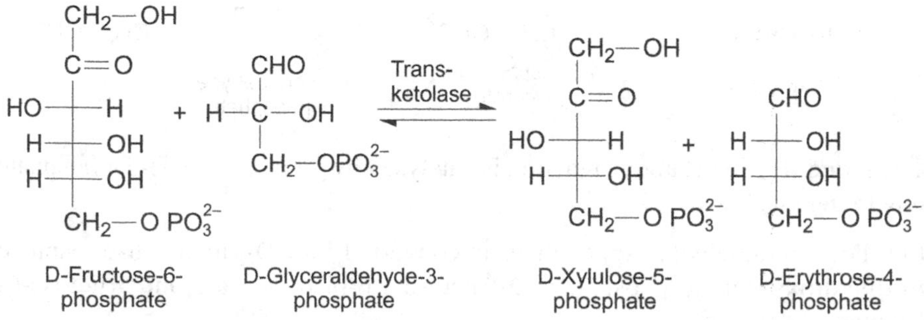

2. D-Erythrose-4-phosphate condenses with dihydroxyacetone phosphate obtained in stage II, to give D-sedoheptulose-1, 7-bisphosphate and then

D-sedoheptulose-7-phosphate. The reacton takes place in the presence of an aldolase.

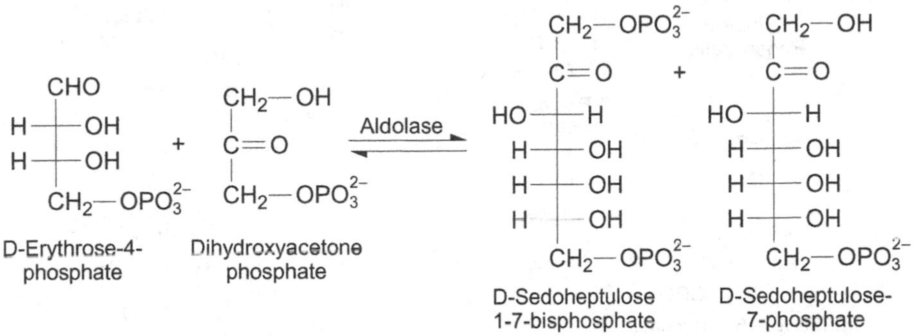
Aldolase
D-Erythrose-4-phosphate
Dihydroxyacetone phosphate
D-Sedoheptulose 1-7-bisphosphate
D-Sedoheptulose-7-phosphate

3. The last step of Calvin cycle is completed by the reaction of D-sedoheptulose-7-phosphate and D-glyceraldehyde-3-phosphate to give D-ribose-5-phosphate and D-xylulose-5-phosphate

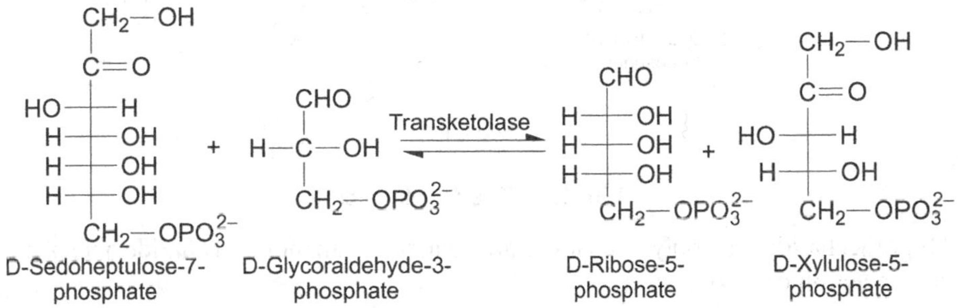
Transketolase
D-Sedoheptulose-7-phosphate
D-Glycoraldehyde-3-phosphate
D-Ribose-5-phosphate
D-Xylulose-5-phosphate

Both D-ribose-5-phosphate and D-xylulose-5-phosphate get converted to D-ribulose-5-phosphate and D-ribulose-1, 5-bisphosphate in the presence of specific enzymes as given below.

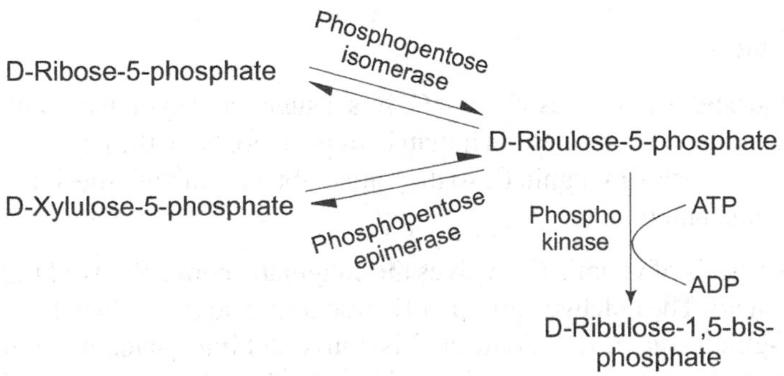
D-Ribose-5-phosphate
Phosphopentose isomerase
D-Ribulose-5-phosphate
D-Xylulose-5-phosphate
Phosphopentose epimerase
Phospho kinase
ATP
ADP
D-Ribulose-1,5-bis-phosphate

The whole Calvin cycle is summarised in Fig. 1.33.

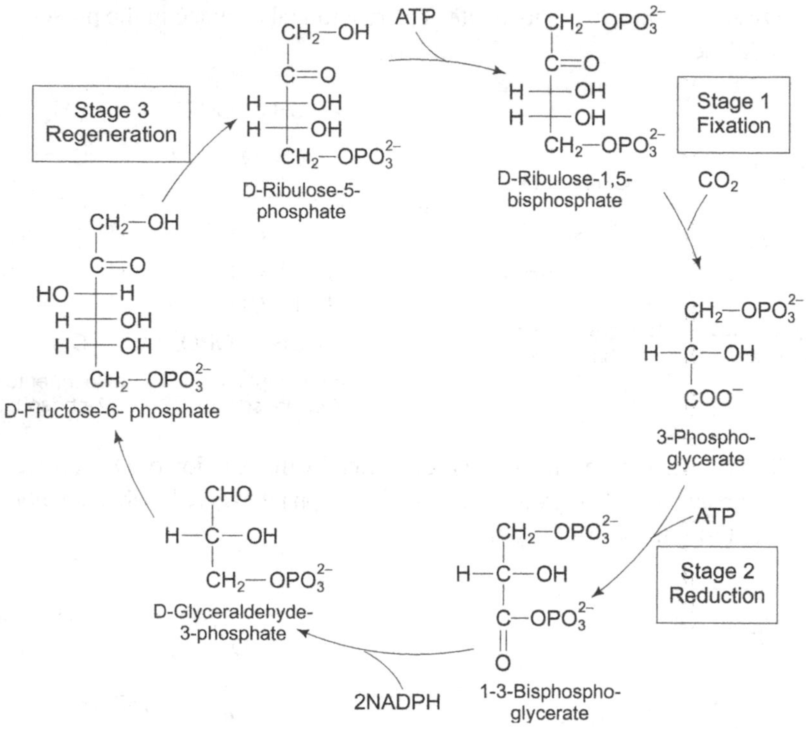

Fig. 1.33 The Calvin cycle.

Oligosaccharides and polysaccharides are obtained from monosaccharide phosphates by the action of enzymes.

1.25 NATURAL PRODUCTS DERIVED FROM CARBOHYDRATES

A. Ascorbic Acid

L-Ascorbic acid is known as vitamin C. It is a sugar acid synthesised in plants and in the livers of most vertebrates. Human beings do not have the necessary enzymes for the biosynthesis of vitamin C, so they must obtain it in their diets. It is abundant in citrus fruits and tomatoes.

The biosynthesis of vitamin C involves the enzymatic conversion of D-glucose into L-gulonic acid. The aldehyde group of D-glucuronic acid is reduced by an enzyme to give L-gulonic acid. L-Gulonic acid is converted into γ-lactone by the enzyme lactonase and then an enzyme called oxidase oxidises the γ-lactone to L-ascorbic acid (Fig. 1.34).

The L-configuration of ascorbic acid refers to the configuration at C-5. which was C-2 in D-glucose.

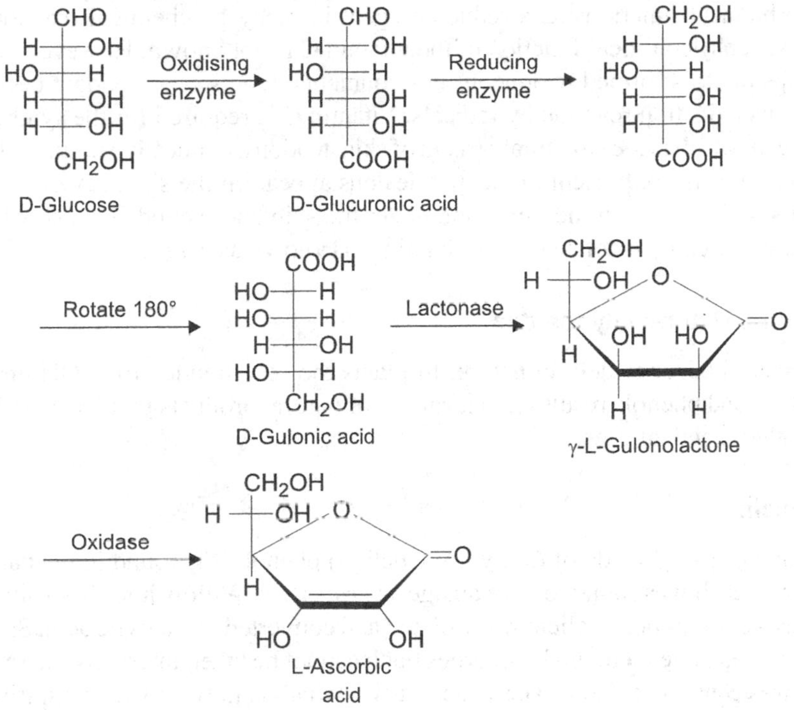

Fig 1.34

Fischer had reduced the same γ-gulonolactone to L-gulose.

Although L-ascorbic acid does not contain a carboxylic acid group, it is an acidic compound because the *pka* of the C-3-OH group is 4.17.

Ascorbic acid is a lactone and an enediol. It is an unstable compound and acts as a reducing agent. It is readily oxidised to L-dehydroascorbic acid, which also has some vitamin C activity. If the lactone ring is opened by hydrolysis, all vitamin C activity is lost. Therefore, vitamin C survives in food that has been thoroughly cooked. If the food is cooked in water and then drained, the water-soluble vitamin goes with the water.

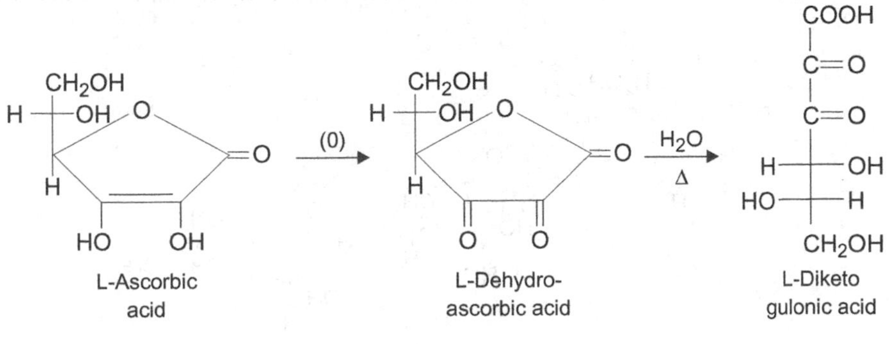

Ascorbic acid functions as a reducing agent in many biochemical reactions but the exact physiological function of the vitamin C is not known. However, vitamin C traps radicals formed in aqueous environments. It acts as an oxidant because it prevents oxidation reaction by radicals. Vitamin C is required for the synthesis of collagen, which is the structural protein of skin, tendons, connective tissue and bone. If vitamin C is not present in the diet, lesions appear on the skin, severe bleeding occurs in the gums, in the joints and under the skin and wounds heal slowly. The disease caused by a deficiency of vitamin C is known as *scurvy*.

B. Some Natural Glycosides

Glycosides occur widely in nature. In plants they are bonded to —OH group of alcohols and phenols resulting in a variety of natural products that have medicinal and other practical uses.

i. Salicin

Salicin is a β-glycoside of *O*-(hydroxymethyl) phenol. It is found in the bark of a willow tree. It was found to have analgesic properties. Willow juice is so bitter that the active component, salicin was isolated and converted to salicylic acid. Salicylic acid has valuable medicinal properties but it cannot be taken internally. So salicylic acid was converted into its acetyl derivative, which is now known as aspirin.

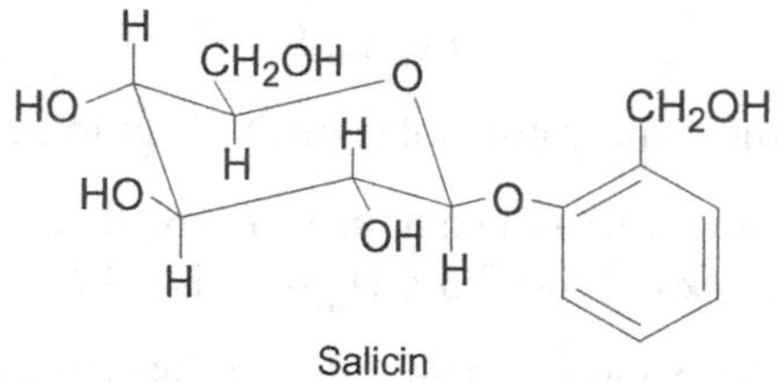

Salicin

ii. Amygdalin

Amygdalin occurs in bitter almonds. It is glycoside formed from the disaccharide gentiobiose and the cyanohydrin of benzaldehyde. Almonds contain an enzyme that catalyses the conversion of amygdalin to HCN, benzaldehyde and two molecules of D-glucose.

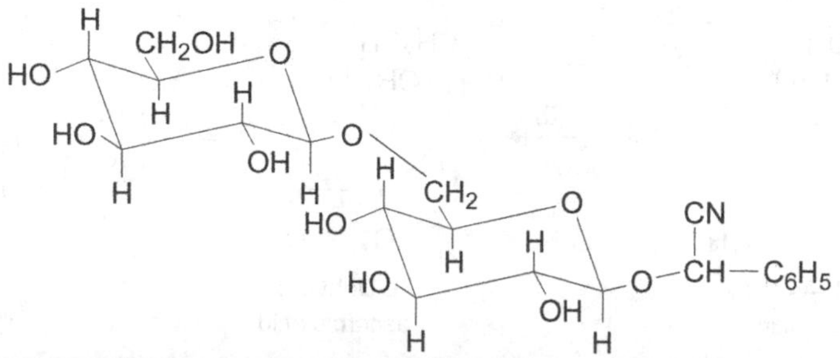

Amygdalin

C. Anthocyanins—Natural Colouring Matters

Anthocyanins are widespread in nature. Much of the colour of plants is due to the presence of anthocyanins. The red colour of rose, purple, crimson to the blue colour of other flowers are all due to these compounds. For example, the blue corn-flower and red rose have the same anthocyanin, cyanin. The blue colour is that of its potassium salt.

Anthocyanins are actually present as glycosides. Hydrolysis gives the corresponding anthocyanidins. Three important anthocyanidins are cyanidin, pelargonidin and delphinidin. They are derivatives of the parent structure flavone. The sugar units are attached at the 3- and 5-positions.

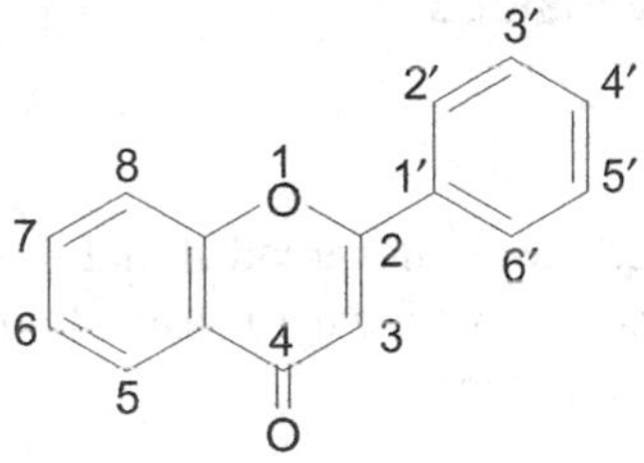

Flavone

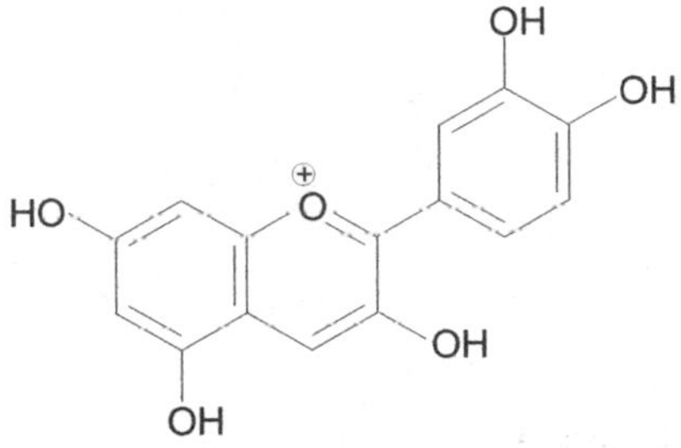

Cyanidin

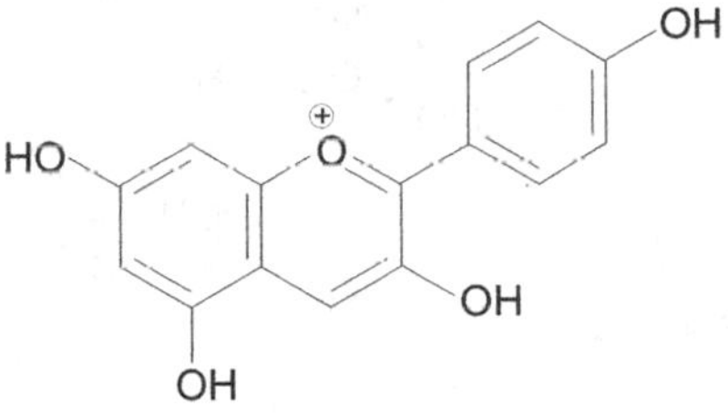

Pelargonidin

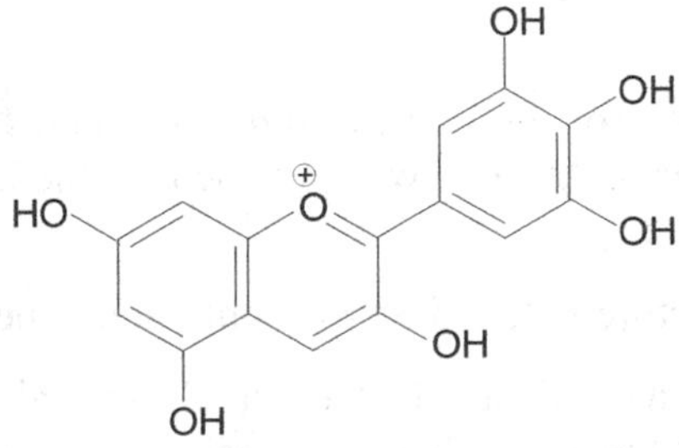

Delphinidin

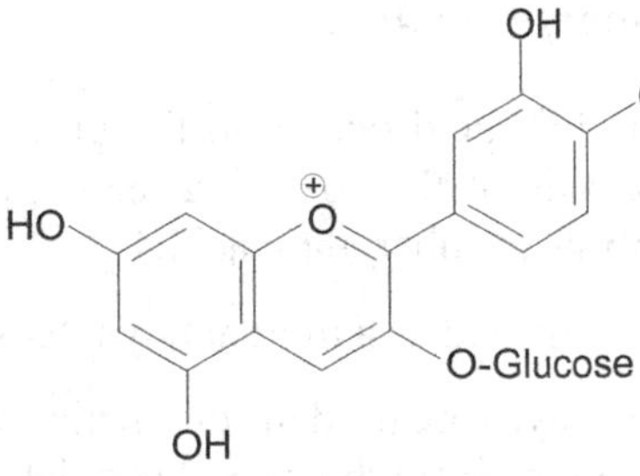

Cyanidin-3-glucoside

Peonin

Another naturally occurring glycoside, which belongs to anthocyanin group, is peonin.

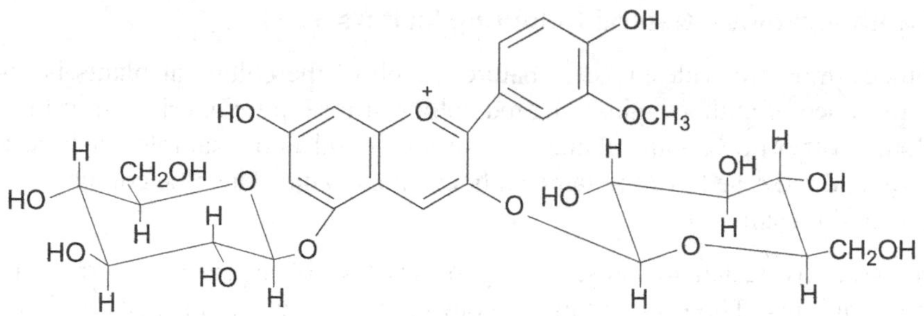

Peonin

D. Anthraquinone Glycosides

i. Carminic Acid

Anthraquinone glycosides are widespread in nature. For example, carminic acid, the principal constituent of cochineal (a dried female insect) is a polyhydroxy anthraquinone attached to glucose.

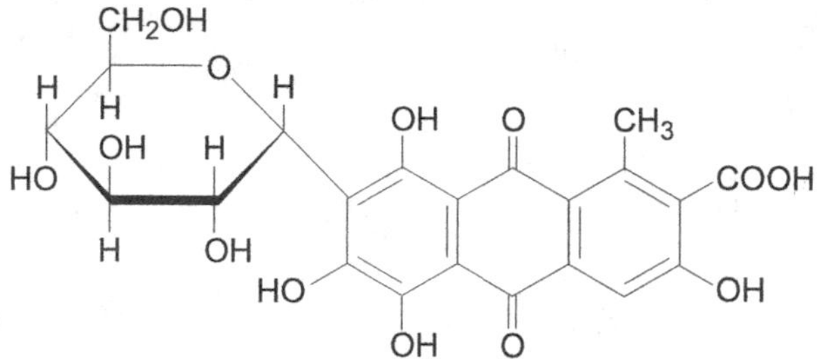

Carminic acid

ii. Ruberythric Acid

Ruberythric acid occurs in the ground root of the plant *Rubia tinctorum*. It is a glucoside of the colouring matter, alizarin. Alizarin was obtained by the acid hydrolysis of ruberythric acid.

Alizarin is used as a mordant dye. Its structure is 1,2-dihydroxy anthraquinone.

Mordant dyes are used in conjunction with a mordant, which is usually a metal salt that forms an insoluble complex or lake with the dye. The dye is applied to the fibre or cloth that has been pretreated with a metal salt in the presence of a mild alkali and a wetting agent. Strongly coloured insoluble metal complexes thereby formed are precipitated within the fibre and resulting dyeings have excellent fastness properties. By varying the metal salt it is possible to obtain several colours from a single dye.

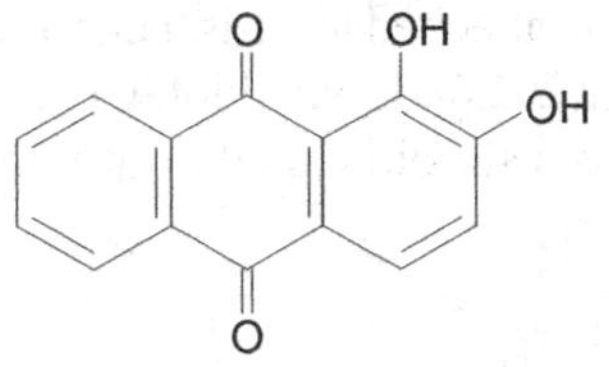

Alizarin

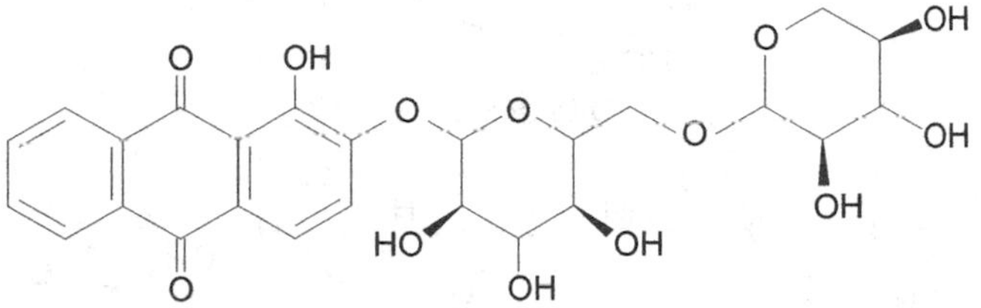

Ruberythric acid

E. Indican–A Natural Colouring Matter

Indigo is an important vat dye which is present as a glucoside of indoxyl. The glucoside is known as *indican*. Enzymatic or acid hydrolysis of indican gives 3-hydroxyindole, known as indoxyl which exists in equilibrium with the corresponding keto form.

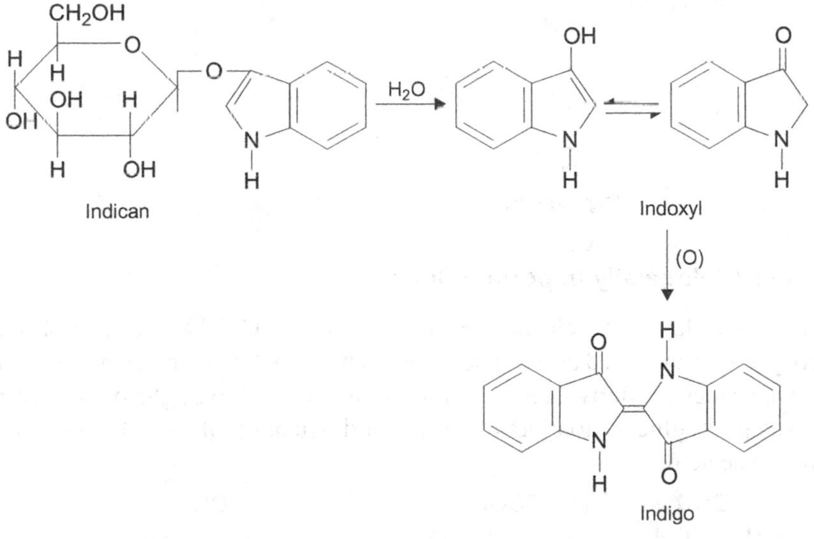

Indican   Indoxyl

Indigo

F. Cardiac Glycosides

The cardiac glycosides have an effect on the actions of the heart and many of them are highly toxic.

A well-known example of a cardiac glycoside is *digitoxin,* which is isolated from *Digitalis purpurea,* known as foxglove. This compound reduces the pulse rate, regularises the rhythm of the heart and strengthens the heartbeat.

Digitoxin consists of a steroid aglycone attached to a trisaccharide. The trisaccharide has three units of D-digitoxose, which is 2,6- deoxyaldohexose. The sugar units are joined by β-linkages. The aglycone is a steroid called digitoxigenin.

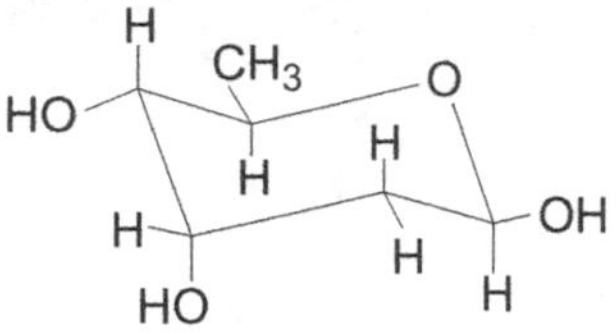

D-Digitoxose

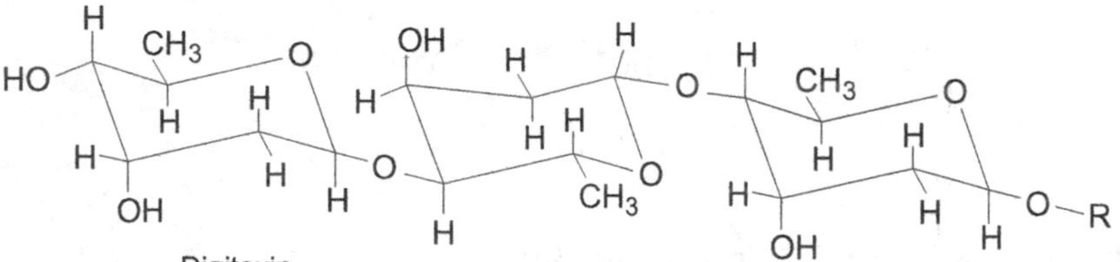

Digitoxin

R* = aglycone
  = digitoxigenin

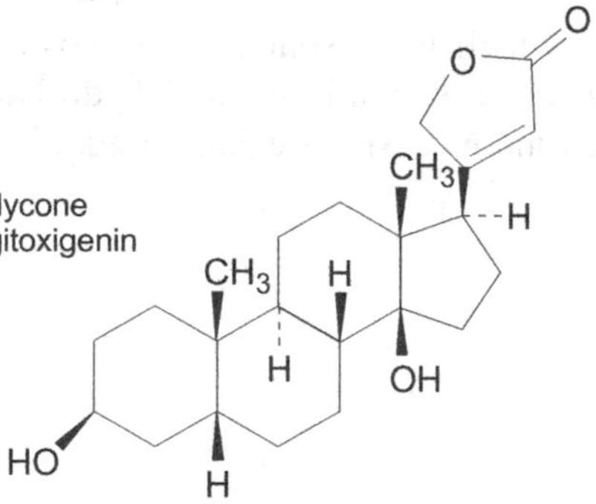

Digitoxigenin

G. Some Biologically Important Sugars

Monosaccharides in which the —CH_2OH group at C-6 has been oxidised to a carboxyl group are called uronic acids. Their names are based on the monosaccharide from which they are derived. For example, oxidation of C-6 of glucose to a carboxyl group converts glucose to glucuronic acid and oxidation of C-6 of galactose gives galacturonic acid.

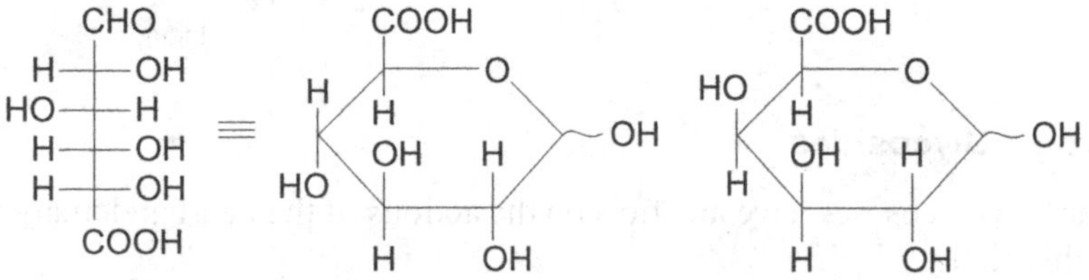

D-Glucuronic acid

D-Galacturonic acid

The D-Glucuronate-2-sulphate unit is present in heparin, a polysaccharide that prevents blood clotting.

H. Glycosylamines

A sugar in which an amino group replaces the anomeric —OH group is called a glycosylamine. For example, we have β-D-glucopyranosyl amine.

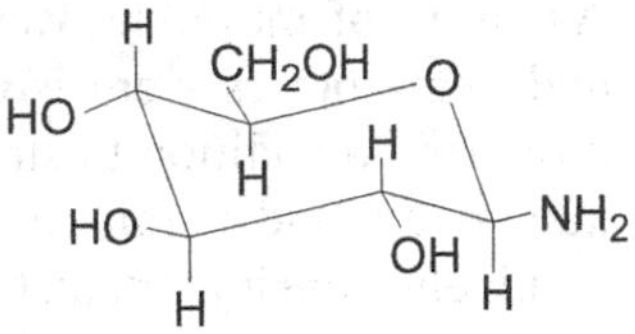

β-D-Glucosylamine

Nucleosides are glycosylamines in which the amino component is a pyrimidine or a purine. For example, adenosine is a glycosylamine, in which D-ribose is the sugar component. The anomeric —OH is replaced by adenine, the amino component.

Nucleosides are the important components of RNA (ribonucleic acid) and DNA (deoxy ribonucleic acid). D-Ribose is the sugar component of RNA while 2-deoxyribose is the sugar component of DNA.

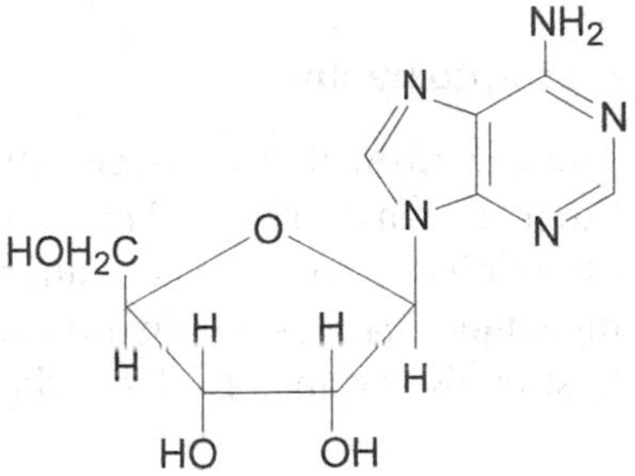

Adenosine

RNA and DNA are *N*-glycosides. Their subunits consist of an amine (base) bonded to the β-position of the anomeric carbon of ribose or 2-deoxyribose. The subunits are linked by a phosphate group between C-3 of one sugar and C-5 of the next sugar. Short segments of RNA and DNA are represented on pages 294, 295.

I. Amino Sugars

When a non-anomeric —OH group in sugars is replaced by an amino group, we get amino sugars. An example is D-glucosamine. In many cases the amino group is acetylated. *N*-Acetyl glucosamine is one of the subunits of chitin, a polysaccharide. *N*-Acetyl muramic acid is an important component of bacterial cell walls.

Amino sugars are the basic units of chitin and heparin, which have been described under polysaccharides.

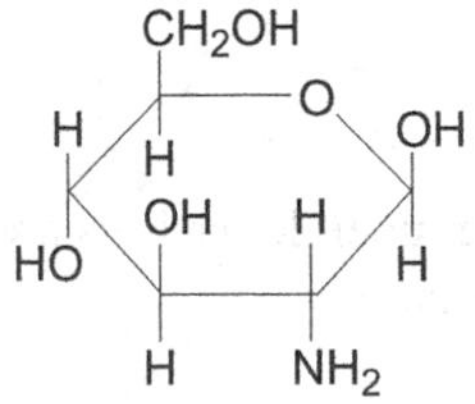

β-D-Glucosamine

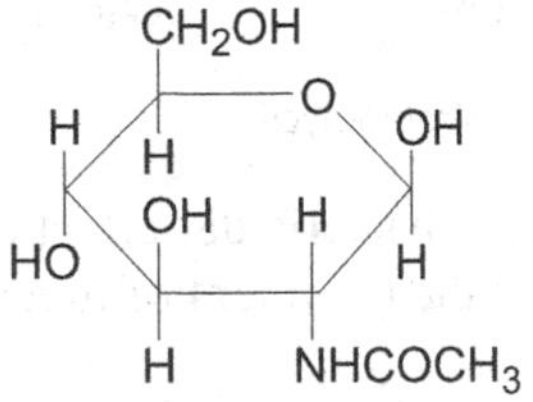

β-*N*-Acetyl-D-glucosamine

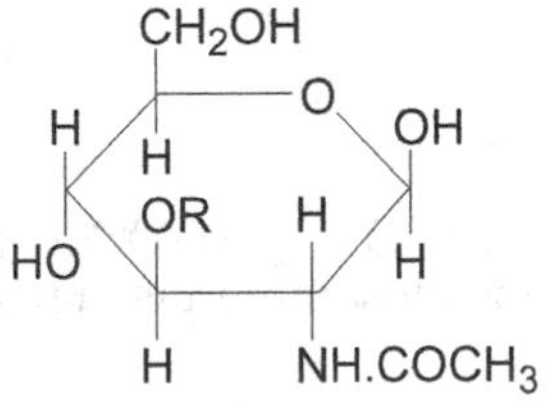

β-*N*-Acetyl muramic acid

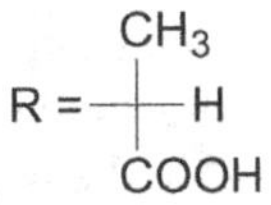

$$R = \begin{matrix} CH_3 \\ | \\ | \text{—H} \\ COOH \end{matrix}$$

J. Carbohydrate Antibiotics

A number of antibiotics were isolated from the genus, *streptomyces*. Streptomycin and many others were isolated from it. They all contain closely related sugar moieties. In addition to streptomycin, other antibiotics of interest are neomycin, kanamycin and paramomycin. Out of these stereptomycin and neomycin are relatively more important. Gentamicin is also an amino glycoside which is produced by *Micromonospora* species. Chloramphenicol contains a derivative of a sugar moiety and therefore, may be grouped with these.

i. Streptomycin

The organism that produces streptomycin is known as *streptomyces griseus*. Streptomycin consists of 3 units, N-methyl-L-glucosamine, streptose and streptidine. Streptidine is not a sugar at all, but a cyclohexane derivative. Its structure is 1,3-diguanido-2,4,5,6- tetrahydroxycyclohexane. N-Methyl-L-glucosamine is linked to streptose by an α-L-glycosidic linkage.

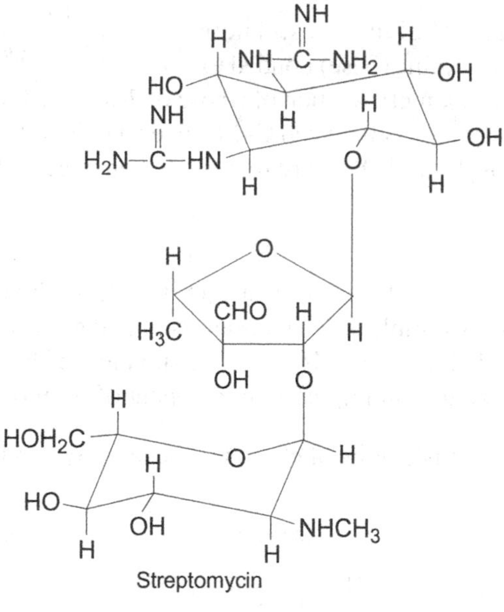

Streptomycin

Streptomycin has a very specific bactericidal effect on the tubercle bacillus and it is the most important agent for the treatment of tuberculosis.

ii. Neomycin

Neomycin was isolated in 1949 from *streptomyces fradiae*. It is a mixture of closely related substances known as Neomycin B, C and A. Neomycin B and C consist of four units, three sugar moieties and one non-sugar unit. Neomycin B differs from Neomycin C in the nature of the sugar attached to D-ribose. Neomycin C has been assigned the following structure:

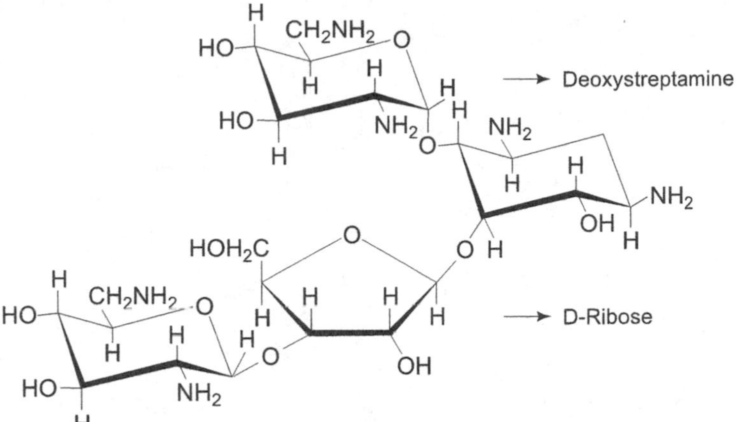

Neomycin

Neomycin has a broad spectrum of activity against a variety of organisms.

iii. Gentamicin

Gentamicin is also known as Garamycin and was isolated in 1958. It is obtained commercially from *Micromonospora purpurea*. Gentamicin is also a mixture of closely related substances known as gentamicins C_1, C_2, and C_1a. Each one of them is composed of three units linked together by glycosidic linkages.

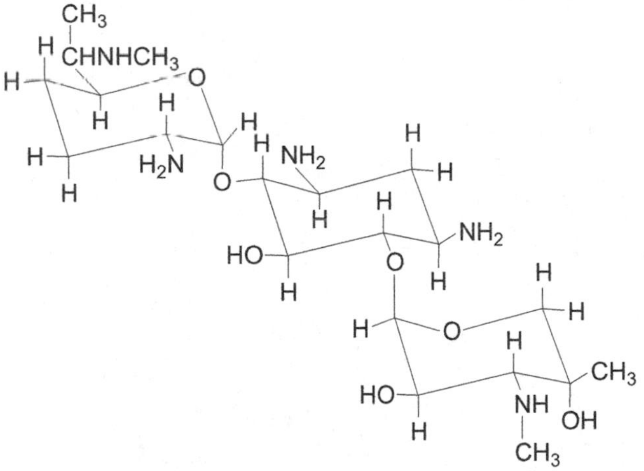

Gentamicin

Gentamicin has broad spectrum activity against a variety of both gram-positive and gram-negative organisms.

2

Amino Acids, Peptides, and Proteins

Learning Objectives

In this chapter we will study

- Amino Acids, Classification and Stereochemistry
- The Essential Amino Acids
- Synthesis and Reactions of Amino Acids
- Resolution of Racemic Mixture of Amino Acids
- Stereoselective Synthesis of Amino Acids
- Biosynthesis of Amino Acids
- Amino Acids as Dipolar Ions, as Acids and Bases
- Isoelectric Point
- Separation of Amino Acids
- Polypeptides: Nomenclature and Synthesis
- Solid Phase Peptide Synthesis
- Determination of Primary Structure of Peptides
- End Group Analysis – Determination of N – terminal and C – terminal Amino Acids
- Some Interesting Peptides
- Importance and Biological Functions of Proteins and Polypeptides
- Molecular Shape and Structure of Proteins
- Denaturation of Proteins

2.1   INTRODUCTION

The three important groups of biopolymers are polysaccharides, proteins and nucleic acids. We already know the importance of polysaccharides. They act as food reserves in animals and human beings and in plants as structural materials. Proteins are the most abundant organic molecules in animals and human beings. We consume proteins primarily for growth and maintenance. Any balanced diet must contain an adequate amount of proteins. They perform a variety of functions in living organisms. They are the principal material of muscles, skin, tendons, nerves and blood. As enzymes and hormones, proteins catalyse and regulate the reactions that occur in the body; as haemoglobins they transfer oxygen to its most remote corners. Even in plants, where carbohydrates are more abundant as structural materials, proteins are present in those parts that are responsible for growth and reproduction.

The fundamental structure of proteins is simple. Proteins are biopolymers of α-amino acids. They consist of long chains of α-amino acids bonded to each other by amide bonds also known as peptide linkages between the carboyxlic group of one amino acid with the amino group of another. Thus, proteins are high molecular weight polypeptides. Their molecular weight may range from 5000–100,000 units. A single protein molecule contains hundreds or even thousands of amino acid units. About twenty different amino acids are the building blocks of proteins. The number of different combinations, that is the number of different protein molecules that are possible, is almost infinite.

2.2   AMINO ACIDS – THE BUILDING BLOCKS OF PROTEINS

We have seen above that α-amino acids are the building blocks of proteins. Hydrolysis of proteins with an acid or base yields a mixture of different amino acids. Naturally occurring proteins on hydrolysis may yield twenty-two different amino acids. These amino acids have an important structural feature in common. All are α-amino acids; *i.e.*, the amino group is bonded to the α-carbon atom next to the carboxyl group. The simplest amino acid is glycine. All others can be derived from glycine.

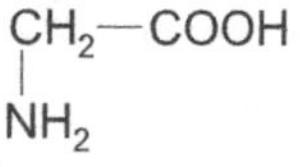

$$CH_2-COOH \qquad\qquad R-CH-COOH$$
$$| \qquad\qquad\qquad\qquad\qquad |$$
$$NH_2 \qquad\qquad\qquad\qquad NH_2$$

    Glycine 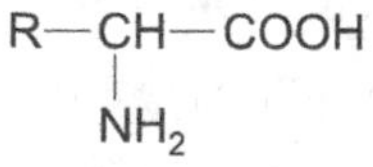   A substituted amino acid

where the R group is different for different amino acids.

With the exception of glycine, all naturally occurring amino acids have the L-configuration at the α-carbon. That means they have the same relative configuration as L-glyceraldehyde.

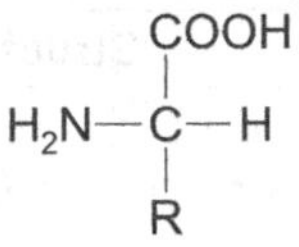

$$
\begin{array}{cc}
\text{COOH} & \text{CHO} \\
| & | \\
\text{H}_2\text{N}-\text{C}-\text{H} & \text{HO}-\text{C}-\text{H} \\
| & | \\
\text{R} & \text{CH}_2\text{OH} \\
\text{An L-amino acid} & \text{L-Glyceraldehyde}
\end{array}
$$

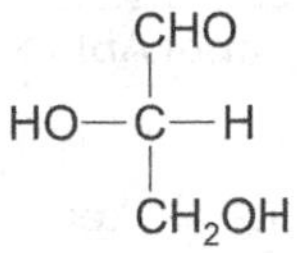

As the two functional groups present in amino acids are acidic and basic, all amino acids are amphoteric and actually exist as zwitterions. Thus, the simplest amino acid glycine exists in the form shown below rather than aminoacetic acid.

$$
\underset{\text{Aminoacetic acid}}{\text{H}_2\text{N}-\text{CH}_2-\text{COOH}} \rightleftharpoons \underset{\text{Zwitterion}}{\overset{+}{\text{H}_3\text{N}}-\text{CH}_2-\overline{\text{COO}}}
$$

2.3 CLASSIFICATION OF AMINO ACIDS

The 22 α-amino acids that are obtained from proteins can be classified into four different groups on the basis of the structures of their side chains:

1. Neutral Amino Acids
2. Acidic Amino Acids
3. Neutral Amino Acids with Polar Side Chains
4. Basic Amino Acids

These are listed with their symbols, abbreviations and structures in Table 2.1.

Only 20 of the 22 amino acids listed in Table 2.1 are actually used by cells when they synthesise proteins. Hydroxyproline is synthesised from proline and cystine is synthesised from cysteine.

Asparagine and glutamine are derived from aspartic acid and glutamic acid.

TABLE 2.1 Amino Acids Commonly Found in Proteins

1. Neutral Amino Acids			
Name	**Symbol**	**Abbreviation**	**Structure**
Glycine	G	Gly	$\begin{array}{l}\text{CH}_2-\text{COOH} \\ | \\ \text{NH}_2\end{array}$
Alanine	A	Ala	$\begin{array}{l}\text{CH}_3-\text{CH}-\text{COOH} \quad \text{e}\\ \phantom{\text{CH}_3-}| \\ \phantom{\text{CH}_3-}\text{NH}_2\end{array}$
Valine	V	Val	$\begin{array}{l}\text{CH}_3 \\ \quad\diagdown \\ \quad\quad\text{CH}-\text{CH}-\text{COOH} \quad \text{e}\\ \quad\diagup \phantom{\text{CH}-}| \\ \text{CH}_3 \phantom{\text{CH}-}\text{NH}_2\end{array}$

Cont'd.

Name	Symbol	Abbreviation	Structure
Leucine	L	Leu	CH_3 > $CH-CH_2-CH-COOH$ with NH_2 e
Isoleucine	I	Ile	CH_3 > $CH-CH-CH-COOH$ with CH_3 NH_2
Proline	P	Pro	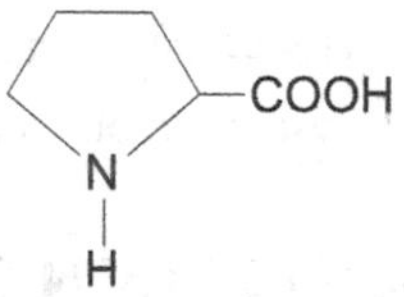 pyrrolidine ring —COOH, N, H
Phenylalanine	F	Phe	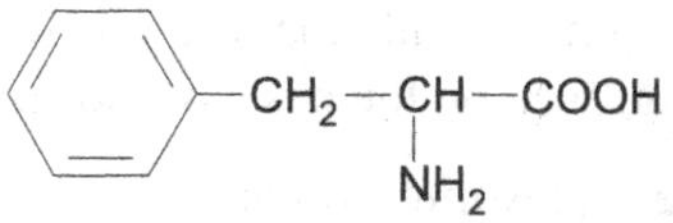 phenyl —$CH_2-CH-COOH$ with NH_2 e
Tryptophan	W	Trp	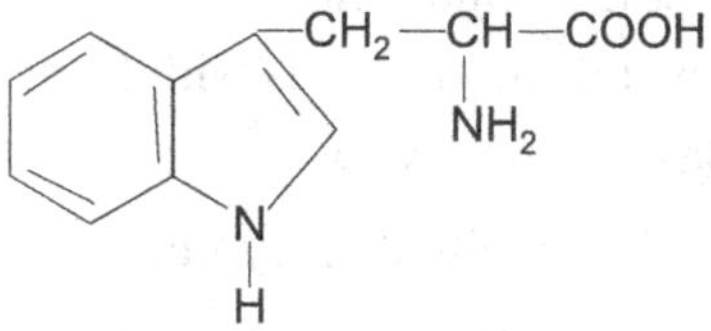 indole —$CH_2-CH-COOH$ with NH_2, N, H e
Methionine	M	Met	$CH_3-S-CH_2-CH_2-CH-COOH$ with NH_2 e
Cystine		$(Cys)_2$	$(S-CH_2-CH-COOH)_2$ with NH_2
2.Acidic Amino Acids			
Aspartic acid	D	Asp	$HOOC-CH_2-CH-COOH$ with NH_2
Glutamic acid	E	Glu	$HOOC-CH_2-CH_2-CH-COOH$ with NH_2
3.Neutral Amino Acids with Polar Side Chains			
Serine	S	Ser	$HO-CH_2-CH-COOH$ with NH_2

Cont'd.

Name	Symbol	Abbreviation	Structure
Threonine	T	Thr	$CH_3-CH-CH-COOH$ e $\qquad\;\;\; OH \;\;\; NH_2$
Cysteine	C	Cys	$HS-CH_2-CH-COOH$ $\qquad\qquad\quad NH_2$
Tyrosine	Y	Tyr	$HO-\langle\bigcirc\rangle-CH_2-CH-COOH$ $\qquad\qquad\qquad\qquad NH_2$
Asparagine	N	Asn	$\qquad\;\; O$ $\qquad\;\; \parallel$ $H_2N-C-CH_2-CH-COOH$ $\qquad\qquad\qquad NH_2$
Glutamine	Q	Gln	$\qquad\;\; O$ $\qquad\;\; \parallel$ $H_2N-C-CH_2-CH_2-CH-COOH$ $\qquad\qquad\qquad\qquad\; NH_2$
Hydroxyproline		Hyp	HO (pyrrolidine ring)—COOH

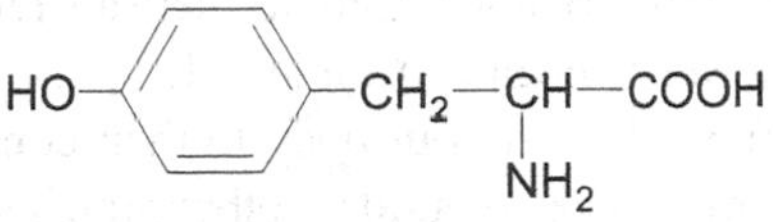

4. Basic Amino Acids

Name	Symbol	Abbreviation	Structure
Lysine	K	Lys	$CH_2-CH_2-CH_2-CH_2-CH-COOH$ e $NH_2 \qquad\qquad\qquad\qquad\;\; NH_2$
Histidine	H	His	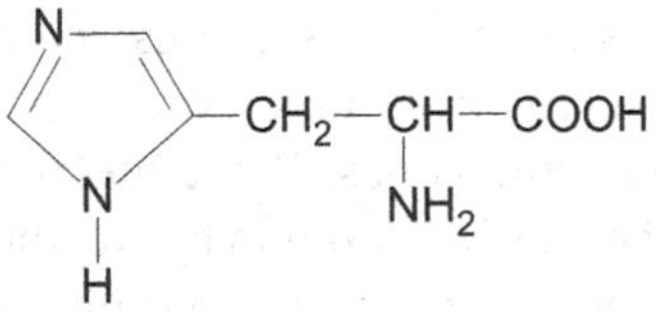 (imidazole ring)$-CH_2-CH-COOH$ e $\qquad\qquad\qquad\qquad NH_2$
Arginine	R	Arg	$H_2N-C-NH-CH_2-CH_2-CH_2-CH-COOH$ $\quad\;\; \parallel \qquad\qquad\qquad\qquad\qquad NH_2$ e $\quad\;\; NH$

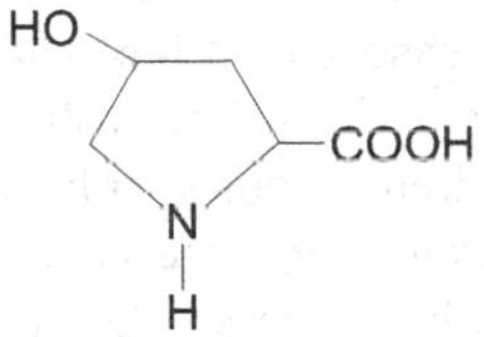

2.4   THE ESSENTIAL AMINO ACIDS

Amino acids are synthesised by all living organisms, animals and plants for their protein requirements. The biosynthesis of proteins requires the presence of all the constitutent amino acids. If one of the 20 amino acids is missing or in short supply, protein biosynthesis is inhibited. Some organisms, such as *E.Coli* can synthesise all the amino acids that they need. However, the human body is unable to synthesise some of these amino acids required for making proteins. These are called **essential amino** acids. For adult humans there are ten amino acids, which have been designated with the superscript **e** in Table 2.1. They must be included in our diet. We eat proteins, break them down in our body to their constituent amino acids and then use some of these amino acids to build up other proteins which we require to maintain good health. So, for the requirement of essential amino acids we depend upon external sources, *i.e.*, the proteins we take. The requirement of each essential amino acid for an adult human body is in the range of 1 to 2 grams per day. An adequate protein supplies all the essential amino acids in the quantities needed for growth and repair of body tissues. Not all proteins contain all the essential amino acids. Amino acids are not stored and dietary sources of essential amino acids are needed at regular intervals.

Most proteins from plant sources are deficient in one or more essential amino acids. Zein, the corn protein, is deficient in two essential amino acids lysine and tryptophan. People whose diet consists chiefly of corn may suffer from malnutrition. A protein deficiency disease, called **Kwashiorkar**, is quite common in parts of Africa where corn is the major food. The problem in this disease, which is particularly severe in growing children, is not simply starvation but the breakdown of the body's own proteins. Protein from rice is short on lysine and threonine. Gliadin, the wheat protein lacks in lysine and soya protein does not have the essential amino acid methionine. A complete protein is one that provides all essential amino acids in appropriate amounts for human survival. These amino acids cannot be synthesised by humans but they are needed for the biosynthesis of proteins. Lysine and methionine are two essential amino acids that are frequently in short supply in plant proteins. Most proteins from animal sources contain all the essential amino acids in adequate amounts. Meat, milk, fish, eggs and cheese are complete proteins. For example, casein, the protein from milk, contains nineteen of the common amino acids and all of the essential ones. Therefore, it is a complete protein. A balanced diet should consist of about 20% protein by weight. It must include all the essential amino acids in sufficient quantity. If we consume proteins from vegetable sources only, we may have deficiency of one or the other essential amino acid. Because grains such as rice and corn are usually poor in lysine and because beans are usually poor in methionine, vegetarians are at risk for malnutrition unless they eat grains and beans together. This leads to the concept of complementary proteins, *i.e.,* mixtures that provide all the essential amino acids. Extreme vegetarianism is, therefore, dangerous. It is best to eat proteins from a variety of sources to ensure an adequate supply of all the essential amino acids in the diet.

2.5   STEREOCHEMISTRY OF α-AMINO ACIDS

With the exception of glycine, all the common amino acids listed in Table 2.1 are chiral molecules. Therefore, there are two (optically active) stereoisomers for each of the amino acids. As with carbohydrates, it is customary to use the D- and L-nomenclature with the amino acids. All the naturally occurring amino acids have the L-configuration, *i.e.*, they have the same relative configuration as L-glyceraldehyde. Applying Fischer representation to these amino acids, D-amino acids are those in which the $-NH_2$ group is on the right side of the chiral centre, $-COO^-$ being at the top and the carbon chain at the bottom. Similarly, L-amino acids have $-\overset{+}{N}H_3$ group on the left of the chiral centre as shown below:

$$
\begin{array}{cc}
\text{COO}^- & \text{COO}^- \\
| & | \\
\text{H} - \overset{|}{\underset{|}{\text{C}}} - \overset{+}{\text{N}}\text{H}_3 \qquad & \overset{+}{\text{N}}\text{H}_3 - \overset{|}{\underset{|}{\text{C}}} - \text{H} \\
\text{R} & \text{R}
\end{array}
$$

D-Amino acid   L-Amino acid

The stereo structures of L-alanine and L-serine in relationship with L-glyceraldehyde are shown in Fig. 2.1.

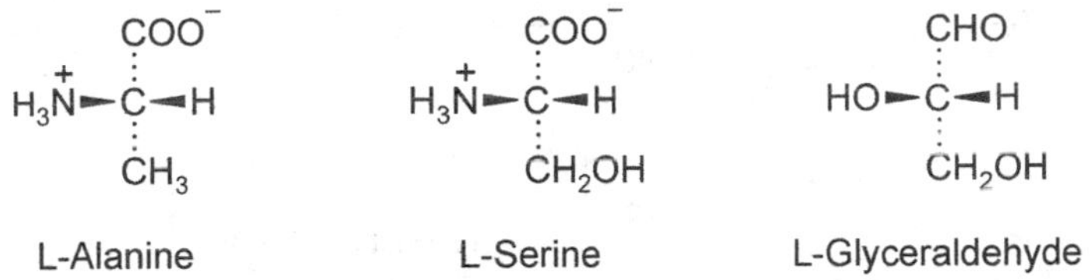

$$
\begin{array}{ccc}
\text{COO}^- & \text{COO}^- & \text{CHO} \\
\overset{+}{\text{H}_3\text{N}} - \overset{|}{\text{C}} - \text{H} & \overset{+}{\text{H}_3\text{N}} - \overset{|}{\text{C}} - \text{H} & \text{HO} - \overset{|}{\text{C}} - \text{H} \\
\text{CH}_3 & \text{CH}_2\text{OH} & \text{CH}_2\text{OH}
\end{array}
$$

L-Alanine   L-Serine   L-Glyceraldehyde

Fig. 2.1. The relationship of L-alanine and L-serine to L-glyceraldehyde.

Some D-amino acids are found as components of the cell walls of bacteria and certain antibiotics. However, only L-amino acids are incorporated in proteins. These D- and L-designations have nothing to do with the actual optical rotation of amino acids, which is determined experimentally.

2.6   SYNTHESIS OF α-AMINO ACIDS

A. Amination of α-Halogeno Acids

This is the first method that has been developed for the synthesis of α-amino acids. With its various modifications, this method is quite useful, although it cannot be applied for the synthesis of all the amino acids.

In this method, α-chloro or α-bromo acids are subjected to direct ammonolysis by treating with a large excess of concentrated aqueous ammonia. The necessary α-halo acids are prepared by the Hell-Volhard Zelinksy reaction of the unsubstituted acids or through the appropriately substituted malonic ester. Generally the bromo compounds show greater reactivity.

$$R-CH_2-COOH \xrightarrow{Br_2/P} R-\underset{\underset{Br}{|}}{C}H-COOH \xrightarrow[\text{excess}]{NH_3} R-\underset{\underset{NH_3}{\overset{|}{\underset{+}{}}}}{C}H-COO^-$$

For example, alanine can be prepared from propionic acid in the following manner:

$$CH_3-CH_2-COOH \xrightarrow{Br_2/P} CH_3-\underset{\underset{Br}{|}}{C}H-COOH \xrightarrow[\substack{25° \\ 4\ days}]{NH_4OH} CH_3-\underset{\underset{NH_3}{\overset{|}{\underset{+}{}}}}{C}H-COO^-$$

The general scheme for the synthesis of α-halo acids from malonic ester is outlined below:

$$\underset{\underset{COOC_2H_5}{\diagdown}}{\overset{\diagup COOC_2H_5}{CH_2}} \xrightarrow{NaOEt} \overset{+}{Na} \overset{-}{C}H \overset{\diagup COOC_2H_5}{\underset{\diagdown COOC_2H_5}{}} \xrightarrow{RX}$$

$$R-CH \overset{\diagup COOC_2H_5}{\underset{\diagdown COOC_2H_5}{}} \xrightarrow[\text{2) HCl}]{\text{1) KOH,}\Delta} R-CH \overset{\diagup COOH}{\underset{\diagdown COOH}{}} \xrightarrow[CCl_4]{Br_2}$$

$$R-\underset{\diagdown COOH}{\overset{\diagup COOH}{C}-Br} \xrightarrow{\Delta} R-\underset{\underset{Br}{|}}{C}H-COOH$$

Phenylalanine may be synthesised starting from malonic ester as given below:

$$\underset{\diagdown COOC_2H_5}{\overset{\diagup COOC_2H_5}{CH_2}} \xrightarrow[\text{2) } C_6H_5CH_2Cl]{\text{1) NaOEt}} C_6H_5CH_2-CH \overset{\diagup COOC_2H_5}{\underset{\diagdown COOC_2H_5}{}} \xrightarrow[\text{2) HCl}]{\text{1) KOH,}\Delta}$$

$$C_6H_5CH_2-CH \overset{\diagup COOH}{\underset{\diagdown COOH}{}} \xrightarrow[\text{reflux}]{Br_2,\ ether} C_6H_5CH_2-\underset{\diagdown COOH}{\overset{\diagup COOH}{C}-Br} \xrightarrow[-CO_2]{\Delta}$$

$$C_6H_5CH_2-\underset{\underset{Br}{|}}{C}H-COOH \xrightarrow[\text{excess}]{NH_3} C_6H_5CH_2-\underset{\underset{NH_3}{\overset{|}{\underset{+}{}}}}{C}H-CO\bar{O}$$

Phenylalanine

This route has also been used for the synthesis of histidine. Other amino acids which can be easily synthesised by the above method are glycine, serine, threonine, valine, leucine and isoleucine.

B. Gabriel Phthalimide Synthesis

Better yields of amino acids are obtained by the Gabriel Phthalimide Synthesis. α-Halo esters are used instead of α-halo acids. The halogen in the α-halo esters is replaced by a phthalimido group either by fusion or by gentle heating with potassium phthalimide in dimethyl formamide. The intermediate ester yields the desired amino acid by drastic hydrolysis or better in two steps by treatment first with 4N alkali followed by acid-hydrolysis. The basic scheme is outlined below:

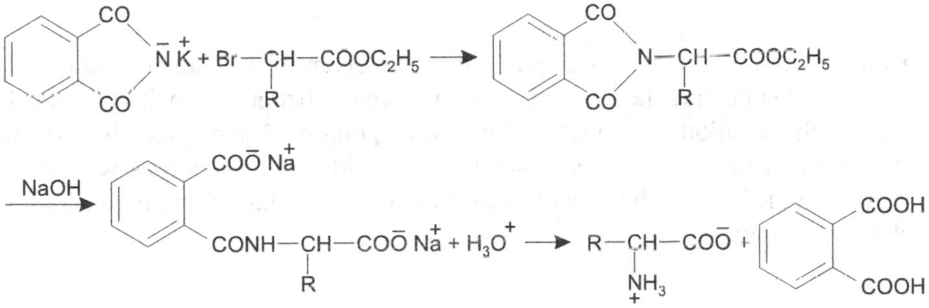

The above method is generally used for the preparation of glycine and leucine.

C. Strecker Synthesis

In this method, α-aminonitriles obtained by the treatment of aldehydes with ammonia and HCN are hydrolysed to α-amino acids. The method was first developed by Strecker to synthesise alanine and is known after his name.

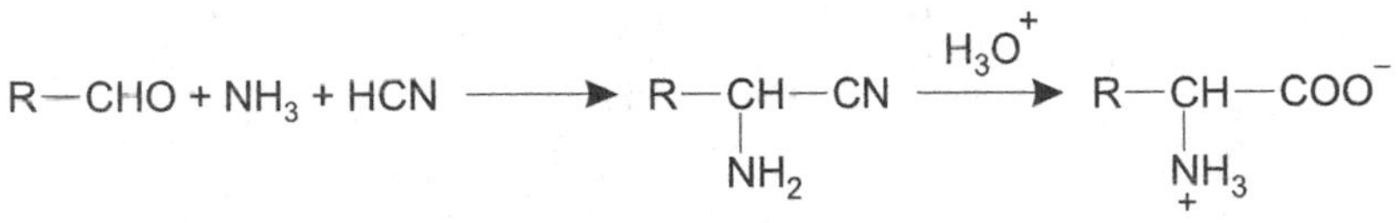

Use of acetaldehyde in the above scheme results in the formation of alanine.

The mechanism of formation of the α-aminonitrile probably involves the addition of HCN to the imine, which is formed by condensation of the aldehyde with ammonia.

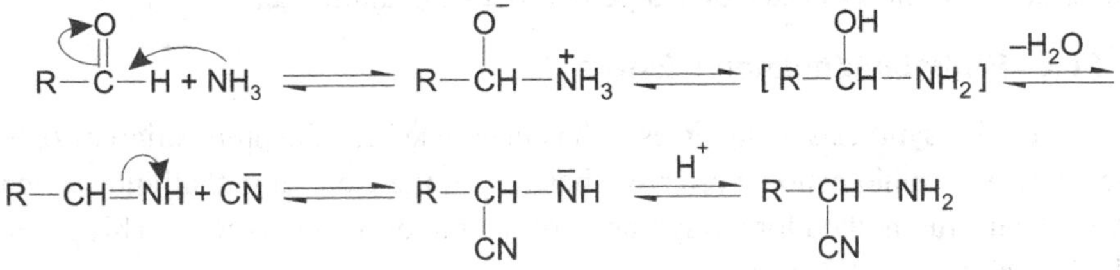

In a modification, an equimolecular mixture of ammonium chloride and potassium cyanide is used instead of the more dangerous HCN. This method is useful for the synthesis of glycine, alanine, serine, valine, methionine, glutamic acid, leucine, isoleucine and phenylalanine. For example, glycine has been synthesised by the following route:

$$HCHO + NH_4CN \xrightarrow{H_2SO_4} H_2N-CH_2-CN \xrightarrow[\substack{H_2O \\ \Delta}]{BaO} H_3\overset{+}{N}-CH_2-COO^-$$

A modification of the chemical procedure was given by Buchrer for improving the yields. On heating the cyanohydrins with urea or better ammonium carbonate in alcoholic solution, high yields of the corresponding 5-substituted hydantoins are formed, which provide the desired amino acids after hydrolysis with alkali or concentrated acid. The cyanohydrin may also be obtained *via* the bisulphite addition compound.

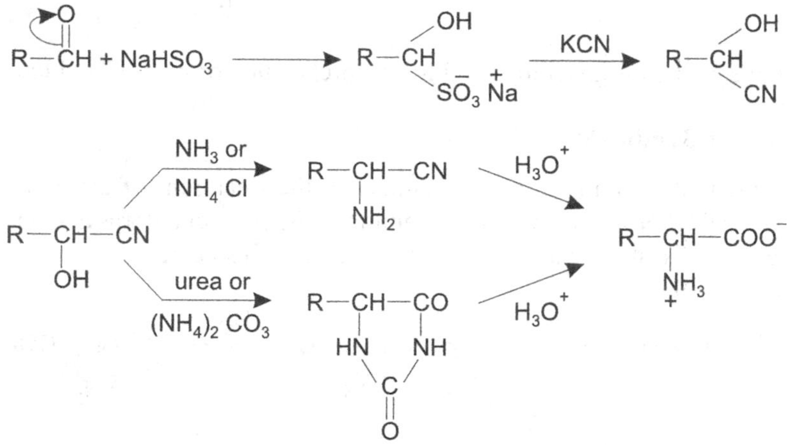

D. Malonic Ester Synthesis

By variations of the malonic ester synthesis, useful general methods have been developed for the synthesis of a large number of α-amino acids.

i. From Phthalimidomalonic Esters

In the earlier synthesis, malonic ester has been used for the preparation of α-halo acids. If we couple malonic ester synthesis with the Gabriel phthalimide scheme, we get a useful method for the synthesis of a number of amino acids. This gives us the Phthalimidomalonic Ester method.

N-Phthalimidomalonic ester is prepared by condensing potassium phthalimide with monobromo malonic ester. The ester may be alkylated by a variety of alkyl halides or condensed with α,β-unsaturated carbonyl compounds. Vigorous acid hydrolysis causes hydrolysis of both ester groups and the phthalimido group. Decarboxylation of the resulting malonic acid gives the desired amino acid. The product is a racemic modification and may be resolved.

The basic scheme is outlined below:

Phthalic acid is formed as a byproduct.

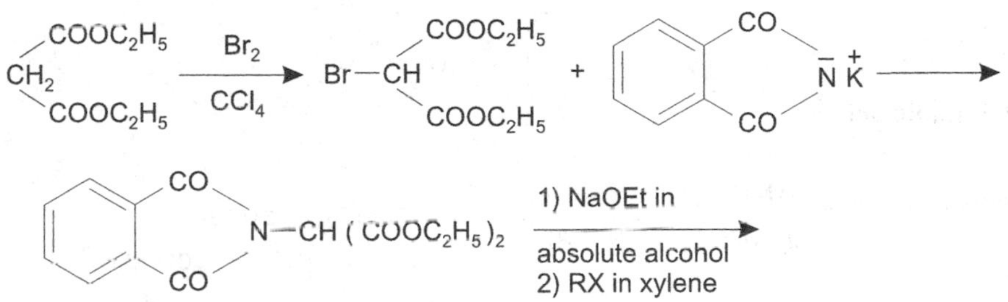

N-Phthalimidomalonic ester

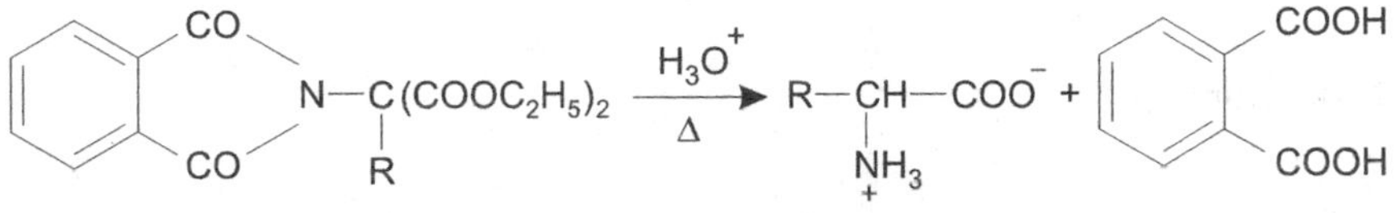

The above method has been successfully used for the synthesis of aspartic acid, phenylalanine, glutamic acid, cysteine, proline, methionine, serine, tyrosine and lysine.

Some specific examples are given below:

Aspartic acid

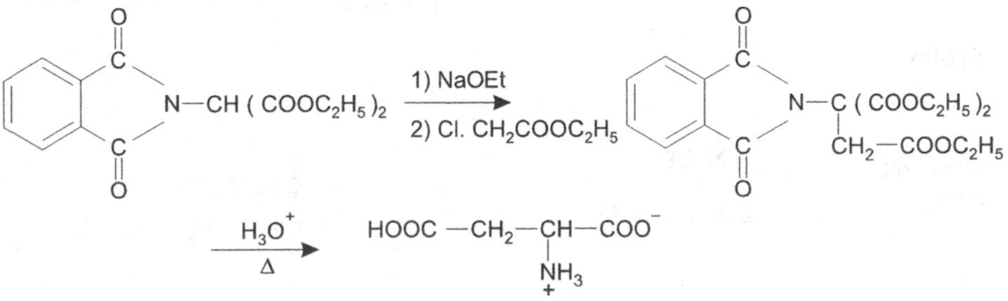

Aspartic acid

Phenylalanine

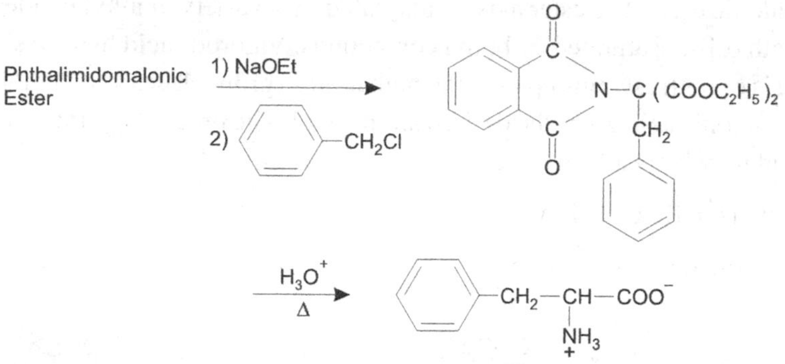

Phthalimidomalonic Ester
1) NaOEt
2) [benzyl chloride]—CH$_2$Cl

$$\xrightarrow[\Delta]{H_3O^+}$$ [phenyl]—CH$_2$—CH—COO$^-$, NH$_3^+$

Phenylalanine

Glutamic acid

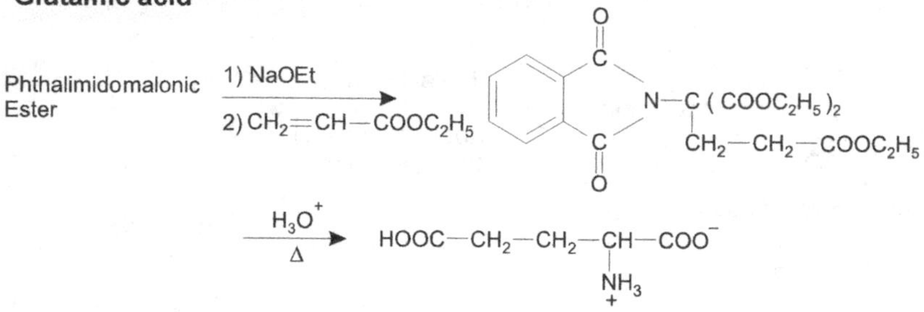

Phthalimidomalonic Ester
1) NaOEt
2) CH$_2$=CH—COOC$_2$H$_5$

$$\xrightarrow[\Delta]{H_3O^+}$$ HOOC—CH$_2$—CH$_2$—CH—COO$^-$, NH$_3^+$

Glutamic acid

Methionine

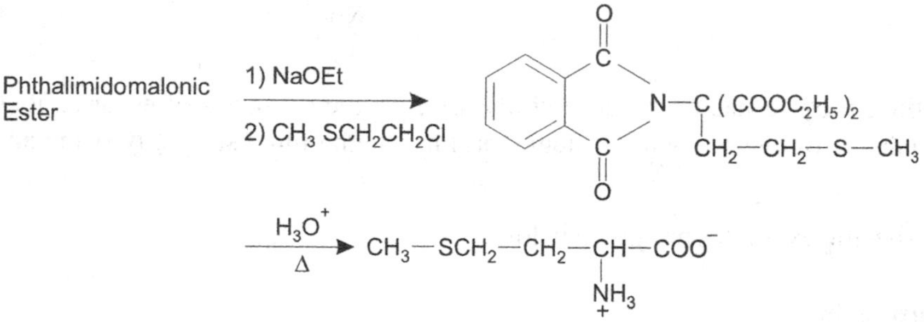

Phthalimidomalonic Ester
1) NaOEt
2) CH$_3$ SCH$_2$CH$_2$Cl

$$\xrightarrow[\Delta]{H_3O^+}$$ CH$_3$—SCH$_2$—CH$_2$—CH—COO$^-$, NH$_3^+$

Methionine

Proline

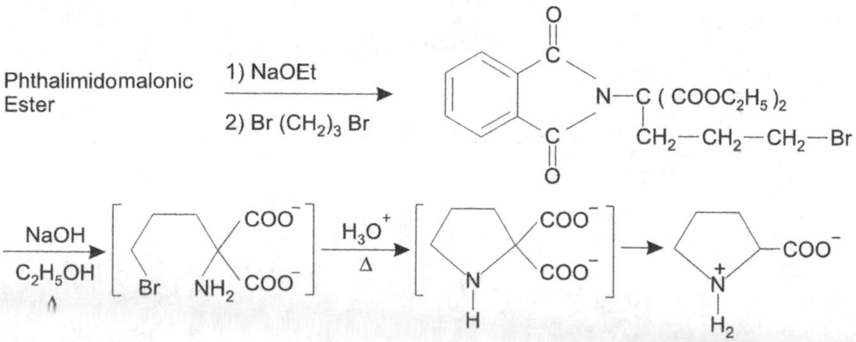

Phthalimidomalonic Ester
1) NaOEt
2) Br (CH$_2$)$_3$ Br

$$\xrightarrow[\text{C}_2\text{H}_5\text{OH}]{\text{NaOH}}$$ $$\xrightarrow[\Delta]{H_3O^+}$$

Proline

Cysteine-Cystine

Benzylthiomethyl chloride is prepared first and condensed with
N-phthalimidomalonic ester. The product is hydrolysed and reduced to obtain
cystenine. Air oxidation of cysteine gives cystine.

$$\text{—CH}_2\text{SH} + \text{HCHO} + \text{HCl} \longrightarrow \text{—CH}_2\text{—S—CH}_2\text{Cl}$$

Benzyl thiol.   Benzylthiomethyl chloride

Phthalimidomalonic   1) NaOEt
Ester $\xrightarrow{\hspace{2cm}}$
                       2) $C_6H_5CH_2SCH_2Cl$

$$\text{N—C (COOC}_2\text{H}_5\text{)}_2$$
$$\text{CH}_2\text{SCH}_2\text{C}_6\text{H}_5$$

1) NaOEt
$\xrightarrow{\hspace{2cm}}$ $C_6H_5CH_2\text{—S—CH}_2\text{—CH—COOH}$ $\xrightarrow[\text{liq. NH}_3]{\text{Na}}$ $\text{H}_3 \text{—CH}_2\text{—CH—COO}$
2) HCl $\quad\quad\quad\quad\text{NH}_2$ $\quad\quad\quad\quad\quad\quad\text{NH}_3 \quad\text{Cysteine}$
                                                                              $+$

$$\text{HOOC—CH—CH}_2\text{—S—S—CH}_2\text{—CH—COOH}$$
$$\quad\quad\text{NH}_2 \quad\quad\quad\quad\quad\quad\quad\quad\text{NH}_2$$
$$\text{Cystine}$$

ii. From Acetamidomalonic Esters

Acetamido derivatives of malonic ester may also be used to synthesise amino
acids. N-Acetamidomalonic ester may be prepared by treating the malonic ester
with nitrous acid to give a nitroso derivative which tautomerises to the oxime and
reduction of the oxime using zinc/acetic acid in the presence of acetic anhydride.
Catalytic hydrogenation of the oxime in acetic anhydride solution also affords
N-acetamidomalonic ester.

$$\text{CH}_2 \overset{\text{COOC}_2\text{H}_5}{\underset{\text{COOC}_2\text{H}_5}{}} \xrightarrow{\text{HNO}_2} \left[\text{ON—CH} \overset{\text{COOC}_2\text{H}_5}{\underset{\text{COOC}_2\text{H}_5}{}} \right] \longrightarrow$$

$$\text{HON=C} \overset{\text{COOC}_2\text{H}_5}{\underset{\text{COOC}_2\text{H}_5}{}} \xrightarrow[\text{Ac}_2\text{O}]{\text{Zn/ HOAc}} \text{CH}_3\text{—CONH—CH} \overset{\text{COOC}_2\text{H}_5}{\underset{\text{COOC}_2\text{H}_5}{}}$$

N-Acetamidomalonic ester is alkylated in the usual manner and the resulting product
is hydrolysed and decarboxylated to give the desired amino acid. Some specific
examples are given below:

Serine

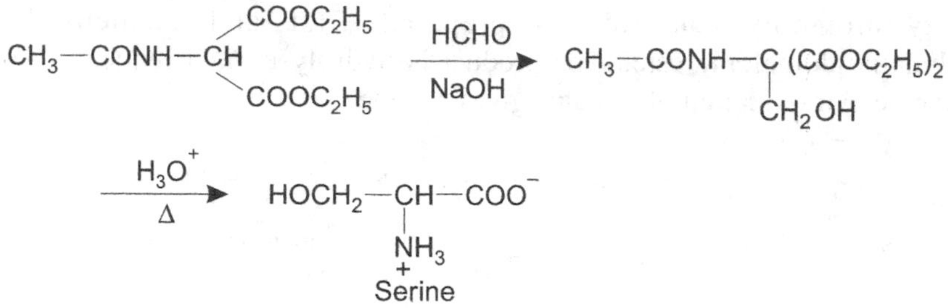

$$CH_3-CONH-CH \begin{matrix} COOC_2H_5 \\ \\ COOC_2H_5 \end{matrix} \xrightarrow[NaOH]{HCHO} CH_3-CONH-\underset{CH_2OH}{C}(COOC_2H_5)_2$$

$$\xrightarrow[\Delta]{H_3O^+} HOCH_2-\underset{\underset{\underset{Serine}{+}}{NH_3}}{CH}-COO^-$$

Leucine

$$\text{Acetamidomalonic ester} \xrightarrow[2)\,(CH_3)_2CHCH_2Br]{1)\,NaOEt} CH_3CONH-\underset{CH_2-CH(CH_3)_2}{C}(COOC_2H_5)_2$$

$$\xrightarrow[\Delta]{H_3O^+} \begin{matrix} CH_3 \\ \\ CH_3 \end{matrix}\!\!\diagup CHCH_2-\underset{\underset{\underset{Leucine}{+}}{NH_3}}{CH}-COO^-$$

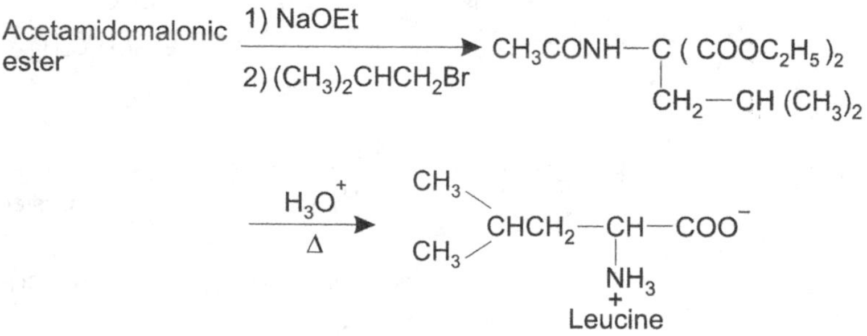

Histidine

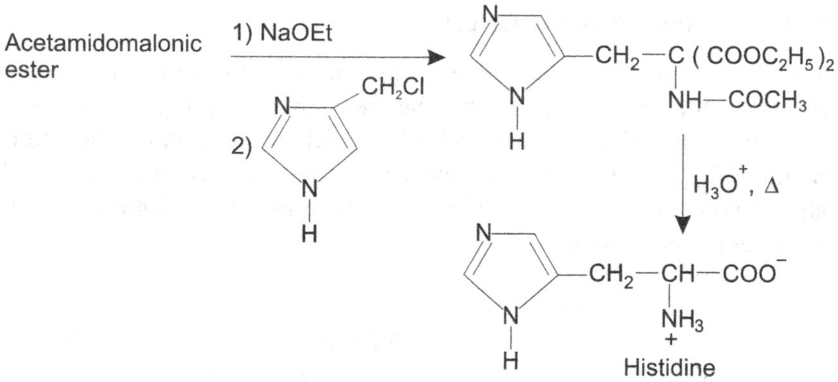

iii. From Benzamidomalonic Esters

Diethylaminomalonate can be prepared from diethyl malonate by the following reactions:

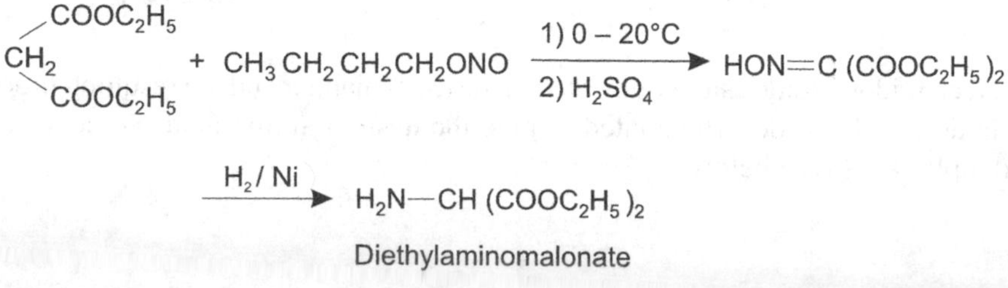

$$CH_2\begin{matrix} COOC_2H_5 \\ \\ COOC_2H_5 \end{matrix} + CH_3CH_2CH_2CH_2ONO \xrightarrow[2)\,H_2SO_4]{1)\,0-20°C} HON{=}C(COOC_2H_5)_2$$

$$\xrightarrow{H_2/Ni} H_2N-CH(COOC_2H_5)_2$$

Diethylaminomalonate

Diethylaminomalonic ester is now benzoylated to give benzamidomalonic ester in good yields which is a useful reagent for the synthesis of amino acids.

$$H_2N-\underset{\underset{\textstyle COOC_2H_5}{|}}{\overset{\overset{\textstyle COOC_2H_5}{|}}{CH}} \quad \xrightarrow[\text{Pyridine}]{C_6H_5COCl} \quad C_6H_5CONH-\underset{\underset{\textstyle COOC_2H_5}{|}}{\overset{\overset{\textstyle COOC_2H_5}{|}}{CH}}$$

Benzamidomalonic ester so obtained is alkylated in the usual manner and the product so obtained on acid hydrolysis and decarboxylation affords the desired amino acid. For example, leucine may be synthesised in the following manner:

$$C_6H_5CONH-\underset{\underset{\textstyle COOC_2H_5}{|}}{\overset{\overset{\textstyle COOC_2H_5}{|}}{CH}} \quad \xrightarrow[\text{2) }(CH_3)_2\text{ CHCH}_2\text{I}]{\text{1) NaOEt}} \quad C_6H_5CONH-\underset{\underset{\textstyle CH_2-CH}{|}}{\overset{\overset{\textstyle }{|}}{C}}(COOC_2H_5)_2 \overset{\textstyle CH_3}{\underset{\textstyle CH_3}{<}}$$

$$\xrightarrow[\Delta]{H_3O^+} \quad \overset{\textstyle CH_3}{\underset{\textstyle CH_3}{>}}CH-CH_2-\underset{\underset{\textstyle \overset{+}{N}H_3}{|}}{CH}-COO^- + C_6H_5COOH$$

E. Aldehyde Condensations with Glycine Derivatives (Erlenmeyer Azlactone Synthesis)

From the structural point of view all the amino acids can be considered as derivatives of glycine in which one of two α-hydrogen atoms is substituted by the side chain.

$$\boxed{H}-\underset{\underset{\textstyle NH_2}{|}}{CH}-COOH \quad \longrightarrow \quad \boxed{R}-\underset{\underset{\textstyle NH_2}{|}}{CH}-COOH$$

It is possible to introduce any group by reacting aldehydes with the α-hydrogens of glycine. For this it is necessary to protect the amino group, which would otherwise react first. In Erlenmeyer azlactone synthesis, hippuric acid (benzoyl glycine) is used for condensing with aromatic aldehydes. The amino group is thus protected by the benzoyl group.

An azlactone is formed by boiling the mixture of an aromatic aldehyde and hippuric acid in the presence of acetic anhydride and sodium acetate. Partial hydrolysis of the azlactone by warming with 1% sodium hydroxide solution opens up the ring. The product on reduction with sodium amalgam followed by acid hydrolysis gives an α-amino acid. Also heating the azlactone with a mixture of red phosphorus and hydriodic acid converts it in one step into the amino acid.

Under the above conditions only aromatic aldehydes condense with hippuric acid. However, it has been found that aliphatic aldehydes may condense with hippuric acid to form azlactones if lead acetate is used instead of sodium acetate.

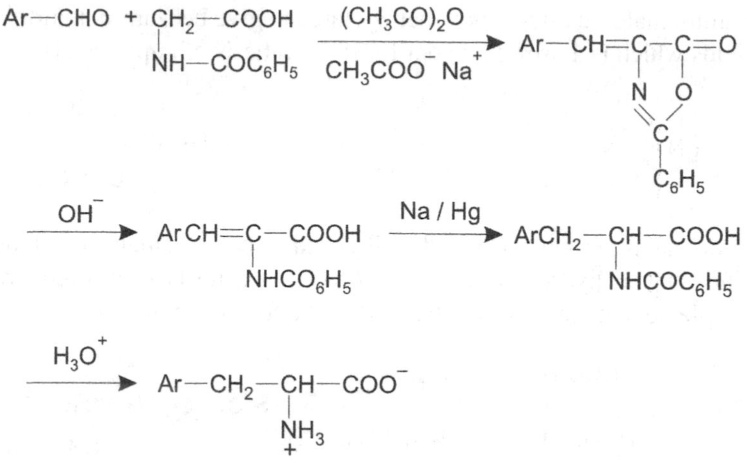

$$Ar—CHO + \underset{\underset{NH—COC_6H_5}{|}}{CH_2—COOH} \xrightarrow[\underset{CH_3COO^-\ Na^+}{}]{(CH_3CO)_2O} Ar—CH{=}\underset{\underset{N}{\|}}{C}—\underset{\underset{O}{}}{C}{=}O$$

$$\xrightarrow{OH^-} \underset{\underset{NHCO_6H_5}{|}}{ArCH{=}C—COOH} \xrightarrow{Na\,/\,Hg} \underset{\underset{NHCOC_6H_5}{|}}{ArCH_2—CH—COOH}$$

$$\xrightarrow{H_3O^+} Ar—CH_2—\underset{\underset{NH_3}{\overset{|}{\underset{+}{}}}}{CH}—COO^-$$

The azlactone synthesis has been used successfully for preparing aromatic amino acids like phenylalanine, tyrosine, tryptophan and thyroxine. For example, tyrosine may be obtained by the following route:

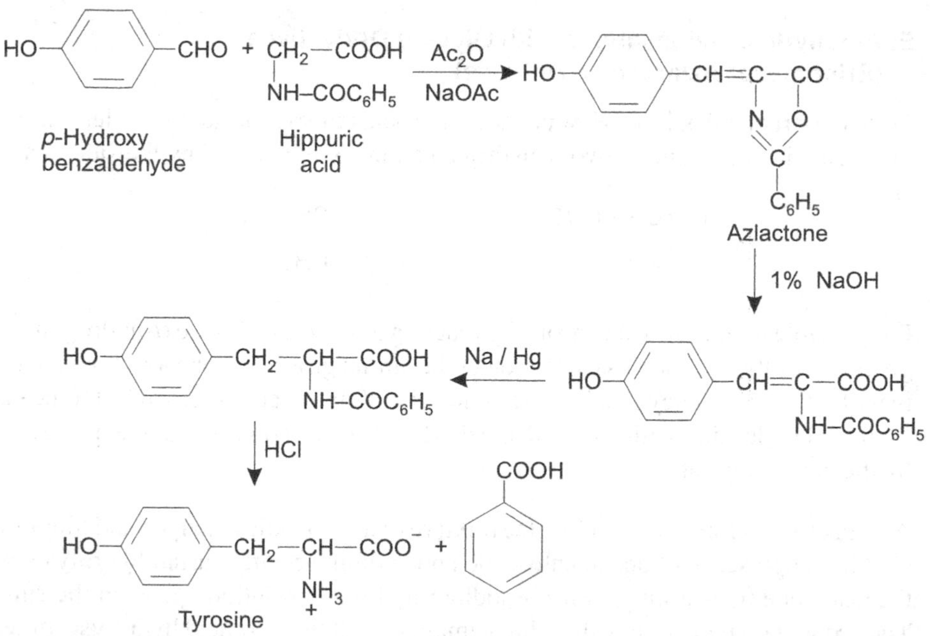

p-Hydroxy benzaldehyde   Hippuric acid

Azlactone

Aromatic aldehydes may also be condensed with 2,5-diketopiperazine in a similar manner. Reductive cleavage of the product by heating with hydriodic acid and red phosphorus gives an α-amino acid. For example,

$$2\ C_6H_5CHO + \underset{\underset{\displaystyle CO-NH}{|}}{\overset{\overset{\displaystyle NH-CO}{|}}{CH_2\ \ CH_2}} \xrightarrow{(CH_3CO)_2O} C_6H_5CH\!=\!\underset{\underset{\displaystyle CO-NH}{|}}{\overset{\overset{\displaystyle NH-CO}{|}}{C}}\ \ C\!=\!CHC_6H_5$$

$$\xrightarrow[\displaystyle (P)]{\displaystyle HI}\ 2\ C_6H_5CH_2\!-\!\underset{\underset{\displaystyle +}{\overset{\displaystyle NH_3}{|}}}{CH}\!-\!COO^-$$

Phenylalanine

Condensation of the aldehydes with glycine or glycine esters provides only β-hydroxy acids as the first step will occur according to the aldol type of condensation.

$$C_6H_5CHO + \underset{\underset{\displaystyle COOR}{|}}{\overset{\overset{\displaystyle NH_2}{|}}{CH_2}} \longrightarrow C_6H_5\!-\!\underset{\underset{\displaystyle OH}{|}}{CH}\!-\!\overset{\overset{\displaystyle N\!=\!CH\!-\!C_6H_5}{|}}{CH}\!-\!COOR$$

$$\downarrow H_2O$$

$$C_6H_5\!-\!\underset{\underset{\displaystyle OH}{|}}{CH}\!-\!\underset{\underset{\underset{\displaystyle +}{\displaystyle NH_3}}{|}}{CH}\!-\!COO^-$$

F. Synthesis via Molecular Rearrangements

A large variety of α-amino acids may be synthesised via molecular rearrangements like the Beckmann rearrangement, Schmidt reaction, Curtius reaction and Hofmann degradation. Basically in all these reactions intramolecular migration of the alkyl group takes place to the more electron deficient nitrogen atom giving intermediates which ultimately yield α-amino acids.

i. By Schmidt Reaction

The Schmidt reaction is used in a variety of ways for the synthesis of amino acids. Simple amino acids may be obtained if this reaction is applied to alkylated malonic acids, e.g.,

$$CH_3\!-\!CH_2\!-\!CH_2\!-\!\underset{\underset{\displaystyle COOH}{|}}{CH}\!-\!COOH \xrightarrow[\displaystyle H_2SO_4]{\displaystyle HN_3} CH_3\!-\!CH_2\!-\!CH_2\!-\!\underset{\underset{\underset{\displaystyle +}{\displaystyle NH_3}}{|}}{CH}\!-\!COO^-$$

By applying the Schmidt reaction to cyclic ketones or keto acids, we get the desired amino acids, e.g., we get lysine by using the Schmidt reaction as given on the next page.

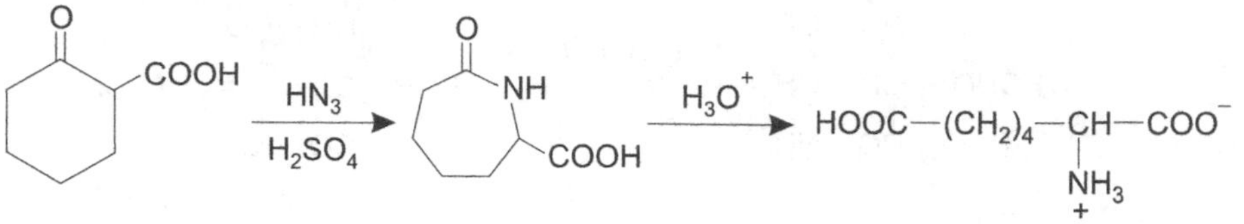

A second application of the Schmidt reaction yields lysine. The carboxyl group that is not α- to the amino group reacts. Actually simple α-amino acids do not react at all in the Schmidt rection.

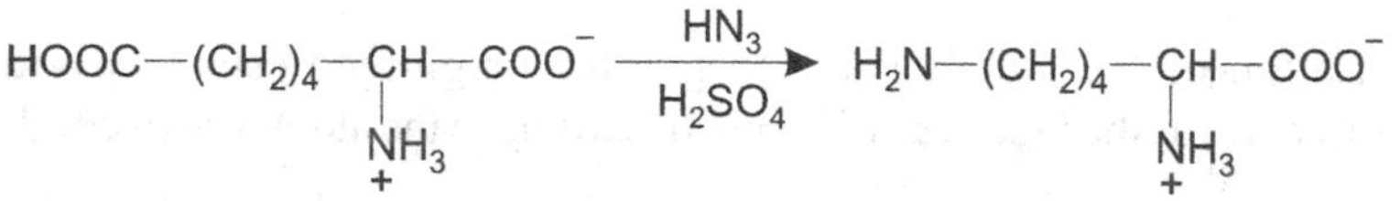

Lysine is also obtained when the Schmidt reaction is applied to cyclohexanone.

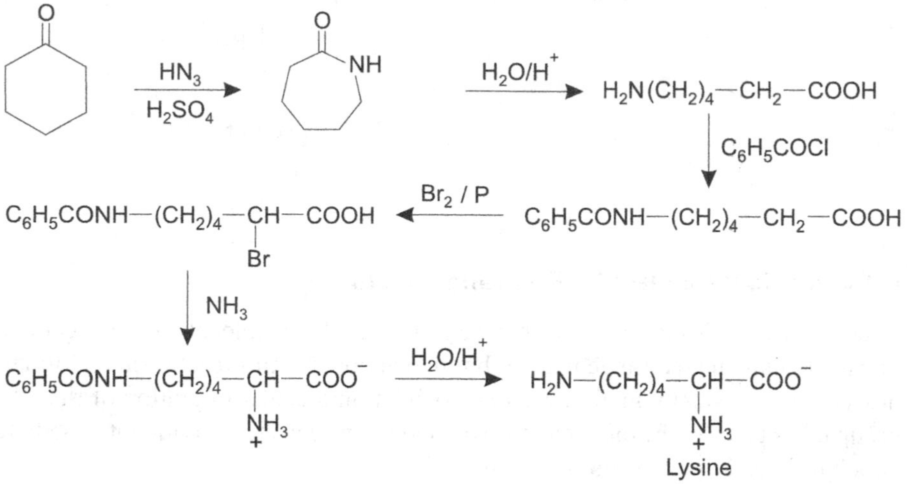

The probable mechanism for the conversion of ketones to amides involves intramolecular migration of the alkyl group to the electron deficient nitrogen atom.

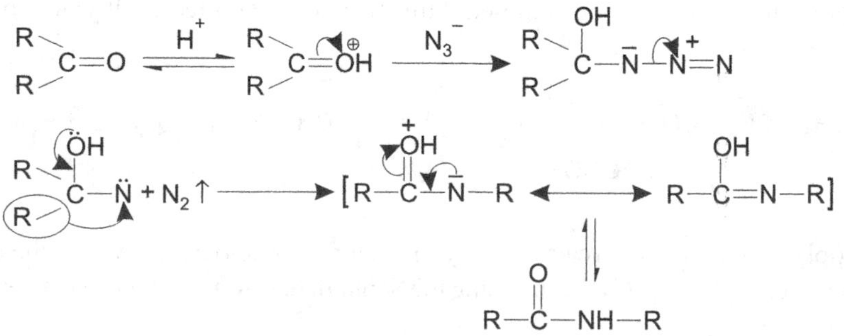

ii. Synthesis through the Beckmann Rearrangement

The Beckmann rearrangement of cyclopentanone oxime gives ω-aminovaleric acid, whch can be easily converted to basic amino acids like proline, ornithine and arginine.

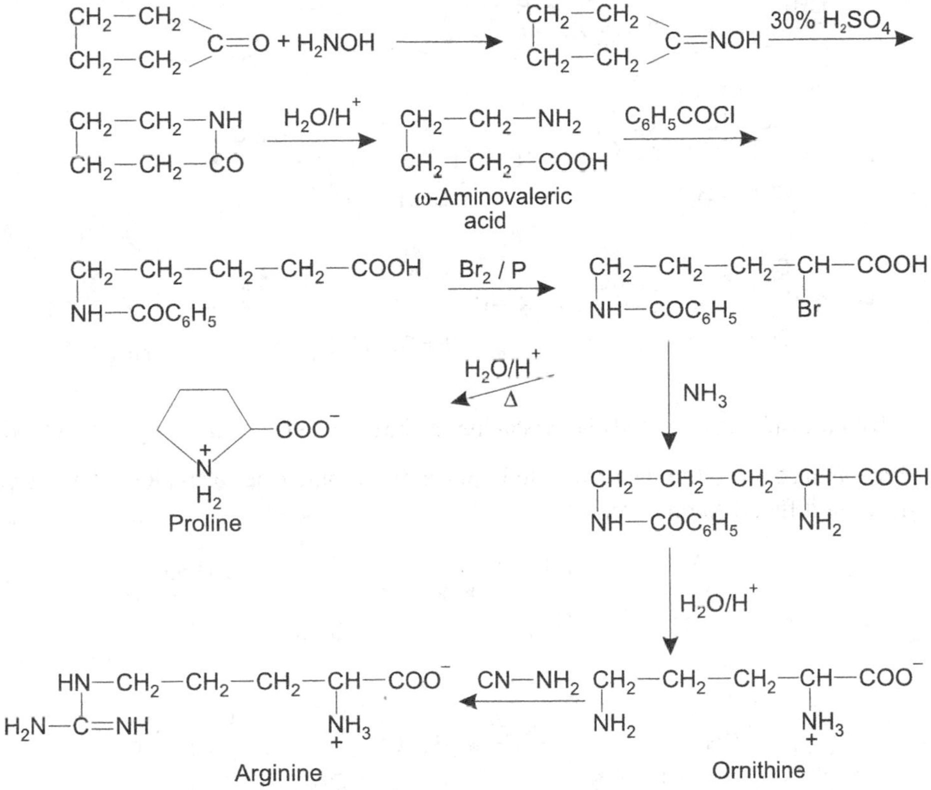

ω-Aminovaleric
acid

Proline

Ornithine

Arginine

The mechanism of the Beckmann rearrangement is as follows:

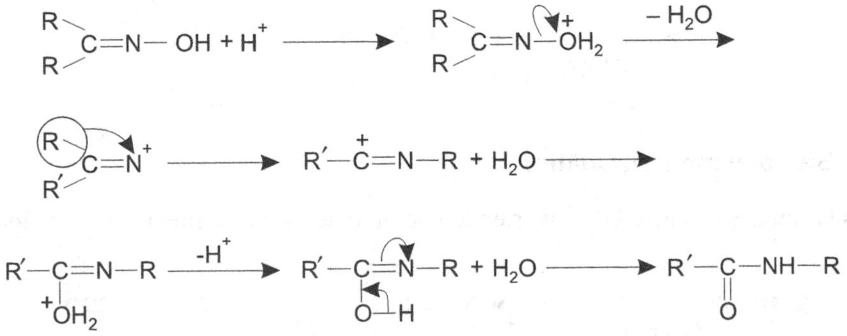

By using cyclohexanone instead of cyclopentanone, we get lysine.

iii. By Curtius Reaction

α-Amino acids may be synthesised by the application of Curtius reaction to alkylated malonic half esters.

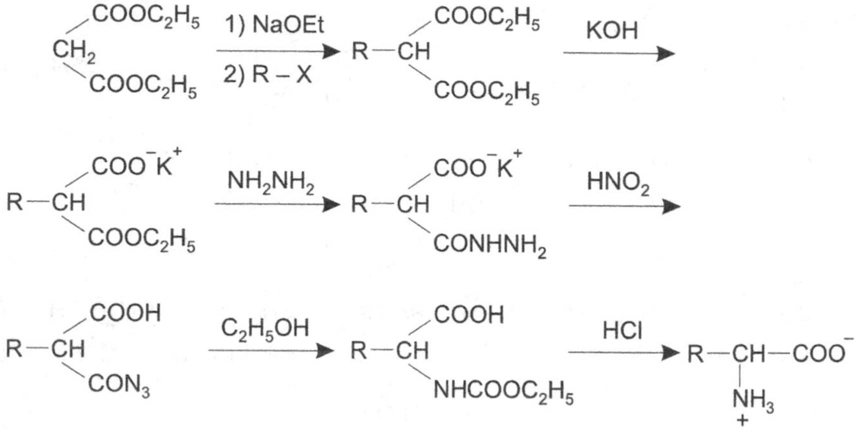

$$CH_2 \begin{matrix} COOC_2H_5 \\ \\ COOC_2H_5 \end{matrix} \xrightarrow[\text{2) R--X}]{\text{1) NaOEt}} R-CH \begin{matrix} COOC_2H_5 \\ \\ COOC_2H_5 \end{matrix} \xrightarrow{\text{KOH}}$$

$$R-CH \begin{matrix} COO^-K^+ \\ \\ COOC_2H_5 \end{matrix} \xrightarrow{NH_2NH_2} R-CH \begin{matrix} COO^-K^+ \\ \\ CONHNH_2 \end{matrix} \xrightarrow{HNO_2}$$

$$R-CH \begin{matrix} COOH \\ \\ CON_3 \end{matrix} \xrightarrow{C_2H_5OH} R-CH \begin{matrix} COOH \\ \\ NHCOOC_2H_5 \end{matrix} \xrightarrow{HCl} R-\underset{\underset{+}{\overset{|}{NH_3}}}{CH}-COO^-$$

Glycine, alanine, valine and phenylalanine can be obtained very easily by this method.

We can also use ethyl cyanoacetate in place of malonic ester as malonic half ester is quite difficult to prepare.

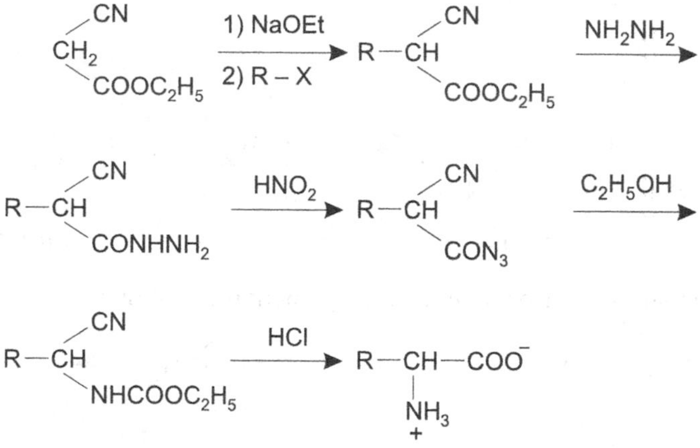

$$CH_2 \begin{matrix} CN \\ \\ COOC_2H_5 \end{matrix} \xrightarrow[\text{2) R--X}]{\text{1) NaOEt}} R-CH \begin{matrix} CN \\ \\ COOC_2H_5 \end{matrix} \xrightarrow{NH_2NH_2}$$

$$R-CH \begin{matrix} CN \\ \\ CONHNH_2 \end{matrix} \xrightarrow{HNO_2} R-CH \begin{matrix} CN \\ \\ CON_3 \end{matrix} \xrightarrow{C_2H_5OH}$$

$$R-CH \begin{matrix} CN \\ \\ NHCOOC_2H_5 \end{matrix} \xrightarrow{HCl} R-\underset{\underset{+}{\overset{|}{NH_3}}}{CH}-COO^-$$

iv. By Hofmann Degradation

Ester amides can also be converted to α-amino acids by Hofmann degradation.

$$R-CH \begin{matrix} CONH_2 \\ \\ COOC_2H_5 \end{matrix} \xrightarrow[\text{NaOH}]{Br_2} R-CH \begin{matrix} NH_2 \\ \\ COOC_2H_5 \end{matrix} \longrightarrow R-\underset{\underset{+}{\overset{|}{NH_3}}}{CH}-COO^-$$

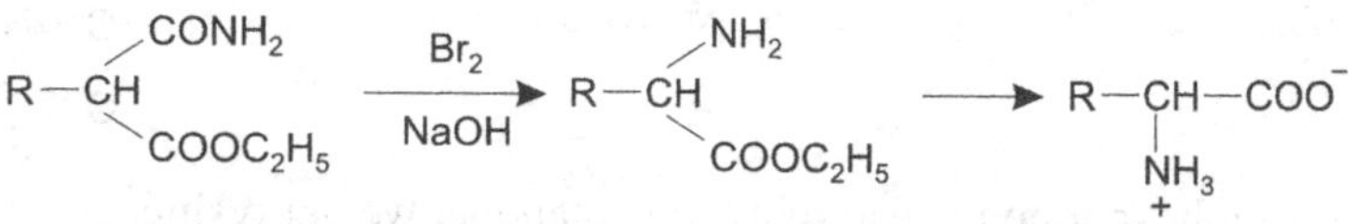

This is a typical conversion of amide to an amine. The formal end-point in the reaction is the isocyanate. Under the reaction conditions, addition of water takes place to yield the unstable carbamic acid, which decarboxylates to give the amine.

$$R-\overset{\overset{\displaystyle O}{\|}}{C}-NH_2 \xrightarrow{\bar{O}Br} R-\overset{\overset{\displaystyle O}{\|}}{C}-NH-Br \xrightarrow{OH^-} (R\overset{\overset{\displaystyle O}{\|}}{\underset{}{C}}-\bar{N}-Br$$

$$R-N{=}C{=}O \longleftarrow [R-\bar{N}-\overset{+}{C}{=}O]$$
Isocyanate

$$R-\overset{\curvearrowleft}{N}{=}C{=}O + H_2O \longrightarrow R-\overset{H}{\underset{\underset{OH}{|}}{N}}-\overset{|}{C}{=}O \xrightarrow{OH^-} R-NH_2 + CO_2 + H_2O$$

The Curtius and Schmidt reactions are actually closely related to that of Hofmann because all of these involve the formation of the same isocyanate as the intermediate. In both these reactions, N_2 is the leaving group from the same azide intermediate. Here again the migration of the alkyl group R takes place in a concerted process, i.e., it is intramolecular. The azide may be obtained either by nitrosation of an acid-hydrazide (Curtius reaction) or by the reaction of hydrazoic acid on a carboxylic acid—The Schmidt reaction.

G. Reductive Amination

The reductive amination of α-keto acids yields α-amino acids. This may be achieved by carrying out reduction catalytically or with sodium and ethanol in the presence of ammonia. The mechanism of the reduction is not certain but it probably occurs via the imino-acid. The oximes and phenylhydrazones of α-keto acids may also be reduced to α-amino acids in a similar manner.

$$
\begin{array}{ccc}
& R-\overset{\overset{\displaystyle O}{\|}}{C}-COOH & \\
\overset{NH_2OH}{\swarrow} & \downarrow NH_3 & \overset{C_6H_5NHNH_2}{\searrow} \\
\overset{N-OH}{\underset{}{\|}} & \overset{NH}{\underset{}{\|}} & \overset{N-NHC_6H_5}{\underset{}{\|}} \\
R-C-COOH & [R-C-COOH] & R-C-COOH \\
\searrow{[H]} & \downarrow{[H]} & \swarrow{[H]} \\
& R-\underset{\underset{NH_2}{|}}{CH}-COOH &
\end{array}
$$

α-Oximino acids are also obtained by the treatment of 2-alkyl or 2-acyl acetoacetic esters with butyl nitrite in sulphuric acid, whereas action of a diazonium salt in the presence of sodium ethoxide affords the corresponding phenyl-hydrazones. Reduction of these can be carried out to get any α-amino acid.

$$\text{CH}_3\text{—CO—CH—COOC}_2\text{H}_5 \xrightarrow[\substack{n\text{-}C_4H_9ONO \\ H_2SO_4}]{} R\text{—}\overset{\displaystyle \overset{NOH}{\|}}{C}\text{—COOC}_2\text{H}_5$$

$$\xrightarrow[\substack{C_6H_5N_2^+Cl^- \\ NaOEt}]{} R\text{—}\overset{\displaystyle \overset{N\text{—NH—}C_6H_5}{\|}}{C}\text{—COOC}_2\text{H}_5 \xrightarrow{H_2O} R\text{—}\overset{\displaystyle \overset{N\text{—NH—}C_6H_5}{\|}}{C}\text{—COOH}$$

The usefulness of these methods depends upon the availability of α-keto acids. Thus, alanine, phenylalanine, leucine, isoleucine and valine may be prepared in this way.

H. Oxidation of Amino Alcohols

Oxidation of the 2-amino alcohols affords a simple route to α-amino acids but is of less importance because of the limited number of amino alcohols which are readily available.

The conversion of the alcoholic group to a carboxyl residue can be carried out by the usual oxidising agents like alkaline potassium permanganate or acid dichromate, *etc.*, but the oxidation has to be preceded by the protection of the labile amino group. Instead of the usual masking with a phthaloyl or bezoyl residue, protonation by addition of a strong acid is quite sufficient for this process.

$$R\text{—}\underset{\underset{NH_2}{|}}{CH}\text{—CH}_2OH \xrightarrow{H^+} R\text{—}\underset{\underset{NH_3^+}{|}}{CH}\text{—CH}_2OH \xrightarrow[2)\ O\bar{H}]{1)\ (O)} R\text{—}\underset{\underset{NH_3^+}{|}}{CH}\text{—COO}^-$$

2.7 RESOLUTION OF RACEMIC MIXTURES OF AMINO ACIDS

When amino acids are synthesised in nature, only the L-enantiomer is formed. However, when amino acids are synthesised in the laboratory, the product is usually a racemic mixture—a mixture of D- and L-enantiomers. When only one isomer is desired, the enantiomers must be separated. In order to obtain the naturally occurring L-amino acid, we must resolve the racemic mixture. Some methods used to resolve the racemic mixture are:

1. We already know that the commonly used method to separate the enantiomers is to convert them into diastereomers, separate the diastereomers and convert the separated diastereomers back to enantiomers.

The amino group is usually converted into an amide so that the material is not amphoteric. For example, alanine reacts with benzoyl chloride in aqueous base to give N-benzoylalanine which is a typical acid.

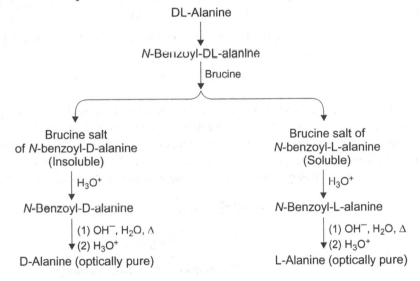

Alanine   N-Benzoylalanine

The racemic N-benzoylalanine is resolved in the normal way with brucine or strychnine. If brucine is used, it is the brucine salt of D-alanine that is less soluble. If strychnine is used, the strychnine salt of L-alanine crystallises. Acidification of the salts yields the D- and L-enantiomers of N-benzoylalanine Basic hydrolysis then affords the pure enantiomeric amino acids. The process is outlined below:

DL-Alanine

↓

N-Benzoyl-DL-alanine

↓ Brucine

Brucine salt of N-benzoyl-D-alanine (Insoluble)	Brucine salt of N-benzoyl-L-alanine (Soluble)
↓ H_3O^+	↓ H_3O^+
N-Benzoyl-D-alanine	N-Benzoyl-L-alanine
↓ (1) OH^-, H_2O, Δ (2) H_3O^+	↓ (1) OH^-, H_2O, Δ (2) H_3O^+
D-Alanine (optically pure)	L-Alanine (optically pure)

In this method, the enantiomer that forms the less soluble salt is usually obtained in an optically pure state. The more soluble enantiomer is obtained in an impure state because some of the less soluble salt invariably remains in solution.

In the above case, N-benzoyl-L-alanine may be treated with strychnine to give the insoluble strychnine salt. In this way both enantiomers may be obtained in an optically pure state.

This method of diastereomeric salts suffers from several severe drawbacks. This is a time consuming method and yields are often low because several recrystallisations may be required in order to purify the product. There is no way to judge which chiral base will give well defined crystals with a given amino acid or which enantiomer of the amino acid will form the less soluble salt.

2. Various biological procedures are more useful for the routine large scale resolution of amino acids.

One important method of resolving amino acids is based on the use of enzymes. Because enzymes are chiral they react at a different rate with each of the enantiomers. These enzymes called *deacylases* catalyse the hydrolysis of N-acylamino acids in living organisms. Since the active site of the enzyme is chiral, it hydrolyses only N-acylamino acids of the L-configuration. When it is exposed to a racemic mixture of N-acylamino acids, only the derivative of the L-amino acid is affected and products are separated easily.

$$DL-R-CH-COO^{-} \xrightarrow{(CH_3CO)_2O} DL-R-CH-COOH$$

with NH_3^{+} below the first structure, and $NH-CHCH_3$ below the second, then $\downarrow$ deacylase

$$H_3\overset{+}{N}-\underset{R}{\overset{COO^{-}}{\underset{|}{\overset{|}{C}}}}-H \quad + \quad H-\underset{R}{\overset{COO^{-}}{\underset{|}{\overset{|}{C}}}}-NH-COCH_3 + CH_3COO^{-}$$

L-Amino acid   D-N-Acylamino acid

An example is the resolution of DL-Leucine by hog renal acylase, an enzyme isolated from hog kidneys. The enzyme catalyses the hydrolysis of amide linkages and is specific for amides of L-amino acids. The enzyme catalyses the hydrolysis of N-acetyl-L-leucine to the amino acid leaving N-acetyl-D-Leucine unchanged. The two enantiomers are easily separable, since one is acidic and other is amphoteric.

$$\text{DL-Leucine} \xrightarrow{(CH_3CO)_2O} \textit{N}\text{-Acetyl-DL-leucine}$$

$\downarrow$ Hog renal acylase

$$\text{L-Leucine} + \textit{N}\text{-Acetyl-D-leucine}$$

2.8 STEREOSELECTIVE SYNTHESIS OF AMINO ACIDS

Various synthethic methods described earlier would give a racemic mixture, *i.e.*, a mixture of D- and L-amino acids. The ideal synthesis of an amino acid would be one that would produce only the naturally occurring L-amino acid. This has been made possible through the use of chiral hydrogenation catalysts derived from transition metals. A variety of catalysts has been used.

We give here the method developed by **B. Bosnich** which is based on a rhodium complex with (R)-1, 2-bis (diphenyl-phosphino) propane. This compound is called "(R)-prophos". When a rhodium complex of norbornadiene (NBD) is treated with (R)-prophos, the (R)-prophos replaces one of the molecules of norbornadiene surrounding the rhodium atom to produce a chiral rhodium complex.

$$(C_6H_5)_2P-CH-CH_2-P(C_6H_5)_2$$

$$CH_3$$

(R)-Prosphos

[RH (NBD)$_2$]ClO$_4$ + (R)-Prophos $\longrightarrow$ [RH (R)-prophos) (NBD)]ClO$_4$ + NBD

Chiral rhodium complex

Treating this rhodium complex with hydrogen in a solvent such as ethanol yields a solution containing an active chiral hydrogenation catalyst, which probably has the composition

$$[\text{Rh (R)}-\text{Prophos(H)}_2\,(\text{EtOH})_2]^+$$

When 2-acetylamino propenoic acid is added to this solution and hydrogenation is carried out, the product of the reaction is the *N*-acetyl derivative of L-alanine in 90% yield. Hydrolysis of the *N*-acetyl group yields L-alanine. Because the hydrogenation catalyst is chiral, it transfers its hydrogen atoms in a stereoselective way. This synthesis is often called an **asymmetric synthesis**.

$$CH_2=C-COOH \xrightarrow[\text{H}_2]{[\text{Rh(R)}-\text{prophos(H}_2)(\text{solvent})_2]^+} CH_3-C-COOH$$

NH—COCH$_3$	NH—COCH$_3$
2-Acetylaminopropenoic acid	*N*-Acetyl-L-alanine

$$\xrightarrow[\text{(2) H}_3\text{O}^+]{\text{(1) OH}^-,\,\text{H}_2\text{O},\,\Delta} \quad H_3C-C-COO$$

$$NH_3$$
$$+$$
L-Alanine

The above procedure has been used to synthesise several other L-amino acids from 2-acetylaminopropenoic acids that have substituents at the 3-position. If we use the (Z)-isomer, it yields the L-amino acid with yields of 87-93%.

$$\underset{R}{\overset{H}{\diagdown}}C=C\underset{NHCOCH_3}{\overset{COOH}{\diagup}} \xrightarrow[\text{(2) OH}^-,\,\Delta \quad \text{(3) H}_3\text{O}^+]{\text{(1) [Rh(R)}-\text{prophos(H}_2)(\text{solvent})_2]^+} \quad R-C-COO^-$$

$$NH_3$$
$$+$$
L-Amino acid

2.9 BIOSYNTHESIS AMINO ACIDS

We have seen in Section 2.4 that every living cell must have a supply of amino acids to carry out various metabolic processes especially protein synthesis. The biosynthesis of proteins requires the presence of all the twenty amino acids. We know some of these are essential. They must be included in our diet because the human body can biosynthesise only half of those required. Although we can obtain all the twenty amino acids from the food we take, still the human body has to biosynthesise to carry out the important metabolic roles. These reactions involve a constant synthesis and degradation of the compounds in quantities much higher than those needed for protein synthesis.

The pathways for the biosynthesis of amino acids are diverse. Some of these may be derived from simple substances like glucose as a carbon source and ammonium ion as a nitrogen source. However, we find that they have an important common feature. Their carbon skeletons come from intermediates in glycolysis, the citric acid cycle (Chapter 3, Section 3) or the pentose phosphate pathway (Chapter 1)

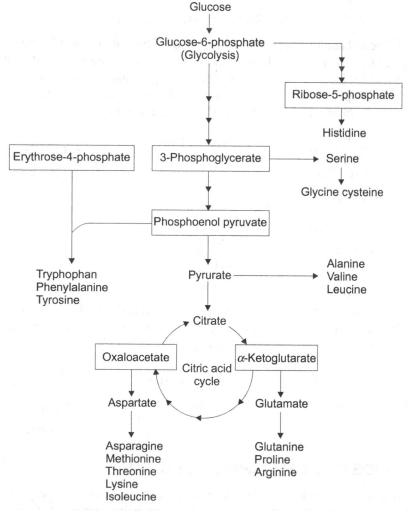

Fig. 2.2. Aminoacid biosynthesis (Precursors are shown in brackets).

It is important to note that catabolic breakdown of amino acids gives intermediates of the citric acid cycle whereas anabolic formation of amino acids utilizes citric acid cycle intermediates as precursors. That means there is a relationship between amino acid metabolism and the citric acid cycle intermediates. We know catabolism means the breakdown of nutrients to provide energy and anabolism is the synthesis of biomolecules from simpler conpounds. Fig. 2.2 shows an overview of the amino acid biosynthesis depicting the precursors from glycolysis, the citric acid cycle and the pentose phosphate pathway.

The non-essential amino acids are synthesised by quite simple reactions whereas the pathways for synthesis of essential amino acids are quite complex. Ten of the amino acids require only one or a few enzymatic steps for their biosynthesis from their precursors. The pathways for other amino acids are more complex. The number of enzymes required by cells to synthesise the essential amino acids is large in relation to the number of enzymes required to synthesise the non-essential amino acids.

We find the biosynthesis of amino acids involves a common set of reactions. For example, glutamate arises form α-ketoglutaric acid, which is an intermediate in the Kreb's cycle. The formation of glutamate is a reductive amination taking place in the presence of reducing enzymes. From glutamate we get glutamine easily by simple amidation.

$$
\begin{array}{ccc}
\underset{\substack{| \\ CH_2 \\ | \\ CH_2-COO^-}}{\overset{\overset{O}{\|}}{C}-COO^-} + NH_3 & \xrightarrow[\text{dehydrogenase}]{\text{Glutamate}} & \underset{\substack{| \\ CH_2 \\ | \\ CH_2-COO^-}}{\overset{+}{H_3N}-CH-COO^-} \longrightarrow \underset{\substack{| \\ CH_2 \\ | \\ CH_2-CONH_2}}{\overset{+}{H_3N}-CH-COO^-} \\
\alpha\text{-Ketoglutarate} & & \text{Glutamic acid} \qquad\quad \text{Glutamine}
\end{array}
$$

The amino group in glutamic acid is transferred to other α-keto acids in the body in another enzymatic reaction. If it reacts with pyruvate (obtained from glycolysis), we get the synthesis of alanine as given below

$$
\begin{array}{cc}
CH_3-\overset{\overset{O}{\|}}{C}-COO^- + \underset{\substack{| \\ CH_2 \\ | \\ CH_2-COO^-}}{\overset{+}{N}H_3-CH-COO^-} & \xrightarrow[\text{amination}]{\text{Trans-}} \quad CH_3-\underset{\substack{| \\ NH_3 \\ +}}{CH-COO^-} + \underset{\substack{| \\ CH_2 \\ | \\ CH_2-COO^-}}{\overset{\overset{O}{\|}}{C}-COO^-} \\
\text{Pyruvate anion} \qquad\quad \text{Glutarate anion} & \qquad\qquad \text{Alanine} \qquad\quad \alpha\text{-Ketoglutarate anion}
\end{array}
$$

Such a transformation is called transamination because an amino group is transferred from one species to another. A keto acid is converted to the corresponding amino acid and glutamic acid becomes α-ketoglutaric acid which then reacts with ammonia to regenerate glutamic acid.

The biosynthesis of amino acids involves a common set of reactions. In addition to transamination reactions, transfer of one-carbon units such as formyl or methyl groups occur frequently. It is not possible here to give details of the reactions for the biosynthesis of all amino acids. A useful way to organise the amino acid biosynthetic pathways is to group them into families corresponding to the metabolic precursor of each amino acid.

We give below biosynthetic families of amino acids based on different metabolic precursors.

1. Glutamate Family-α-Ketoglutarate obtained from Kreb's cycle in the preeursor

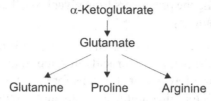

α-Ketoglutarate

Glutamate

Glutamine   Proline   Arginine

2. Asparate Family-Oxaloacetate is the precursor

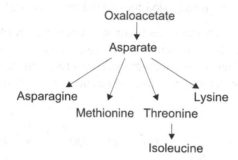

Oxaloacetate

Asparate

Asparagine   Lysine
Methionine Threonine

Isoleucine

3. Serine Family-3-Phosphoglycerate is the precursor

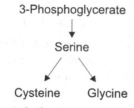

3-Phosphoglycerate

Serine

Cysteine   Glycine

4. Pyruvate Family-Pyruvate is the precursor

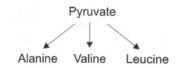

Pyruvate

Alanine   Valine   Leucine

5. Aromatic Family

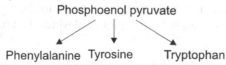

Phosphoenol pyruvate

Phenylalanine Tyrosine   Tryptophan

6. Histidine Family

Ribose-5-phosphate

Histidine

2.10 AMINO ACIDS AS DIPOLAR IONS

Amino acids have relatively high melting points. They usually decompose above 200°C. They have high solubility in water and low solubility in non-polar solvents. They have large dipole moments and these solutions in water have high dielectric constants. All these properties show that amino acids exist as dipolar ions.

As amino acids contain both an acidic group (—COOH) and a basic group (—NH$_2$), they are amphoteric. The acidic group in an amino acid is a substituted ammonium ion and the basic group is the carboxylate ion. Thus, in the crystalline state, amino acids exist as dipolar ions, also called *zwitterions*, in which the amino group is protonated and the carboxyl group exists as a carboxylate anion. In aqueous solution an equilibrium exists between the dipolar ion and the anionic and cationic form of an amino acid.

$$\overset{+}{H_3N}-CH-COOH \underset{+H^+}{\overset{-H^+}{\rightleftharpoons}} \overset{+}{H_3N}-\underset{\underset{R}{|}}{C}H-COO^- \underset{+H^+}{\overset{-H^+}{\rightleftharpoons}} H_2N-\underset{\underset{R}{|}}{C}H-COO^-$$

Cationic form (predominant in strongly acidic solution)	Dipolar ion	Anionic form (predominant in strongly basic solution)

The predominant form of the amino acid present in a solution depends on the pH of the solution. In strongly acidic solution, all amino acids are present as cations and in strongly basic solutions they are present as anions. The pK_a for the carboxyl group of cationic form and pK_a of the dipolar ion form will depend upon the pH of the solution.

To correlate the relationship between pH of a solution to the pK_a of a weak acid, let us consider the dissociation of an acid AH^+.

According to the Bronsted Theory of acids and bases, we know that an acid dissociates to give a proton and a conjugate base. For the dissociation of acid AH^+, we have

$$AH^+ \rightleftharpoons B + H^+ \tag{2.10.1}$$

where B is the conjugate base of acid AH^+.

The extent of dissociation is determined from its K_a by applying the Law of Mass Action. Thus, K_a for the above equation (1) is given by

$$K_a = \frac{[H^+][B]}{[AH^+]} \tag{2.10.2}$$

where K_a is the dissociation constant for the acid AH^+. The term in brackets refers to the concentrations of the species in moles per litre, which we consider equal to the activities of the species.

From the above equation (2.10.2), we have

$$[H^+] = \frac{K_a[AH^+]}{[B]}$$

By taking the common logarithm of both sides

$$\log [H^+] = \log K_a + \log \frac{[AH^+]}{[B]}$$

$$pH = pK_a + \log \frac{[B]}{[AH^+]} \qquad (2.10.3)$$

$$pH = pK_a + \log \frac{[\text{Conjugate base}]}{[\text{Acid}]}$$

This is the Henderson-Hasselbach equation

or

$$pK_a = pH + \log \frac{[\text{Acid}]}{[\text{Conjugate base}]}$$

Thus, the pH of a weak acid is related to the pK_a of the acid and the ratio of the concentration of the conjugate base and concentration of the acid.

When the term, $\log \dfrac{[\text{Conjugate base}]}{[\text{Acid}]}$ is equal to zero, the Henderson-Hasselbach equation becomes

$$pH = pK_a \qquad (2.10.4)$$

i.e., equation (2.10.4) holds good when the molar concentrations of the acid and its conjugate base are equal.

In amino acids, there are at least two ionisable groups —COOH and $\overset{+}{N}H_3$. There are, therefore, at least two pH values for a solution in which equation (2.10.4) holds true. For those amino acids which have a third ionisable group on the side chain, there is a third value of the pH where $pH = pK_a$. That is true for acidic and basic amino acids.

2.11 AMINO ACIDS AS ACIDS AND BASES

We have seen above that the exact structure of an amino acid is a function of the pH, which is related to its pK_a. In acidic solutions, the amino acid is completely protonated and exists as the conjugate acid. For example, glycine exists as glycine hydrochloride in an acidic medium.

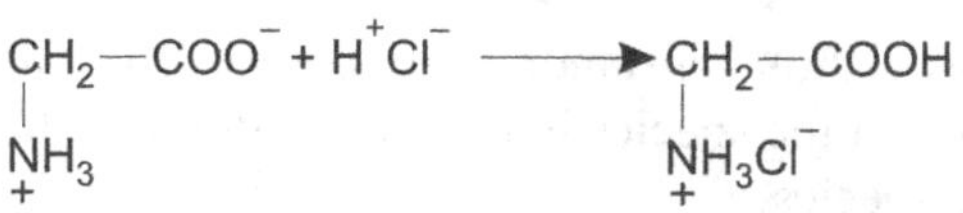

$$\underset{\underset{+}{NH_3}}{CH_2}{-}COO^- + H^+Cl^- \longrightarrow \underset{\underset{+}{NH_3Cl^-}}{CH_2}{-}COOH$$

To understand the relationship between pK_as and structure, let us study the titration of any amino acid, say alanine.

We start with 1 equivalent of the cationic form of alanine and titrate it with standardised sodium hydroxide solution. We shall proceed by steps and see the structures of the species each time after $\frac{1}{2}$ equivalent of the base has been added.

I. Before the actual titration begins, all the species in solution are in the cationic form and there is a net charge of +1. There are two ionisable groups —COOH and $-NH_3$. The carboxyl group is the more acidic of the two and will give a proton first.

Step 1: No addition of base

Structures in solution

$$CH_3-CH-COOH \quad + \quad CH_3-CH-COO^-$$
$$\overset{|}{\underset{+}{NH_3}} \qquad\qquad\qquad \overset{|}{\underset{+}{NH_3}}$$

Cationic form   Zwittorion form
(1/2 equivalent)   (1/2 equivalent)

Net charge = + 1

II. After the addition of $\frac{1}{2}$ equivalent of base, *i.e.*, when the hydrochloride has been half neutralised we will have an equal amount of the cationic form, $CH_3-CH-COOH$

$$\overset{|}{\underset{+}{NH_3}}$$

and its conjugate base, $CH_3-CH-COO^-$

$$\overset{|}{\underset{+}{NH_3}}$$

The pH of the solution at this point is equal to $pK_{a_1} = 2.34$

The net charge of all the species is $+\frac{1}{2}$.

Step 2: After addition of $\frac{1}{2}$ equivalent of base

Structures in solution

$$CH_3-CH-COOH \quad + \quad CH_3-CH-COO^-$$
$$\overset{|}{\underset{+}{NH_3}} \qquad\qquad\qquad \overset{|}{\underset{+}{NH_3}}$$

Cationic form   Zwitterion form
(1/2 equivalent)   (1/2 equivalent)

Net charge = $+\frac{1}{2}$

$$pH = pK_{a_1} = 2.34$$

This dissociaton constant refers to the ionisation of the —COOH group, which is the more acidic of the two acidic groups.

$$CH_3-CH-COOH \rightleftharpoons CH_3-CH-COO^-$$
$$\underset{+}{NH_3} \qquad\qquad \underset{+}{NH_3}$$

III. After one equivalent of the base has been added, all the —COOH groups have become ionised. The main species in solution is the zwitterionic form of alanine with a net charge of zero. The pH of the solution at this point is simply the pH of the solution of the amino acid in water.

Step 3: After addition of 1 equivalent to base

Structure in solution

$$CH_3-CH-COO^-$$
$$\underset{+}{NH_3}$$

Net charge = 0

IV. Addition of a further $\dfrac{1}{2}$ equivalent of base corresponds to half neutralisation of the zwitterionic form.

At this point we have equal amounts of $CH_3-CH-COO^-$ acid and its conjugate base, $CH_3-CH-COO^-$
$$\underset{+}{NH_3}$$
$$NH_2$$

The pH of the solution at this point is equal to pK_{a_2} and for alanine this value is 9.87. The net charge of all the species in solution is $-\dfrac{1}{2}$.

Step 4: After addition of $1\dfrac{1}{2}$ equivalents of base

Structures in solution

$$CH_3-CH-COO^- + CH_3-CH-COO^-$$
$$\underset{+}{NH_3} \qquad\qquad NH_2$$

Zwitterion form   Anionic form
(½ equivalent)   (½ equivalent)

Net charge $= -\dfrac{1}{2}$

V. When two equivalents of NaOH have been added, all the $-\overset{+}{N}H_3$ groups are neutralised. All the species exist in the anionic form with a net charge of –1.

Step 5: After addition of 2 equivalents of base

Structures in solution

$$CH_3-CH-COO^-$$
$$NH_2$$

Anionic Form
Net charge   1

Further addition of base causes no change in the composition of the solution because no ionisable group is present.

The titration curve for alanine hydrochloride is shown in Fig. 2.3

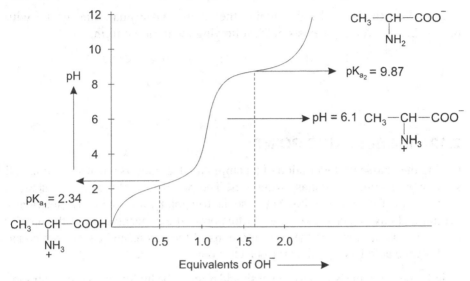

Fig. 2.3. Titration curve for alanine.

Most of the amino acids show similar pK_{a_1} and pK_{a_2} values. These are listed in **Table 2.2.**

Acidic amino acids, aspartic acid and glutamic acid, have an additional carboxyl group with pK_a of 3.87 and 4.28 respectively.

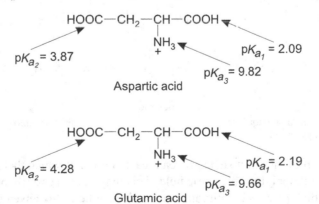

Lysine has two amino groups with pK_a of 8.95 and 10.53. The more basic group is probably the one more remote from the carboxyl group. It has the following zwitterionic structure.

$$H_3\overset{+}{N}-CH_2-CH_2-CH_2-CH_2-\underset{\underset{NH_2}{|}}{CH}-COO^-$$

Lysine

Arginine on the other hand contains the strong basic guanidino group with $pK_a = 12.2$. Therefore, it exists in the following zwitterionic form.

$$H_2N-\underset{\underset{+}{\overset{NH_2}{\underset{|}{\overset{||}{C}}}}}{C}-HN-CH_2-CH_2-CH_2-\underset{\underset{NH_2}{|}}{CH}-COO^-$$

Arginine

2.12   ISOELECTRIC POINT

During the course of the titration of an amino acid, a point is reached at which all structures in solution exist as zwitterions. The pH of the solution at this stage is known as the **isoelectric point**. At the isoelectric point amino acids are electrically neutral and have a zero net charge. At this point the concentration of the dipolar ion is at its maximum and the concentrations of the anions and cations are equal. Each amino acid has a particular isoelectric point.

Strictly speaking, the isoelectric point should refer to the hydrogen ion concentration of the solution in which a particular amino acid does not migrate toward either the anode or the cathode under the influence of an electric field. When a solution of an amino acid is placed in an electric field, the migration toward anode or cathode depends upon the acidity or basicity of the solution. In a quite alkaline solution, there is a net migration of the amino acid toward the anode and in an acidic solution, there is net migration of the amino acid toward the cathode.

$$H_3\overset{+}{N}-\underset{\underset{R}{|}}{CH}-COOH \underset{+H^+}{\overset{-H^+}{\rightleftharpoons}} \overset{+}{N}H_3-\underset{\underset{R}{|}}{CH}-COO^- \underset{+H^+}{\overset{-H^+}{\rightleftharpoons}} H_2N-\underset{\underset{R}{|}}{CH}-COO^-$$

Dipolar ion

← Cathodic migration   Anodic migration →

Cathode   Lower pH   High pH   Anode

In the titration of the amino acids we can reach a point at which there is a zero net charge without applying an electric field. This may not necessarily be the same as the pH at which there is no migration in an electric field. **Sorensen** used the term isoionic to refer to the pH at which there is no net charge. Thus, the term isoionic is more appropriate than isoelectric. As there is no difference between the two values for simple amino acids, it has become customary to use the term isoelectric for isoionic. However, the distinction between the two is very important in the case of proteins.

The isoelectric point for an amino acid is denoted by pI, *i.e.*, the pH at which an amino acid is isoionic. It can be calculated from its pK_a values. We have seen that for monoamino carboxylic acids there are two ionisable groups. In a generalised way, the two equilibria can be written as follows:

$$X^+ \xrightleftharpoons{K_{a_1}} X^{\pm} + H^+ \tag{2.12.1}$$

$$X^{\pm} \xrightleftharpoons{K_{a_2}} X^- + H^+ \tag{2.12.2}$$

where X^+ is the cationic form of any amino acid

$X^{\pm}$ is the zwitterionic form.

X^- is the anionic form.

The corresponding equilibrium constants for (2.12.1) and (2.12.2) are given by:

$$K_{a_1} = \frac{[X^{\pm}][H^+]}{[X^+]} \tag{2.12.3}$$

$$K_{a_2} = \frac{[X^-][H^+]}{[X^{\pm}]} \tag{2.12.4}$$

From equation (2.12.3),

$$[X^+] = \frac{[X^{\pm}][H^+]}{K_{a_1}}$$

From equation (2.12.4),

$$[X^-] = \frac{[X^{\pm}]K_{a_2}}{[H^+]}$$

At the isoelectric point when there is a zero net charge, the concentration of the acid and its conjugate base must be equal

i.e., $\qquad\qquad\qquad [X^+] = [X^-]$

Hence,

$$\frac{[X^{\pm}][H^+]}{K_{a_1}} = \frac{[X^{\pm}]K_{a_2}}{[H^+]}$$

$$[H^+]^2 = K_{a_1} \times K_{a_2}$$

Taking a common logarithm on both sides, we get

$$2\log [H^+] = \log K_{a_1} + \log K_{a_2}$$

$$2pH = pK_{a_1} + pK_{a_2}$$

$$pH = \frac{pK_{a_1} + pK_{a_2}}{2}$$

At the isoelectric point pH is denoted by pI. Therefore,

$$pI = \frac{pK_{a_1} + pK_{a_2}}{2}$$

The pK_a and pI values for different amino acids are given in **Table 2.2.**

As shown above the isoelectric point of an amino acid can be calculated from its titration for simple amino acids with no additional acidic or basic group; the isoelectric point lies exactly half way between the two pK_a values. Let us consider the example of alanine. In strongly acidic solution (*e.g.*, at pH 0), it is present mainly in the cationic form. The pK_{a_1} for the carboxyl group of the cationic form is 2.34. This value is quite lower than the pK_a of corresponding carboxylic acid (propanoic acid, $pK_a = 4.89$). This indicates that the cationic form of alanine is the stronger acid because it is a positively charged species. It should lose a proton more readily.

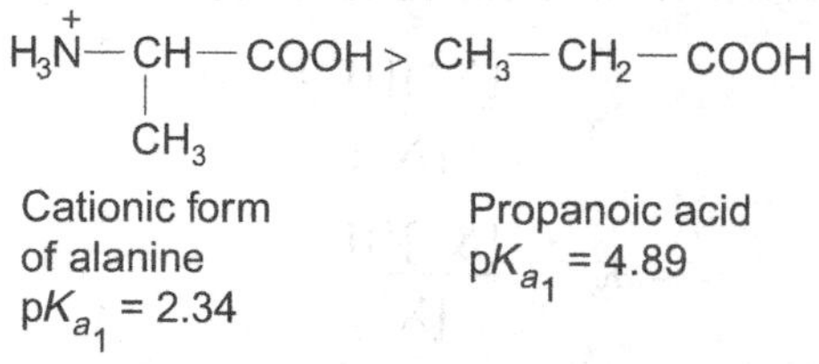

$$\overset{+}{H_3N}-\underset{\underset{CH_3}{|}}{CH}-COOH > CH_3-CH_2-COOH$$

Cationic form
of alanine
$pK_{a_1} = 2.34$

Propanoic acid
$pK_{a_1} = 4.89$

The pK_{a_2} for the dipolar ion of alanine is 9.87. Therefore, the isoelectric point, pI for alanine is obtained by taking the average of the two pK_a values

$$pI = \frac{2.34 + 9.87}{2} = 6.1$$

When the amino acid contains an extra amino group, the situation is going to be different. For example, lysine has an extra amino group on its 2nd carbon. In strongly acidic solution, lysine will be present as a dication because both amino groups will be protonated. The first proton to be lost as the pH is raised is a proton of the carboxyl group $pK_{a_1} = 2.2$).

Lysine exists as a zwitterion with zero net charge between pK_{a_2} and pK_{a_3} as shown below:

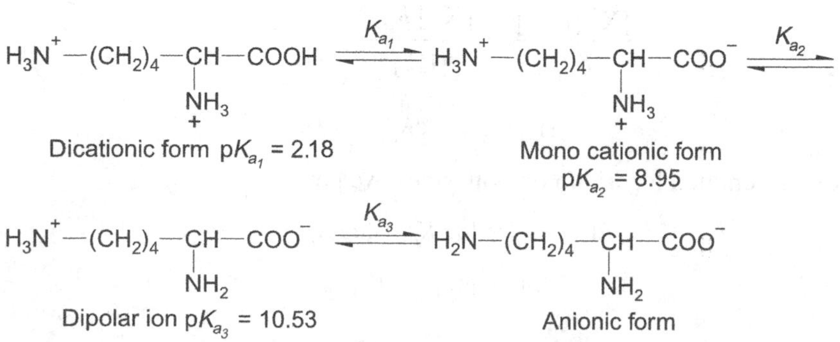

$$H_3\overset{+}{N}-(CH_2)_4-\underset{\underset{+}{\underset{NH_3}{|}}}{CH}-COOH \overset{K_{a_1}}{\rightleftharpoons} H_3\overset{+}{N}-(CH_2)_4-\underset{\underset{+}{\underset{NH_3}{|}}}{CH}-COO^- \overset{K_{a_2}}{\rightleftharpoons}$$

Dicationic form $pK_{a_1} = 2.18$

Mono cationic form
$pK_{a_2} = 8.95$

$$H_3\overset{+}{N}-(CH_2)_4-\underset{\underset{NH_2}{|}}{CH}-COO^- \overset{K_{a_3}}{\rightleftharpoons} H_2N-(CH_2)_4-\underset{\underset{NH_2}{|}}{CH}-COO^-$$

Dipolar ion $pK_{a_3} = 10.53$

Anionic form

The isoelectric point for lysine is, therefore, the average of pK_{a_2} and pK_{a_3}

$$pI = \frac{8.95 + 10.53}{2} = 9.7$$

For acidic amino acids, the zwitterion structure exists between K_{a_1} and K_{a_2}. Therefore, pI is calculated as the average of the first two pK_a values.

For aspartic acid, we have

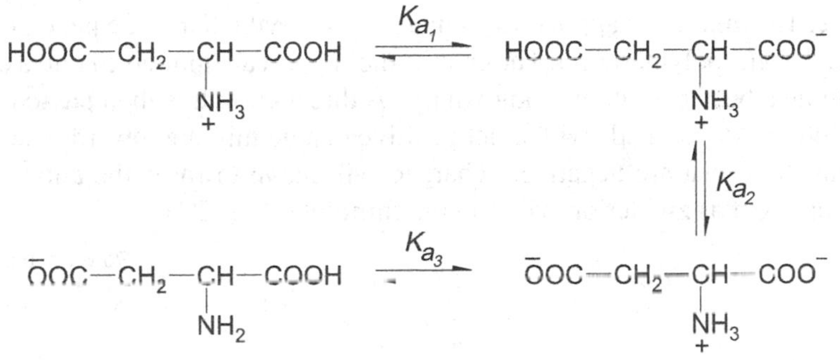

pI is, therefore, given as

$$= \frac{pK_{a_1} + pK_{a_2}}{2}$$

$$= \frac{2.09 + 3.87}{2}$$

$$= 3.0$$

TABLE 2.2   *The pK$_a$ and pI values for amino acids*

S. No.	Amino acid	pK_{a_1}	pK_{a_2}	pK_{a_3}	pI
1.	Glycine	2.35	9.78		6.1
2.	Alanine	2.34	9.87		6.1
3.	Valine	2.32	9.62		6.0
4.	Leucine	2.36	9.60		6.0
5.	Isoleucine	2.36	9.66		6.0
6.	Proline	2.00	10.60		6.3
7.	Phenylalanine	2.58	9.24		5.9
8.	Tryptophan	2.38	9.39		5.9
9.	Methionine	2.28	9.21		5.7
10.	Serine	2.21	9.15		5.7
11.	Threonine	2.63	10.43		6.5
12.	Cysteine	1.96	8.18	10.28	5.1
13.	Aspartic acid	2.09	3.87	9.82	3.0
14.	Glutamic acid	2.19	4.28	9.66	3.2
15.	Lysine	2.18	8.95	10.53	9.7
16.	Arginine	2.02	9.04	12.48	10.8
17.	Tyrosine	2.20	9.10	10.10	5.7
18.	Histidine	1.77	6.10	9.18	7.6

2.13   SEPARATION OF AMINO ACIDS

A. Electrophoresis

Electrophroesis is the migration of charged molecules such as amino acids, proteins, *etc.*, when placed in an electric field. The process is used for the separation of different amino acids from a mixture.

The separation is based on the difference in net charges of the components in the mixture. The mixture is applied to an inert support, which may be paper, cellulose acetate or some polymeric gel. The ends of the support are immersed into a chamber containing a buffer solution of known pH. A direct current is then passed through the solution. Amino acids with a net positive charge migrate towards the cathode whereas those that are negatively charged will move towards the anode. Amino acids that exist as zwitterion will remain immobile (Fig. 2.4).

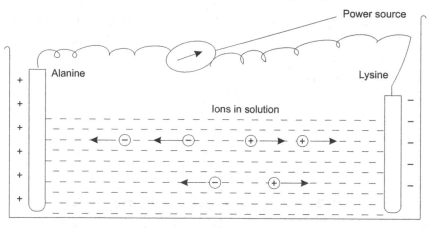

Fig. 2.4.   Principle of electrophoresis.

Detection of the amino acids can be carried out by ninhydrin. The paper is sprayed with a dilute solution of ninhydrin and heated at 100°C. Purple coloured spots appear on the paper which can be compared with R_f values of known compounds.

B. Paper Chromatography

At one time paper chromatography played an important role in biochemical analysis because it provided a method to separate amino acids using very simple equipment. More modern techniques are now commonly used. We describe here the principle involved in paper chromatography because the same principles are employed in modern separation techniques.

The method of paper chromatography separates amino acids on the basis of polarity. A few drops of a solution of an amino acid mixture are applied to the bottom of a strip of the paper. This point is called the origin. The paper strip is immersed in a solvent in a jar. The solvent used is usually a mixture of *n*-butanol, acetic acid and water. The solvent moves up the paper by capillary action, carrying the amino acids with it. Depending on the polarities, the amino acids have affinities for the mobile (solvent) and stationary (paper) phases and therefore, travel up the paper at different rates. The more polar the amino acid, the more strongly it is adsorbed onto the paper. The less polar amino acids move up the paper more rapidly since they have a greater affinity for the mobile phase. The solvent front is marked when the solvent has moved to sufficient height. Now the paper is developed with ninhydrin. Purple-violet coloured spots appear on the chromatogram. The spot closest to the origin is the most polar amino acid and the spot farther away from the origin is the least polar amino acid (Fig. 2.5).

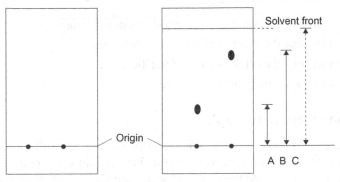

A = Distance moved by component 1
B = Distance moved by component 2
C = Distance moved by solvent

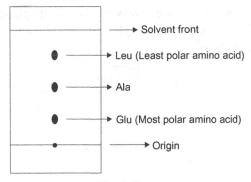

Fig. 2.5.

The most polar amino acids are those with charged side chains, the next most polar are those with side chains that can form hydrogen bonds and the least polar are those with hydrocarbon side chains. For amino acids with alkyl side chains, the larger the alkyl group the less polar the amino acid. For example, leucine is less polar than valine.

C. Thin Layer Chromatography (TLC)

TLC is similar to paper chromatography. In this an adsorbent is coated on the glass plate or a thin sheet of plastic. The adsorbent is usually silica gel or alumina. The adsorbent is coated on the plate with the help of a binder, mostly calcium sulphate.

Here, the adsorbent serves as a stationary phase and a suitable solvent is used, which acts as a mobile phase. The spotting of the mixture of the amino acids is carried out in a similar way as for paper chromatography. The plate is immersed in a jar containing the solvent. After the chromatogram has developed, the plate is sprayed with ninhydrin solution in acetone and heated in an oven at 100°C.

TLC is a convenient and simple method of separating a mixture of amino acids. It is a sensitive and inexpensive analytical technique. The speed of this technique makes it useful for monitoring large scale column chromatography. Important uses of thin layer chromatography are:

1. To determine the number of components in a mixture.
2. To determine the identity of the two substances.
3. To determine the effectiveness of purification.
4. To monitor column chromatography.

D. Column Chromatography

We have seen that chromatography is based on the fact that different compounds can distribute themselves to varying extents between different phases. One phase is the stationary phase and the other is the mobile phase. The mobile phase flows over the stationary material and carries the sample to be separated along with it. The components of the sample interact with the stationary phase to different extents. Some components interact more strongly with the stationary phase and are, therefore, carried away more slowly by the mobile phase. On the other hand, components which interact less strongly are carried away by the mobile phase. Thus, a separation is possible.

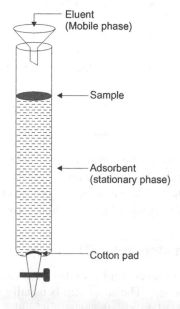

Fig. 2.6.

Many chromatographic techniques used for the separation of amino acids are forms of column chromatography. In typical column chromatography the column is packed with the adsorbent, the material which makes up the stationary phase. Normally silica gel or alumina are used as adsorbents. The sample is a small volume of concentrated solution that is applied to the top of the column. The mobile phase is a suitable solvent which is called eluent. It is passed through the column. The sample is diluted by the eluent and the separation process increases the volume occupied by the sample. Eluents are sometimes checked at regular intervals resulting in the separation of various components in the mixture. The entire sample eventually comes off the column (Fig. 2.6).

E. Ion-Exchange Chromatography

This is a common method used for the quantitative analysis of amino-acid mixtures. The technique is similar to column chromatography. Here the column consists of a resin which may be a strongly acidic cation or a strongly basic anion exchanger. One commonly used resin is a copolymer of styrene and divinylbenzene with negatively charged sulphonic acid groups on some of the benzene rings. When an acid solution of a mixture of amino acids is added, the most basic acids are most tightly bound and the most acidic acids are most weakly bound. Elution is carried out with a series of buffer solutions. The amino acids separated because they flow through the column at different rates.

If a mixture of lysine and glutamate in a solution with pH 6 is loaded onto the above column, glutamate travels down the column rapidly because its negatively charged side chain is repelled by the negatively charged sulphonic acid groups of the resin. The positively charged side chains of the lysine are retained on the column. Such a resin is called a cation-exchange resin. In this Na^+ of the SO_3^- groups are exchanged by the positively charged amino acids.

Resins with positively charged groups are called anion exchange resins because they impede the flow of anions. A common anion exchange resin has $R—N(CH_3)_3^+Cl^-$ groups in place of $SO_3^- Na^+$ groups.

When a mixture of amino acids possessing different charges passes through a charged column, the amino acids possessing opposite charges get associated with the column while those with the same charge or no charge pass through the column. Thus, separation of a mixture of amino acids is achieved.

This method has now been automated. The instrument that automates ion-exhange chromatography is known as an amino acid analyser. When a solution of amino acid mixture passes through the column of an amino acid analyser containing a cation exchange resin, the amino acids move through the column at different rates depending on their overall charge. Fractions of the effluent are collected and analysed for their amino acid content using ninhydrin for formation of the complex with different amino acids. In this way, the identity and the relative amount of each amino acid can be determined.

2.14 CHEMICAL REACTIONS OF AMINO ACIDS

The reactions of amino acids are in general the ones we would expect of compounds containing —NH_2 and —COOH groups. In addition, other groups that may be present undergo their own characteristic reactions.

A. Reactions due to the Amino Group

i. Alkylation

Alkylation of amino acids is carried out by treating the amino acids with alkyl iodides. Mono alkyl amino acids are obtained by this method.

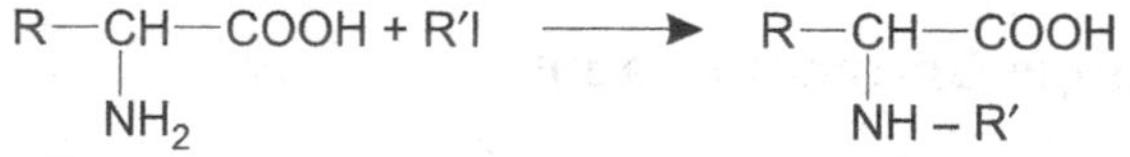
R—CH—COOH + R′I
NH2
R—CH—COOH
NH – R′

In the presence of excess alkyl halide, these form internal quaternary ammonium salts and are called betaines. The trimethyl derivative of glycine, $(CH_3)_3—\overset{+}{N}—CH_2—COO^-$, known as betaine, occurs naturally.

Amino acids may also be arylated by 2,4-dinitrochlorobenzene or 2,4-dinitrofluorobenzene (DNFB) known as **Sanger's** reagent. These derivatives are useful for the analysis of peptides. Barger, however, used 2,4,5-trinitrotoluene for arylation of amino acids. The reaction is carried out by warming an alcoholic solution of the amino acid with trinitrotoluene. Nitrous acid is eliminated.

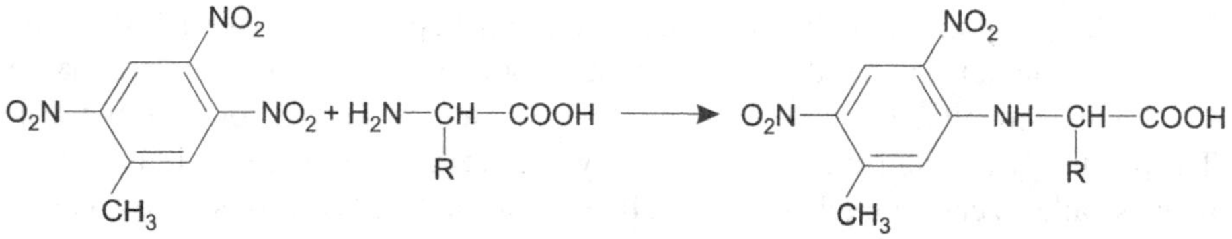
NO2
O2N
NO2 + H2N—CH—COOH
R
CH3
NO2
O2N
NH—CH—COOH
R
CH3

Sanger's reagent is more reactive. It readily gives dinitrophenyl (DNP) derivatives of amino acids.

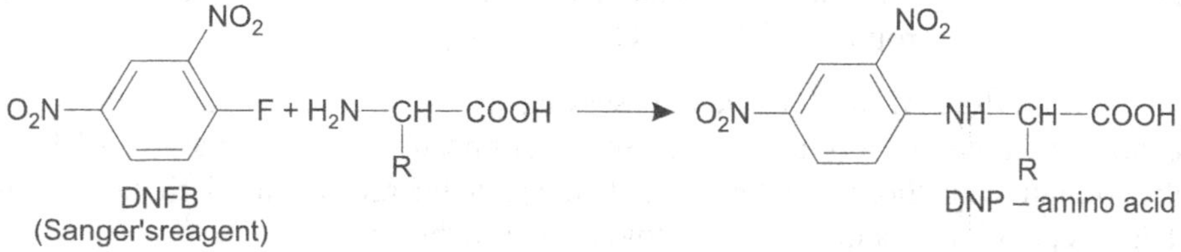
NO2
O2N
F + H2N—CH—COOH
R
DNFB
(Sanger'sreagent)
NO2
O2N
NH—CH—COOH
R
DNP – amino acid

ii. Acylation

Amino acids may be acylated in a general way by using acyl halides in aqueous alkali. For acetylation, we can use acetyl chloride.

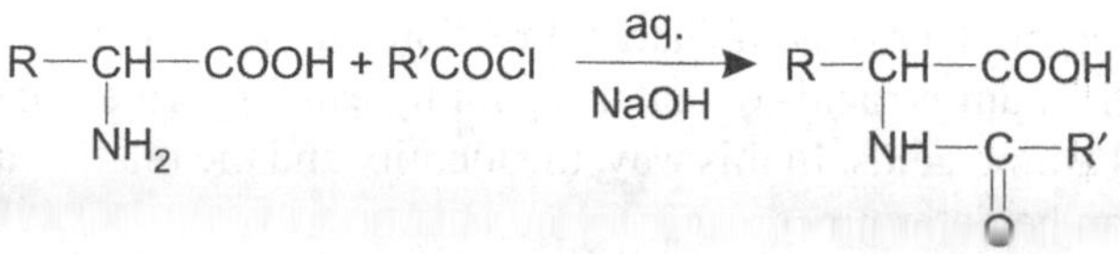
R—CH—COOH + R′COCl
NH2
aq.
NaOH
R—CH—COOH
NH—C—R′
O

Amides may also be prepared by reaction with acetic anhydride. For example,

$$\underset{\text{Glycine}}{\text{H}_2\text{N}-\text{CH}_2-\text{COOH}} \xrightarrow{\text{(CH}_3\text{CO)}_2\text{O}} \underset{N-\text{Acetyl glycine}}{\text{CH}_3-\text{CONH}-\text{CH}_2-\text{COOH}}$$

Similarly, benzoylation may be carried out by using benzoyl chloride under Schotten—Baumann conditions (aq. sodium hydroxide at room temperature). However, it is necessary to acidify the aqueous solution to obtain the acidic product.

$$\underset{\substack{|\\ \text{CH (CH}_3)_2 \\ \text{Valine}}}{\text{H}_2\text{N}-\text{CH}-\text{COOH}} \xrightarrow[\text{2) HCl 2hr, 4°C}]{\text{1) C}_6\text{H}_5\text{COCl / NaOH}} \underset{\substack{|\\ \text{CH (CH}_3)_2 \\ \text{Benzoyl valine}}}{\text{C}_6\text{H}_5-\text{CONH}-\text{CH}-\text{COOH}}$$

Histidine may be acylated with the following method:

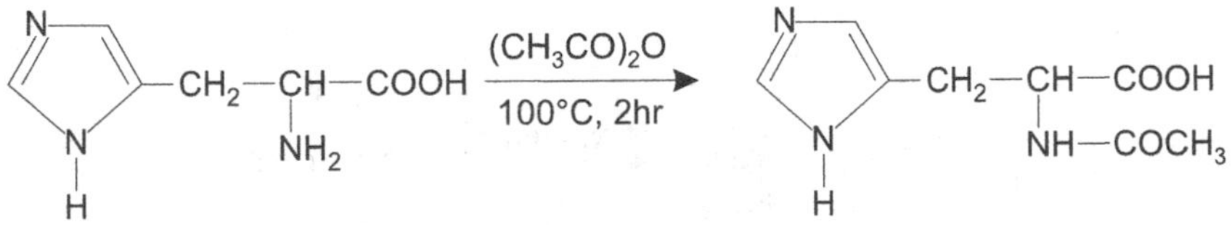

$$\xrightarrow[\text{100°C, 2hr}]{\text{(CH}_3\text{CO)}_2\text{O}}$$

N – Acetylhistidine

The acylated derivatives of amino acids are quite important as they possess sharp and characteristic melting points. The lone pair of electrons in the amino group is withdrawn by the acyl group attached to the nitrogen atom. The benzoyl derivatives are especially useful for characterisation.

Condensation of amino acids with phthalic anhydride gives phthaloyl derivatives, which are of value in peptide synthesis.

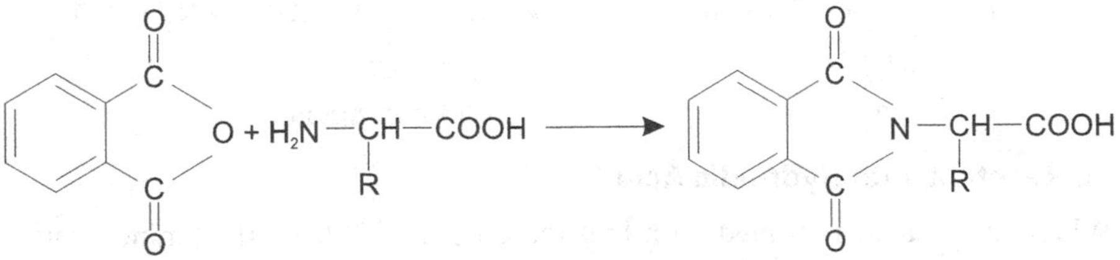

Sulphonamides result from the reaction with sulphonyl chloride. For example,

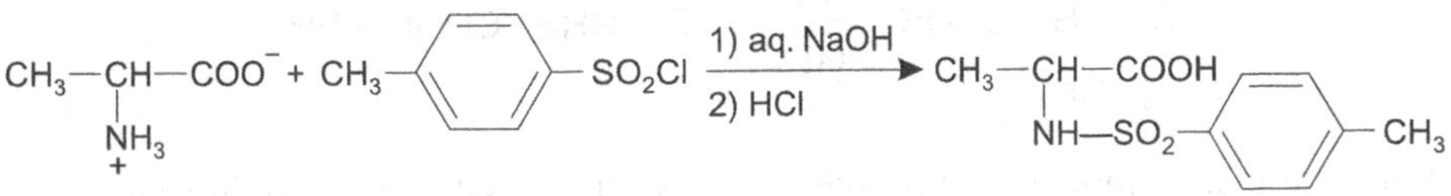

iii. Nitrous Acid Reaction

Nitrous acid reacts with amino acids to liberate one mole of nitrogen gas per mole of amino acid and gives the corresponding hydroxy acids.

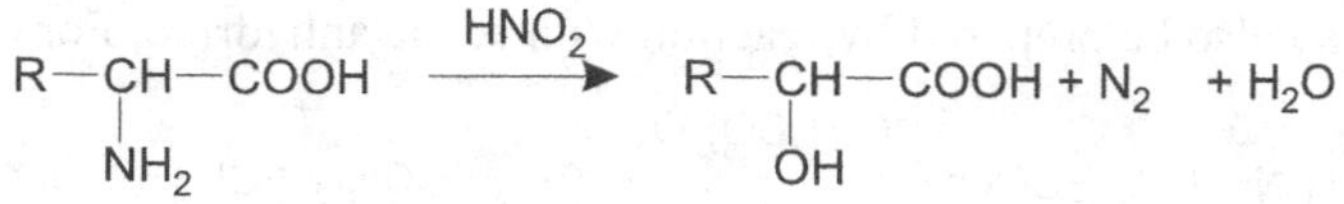

$$R-\underset{\underset{NH_2}{|}}{CH}-COOH \xrightarrow{HNO_2} R-\underset{\underset{OH}{|}}{CH}-COOH + N_2 + H_2O$$

The reaction takes place with α–amino acids at room temperature in less than ten minutes and is of great utility for identification and assay of the purity of amino acids. The α-amino groups of lysine and arginine react in thirty minutes. The method and apparatus were developed by van Slyke and is known by his name.

iv. Formol Reaction

The H-atom of the amino group is substituted by the methylol group by treating with formaldehyde at room temperature.

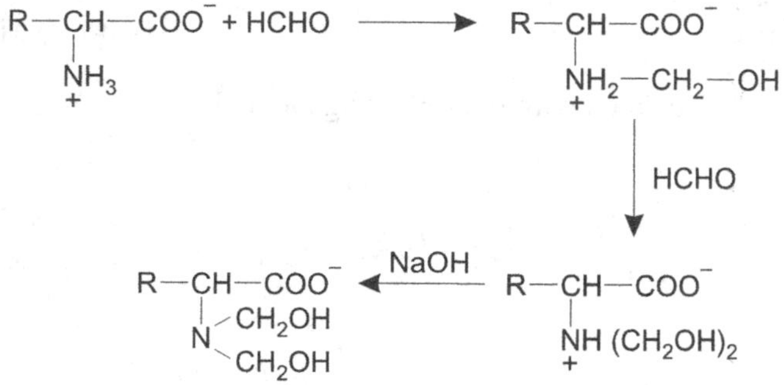

$$R-\underset{\underset{+}{\underset{NH_3}{|}}}{CH}-COO^- + HCHO \longrightarrow R-\underset{\underset{+}{\underset{NH_2-CH_2-OH}{|}}}{CH}-COO^-$$

$$\Big\downarrow HCHO$$

$$R-\underset{\underset{N}{|}}{CH}-COO^- \underset{}{\overset{NaOH}{\longleftarrow}} R-\underset{\underset{+}{\underset{NH(CH_2OH)_2}{|}}}{CH}-COO^-$$

The formol reaction has the same analytical utility as the nitrous acid reaction. Aldehydes other than formaldehyde react in a similar manner.

v. Reaction with Nitrosyl Chloride

Nitrosyl chloride (or bromide) reacts with amino acids to form chloro (or bromo) acids, *e.g.,*

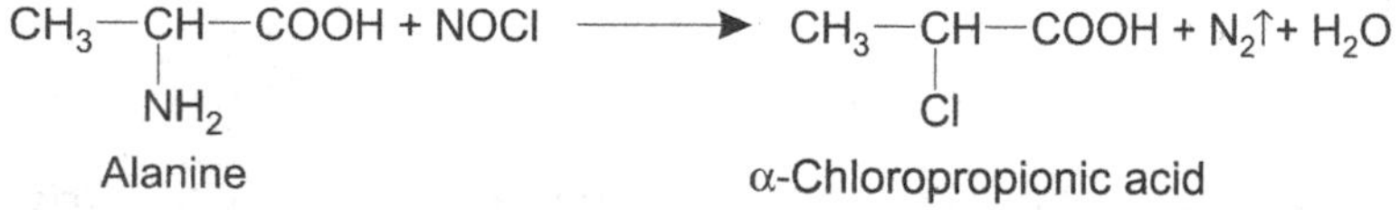

$$\underset{\text{Alanine}}{CH_3-\underset{\underset{NH_2}{|}}{CH}-COOH} + NOCl \longrightarrow \underset{\text{α-Chloropropionic acid}}{CH_3-\underset{\underset{Cl}{|}}{CH}-COOH} + N_2\uparrow + H_2O$$

vi. Reaction with Hydriodic Acid

When amino acid is heated with hydriodic acid at 200°C, the amino group is eliminated with the formation of a fatty acid.

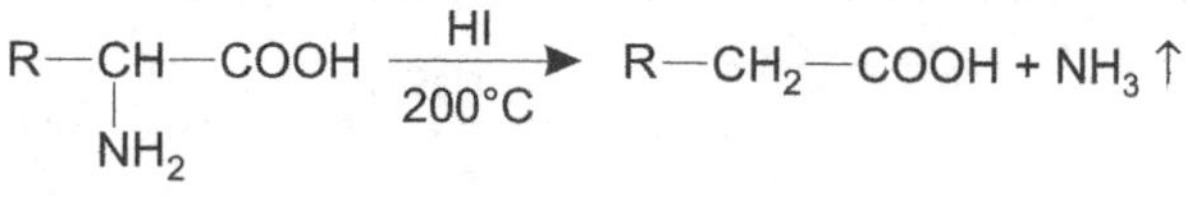

$$R-\underset{\underset{NH_2}{|}}{CH}-COOH \xrightarrow[200°C]{HI} R-CH_2-COOH + NH_3 \uparrow$$

For example, alanine gives propionic acid when heated with HI at 200°C.

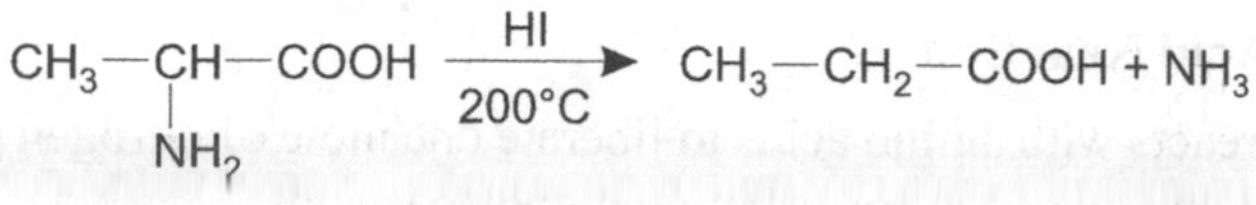

$$CH_3-\underset{\underset{NH_2}{|}}{CH}-COOH \xrightarrow[200°C]{HI} CH_3-CH_2-COOH + NH_3$$

vii. Oxidation to Keto Acid

The deamination of amino acids to keto acids is of considerable biological significance. Chemically an amino acid can be converted to a keto acid by treating with hydrogen peroxide. In the biological systems, such oxidations are brought about by L-amino acid oxidases and D-amino acid oxidases. The keto acids, which are obtained by deamination of the amino group of the amino acids are identical to some of the keto acids, which result from carbohydrate metabolism.

$$R-\underset{\underset{NH_2}{|}}{CH}-COOH \xrightarrow[\substack{or\\ L-amino\ acid\\ oxidase}]{H_2O_2} R-\underset{\underset{O}{||}}{C}-COOH$$

B. Reactions due to the Carboxyl Group

i. Esterification

The carboxyl group of an amino acid may be esterified in the usual manner. The methyl and the ethyl esters are obtained by refluxing the suspension of the amino acid in appropriate alcohol with anhydrous hydrogen chloride. The ester is separated as the crystalline hydrochloride salt.

$$R-\underset{\underset{\overset{+}{N}H_3}{|}}{CH}-COO^- \xrightarrow[HCl]{C_2H_5OH} R-\underset{\underset{\overset{+}{N}H_3\ \ \overset{-}{C}l}{|}}{CH}-COOC_2H_5$$

The free ester may be obtained by the action of aqueous sodium carbonate on the ester salt.

Benzyl esters are often prepared by using benzene sulphonic acid as the catalyst. The water produced in the reaction is removed by azeotropic distillation. For example, glycine gives the following benzyl ester derivative.

$$\underset{\underset{\overset{+}{N}H_3}{|}}{CH_2}-COO^- + \text{⟨benzene ring⟩}-CH_2OH \xrightarrow{C_6H_5SO_3H} \underset{\underset{\overset{+}{N}H_3\ \ C_6H_5\overset{-}{S}O_3}{|}}{CH_2}-COOCH_2-\text{⟨benzene ring⟩}$$

Glycine benzyl ester benzene sulphonate

The esters do not possess the dipolar ion structure found in the unsubstituted amino acids and are, therefore, more volatile. So, it is possible to distill the esters without causing extensive decomposition.

Methyl, ethyl and benzyl esters are employed extensively as intermediates in the synthesis of peptides. However, benzyl esters are more useful because they can be converted back to acids by non-hydrolytic methods. For example, glycine benzyl ester reacts with hydrogen in the presence of palladium to give glycine salt from which the free amino acid can be obtained.

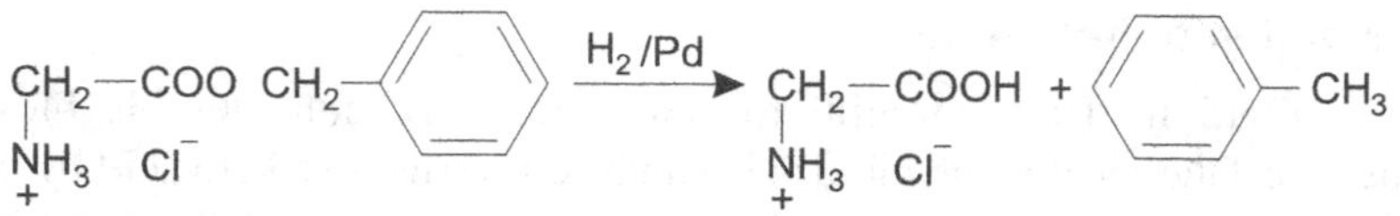

$$CH_2{-}COO\ CH_2{-}\bigcirc \xrightarrow{H_2/Pd} CH_2{-}COOH + \bigcirc{-}CH_3$$

with $\overset{+}{NH_3}\ Cl^-$ on the left and $\overset{+}{NH_3}\ Cl^-$ on the product.

ii. Acyl Halide Formation

It is not possible to convert the carboxyl group of an amino acid directly to an aminoacyl halide because of the reactivity of the amino group in the presence of halogenating agents. The aminoacyl halides are desirable intermediates in the synthesis of peptides. The amino group is protected first and then treated with phosphorus pentachloride or refluxed with thionyl chloride. The benzoyl group is frequently used as the protecting group.

$$R{-}\underset{NH_2}{\overset{|}{CH}}{-}COOH \xrightarrow{C_6H_5COCl} R{-}\underset{NH\ COC_6H_5}{\overset{|}{CH}}{-}COOH$$

$$\downarrow PCl_5\ or\ SOCl_2$$

$$C_6H_5CO{-}NH{-}\underset{R}{\overset{|}{CH}}{-}COCl$$

Deprotection of the benzoyl group is carried out in the usual manner. When amino acids are suspended in acetyl chloride and then treated with phosphorus pentachloride, they form hydrochlorides of the acid chloride.

$$R{-}\underset{NH_2}{\overset{|}{CH}}{-}COOH \xrightarrow[2)\ PCl_5]{1)\ CH_3COCl} R{-}\underset{\overset{+}{NH_3}\ Cl^-}{\overset{|}{CH_2}}{-}COCl + POCl_3$$

iii. Decarboxylation

Decarboxylation of amino acids may be achieved by dry distillation or by heating with bases such as barium oxide. There are specific enzymes which catalyse the decarboxylation of specific amino acids.

$$R{-}\underset{NH_2}{\overset{|}{CH}}{-}COOH \xrightarrow{\Delta} R{-}CH_2{-}NH_2 + CO_2$$

Decarboxylation of amino acids is quite significant biologically. The decarboxylation of histidine to histamine has great physiological significance.

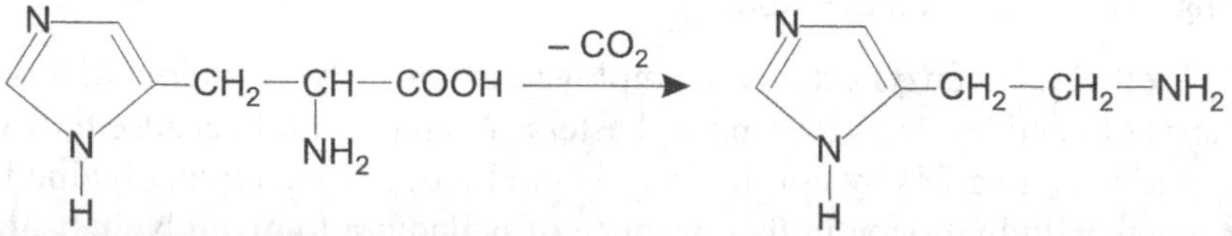

Histidine   Histamine

Histamine is a direct causative agent in allergy, shock and similar states and is believed to arise by decarboxylation.

iv. Cyclodehydration

Two molecules of an amino acid may react with the loss of water to form an amino acid anhydride, *i.e.*, disubstituted piperazine. The reaction may be brought about by heating the amino acid in high boiling solvents such as ethylene glycol. The esters of amino acids react more readily with the splitting out of two molecules of alcohol.

$$R—\underset{\underset{NH_2}{|}}{\overset{\overset{COOH}{|}}{CH}} \quad + \quad \underset{\underset{HOOC}{|}}{\overset{\overset{H_2N}{|}}{CH}}—R' \longrightarrow R—\underset{\underset{NH—CO}{|}}{\overset{\overset{CO—NH}{|}}{CH}} \quad \overset{}{CH}—R'$$

The diketopiperazines are of considerable interest because they have frequently been found in the protein hydrolysates.

v. The Dakin-West Reaction

Amino acids on heating with acetic anhydride and pyridine are converted into methyl α-acetamido ketones. The reaction is often referred to as the Dakin-West reaction.

$$R—\underset{\underset{NH_2}{|}}{CH}—COOH \xrightarrow[\text{Pyridine}]{(CH_3CO)_2O} R—\underset{\underset{NH—COCH_3}{|}}{CH}—COCH_3$$

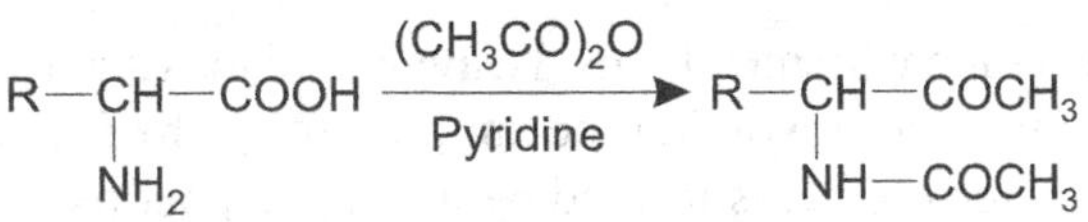

vi. Reduction

Amino acids form amino alcohols on reduction with $LiAlH_4$. The optical activity of the alcohols is retained.

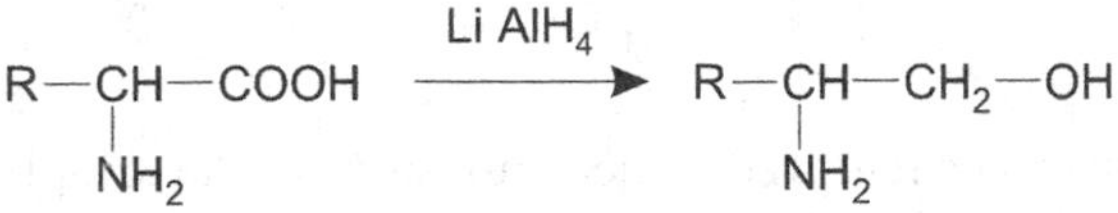

$$R—\underset{\underset{NH_2}{|}}{CH}—COOH \xrightarrow{Li\,AlH_4} R—\underset{\underset{NH_2}{|}}{CH}—CH_2—OH$$

C. Reactions due to both the Amino and Carboxyl Groups

i. Oxidation

The oxidation of amino acids to the aldehyde, in which the carbon containing the amino group is used up, may be carried out by sodium hypochlorite.

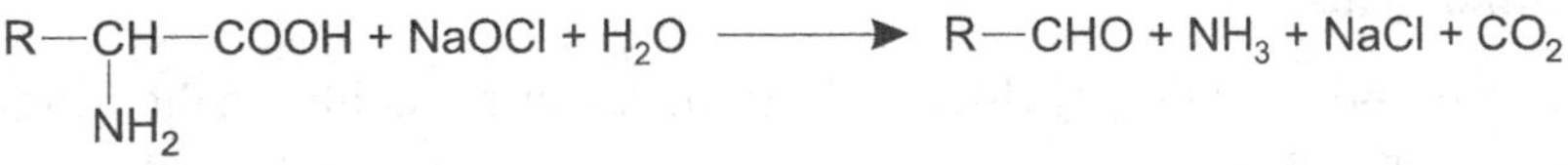

$$R—\underset{\underset{NH_2}{|}}{CH}—COOH + NaOCl + H_2O \longrightarrow R—CHO + NH_3 + NaCl + CO_2$$

Instead of $NaOCl$, chloramine-T may be employed.

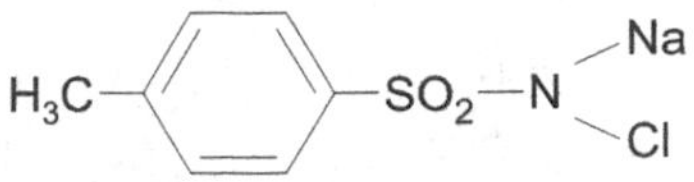

$$H_3C—\underset{}{\overset{}{\bigcirc}}—SO_2—N\overset{\diagup Na}{\diagdown Cl}$$

Chloramine – T

Periodic acid or its sodium salt may be used for cleaving β-hydroxy amino acids.

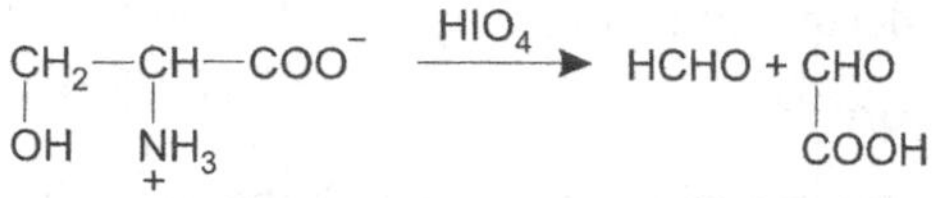

$$CH_2-CH-COO^- \xrightarrow{HIO_4} HCHO + \underset{\underset{COOH}{|}}{CHO}$$
$$\underset{OH}{|} \quad \underset{\underset{+}{NH_3}}{|}$$

When two moles of chloramine-T are used as the oxidising agent, a nitrile instead of an aldehyde is obtained.

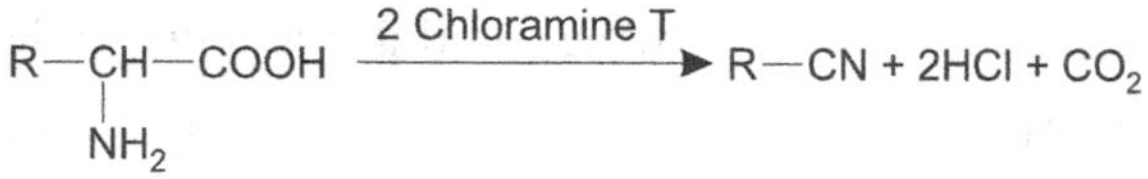

$$R-\underset{\underset{NH_2}{|}}{CH}-COOH \xrightarrow{2 \text{ Chloramine T}} R-CN + 2HCl + CO_2$$

ii. Reaction with Organic Acids

Strong organic acids convert the amino acids to cationic form such as

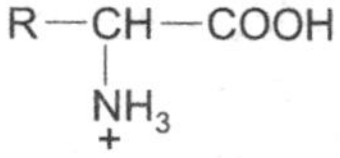

$$R-\underset{\underset{\underset{+}{NH_3}}{|}}{CH}-COOH$$

With picric acid (2,4,6-trinitrophenol) an insoluble salt precipitates out because of the affinity of the acid anion for the amino acid cation.

iii. Reaction with Metal Ions and Complex Salts

Amino acids form salts with metal ions. Heavy metal ions such as Cu^{2+}, Ag^+, Hg^{2+} form insoluble complexes involving the amino group. The copper salt of glycine, which has a deep blue colour, is formed by heating copper oxide with an aqueous solution of glycine. These metal ions also precipitate proteins. The structure of the complex of copper with glycine is given below, which is obtained by mixing glycine with an aqueous solution of $CuSO_4$.

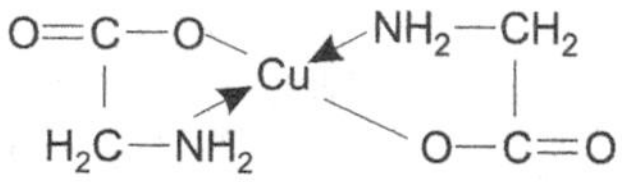

$$O=C-O \diagdown \underset{\diagup}{Cu} \diagup NH_2-CH_2$$
$$\underset{H_2C-NH_2}{|} \qquad \underset{O-C=O}{|}$$

The heavy metal can be removed by decomposition of the complex with hydrogen sulphide and the amino acid is recovered.

Some inorganic complex salts have also been used for selective precipitation of amino acids from protein hydrolysates.

iv. Action of Heat

α-Amino Acids

As given above under cyclodehydration, α-amino acids on heating form 2,5-diketopiperazines.

For example, glycine forms 2,5-diketopiperazine

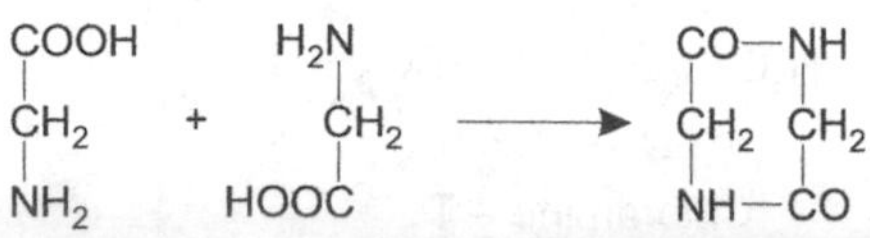

$$\underset{\underset{NH_2}{|}}{\overset{\overset{COOH}{|}}{\underset{CH_2}{|}}} + \underset{\underset{HOOC}{|}}{\overset{\overset{H_2N}{|}}{\underset{CH_2}{|}}} \longrightarrow \underset{\underset{NH-CO}{|}}{\overset{\overset{CO-NH}{|}}{CH_2 \ CH_2}}$$

2,5 Diketopiperazine

β-Amino acids

On heating, β-amino acids eliminate a molecule of ammonia to give α,β-unsaturated acids.

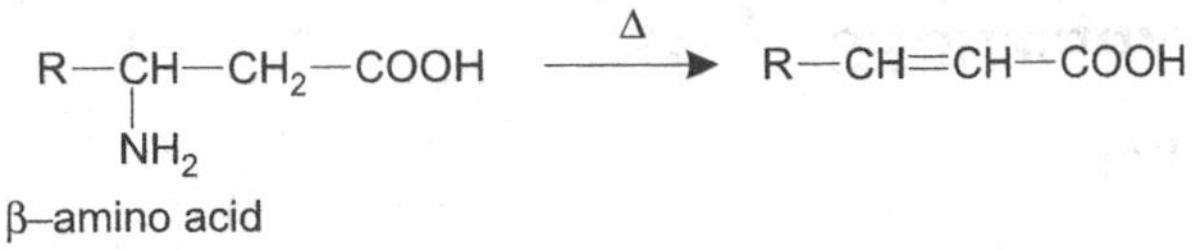
R—CH—CH2—COOH
NH2
Δ
R—CH=CH—COOH
β–amino acid

γ-Amino acids and δ-Amino acids

γ- and δ-amino acids on heating form cyclic amides called lactams.

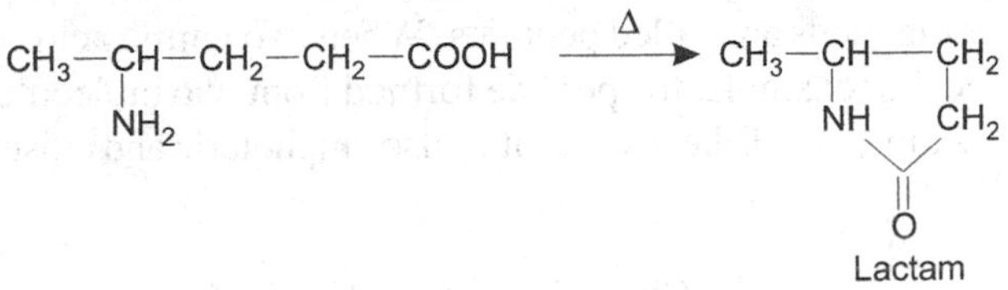
CH3—CH—CH2—CH2—COOH
NH2
Δ
CH3—CH—CH2
NH
CH2
O
Lactam

v. The Ninhydrin Reaction

All α-amino acids give a characteristic violet coloured product when heated with ninhydrin solution. This reaction is the basis of detection of α-amino acids in chromatography techniques. The chromatogram is sprayed with ninhydrin solution in acetone. On heating slowly violet coloured spots develop.

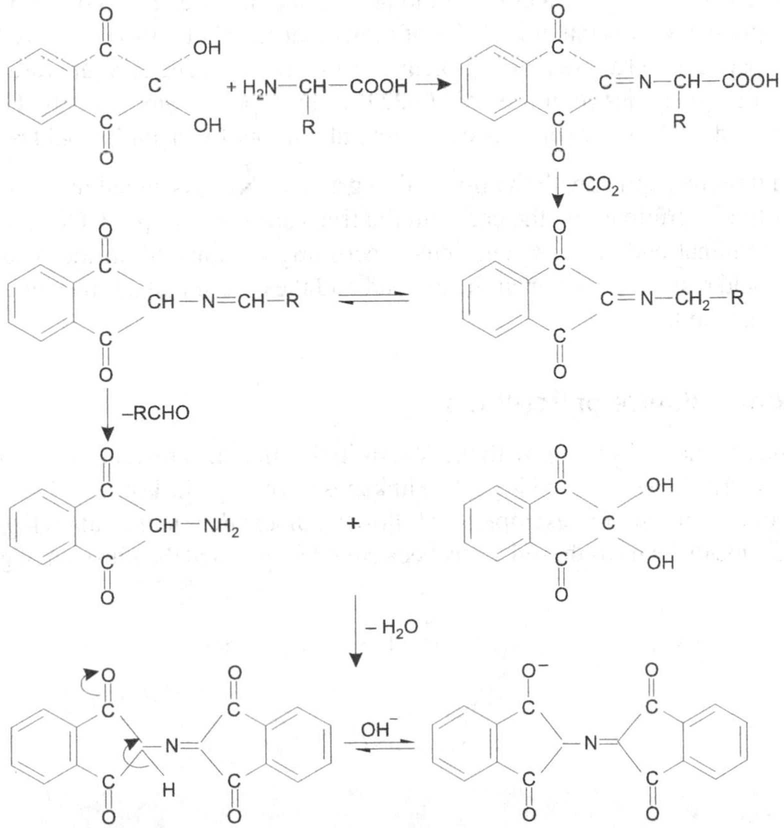
+ H2N—CH—COOH
R
C=N—CH—COOH
- CO2
CH—N=CH—R
C=N—CH2—R
-RCHO
CH—NH2
+
– H2O
OH⁻
N=

On treating with ninhydrin, which is triketohydrindene hydrate, α-amino acids get converted into an amino derivative which condenses with another mole of ninhydrin to finally give the coloured product. The sequence of reactions is given above:

2.15   POLYPEPTIDES

A. Introduction

We have already seen that amino acids may form amide bonds between two molecules, *i.e.*, a carboxyl group of one amino acid condenses with the amino group of another. The amide linkage between the amino acids is called a peptide bond and the resulting compounds are called peptides. When two amino acids condense they give a dipeptide. For example, the peptide formed from two molecules of glycine is glycylglycine, a dipeptide. Like glycine, it is also amphoteric and exists as *zwitterion*.

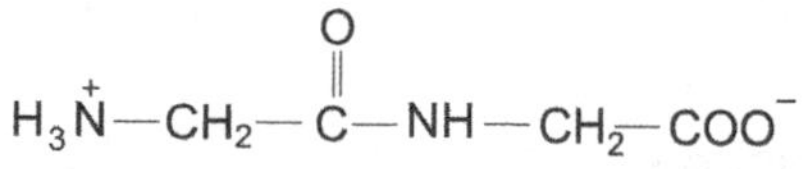

$$H_3\overset{+}{N}-CH_2-\overset{\overset{\displaystyle O}{\|}}{C}-NH-CH_2-COO^-$$

Glycylglycine
(a dipeptide)

Similarly, a tripeptide contains three amino acid units and a tetrapeptide four, and so on. When the number of amino acid units increases beyond ten we get a polypeptide. By convention peptides of molecular weight up to 10,000 are known as polypeptides and above that as proteins. As mentioned earlier, proteins are special types of polypeptides composed of 20–22 different specific amino acids. The term oligopeptide is used for peptides containing about four to ten amino acid residues.

The end of the peptide with the free amino group $(-\overset{+}{N}H_3)$ is called the N-terminal end or the N-terminus and the end with the free carboxyl group $(-COO^-)$ is called the C-terminal end or the C-terminus. According to convention, the N-terminal amino acid residue is written at the left end and the C-terminal amino acid residue at the right end.

B. Nomenclature of Peptides

Peptides are named starting with the N-terminal amino acid residue. The names of amino acid residues involved in peptide linkages (all except the last) are given the –yl suffix of acyl groups. For example, the following dipeptide is named alanyl- glycine. The alanine residue has the –yl suffix because it has acylated the nitrogen of glycine

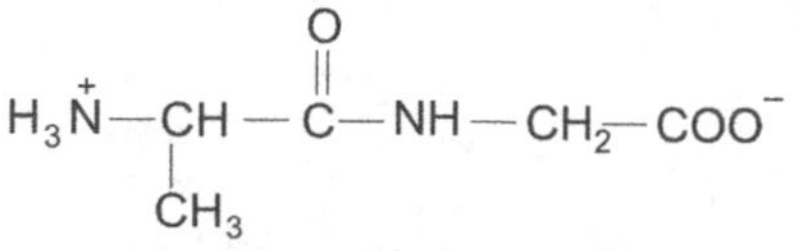

$$H_3\overset{+}{N}-\underset{\underset{\displaystyle CH_3}{|}}{CH}-\overset{\overset{\displaystyle O}{\|}}{C}-NH-CH_2-COO^-$$

Alanylglycine (Ala–Gly)

The abbreviations given in Table 2.1 are used for different amino acids.

Similarly, a tripeptide (Gly-Ala-Phe) is named

$$
\overset{+}{H_3N}-CH_2-\overset{\overset{\displaystyle O}{\|}}{C}-HN-\underset{\underset{\displaystyle CH_3}{|}}{CH}-\overset{\overset{\displaystyle O}{\|}}{C}-HN-\underset{\underset{\displaystyle CH_2-C_6H_5}{|}}{CH}-COO^{-}
$$

Glycylalanylphenylalanine

2.16 SYNTHESIS OF PEPTIDES

A. The Basic Scheme

The synthesis of a peptide requires the formation of a peptide bond between the carboxyl group of one amino acid and the amino group of another one. The reaction involves a nucleophilic substitution by an amino group at a carbonyl group of a carboxylic acid. It is quite likely that reaction may occur between the carboxyl group and the amino group of the same amino acid. For example, the synthesis of simple dipeptide alanylglycine involves the conversion of alanine to its chloride and then allowing it to react with glycine. However, we cannot prevent alanyl chloride from reacting with itself. So, our reaction would give not only Ala-Gly but also Ala-Ala. We may also get tripeptides and so on.

$$
\underset{\underset{+}{\underset{NH_3}{|}}}{CH_3-CH-}COO^{-} \xrightarrow[\underset{+}{2)\ H_3N-CH_2-COO^{-}}]{1)\ SOCl_2} \underset{\underset{+}{\underset{NH_3}{|}}}{CH_3-CH-CONH-CH_2-COO^{-}}
$$

Ala – Gly

+

$$
\underset{\underset{+}{\underset{NH_3}{|}}}{CH_3-CH-CONH-}\underset{\underset{CH_3}{|}}{CH-COO^{-}}
$$

Ala – Ala

Moreover, there are other nucleophilic groups present in amino acids such as other amino groups, —OH and —SH groups on the side chains of amino acids. These groups are also expected to react with carboxylic acids. Therefore, we must protect these groups so that the reaction can be directed to give the desired product.

The first step in the basic scheme of peptide synthesis is to protect the amino group of the first amino acid before we activate it and allow it to react with the second. The second step involves the activation of the —COOH group of protected amino acid and the formation of the peptide bond with the next amino acid under conditions that do not destroy peptide bonds and other functional groups that may be present in the molecule. After getting the desired peptide the last step is the removal of the protecting group by using a reagent that does not affect the peptide bonds. The steps involved in the peptide synthesis are given below:

1. Protection of the amino group

$$Z + \overset{+}{N}H_3-\underset{\underset{R}{|}}{C}H-COO^- \longrightarrow Z-NH-\underset{\underset{R}{|}}{C}H-COOH$$

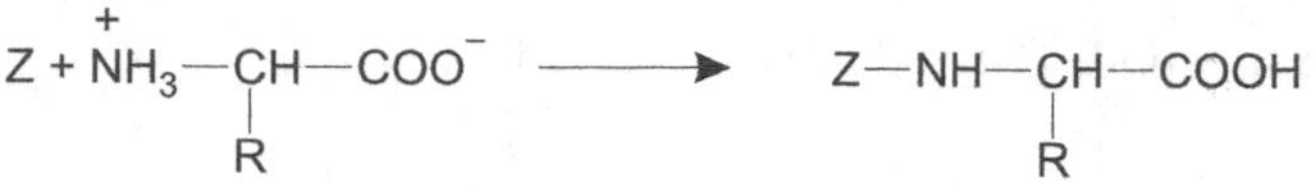

2. Activation of —COOH group and formation of peptide bond

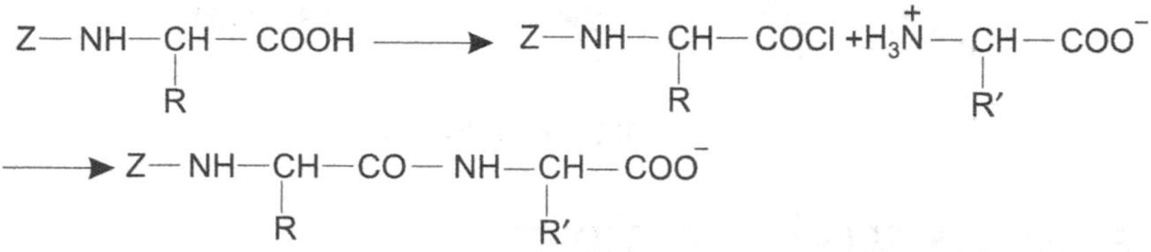

$$Z-NH-\underset{\underset{R}{|}}{C}H-COOH \longrightarrow Z-NH-\underset{\underset{R}{|}}{C}H-COCl + \overset{+}{H_3}N-\underset{\underset{R'}{|}}{C}H-COO^-$$

$$\longrightarrow Z-NH-\underset{\underset{R}{|}}{C}H-CO-NH-\underset{\underset{R'}{|}}{C}H-COO^-$$

3. Removal of the protecting group

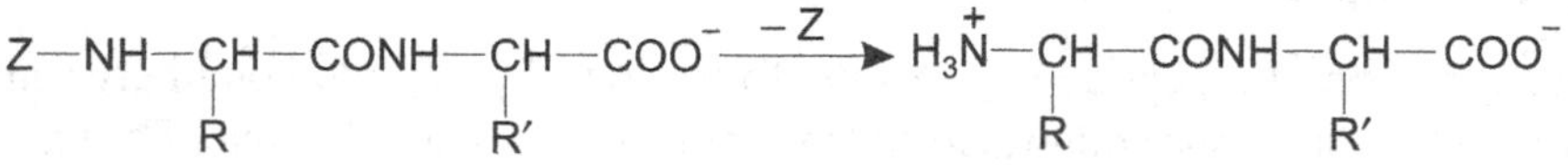

$$Z-NH-\underset{\underset{R}{|}}{C}H-CONH-\underset{\underset{R'}{|}}{C}H-COO^- \xrightarrow{-Z} \overset{+}{H_3}N-\underset{\underset{R}{|}}{C}H-CONH-\underset{\underset{R'}{|}}{C}H-COO^-$$

B. Protection of the Amino Group

Numerous methods have been developed to protect an amino group. The protecting group must be carefully chosen because after we have synthesised the desired peptide the protecting group should be removed easily without disturbing the new peptide bonds. Moreover, it should not result in racemisation of the neighbouring chiral atom. We give below some important protecting groups:

i. Benzyloxycarbonyl Group

This method was introduced in 1932 by **Bergmann** and **Zervas** of the University of Berlin (later of the Rockefeller Institute). In this method, we do acylation of the amino group by benzylchloroformate earlier known as carbobenzoxy chloride. (Cbz or simply Z group). This is prepared in toluene solution from benzyl alcohol and phosgene.

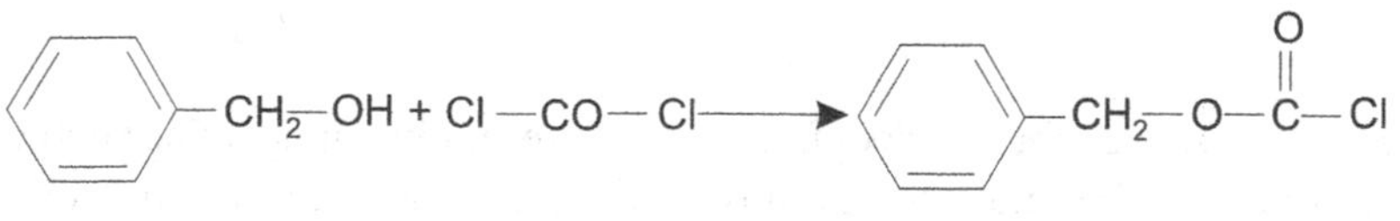

Benzylchloroformate

This may be condensed with the amino acid by the Schotten Baumann reaction. The pH of the reaction mixture is kept high so as to neutralise the HCl so generated in the reaction.

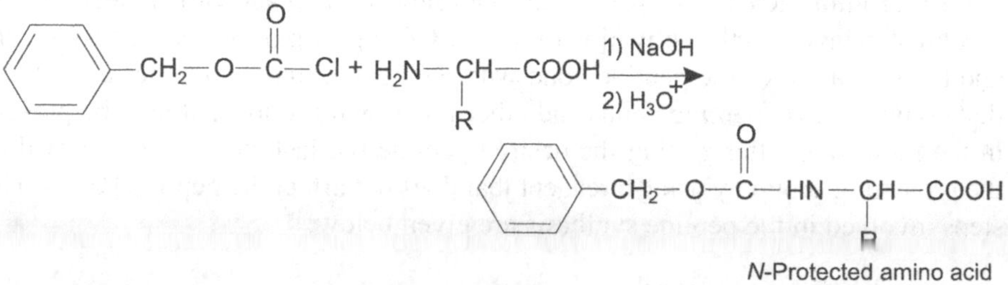

N-Protected amino acid

The benzyloxycarbonyl group can be removed by reagents that do not affect peptide linkages, *i.e.*, by catalytic hydrogenation or by using cold HBr in acetic acid.

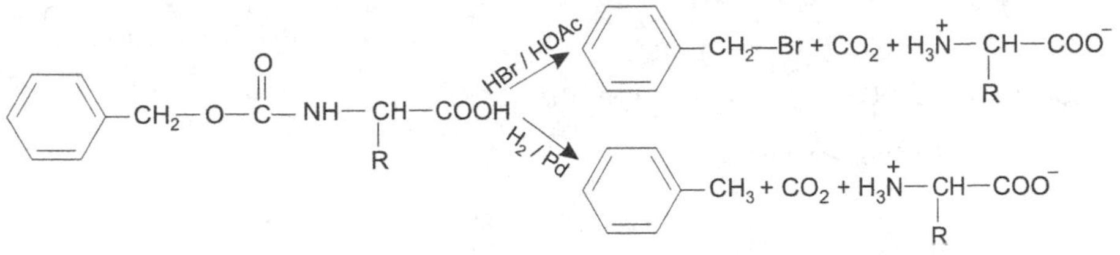

$$\text{C}_6\text{H}_5\text{—CH}_2\text{—O—C(=O)—NH—CH(R)—COOH} \xrightarrow[\text{H}_2/\text{Pd}]{\text{HBr}/\text{HOAc}} \begin{cases} \text{C}_6\text{H}_5\text{—CH}_2\text{—Br} + \text{CO}_2 + \overset{+}{\text{H}_3\text{N}}\text{—CH(R)—COO}^- \\ \text{C}_6\text{H}_5\text{—CH}_3 + \text{CO}_2 + \overset{+}{\text{H}_3\text{N}}\text{—CH(R)—COO}^- \end{cases}$$

The cleavage of the Z-group by HBr in acetic acid proceeds by the S_N2 mechanism as given below:

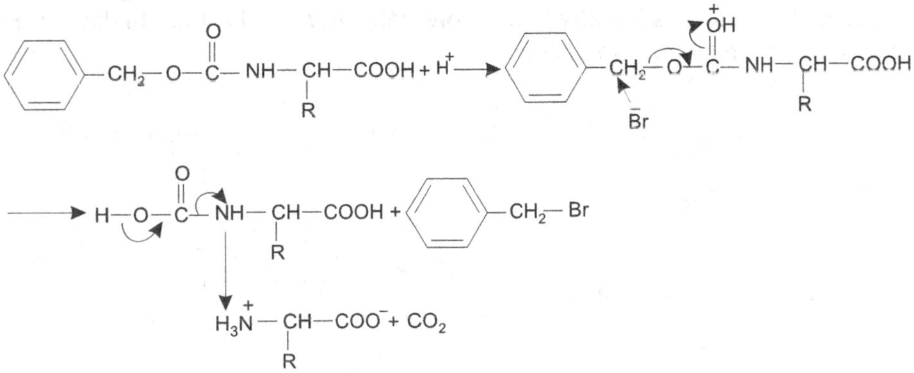

$$\text{C}_6\text{H}_5\text{—CH}_2\text{—O—C(=O)—NH—CH(R)—COOH} + \text{H}^+ \longrightarrow \text{C}_6\text{H}_5\text{—CH}_2\text{—O}^-\text{—C(}\overset{+}{\text{O}}\text{H)—NH—CH(R)—COOH}, \quad \overset{-}{\text{Br}}$$

$$\longrightarrow \text{H—O—C(=O)—NH—CH(R)—COOH} + \text{C}_6\text{H}_5\text{—CH}_2\text{—Br}$$

$$\downarrow$$

$$\overset{+}{\text{H}_3\text{N}}\text{—CH(R)—COO}^- + \text{CO}_2$$

Removal of the Z-group with H_2/Pd depends on the fact that the C-O bond is weak and hydrogenolysis takes place at low temperatures.

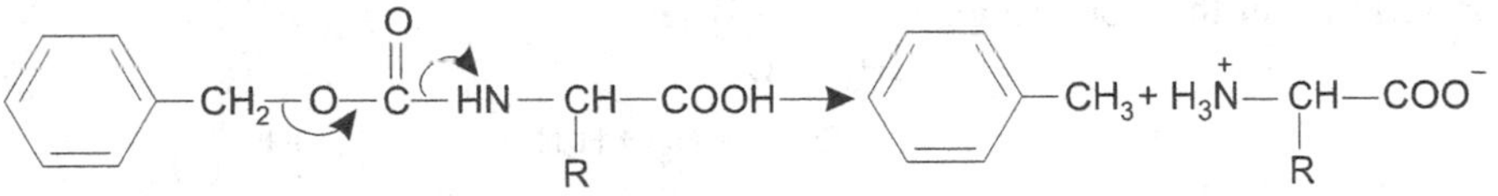

$$\text{C}_6\text{H}_5\text{—CH}_2\text{—O—C(=O)—HN—CH(R)—COOH} \longrightarrow \text{C}_6\text{H}_5\text{—CH}_3 + \overset{+}{\text{H}_3\text{N}}\text{—CH(R)—COO}^-$$

The benzyloxycarbonyl group is normally not affected by mildly basic or other nucleophilic reagents. Therefore, it is quite useful for carrying out peptide synthesis.

ii. *t*-Butyloxycarbonyl Group (Boc Group)

This is the most popular group used for protecting α-amino groups. We can introduce this group by using *t*-butylazidoformate or di-tert-butyl carbonate as tert-butylchloroformate is not very stable. When we use azido-formate, the azide ion is the leaving group. Formation of the Boc-amino acid is shown below.

The Boc group is stable to hydrogenation and general reducing conditions and is more labile to acid than the Z-group. Therefore, the Boc group can be removed easily under mild acidic conditions such as dry trifluoroacetic acid in dichloromethane or dry hydrogen chloride in ether.

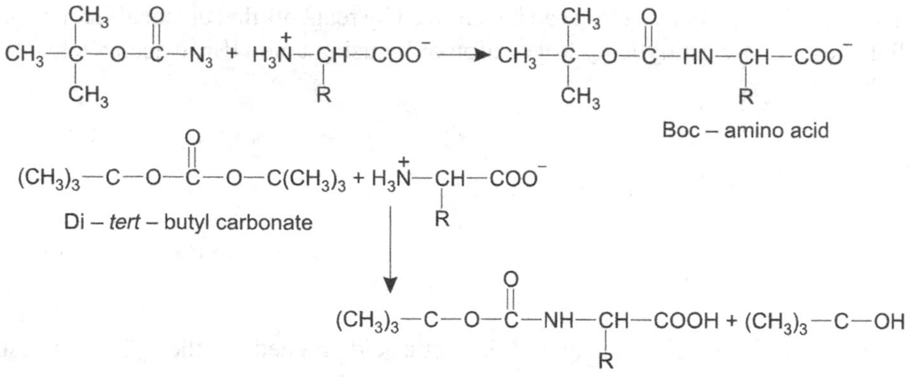

Boc – amino acid

Di – *tert* – butyl carbonate

The mechanism of deprotection of the Boc group is similar to the one given for the Z-group. It also results initially in the more stable *tert*-butyl cation. In the process it gives the formation of isobutene.

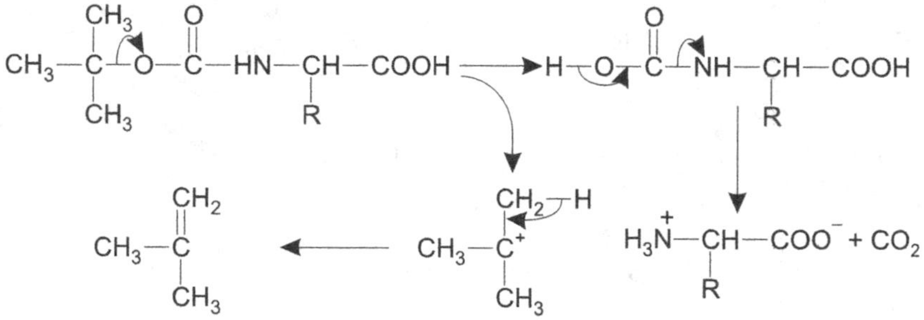

iii. 2-(4-Biphenyl)-isopropoxycarbonyl Group-(Bpoc)

This group is introduced *via* the azide. It is more acid labile than the Boc group because of the formation of more stable intermediate carbocation and hence, it is preferred over the Boc group.

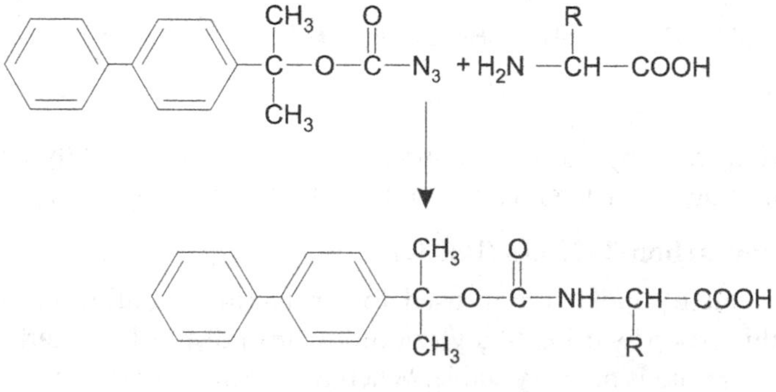

Bpoc Amino acid

Deprotection of the Bpoc group can be carried out by treating the protected amino acid with chloroacetic acid in dichloromethane. The Boc and Z group remain unaffected under these conditions. Therefore, the Bpoc group can be used along with these groups and can be selectively removed.

iv. Phthaloyl Group

Preparation of *N*-phthaloyl amino acids by direct interaction of phthalic anhydride and amino acids in the fused state or in organic solvents frequently leads to partial or complete racemisation. Therefore, *N*-phthaloyl amino acids are obtained by condensing *N*-ethoxy carbonyl phthalimide with amino acids in alkaline media. The group can be cleaved by treating with hydrazine under refluxing conditions as the *N*-phthaloyl group is sensitive to alkaline conditions.

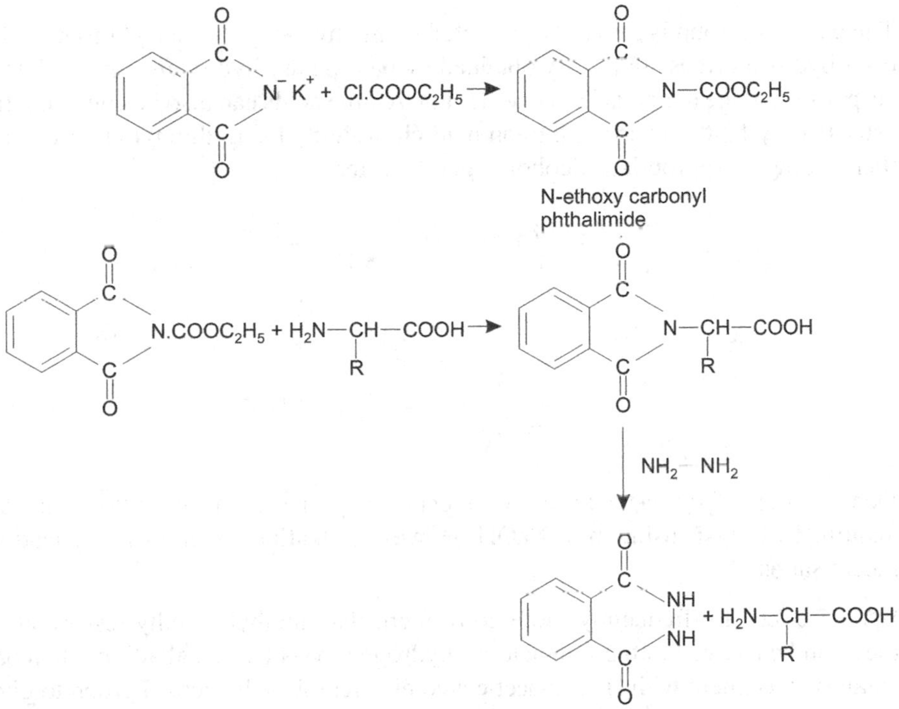

N-ethoxy carbonyl phthalimide

v. Toluene-*p* Sulphonyl Group (Tosyl)

The tosyl group was used for blocking α-amino groups at an early stage in the development of methods of peptide synthesis. More recently, it has proved useful for protecting the ε-amino group of lysine and arginine. It can be cleaved easily by the reducing action of sodium in liquid ammonia.

C. Protection of the Carboxyl Group

In general, carboxyl groups of amino acids are esterified before the peptide bond is formed. This is necessary because amino acids occur as dipolar ions while esters are simple bases which are soluble in aprotic solvents and formation of the peptide bond may have to be conducted in an aprotic solvent. Peptide bond formation in protic solvents is prone to racemisation. Moreover, it is easier to detach a proton from $\overset{+}{N}H_3-\underset{\underset{R}{|}}{C}-COOR'$ than from $\overset{+}{N}H_3-\underset{\underset{R}{|}}{C}-COO^-$.

Esters of amino acids can be liberated from their hydrochlorides by shaking with a solution of ammonia in chloroform before peptide synthesis or *in situ* during peptide synthesis by the addition of a tertiary base. The former method is preferable because the presence of tertiary base can sometimes lead to racemisation.

The carboxyl group is generally protected by methyl, ethyl or benzyl groups. The ester hydrochlorides are readily obtained by passing dry hydrogen chloride into a suspension of the amino acid in dry alcohol. Amino acids can also be conveniently esterified by first converting into an acid chloride by using thionyl chloride and then adding an appropriate alcohol to get the ester.

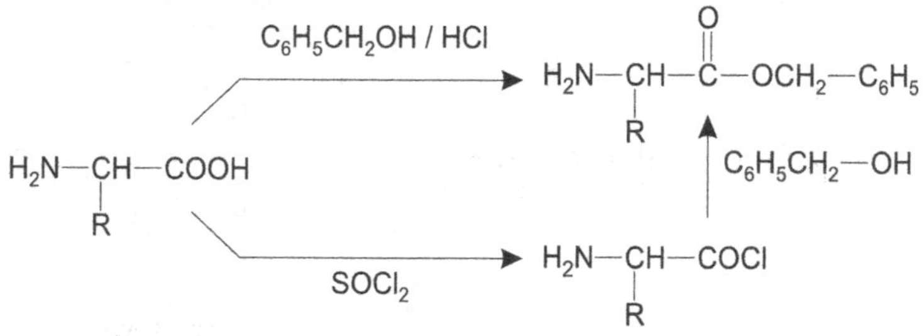

Deprotection of the ester group after peptide synthesis can be carried out by alkaline hydrolysis using mild NaOH. However, alkaline conditions may lead to racemisation.

Benzyl esters are frequently more convenient than methyl or ethyl esters since they can be cleaved to the free acid by hydrogenolysis using palladium-charcoal catalyst. Treatment with HBr in acetic acid also removes the benzyl group to give the free acid.

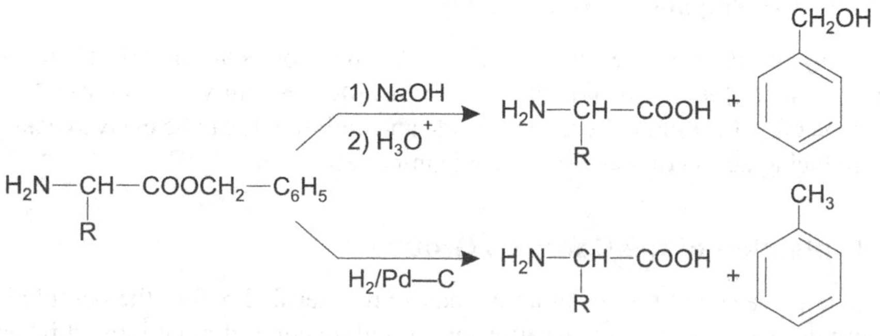

However, *t*-butyl esters, like *t*-butyloxycarbonyl derivatives are resistant to hydrogenolysis but are very sensitive to anhydrous acids. Dry hydrogen chloride in a suitable solvent or trifluoroacetic acid are effective mild reagents for this purpose.

t-Butyl esters are formed when amino acids are allowed to react with isobutylene in the presence of concentrated sulphuric acid.

$$\underset{R}{H_2N-CH-COOH} + \underset{CH_3}{CH_3-C=CH_2} \xrightarrow{\overset{+}{H}} \underset{R}{H_2N-CH-COO-}\underset{CH_3}{\overset{CH_3}{C}-CH_3}$$

D. Protection of Side-Chains of Amino acids

It is not possible to generalise the protecting groups for the side chains because the choice of protecting group depends upon the nature of the side chain.

The ε-amino group in the side chain of lysine can be protected by any of the methods described for α-amino groups. However, it is necessary to employ separate protecting groups of different stability for the α- and ε-amino groups. The protecting group on the α-position can then be selectively removed to permit the coupling of further amino acid residues. The protecting group on the ε-position is left intact until the end of the synthesis.

Lysine chelates with Cu^{2+} ions by involving the α-amino group and $-COO^-$ anion; the ε-amino group can thus be selectively protected. Decomposition of the copper chelate with hydrogen sulphide liberates the α-amino group which can then be protected by using a different reagent. For example, the ε-amino group can first be protected by the Boc group and the α-amino group can then be blocked with the Z-group. After formation of the peptide bond between the $-COOH$ group of lysine and the amino group of the C-protected amino acid, the Z-group can be removed by hydrogenolysis and the liberated amino group can be coupled with another protected amino acid. Finally, all the protecting groups can be removed with anhydrous hydrogen chloride or trifluoroacetic acid to give the free tripeptide.

The β-carboxyl group of aspartic acid and γ-carboxyl group of glutamic acid are usually protected by benzyl or *t*-butyl groups. It is necessary to choose a protecting group for the *N*-terminus which can be removed without affecting the ester groups in the side chains of aspartic or glutamic acids.

The thiol group in the side-chain of cysteine is a powerful nucleophile and it must, therefore, be protected during peptide synthesis. The *S*-benzyl group is the most commonly used. For example, *S*-benzyl-cysteine is prepared by dissolving cysteine in liquid ammonia, adding sufficient sodium to convert the thiol into the anion and then treating with benzyl chloride. The benzyl group is removed by reduction with excess sodium in liquid ammonia.

The guanidino group in the side-chain of arginine is strongly basic. Mere protonation prevents it from entering into the side reactions during peptide synthesis.

The alcoholic group in the side chain is usually protected either as benzyl or *t*-butyl ethers.

2.17   METHODS OF PEPTIDE SYNTHESIS (COUPLING METHODS)

A. Acid Chloride Method

For formation of a peptide bond it is necessary that the hydroxyl group present in the carboxylic acid function is converted into a good leaving group. The most obvious and simple approach to activate a —COOH group is to convert it to an acyl chloride. In this method, the protected amino acid is converted to the corresponding acyl chloride followed by reaction with amino acids or esters under Schotten-Baumann conditions or treatment with amino esters in organic solvents to give the N-protected peptide. Deprotection under hydrolytic conditions gives the free peptide.

This approach played an important role in the early days of peptide synthesis. The reagents normally used for acyl chloride formation, *e.g.*, PCl_5 or $SOCl_2$ are too vigorous to be used in complex or sensitive substrates and may lead to racemisation.

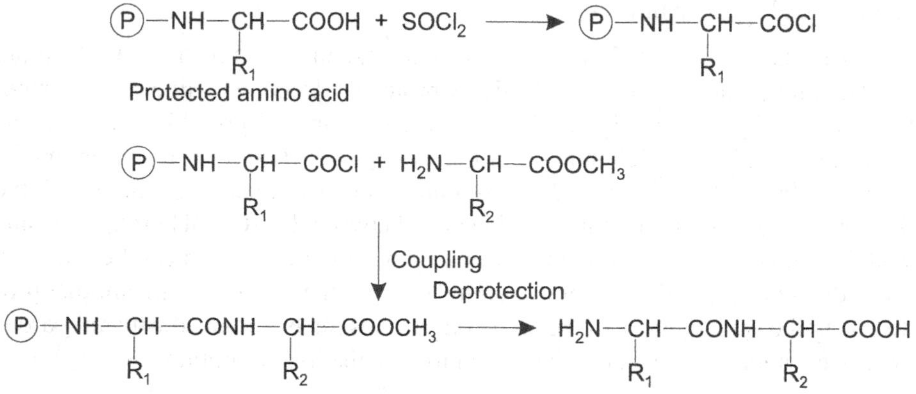

$$\text{P}-NH-\underset{\underset{R_1}{|}}{CH}-COOH + SOCl_2 \longrightarrow \text{P}-NH-\underset{\underset{R_1}{|}}{CH}-COCl$$

Protected amino acid

$$\text{P}-NH-\underset{\underset{R_1}{|}}{CH}-COCl + H_2N-\underset{\underset{R_2}{|}}{CH}-COOCH_3$$

Coupling

Deprotection

$$\text{P}-NH-\underset{\underset{R_1}{|}}{CH}-CONH-\underset{\underset{R_2}{|}}{CH}-COOCH_3 \longrightarrow H_2N-\underset{\underset{R_1}{|}}{CH}-CONH-\underset{\underset{R_2}{|}}{CH}-COOH$$

This method has become completely obsolete as many more convenient and efficient methods are available.

B. Curtius Azide Method

Difficulties observed in the use of acyl chloride were overcome by Curtius. He used the acid azide for coupling with another amino acid.

In this method, N-protected amino acid is converted to its hydrazide which gives the corresponding azide by reaction with nitrous acid. N-protected amino acid azide is then reacted with amino acid ester to give the desired peptide. The scheme is given below:

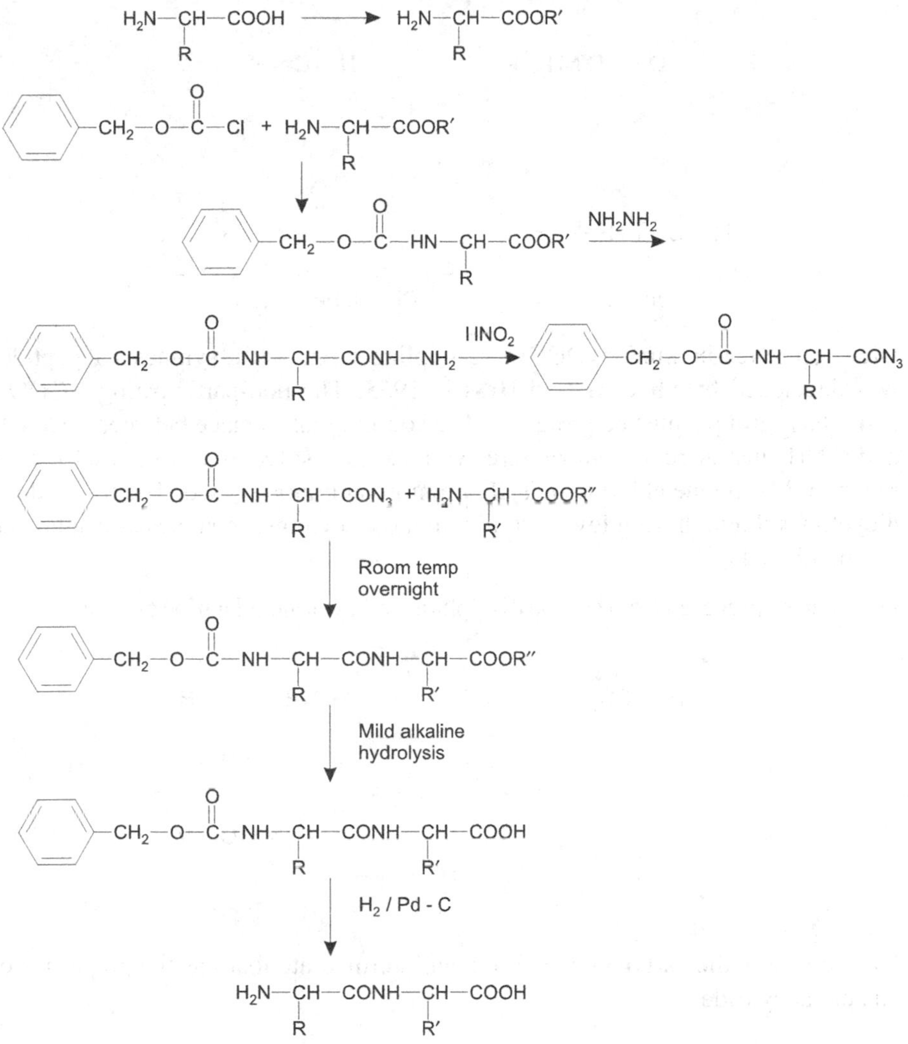

$$H_2N-\underset{\underset{R}{|}}{CH}-COOH \longrightarrow H_2N-\underset{\underset{R}{|}}{CH}-COOR'$$

$$\langle\bigcirc\rangle-CH_2-O-\underset{\underset{O}{\parallel}}{C}-Cl + H_2N-\underset{\underset{R}{|}}{CH}-COOR'$$

$$\downarrow$$

$$\langle\bigcirc\rangle-CH_2-O-\underset{\underset{O}{\parallel}}{C}-HN-\underset{\underset{R}{|}}{CH}-COOR' \xrightarrow{NH_2NH_2}$$

$$\langle\bigcirc\rangle-CH_2-O-\underset{\underset{O}{\parallel}}{C}-NH-\underset{\underset{R}{|}}{CH}-CONH-NH_2 \xrightarrow{I\,NO_2} \langle\bigcirc\rangle-CH_2-O-\underset{\underset{O}{\parallel}}{C}-NH-\underset{\underset{R}{|}}{CH}-CON_3$$

$$\langle\bigcirc\rangle-CH_2-O-\underset{\underset{O}{\parallel}}{C}-NH-\underset{\underset{R}{|}}{CH}-CON_3 + H_2N-\underset{\underset{R'}{|}}{CH}-COOR''$$

$$\downarrow \quad \text{Room temp overnight}$$

$$\langle\bigcirc\rangle-CH_2-O-\underset{\underset{O}{\parallel}}{C}-NH-\underset{\underset{R}{|}}{CH}-CONH-\underset{\underset{R'}{|}}{CH}-COOR''$$

$$\downarrow \quad \text{Mild alkaline hydrolysis}$$

$$\langle\bigcirc\rangle-CH_2-O-\underset{\underset{O}{\parallel}}{C}-NH-\underset{\underset{R}{|}}{CH}-CONH-\underset{\underset{R'}{|}}{CH}-COOH$$

$$\downarrow \quad H_2\,/\,Pd-C$$

$$H_2N-\underset{\underset{R}{|}}{CH}-CONH-\underset{\underset{R'}{|}}{CH}-COOH$$

This synthesis is accomplished without the risk of racemisation. However, some disadvantages are there. *Firstly,* more steps are involved. *Secondly,* the acyl azide may undergo the Curtius rearrangement to give the formation of isocyanate. Therefore, the desired product is difficult to purify.

C. DCC Method

In the presence of dicyclohexylcarbodiimide, an acid reacts with an amine to give an amide.

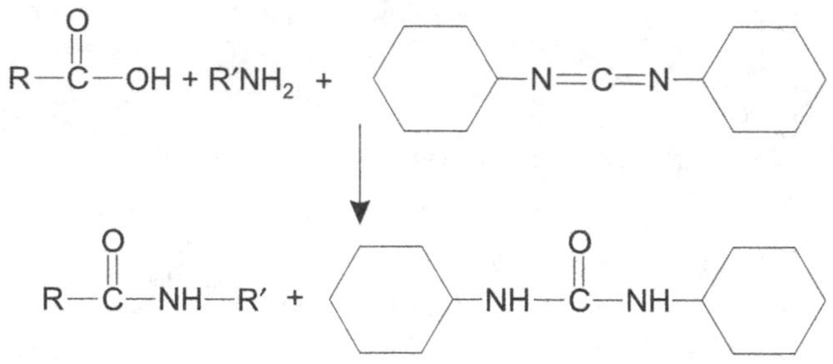

An amide   Dicyclohexylurea

Dicyclohexylcarbodiimide (DCC) as a coupling agent for the synthesis of peptides was developed by **Sheehan** and **Hess** in 1955. The non-participating —COOH and —NH₂ groups must be protected. The coupling takes place between —COOH and —NH₂ just at room temperature. Various inert solvents can be used for the reaction. Methylene chloride or THF are more commonly used. The by-product, dicyclohexylurea, having low solubility in most solvents, separates as a solid and can be filtered off.

DCC can be prepared from carbon disulphide and cyclohexyl amine as given below:

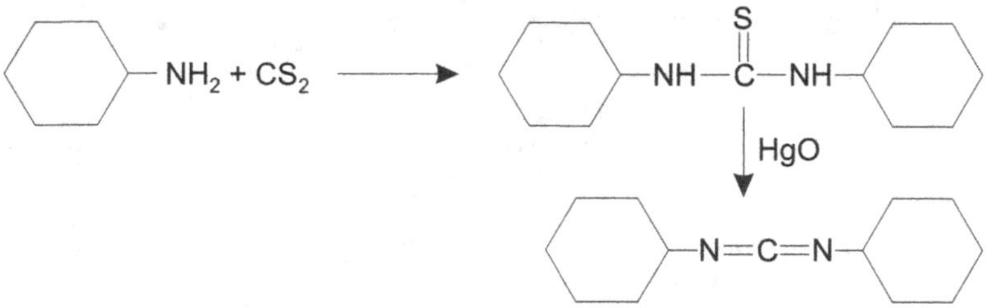

DCC

DCC converts the carboxylic acid into an intermediate that has the properties of an acid anhydride

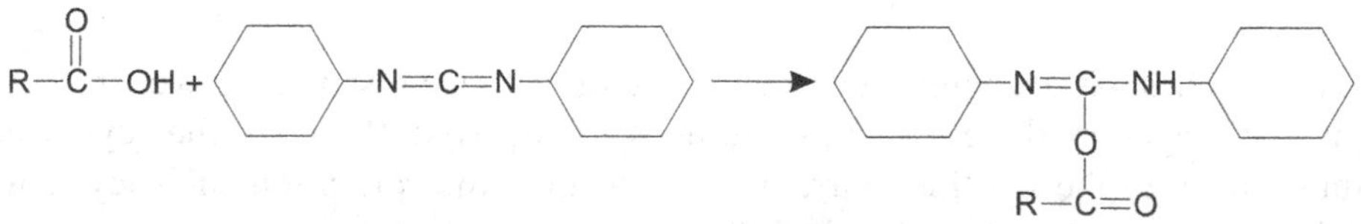

The scheme of reactions for the synthesis of Gly-Ala is given below:

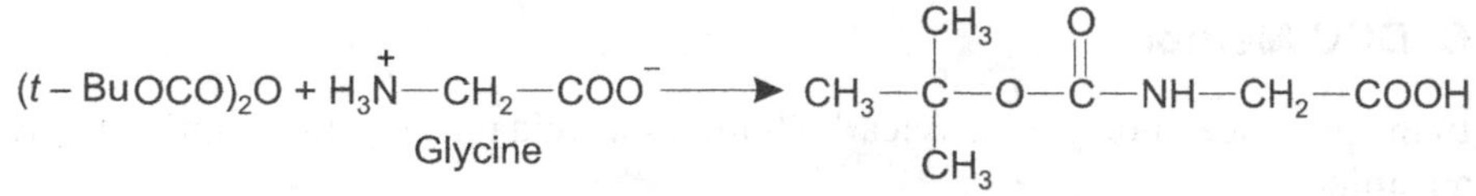

Boc - Glycine

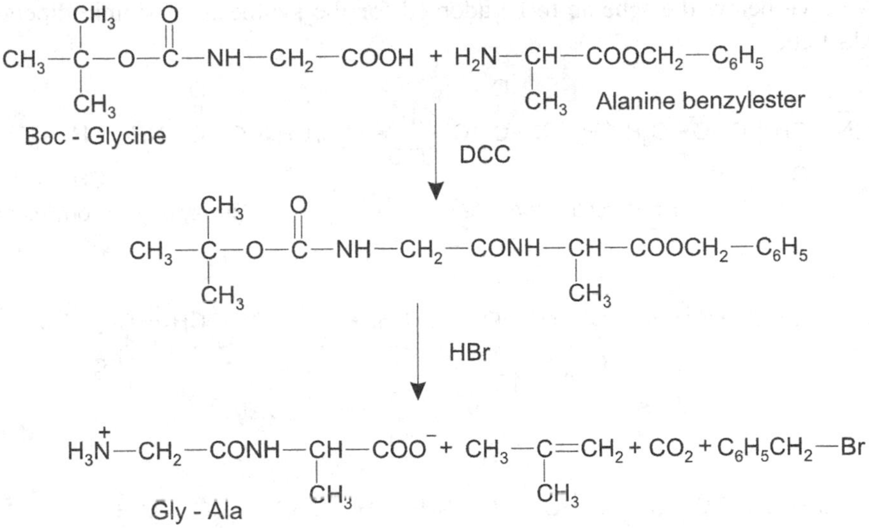

$$CH_3-\underset{\underset{CH_3}{|}}{\overset{\overset{CH_3}{|}}{C}}-O-\overset{\overset{O}{\|}}{C}-NH-CH_2-COOH \ + \ H_2N-\underset{\underset{CH_3}{|}}{CH}-COOCH_2-C_6H_5$$

Boc - Glycine

Alanine benzylester

DCC

$$CH_3-\underset{\underset{CH_3}{|}}{\overset{\overset{CH_3}{|}}{C}}-O-\overset{\overset{O}{\|}}{C}-NH-CH_2-CONH-\underset{\underset{CH_3}{|}}{CH}-COOCH_2-C_6H_5$$

HBr

$$\overset{+}{H_3N}-CH_2-CONH-\underset{\underset{CH_3}{|}}{CH}-COO^- \ + \ CH_3-\underset{\underset{CH_3}{|}}{C}=CH_2 + CO_2 + C_6H_5CH_2-Br$$

Gly - Ala

The repetition of the above reactions can give us synthesis of any desired peptide. By following the above approach oxytocin was synthesised by **Vincent du Vigneaud** in 1955 and synthesis of insulin with 51 amino acids was reported by Sanger in 1963.

D. Mixed Anhydride Method

In this method, the carboxyl group of the protected amino acid is converted to a mixed anhydride using ethyl chloroformate in an organic solvent in the presence of a tertiary base like triethylamine. This method is used in such cases where the separation of the by-product dicyclohexylurea in the DCC method is difficult.

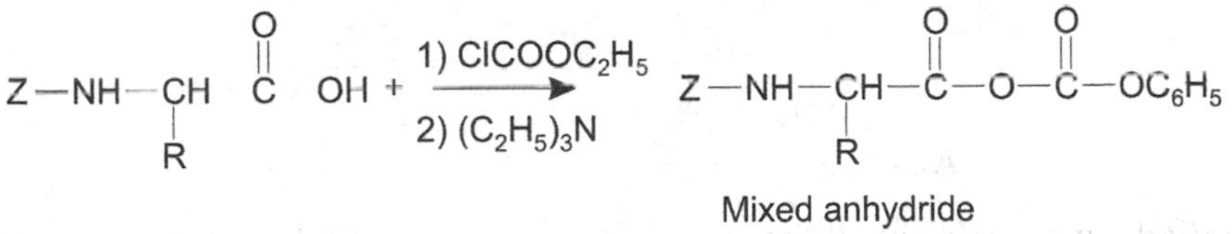

$$Z-NH-\underset{\underset{R}{|}}{CH} \ \overset{\overset{O}{\|}}{C} \ OH \ + \ \xrightarrow[\text{2) }(C_2H_5)_3N]{\text{1) ClCOOC}_2H_5} \ Z-NH-\underset{\underset{R}{|}}{CH}-\overset{\overset{O}{\|}}{C}-O-\overset{\overset{O}{\|}}{C}-OC_6H_5$$

Mixed anhydride

The mixed anhydride can then be used to acylate another amino acid or its ester to give the desired product. The reaction takes place at low temperatures giving high yields and pure products.

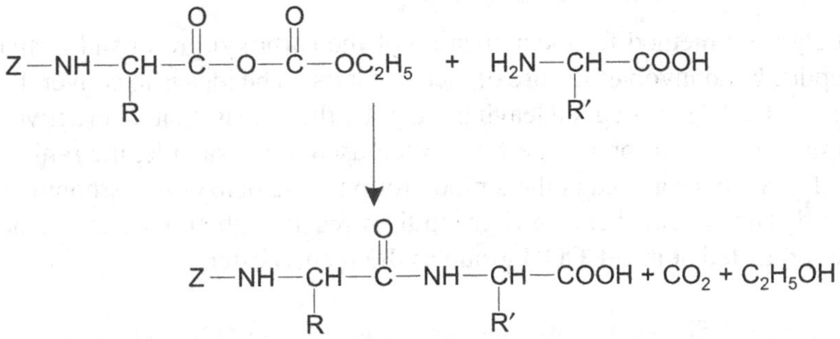

$$Z-NH-\underset{\underset{R}{|}}{CH}-\overset{\overset{O}{\|}}{C}-O-\overset{\overset{O}{\|}}{C}-OC_2H_5 \ + \ H_2N-\underset{\underset{R'}{|}}{CH}-COOH$$

$$Z-NH-\underset{\underset{R}{|}}{CH}-\overset{\overset{O}{\|}}{C}-NH-\underset{\underset{R'}{|}}{CH}-COOH + CO_2 + C_2H_5OH$$

We give below the scheme to be adopted for the synthesis of simple dipeptide Ala-Leu:

$$H_3\overset{+}{N}-CH-COO^{\pm} + C_6H_5CH_2-O-\overset{\overset{\displaystyle O}{\|}}{C}-Cl \xrightarrow[25°C]{OH^{\pm}} C_6H_5CH_2-O-\overset{\overset{\displaystyle O}{\|}}{C}-NH-CH-COOH$$

$$\qquad\quad \overset{|}{CH_3}\qquad\qquad\qquad\qquad\qquad\qquad\qquad\qquad\qquad\qquad\qquad\qquad \overset{|}{CH_3}$$

Benzylchloroformate   Benzyloxycarbonylalanine

$$C_6H_5CH_2-O-\overset{\overset{\displaystyle O}{\|}}{C}-NH-CH-\overset{\overset{\displaystyle O}{\|}}{C}-O-\overset{\overset{\displaystyle O}{\|}}{C}OC_2H_5 + (CH_3)_2CH-CH_2-CH-COO^-$$

$$\qquad\qquad\qquad\qquad\qquad \overset{|}{CH_3}\qquad\qquad\qquad\qquad\qquad\qquad\qquad\qquad \overset{|}{\underset{+}{NH_3}}$$

Z - Ala   Leu

$$\Big\downarrow$$

$$C_6H_5CH_2-O-\overset{\overset{\displaystyle O}{\|}}{C}-NH-CH-\overset{\overset{\displaystyle O}{\|}}{C}-NH-CH-COOH + CO_2 + C_2H_5OH$$

$$\qquad\qquad\qquad\qquad\qquad \overset{|}{CH_3}\qquad\qquad \overset{|}{CH_2}$$

Z - Ala - Leu   CH

$$\qquad\qquad\qquad\qquad\qquad\qquad\qquad\qquad H_3C\quad CH_3$$

$$\Big\downarrow H_2 / Pd$$

$$CH_3-CH-\overset{\overset{\displaystyle O}{\|}}{C}-NH-CH-COO^- \quad + \quad \bigcirc\!\!\!-CH_3 + CO_2$$

$$\qquad\quad \overset{|}{\underset{+}{NH_3}}\qquad\qquad \overset{|}{CH_2}$$

$$\qquad\qquad\qquad\qquad\qquad CH$$

$$\qquad\qquad\qquad\qquad H_3C\quad CH_3$$

Ala - Leu

Similarly, we can show all the steps in the synthesis of Gly-Val-Ala using Boc-group as a protecting group.

E. Use of Other Activating Groups

A widely used method for the activation of the carboxyl group and formation of the peptide bond involves the use of 'active esters'. The idea is to convert the —OH group of —COOH into a good leaving group by the preparation of an active ester of the amino acid. p-Nitrophenyl ester is often used. For example, the p-nitrophenyl ester of glycine, protected at the amino group by the benzyloxycarbonyl group, is prepared. The activated carboxyl group then reacts with another amino acid, say valine, protected at its —COOH group as the methyl ester.

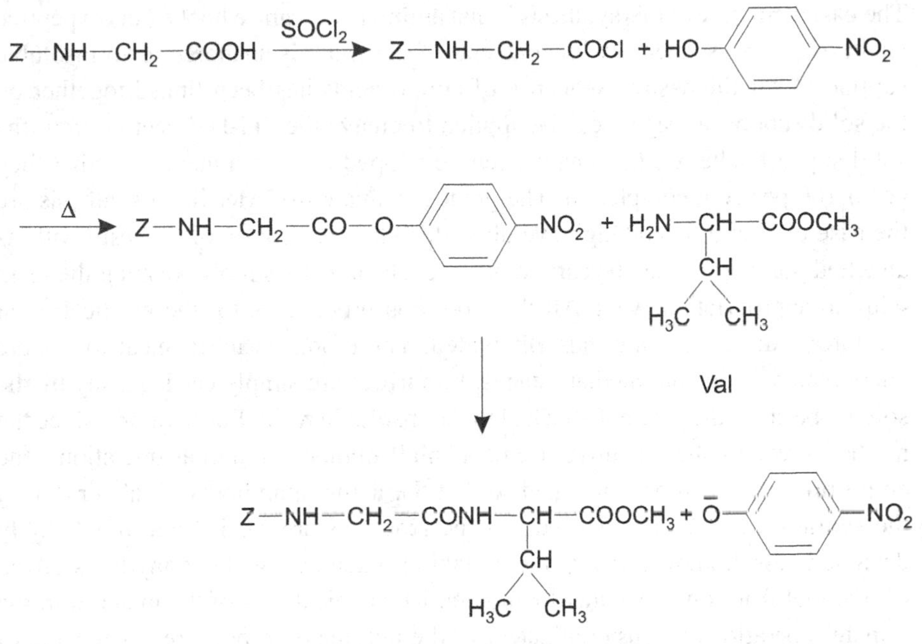

$$Z-NH-CH_2-COOH \xrightarrow{SOCl_2} Z-NH-CH_2-COCl + HO--NO_2$$

$$\xrightarrow{\Delta} Z-NH-CH_2-CO-O--NO_2 + H_2N-\underset{\underset{H_3C \quad CH_3}{\underset{|}{CH}}}{\overset{|}{CH}}-COOCH_3$$

Val

$$Z-NH-CH_2-CONH-\underset{\underset{H_3C \quad CH_3}{\underset{|}{CH}}}{\overset{|}{CH}}-COOCH_3 + \bar{O}--NO_2$$

Z - Gly – Val – O – CH$_3$

The conditions used for the formation of the *p*-nitrophenyl ester are quite harsh and might not be suitable for activating the —COOH group of larger peptides. So other types of esters, formed in the presence of DCC, are used for many peptide syntheses.

Esters of *N*-hydroxy succinimide are often used where *N*-oxysuccinimide anion is a good leaving group.

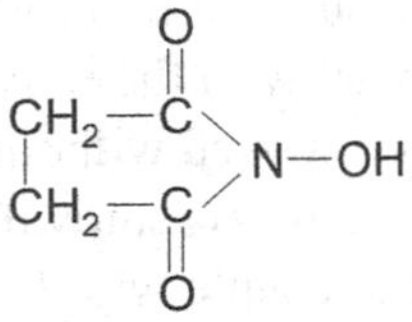

$$\begin{array}{c} CH_2-\overset{O}{\overset{\|}{C}} \\ \diagdown \\ N-OH \\ \diagup \\ CH_2-\underset{\|}{\overset{}{C}} \\ O \end{array}$$

N-Hydroxy-
succinimide

2.18 SOLID PHASE PEPTIDE SYNTHESIS

A. Introduction

Solid phase peptide synthesis was developed and introduced by R.B. Merrifield of Rockfeller University in order to simplify and accelerate peptide synthesis in a way that would make the synthesis of long peptides practical.

The basic principle of this synthesis is that amino acids can be hooked into a peptide chain of desired sequence while one end of the chain is anchored to an insoluble support. After the desired sequence of amino acids has been linked together on the solid support a reagent can be applied to cleave the finished peptide from the solid support. The reactions have been developed in such a manner so that they go to 100 per cent completion. The great advantages of Merrifield synthesis are the ease of operation and high overall yield. The purification of the resin with its attached polypeptide can be carried out at each stage by simply washing the resin with an appropriate solvent. All the laborious procedures for the purification at the intermediate steps are thus eliminated. Therefore, no mechanical losses are encountered in the intermediate stages. Impurities are simply carried away by the solvent because they are not attached to the insoluble resin. Furthermore, since the method involves the repetitive use of a small number of similar operations, the entire procedure can be automated so that the actual man hours of labour during the synthesis are greatly reduced. All the reactions are carried out in a suitably designed vessel, known as a "protein making machine", without any transferring of material from one container to another. Mechanical loss of the material in the transfer operations is thus eliminated. In the automated procedure, each cycle of the Merrifield synthesis requires only four hours and attaches one new amino acid residue.

B. The Basic Scheme of Solid Phase Peptide Synthesis

The solid support is a synthetic polymer that bears reactive groups (X). These groups are made to react with the carboxyl groups of an amino acid in such a way that the amino acid is bound covalently to the polymer support. During this operation the amino group of the amino acid must be protected by a suitable group (Y) so that the amino group will not react with the polymer. The protecting group must be such that it can be selectively removed without damage to the bond holding the amino acid to the polymer support. After removal of the protecting group a second N-protected amino acid is used to couple with the amino group of the first amino acid, thus forming the first peptide chain. By repeating these two steps, deprotection and coupling, the peptide of desired sequence is assembled on the polymer support. At the end a different reagent is applied to cleave the peptide chain from the polymer and also to carry out the deprotection. The basic plan of solid phase peptide synthesis is outlined ahead (Fig. 2.7).

C. The Merrifield Method

In the Merrifield method, the solid support used is a copolymer of styrene and 2% divinyl benzene. The polymer is composed of long alkyl chains bearing a phenyl ring on every second carbon atom. These chains are cross-linked at every fiftieth carbon by *p*-diethylphenyl residues derived from divinyl benzene. This cross-linking makes the polymer completely insoluble in all ordinary solvents. The polymer also swells in certain organic solvents but is chemically inert during peptide synthesis. The resin is used in the form of fine beads 20–70 microns in diameter.

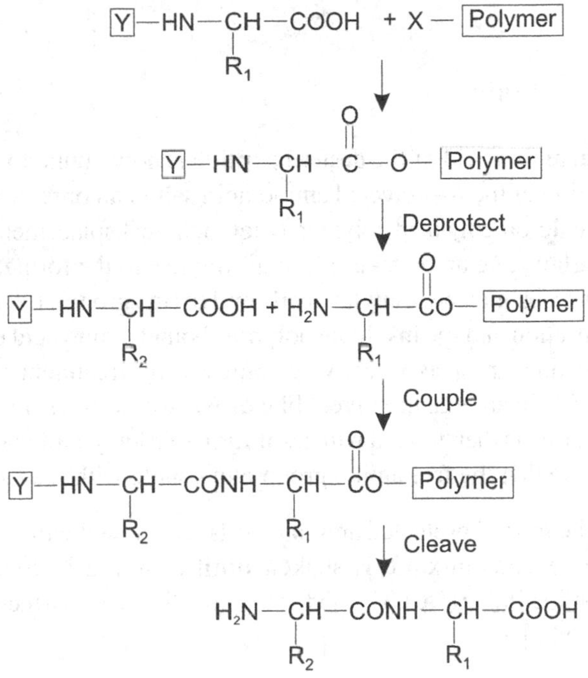

Fig. 2.7.   The basic plan of solid phase peptide synthesis.

The first step in the automated peptide synthesis (Fig. 2.7) involves the attachment of the *N*-protected amino acid residue to the resin support. In order to effect this, the polymer must first be made reactive. That is done by chloromethylating the benzene rings in the chain by a Friedel-Crafts reaction using chloromethyl methyl ether and stannic choride.

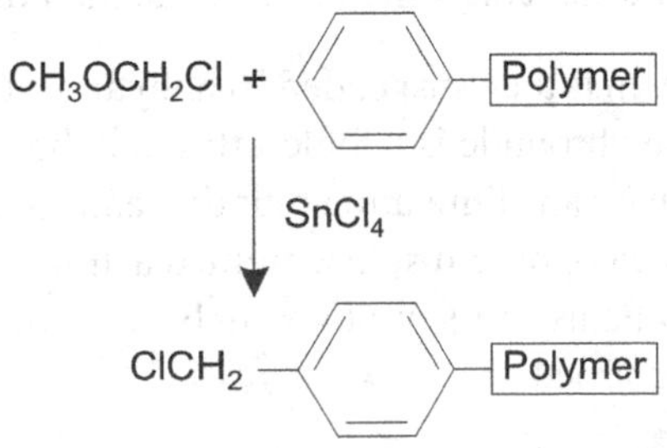

Thus, about one out of 10–100 phenyl groups is chloromethylated giving

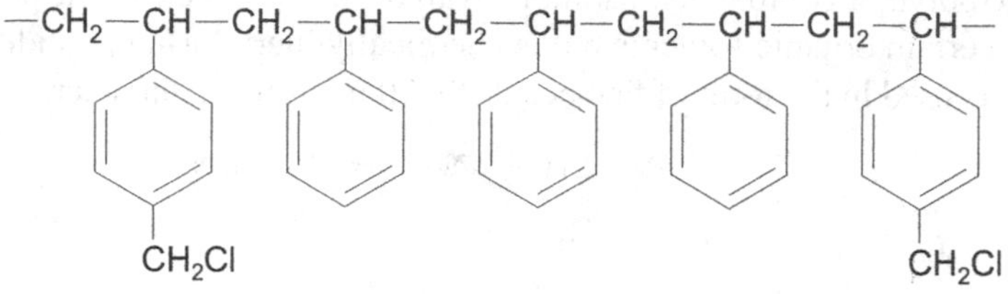

$$-CH_2-CH-CH_2-CH-CH_2-CH-CH_2-CH-CH_2-CH-$$

$$CH_2Cl \qquad\qquad\qquad\qquad\qquad\qquad CH_2Cl$$

The C-terminal amino acid of the desired peptide is now bound to the polymer by shaking a solution of the N-protected amino acid salt in an organic solvent such as DMF with the chloromethylated polymer. Nucleophilic displacement of the halogen atom by the carboxylate anion takes place giving rise to the formation of an ester. The protecting group may be t-butyloxycarbonyl (a Boc group). Excess reagents are removed by filtration and the insoluble polymer bound amino acid ester is washed. The Boc-protecting group is selectively removed by treatment with anhydrous hydrogen chloride in an organic solvent like dioxane or acetic acid. The aminoacyl resin hydrochloride is then treated with a solution of triethylamine to neutralise the hydrochloride, so that the free amino group can couple with the next amino acid.

A solution of the next N-protected amino acid is then added with a coupling agent and the heterogeneous mixture is shaken until coupling is complete. Usually dicyclohexylcarbodiimide (DCC) is used as the coupling agent, which gets converted into dicyclohexylurea.

The polymer now is bound to the N-protected dipeptide. Deprotection and neutralisation of the protected dipeptide resin make the amino group ready for coupling with the next amino acid. The three steps—deprotection, neutralisation and coupling are repeated as many times as desired or until the desired peptide chain is linked together on the resin. Throughout all these operations, the reagents and reaction conditions are chosen so that all reactions go to 100 per cent completion. The only purification used at any step is the washing of the resin and its attached peptide with appropriate solvents which can be handled rapidly and effectively.

The finished peptide resin is then suspended in anhydrous trifluoroacetic acid and a slow stream of hydrogen bromide is bubbled through the suspension. This reagent removes the Boc-group by an elimination reaction and cleaves the finished peptide from the resin by a nucleophilic displacement reaction. Certain other protecting groups used for side chains are simultaneously removed from the peptide, but peptide bonds are not harmed.

The complete Merrifield method is outlined on the next page (Fig. 2.8):

As Merrifield solid phase peptide synthesis is a stepwise synthesis of peptides from the C-terminus, it is free from the danger of racemisation. Only the carboxyl groups of *N*-protected amino acids are activated for coupling, not the carboxyl groups of peptides.

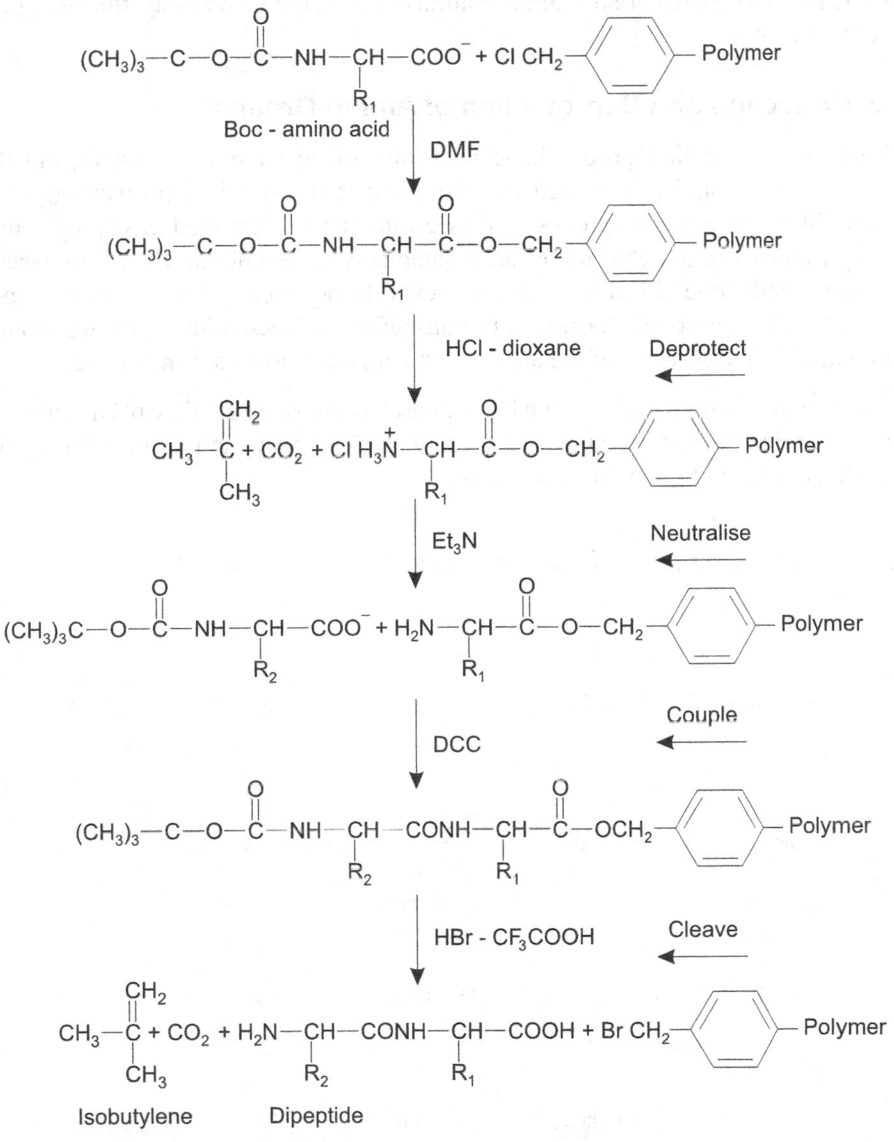

$(CH_3)_3-C-O-\overset{O}{\overset{\|}{C}}-NH-\underset{R_1}{CH}-COO^- + Cl\,CH_2-\langle\text{benzene ring}\rangle-\text{Polymer}$

Boc - amino acid

| DMF

$(CH_3)_3-C-O-\overset{O}{\overset{\|}{C}}-NH-\underset{R_1}{CH}-\overset{O}{\overset{\|}{C}}-O-CH_2-\langle\text{benzene ring}\rangle-\text{Polymer}$

| HCl - dioxane   Deprotect ←

$\underset{CH_3}{\overset{CH_2}{\overset{\|}{CH_3-C}}} + CO_2 + Cl\,H_3\overset{+}{N}-\underset{R_1}{CH}-\overset{O}{\overset{\|}{C}}-O-CH_2-\langle\text{benzene ring}\rangle-\text{Polymer}$

| Et₃N   Neutralise ←

$(CH_3)_3C-O-\overset{O}{\overset{\|}{C}}-NH-\underset{R_2}{CH}-COO^- + H_2N-\underset{R_1}{CH}-\overset{O}{\overset{\|}{C}}-O-CH_2-\langle\text{benzene ring}\rangle-\text{Polymer}$

| DCC   Couple ←

$(CH_3)_3-C-O-\overset{O}{\overset{\|}{C}}-NH-\underset{R_2}{CH}-CONH-\underset{R_1}{CH}-\overset{O}{\overset{\|}{C}}-OCH_2-\langle\text{benzene ring}\rangle-\text{Polymer}$

| HBr - CF₃COOH   Cleave ←

$\underset{CH_3}{\overset{CH_2}{\overset{\|}{CH_3-C}}} + CO_2 + H_2N-\underset{R_2}{CH}-CONH-\underset{R_1}{CH}-COOH + Br\,CH_2-\langle\text{benzene ring}\rangle-\text{Polymer}$

Isobutylene   Dipeptide

Fig. 2.8.   The complete Merrifield Method.

Finally, when the Boc-group is used as the protecting group, deprotection and cleavage are carried out in non-hydrolytic solvents, the danger of racemisation is minimised.

The Merrifield technique has been applied successfully to the synthesis of ribonuclease, a protein having 124 amino acid residues. The synthesis involved 369 different chemical reactions and 11,931 automated steps including injection of reagents, filtrations, washings and so on. All these operations were carried out without isolating an intermediate. The synthetic ribonuclease not only had the same physical characteristics of the natural enzyme, but it possessed the biological activity as well.

D. Protection and Deprotection of Amino Groups

Various N-protecting groups have been developed by other workers, but the t-butyloxycarbonyl group used by Merrifield is the standard protecting group in solid phase peptide synthesis. This group can be removed easily by either anhydrous hydrogen chloride in an organic solvent or anhydrous trifluoroacetic acid in methylene chloride. These reagents do not cleave benzyl esters. Since the t-butyloxycarbonyl chloride is too unstable, the Boc-amino acids are usually prepared by the reaction of the amino acid with t-butyloxycarbonyl azide.

The amino acid is treated with the Boc-azide at a controlled alkaline pH in a pH-stat or by a mixture of magnesium oxide, sodium bicarbonate and triethylamine. The azide is obtained by the following route:

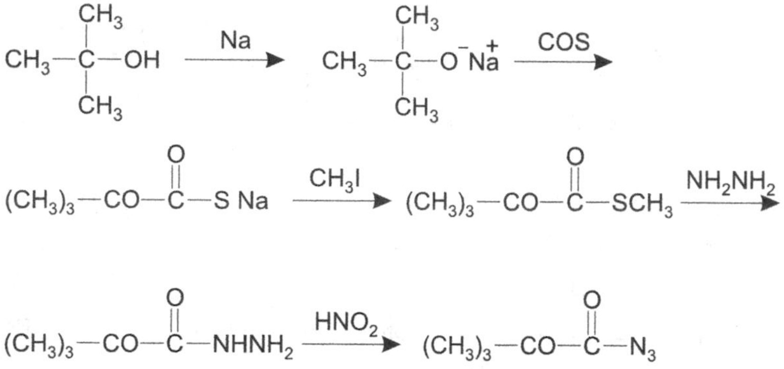

$$CH_3-\underset{\underset{CH_3}{|}}{\overset{\overset{CH_3}{|}}{C}}-OH \xrightarrow{Na} CH_3-\underset{\underset{CH_3}{|}}{\overset{\overset{CH_3}{|}}{C}}-O^-Na^+ \xrightarrow{COS}$$

$$(CH_3)_3-CO-\overset{\overset{O}{||}}{C}-S\,Na \xrightarrow{CH_3I} (CH_3)_3-CO-\overset{\overset{O}{||}}{C}-SCH_3 \xrightarrow{NH_2NH_2}$$

$$(CH_3)_3-CO-\overset{\overset{O}{||}}{C}-NHNH_2 \xrightarrow{HNO_2} (CH_3)_3-CO-\overset{\overset{O}{||}}{C}-N_3$$

t – Butyloxycarbonyl azide

$$(CH_3)_3-CO-\overset{\overset{O}{||}}{C}-N_3 + H_2N-\underset{\underset{R}{|}}{CH}-COOH \xrightarrow{Base}$$

$$(CH_3)_3-CO-\overset{\overset{O}{||}}{C}-NH-\underset{\underset{R}{|}}{CH}-COOH$$

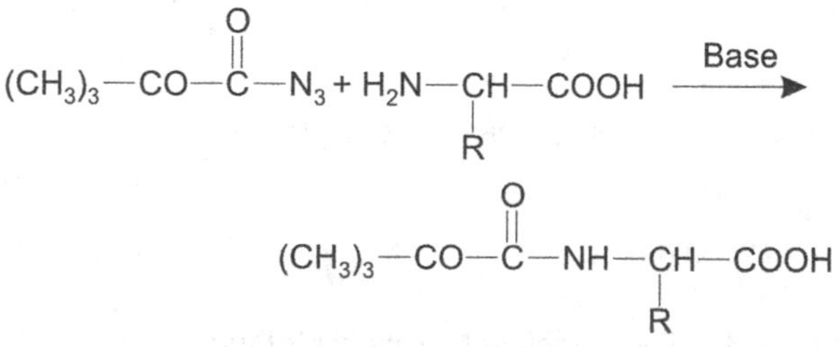

Boc - amino acid

Deprotection of the Boc-amino acid takes place by simple elimination reaction giving rise to the formation of isobutylene and a free amino acid with elimination of CO_2.

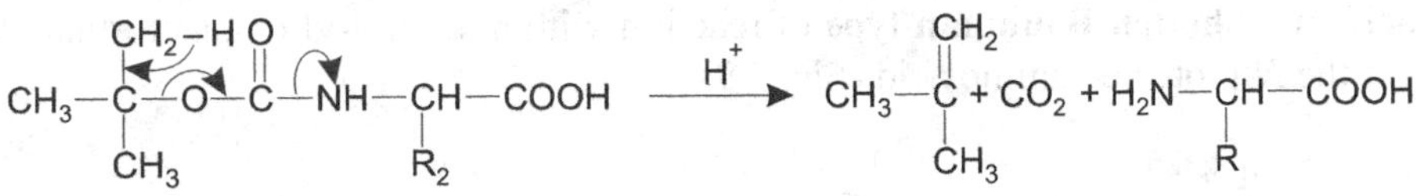

$$CH_3-\underset{\underset{CH_3}{|}}{\overset{\overset{CH_2-H}{|}}{C}}-O-\overset{\overset{O}{||}}{C}-NH-CH-COOH \xrightarrow{H^+} CH_3-\underset{\underset{CH_3}{|}}{\overset{\overset{CH_2}{||}}{C}} + CO_2 + H_2N-\underset{\underset{R}{|}}{CH}-COOH$$

The *t*-Amyloxy carbonyl group (Aoc-group) has replaced to some extent the Boc-group as the NH_2-protecting group because *t*-amyloxycarbonyl chloride is sufficiently more stable than the *t*-butyloxycarbonyl chloride. Therefore, it can be used directly for the synthesis of protected amino acids without recourse to the much more expensive azide method.

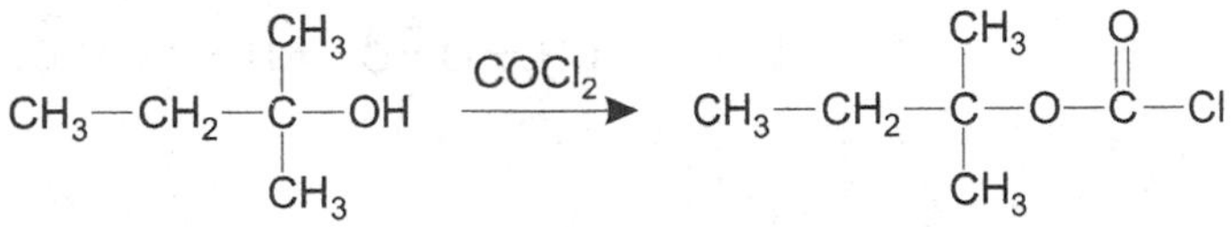

$$CH_3-CH_2-\underset{\underset{CH_3}{|}}{\overset{\overset{CH_3}{|}}{C}}-OH \xrightarrow{COCl_2} CH_3-CH_2-\underset{\underset{CH_3}{|}}{\overset{\overset{CH_3}{|}}{C}}-O-\overset{\overset{O}{||}}{C}-Cl$$

t-Amyloxycarbonyl chloride

$$CH_3-CH_2-\underset{\underset{CH_3}{|}}{\overset{\overset{CH_3}{|}}{C}}-O-\overset{\overset{O}{||}}{C}-Cl + H_2N-\underset{\underset{R}{|}}{CH}-COOH$$

$$\downarrow \text{Base}$$

$$CH_3-CH_2-\underset{\underset{CH_3}{|}}{\overset{\overset{CH_3}{|}}{C}}-O-\overset{\overset{O}{||}}{C}-NH-\underset{\underset{R}{|}}{CH}-COOH$$

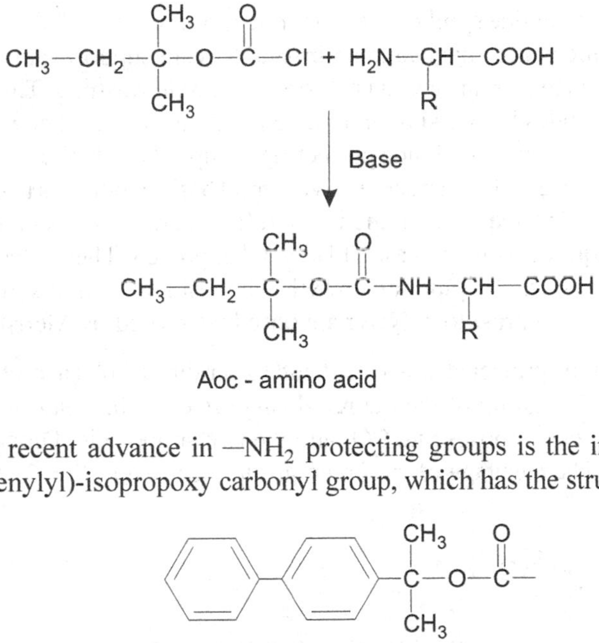

Aoc - amino acid

The most recent advance in $-NH_2$ protecting groups is the introduction of the 2-(4-biphenylyl)-isopropoxy carbonyl group, which has the structure

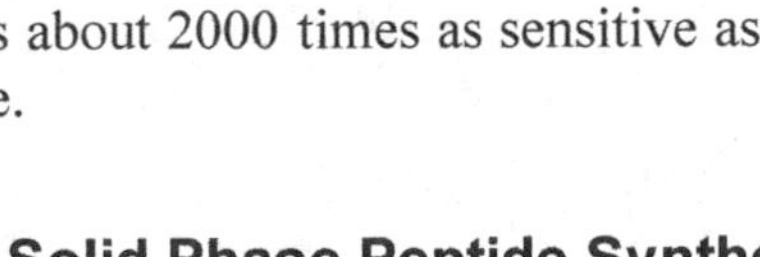

$$\text{(biphenyl)}-\underset{\underset{CH_3}{|}}{\overset{\overset{CH_3}{|}}{C}}-O-\overset{\overset{O}{||}}{C}-$$

This protecting group is about 2000 times as sensitive as the *t*-butyloxycarbonyl group to acidic cleavage.

E. Fmoc Method of Solid Phase Peptide Synthesis

A new solid phase peptide synthesis was developed by **Sheppar** *et al*. using Fmoc for protecting the *N*-terminal amino group, Fmoc is 9-fluorenyl methoxy carbonyl used as a protecting group. The Fmoc group is introduced by reacting the amino acid in a **Sholten Baumann** type of reaction with fluoren-9-yl chloro formate to get the N-protected amino acid (Fig. 2.9).

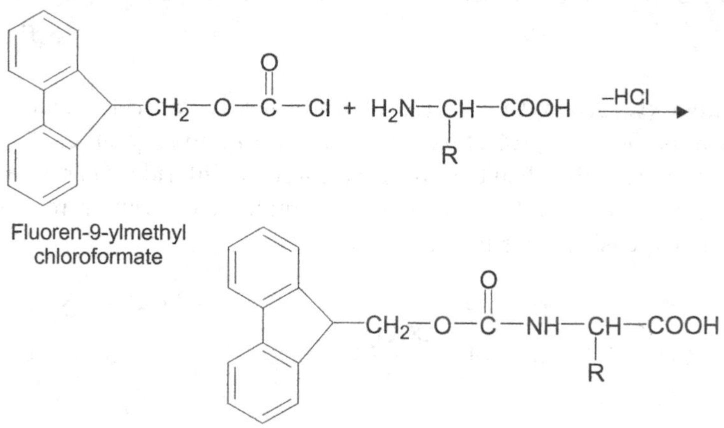

Fluoren-9-ylmethyl chloroformate

Fig. 2.9.   Use of Fmoc for protecting-NH_2 group.

In Fmoc solid phase peptide synthesis, the peptide chain is assembled stepwise, one amino acid at a time while attached to an insoluble resin support. This allows the reaction by products to be removed at each step by simple washing. The procedure for building the peptide chain is similar to the one which we use when t-BOC is the protecting group. However, the Fmoc protecting group is base labile. The cleavage of the Fmoc group is carried out under relatively mild basic conditions using 20-50% piperidine in DMF. The exposed amine is, therefore, neutral; no neutralisation of peptide resin is required as in the case of Boc/Bzl approach. The protecting group is removed at every cycle by plperidine/DMF. The resin used in this proceduce is a polyamide resin in contrast to polystyrene based resin used by Merrifield.

Treatment of the Fmoc protected amino acid with piperidine results in proton abstraction from the methine group of the fluorenyl ring system. This leads to release of a carbamate which decomposes into CO_2 and the free amino acid. Dibenzofulvene is also generated. The cleavage of the Fmoc group is shown in Fig. 2.10.

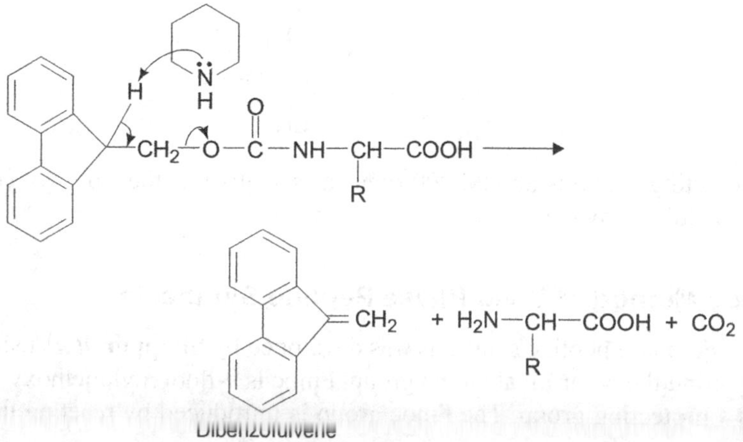

Dibenzofulvene

Fig. 2.10.   Cleavage of the Fmoc group.

This reaction is able to occur due to the acidity of the fluorenyl proton. The dibenzo fulvene by-product can react with nucleophiles such as piperidine which is in large excess.

The Fmoc method is easy to automate because there is no need is use corrosive trifluoro acetic acid in the synthetic cyles. Moreover, the fluorenyl group is a chromophore; Fmoc deprotection can be monitored by UV absorbance of the reaction mixture.

The Fmoc method is a gentler method than *t*-Boc since the peptide chain is not subjected to acid at each cycle and has become the major method employed in commercial automated peptide synthesis.

Today Fmoc is the method of choice for peptide synthesis Very high quality building blocks are available at low cost. Many modified derivatives are commercially available as Fmoc building blocks, making synthetic access to a broad range of peptide derivatives. The majority of the synthetic peptides are now prepared by Fmoc solid-phase peptide synthesis. The *t*-Boc method is now generally used for specialist applications.

2.19 DETERMINATION OF PRIMARY STRUCTURE OF PEPTIDES

A. Introduction

Structure determination of peptides means arriving at the sequence of amino acids in a polypeptide chain. When the polypeptide is a protein, this sequence is referred to as its primary structure. In order to determine the structure of a peptide, we must know:

1. The kind of amino acids that make up the peptide.
2. The number of amino acids of each type.
3. The order in which these amino acids are arranged in the peptide chain.

Therefore, it is necessary to have a qualitative and quantitative analysis of the amino acids present. In the first step, the peptide chain is converted into its amino acid components by complete hydrolysis. In the second step, the hydrolysate that contains the mixture of amino acids from the first step is analysed to determine the amino acid composition. In order to calculate the number of residues of each amino acid present, we must know the molecular weight of the polypeptide, which is determined by suitable physical methods. Finally, in the third step, the sequence of amino acids is determined.

B. Complete Hydrolysis

Hydrolysis of the peptides may be carried out with acids, bases or with certain enzymes to give a mixture of amino acids. For example, a dipeptide on hydrolysis gives a mixture of two amino acids.

$$H_2N-\underset{\underset{R}{|}}{CH}-CONH-\underset{\underset{R'}{|}}{CH}-COOH$$

$$\downarrow H_2O$$

$$H_2N-\underset{\underset{R}{|}}{CH}-COOH \ + \ H_2N-\underset{\underset{R'}{|}}{CH}-COOH$$

When the hydrolysis is carried out with an acid or with a base, all the peptide bonds are split giving a mixture of amino acids in the hydrolysate; *i.e.*, under the action of an acid or alkali complete hydrolysis takes place. Enzymatic hydrolysis, however, is selective and incomplete; only specific peptide linkages are broken. Enzymatic hydrolysis is usually employed in the third step, *i.e.*, for determining the sequence of amino acids in peptides.

In the acid hydrolysis, the peptide or protein is heated with 6N HCl in a sealed tube at 110°C for 10 to 100 hours, depending on the peptide. The hydrolysate is analysed for the amino acid content. Tryptophan is largely destroyed by acid hydrolysis. Asparagine and glutamine are completely converted into aspartic and glutamic acids respectively. Cysteine and cystine are partially destroyed and better results are obtained by previously oxidising them with performic acid to cysteic acid which is stable. Serine and threonine are slowly destroyed during acid hydrolysis. Finally, valine and isoleucine are slowly liberated because their bulky side chains cause steric hinderance.

Alkaline hydrolysis is effected by heating the peptide at 100°C with 2 to 4 N NaOH for 1 to 10 hours. It is, however, useless for complete analysis of amino acids in a peptide or protein because several amino acids are destroyed by alkaline hydrolysis. Amino acids which are destroyed by alkaline hydrolysis are arginine, cysteine, cystine, threonine and serine. For this reason, acid hydrolysis is usually preferred over alkaline hydrolysis. Tryptophan, however, largely survives alkaline hydrolysis. So, to obtain tryptophan content, a peptide or protein is hydrolysed under alkaline conditions.

C. Analysis of the Hydrolysate

The hydrolysate obtained after complete hydrolysis is examined for its amino acid content. The relative abundance of each amino acid in the hydrolysate is identified and measured. This may be done by any of the suitable methods mentioned earlier. The commonly used methods are paper and thin layer chromatography, electrophoresis and ion-exchange chromatography.

Ion-exchange chromatography is the most popular method for the qualitative analysis of a mixture of amino acids. There are several technical variations, but basically the mixture of amino acids, derived from the hydrolysis of polypeptide or protein, is resolved on a column of a sulphonated polystyrene resin by elution with buffers. More recently the process has been mechanised and the commercial instrument is called an amino acid analyser. The buffer is pumped through the column at a fixed rate, the effluent from the column is automatically mixed with a controlled flow of ninhydrin solution and heated by passage through a reaction coil at 100°C. The developed colour is measured with a visible spectrometer and the absorbance is plotted by a recorder as a function of time. The results have a precision of 100 ± 3 per cent. The amino acids can be identified from their elution positions. By comparison of the chromatogram of an unknown mixture with that of a mixture of known composition, we can arrive at quantitative analysis of the mixture.

D. Cleavage of Disulphide Bridges

After determining the total amino acid content and the kind of amino acids in a polypeptide chain, the next step is to cleave any disulphide bridge that may be present before carrying out sequence studies. The reaction is usually carried out when the peptide or protein contains a cystine irrespective of whether this forms a loop within a chain or links two chains together. This can be achieved by oxidising the substance with performic acid, which converts the two cysteine units into cysteic acid units. This oxidation converts the disulphide bridge to sulphonic acid $(-SO_3H)$ groups.

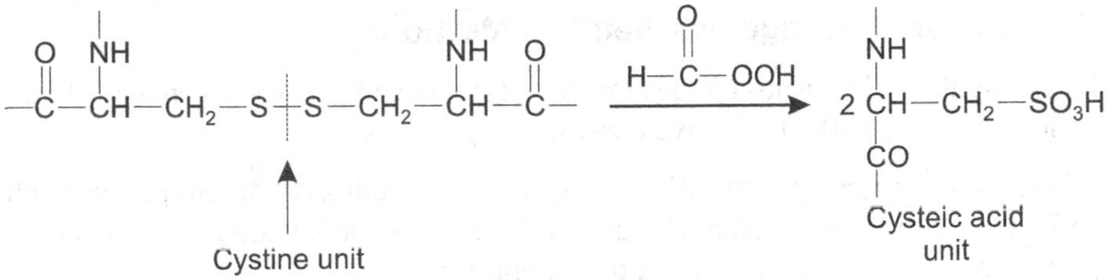

Cystine unit   Cysteic acid unit

If the compound contains no disulphide bridges this step is not necessary.

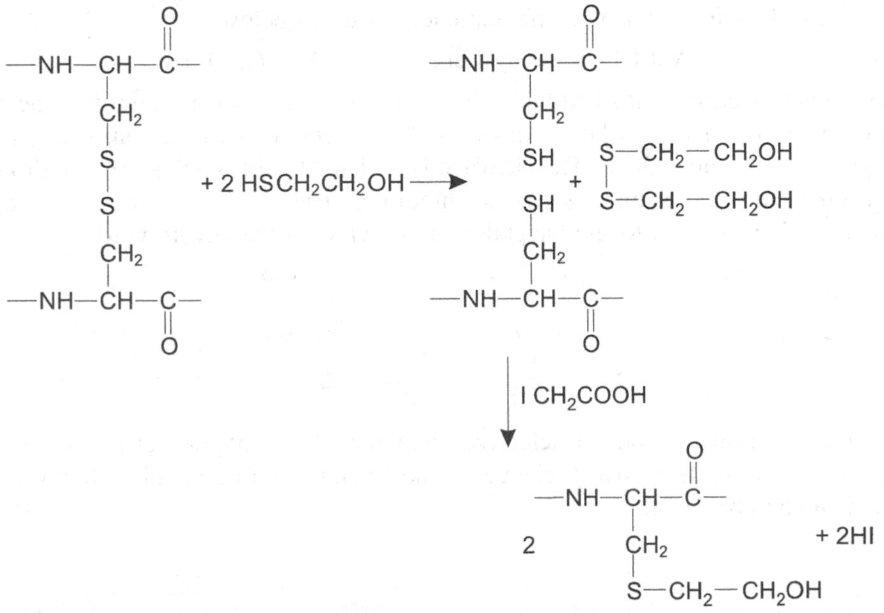

Alternatively, the disulphide bridges can be reductively cleaved. Several reagents for reduction have been used, but the best reagent is perhaps 2-mercaptoethanol. When reduction is complete, the liberated thiol groups in the peptide are alkylated with iodoacetic acid. Cysteine is thus converted into S-carboxymethyl cysteine.

E. Determination of the Amino Acid Sequence

The determination of the correct amino acid sequence of a peptide or protein is carried out in two parts. In the first part, partial degradation of the polypeptide chain is carried out by means of certain enzymes. The fragments so obtained are analysed individually for total amino acid content. The second part involves determination of the amino acids that are present at the N-terminus and at the C-terminus in the original peptide as well as in the fragments obtained through enzymatic hydrolysis. This is called **End Group Analysis**. The pieces are then fitted together to arrive at the complete sequence of the peptide chain.

The process is then repeated using a different cleavage method usually cyanogen bromide. The second set of peptide fragments are analysed separately and the amino acid sequence is determined for each fragment. The complete amino acid sequence is arrived at sometimes by using both sets of data.

F. Selective Cleavage by Chemical Methods

Partial hydrolysis by acids has been widely used and efforts have been made to find conditions which effect selective cleavage of peptides.

Hydrolysis of a peptide with 0.25N acetic acid under reflux conditions preferentially splits the peptide bonds on both sides of the aspartic acid residues. For example, partial hydrolysis of the following pentapeptide:

Val-Glu-Asp-Gly-Leu

will give four fragments with the sequences as given below:

Val-Glu, Asp-Gly-Leu, Val-Glu-Asp, Gly-Leu

However, peptide bonds involving the amino groups of serine and threonine are preferentially cleaved when hydrolysis of the peptide is carried out using 10N hydrochloric acid at 30°C. The cleavage is assisted by the neighbouring hydroxyl group in the side-chain of serine or threonine. The peptide rearranges through oxazolidine tautomer to an O-acyl derivative which is then hydrolysed.

Another useful method for selective cleavage of polypeptide chains employs cyanogen bromide, CNBr. It cleaves dialkyl sulphides to give alkyl thiocyanate and an alkyl bromide.

$$CN—Br + R—CH_2—S—CH_2—R \longrightarrow R—CH_2—S—CN + R—CH_2—Br$$

Cyanogen bromide has been used to cleave the peptide bonds containing the carboxyl group of methionine units; the methionine is converted into a C-terminal homoserine lactone unit. The reaction is quite complicated since ring closure and fission of the peptide bond occur simultaneously.

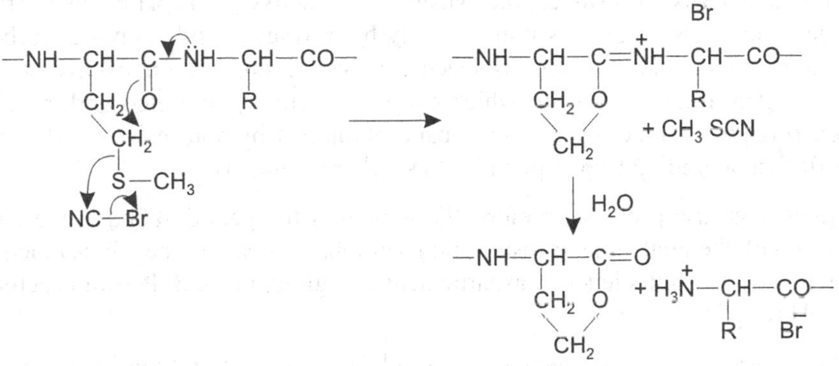

Homoserine Lactone
unit

G. Partial Hydrolysis by Enzymes

The partial hydrolysis of peptides is carried out by means of enzymes generally known as proteases. They readily hydrolyse peptide bonds only between specific amino acid residues. The commonly used proteases are trypsin, chymotrypsin and pepsin, all hydrolyse specific peptide linkages to give smaller fragments.

Trypsin, the pancreatic enzyme occurs in the intestine of mammals. This preferentially hydrolyses the peptide bond at the carboxyl position of basic amino acids such as lysine and arginine.

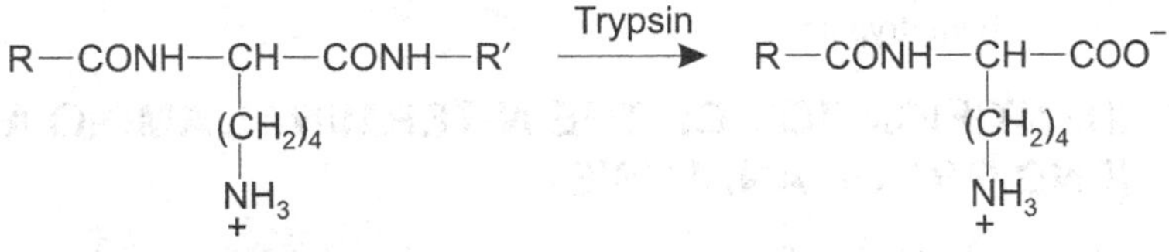

Trypsin is most active in the pH range 7–9, but is fairly rapidly inactivated by self-digestion. Addition of a low concentration of calcium ions markedly inhibits self-digestion. It is possible both to extend and restrict the action of trypsin on proteins. If an amino group of lysine is blocked the action of trypsin can be restricted to the hydrolysis of arginyl peptide bonds.

As the action of trypsin is highly specific it has been extensively used for comparing the primary structures of closely related proteins. The method is known as finger printing. It involves hydrolysis of the proteins with trypsin and comparison of the resulting peptides. Separation of these peptides is achieved by electrophoresis on filter paper followed by chromatography at right angles to the direction of electrophoresis. Peptides are spotted by spraying with ninhydrin.

Chymotrypsin, another intestinal enzyme selectively splits the peptide bond at the carboxyl position of phenylalanine, tyrosine and tryptophan (all aromatic amino acids). Like trypsin, chymotrypsin is most active in the pH range 7–9 but is considerably less specific as it also slowly hydrolyses peptide bonds at carboxyl positions of methionine, leucine, isoleucine and valine. Preparations of chymotrypsin may contain traces of trypsin which can be selectively inhibited with naturally occurring proteins like soya-bean trypsin inhibitor. Chymotrypsin is widely used for further degrading tryptic peptides to smaller fragments.

Pepsin, a gastric protease, preferentially cleaves the peptide bond at the amino position of phenylalanine, tyrosine and tryptophan. However, pepsin is much less specific. It also splits leucine, aspartic acid and glutamic acid. Pepsin is active in the pH range 1–2.

Other proteolytic enzymes which are useful for degrading proteins include papain (active at pH 5–7) and a bacterial enzyme subtilisin (active at pH 6–8) from *Bacillus subtilis*. These enzymes have broad *specificities* and are mainly used for further degradation of large fragments which are produced by tryptic hydrolysis.

The specific action of three proteases on a hepta peptide is illustrated below:

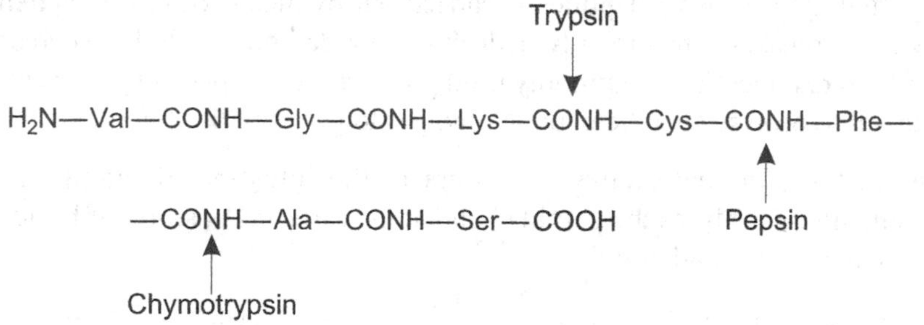

Trypsin

H_2N—Val—CONH—Gly—CONH—Lys—CONH—Cys—CONH—Phe—

Pepsin

—CONH—Ala—CONH—Ser—COOH

Chymotrypsin

2.20 IDENTIFICATION OF THE *N*-TERMINAL AMINO ACID (END GROUP ANALYSIS)

There are two main types of methods for identifying the *N*-terminal residue of a peptide or protein.

In the first type, a suitable group is attached to the α-amino group of the *N*-terminal residue so that after hydrolysis with acid or enzymes the labelled residue can be detected by spectrophotometry, fluorescence or radioactive counting and identified by chromatography.

A number of reagents have been developed for condensation of the amino group of the *N*-terminus.

A. Barger's Method

The earlier method which was most popular for a number of years is that of Barger. As mentioned earlier, trinitrotoluene is condensed with the amino end of peptides with the elimination of one of the nitro groups. The peptide is then hydrolysed and the resultant yellow dinitrotolyl derivative of the N-terminus amino acid separates out readily, because its solubility properties differ from those of the unsubstituted amino acids obtained by hydrolysis.

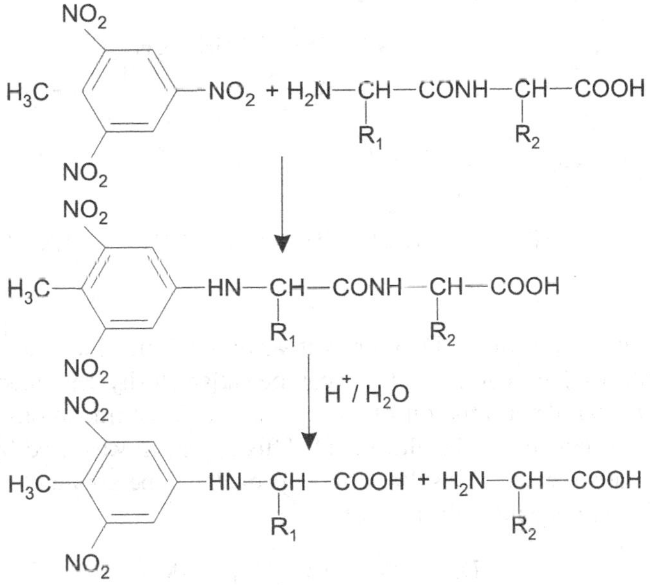

B. The DNP Method (Sanger Method)

A very successful method of identifying the N-terminal residue was introduced by F. **Sanger** of Cambridge University. The method makes use of 2,4-dinitrofluorobenzene (DNFB) which undergoes nucleophilic substitution by the free amino group to give an N-dinitrophenyl (DNP) derivative. The DNP derivative of the peptide is obtained by the reaction with 2,4-dinitrofluorobenzene in a buffer at about pH 8. Reaction is likely to occur also with the ε-amino groups of lysine, the phenolic hydroxyl group of tyrosine, the thiol group of cysteine and the imidazole nucleus of histidine.

Complete hydrolysis of the DNP-peptide with acid gives the yellow DNP-derivative of the N-terminal residue together with ε-DNP-lysine, o-DNP tyrosine and free amino acids, which can be identified by chromatography. Thin layer chromatography is particularly useful for rapid qualitative work.

The DNP derivatives of most of the amino acids have been prepared and characterised. The DNP method cannot be used repeatedly since its use requires complete hydrolysis of the DNP derivative.

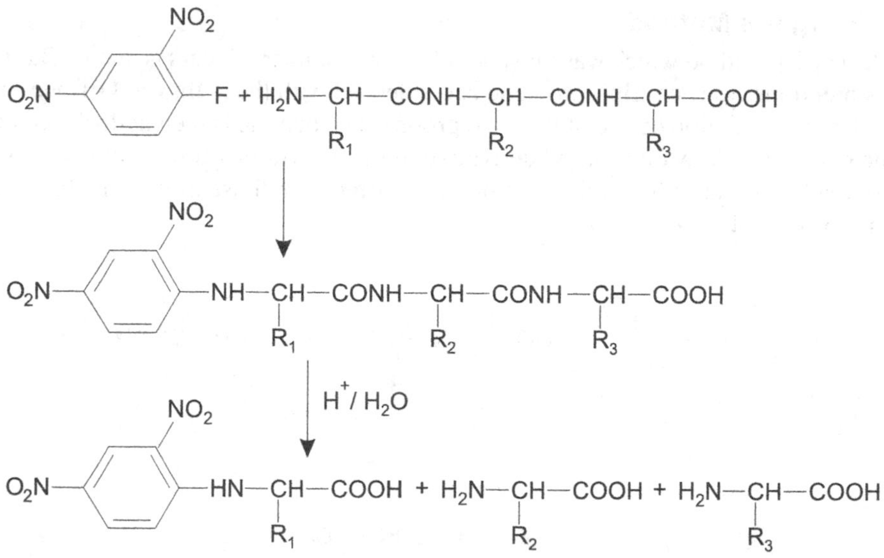

$$O_2N-\underset{\overset{|}{NO_2}}{\bigcirc}-F + H_2N-\underset{\overset{|}{R_1}}{CH}-CONH-\underset{\overset{|}{R_2}}{CH}-CONH-\underset{\overset{|}{R_3}}{CH}-COOH$$

$$\downarrow$$

$$O_2N-\underset{\overset{|}{NO_2}}{\bigcirc}-NH-\underset{\overset{|}{R_1}}{CH}-CONH-\underset{\overset{|}{R_2}}{CH}-CONH-\underset{\overset{|}{R_3}}{CH}-COOH$$

$$\downarrow H^+/H_2O$$

$$O_2N-\underset{\overset{|}{NO_2}}{\bigcirc}-HN-\underset{\overset{|}{R_1}}{CH}-COOH + H_2N-\underset{\overset{|}{R_2}}{CH}-COOH + H_2N-\underset{\overset{|}{R_3}}{CH}-COOH$$

Partial acid hydrolysis of the DNP-derivative of a polypeptide may give rise to a series of yellow DNP-peptides. If these can be satisfactorily separated (by column chromatography), determination of their amino acid compositions may permit the N-terminal sequence to be elucidated. This approach was used by Sanger for elucidating the structure of insulin. This approach can be applied to any peptide. For example, partial acid hydrolysis of

DNP—Phe—Val—Asp—Glu

will give rise to the following

DNP—Phe—Val—Asp
DNP—Phe—Val
DNP—Phe

From these products the original sequence of the tetrapeptide can be reconstructed.

C. The Dansyl Method

A modification of the DNP method is the use of 5-dimethylamino-1-naphthalenesulphonyl chloride (dansyl chloride) since these derivatives are more resistant to acid hydrolysis and give strong fluoroscence in ultraviolet light. Therefore, very minute quantities (10^{-10} to 10^{-9} moles) can be easily detected. The method is about 100 times more sensitive than the DNP-technique. The dansyl amino acids can also be analysed spectroscopically.

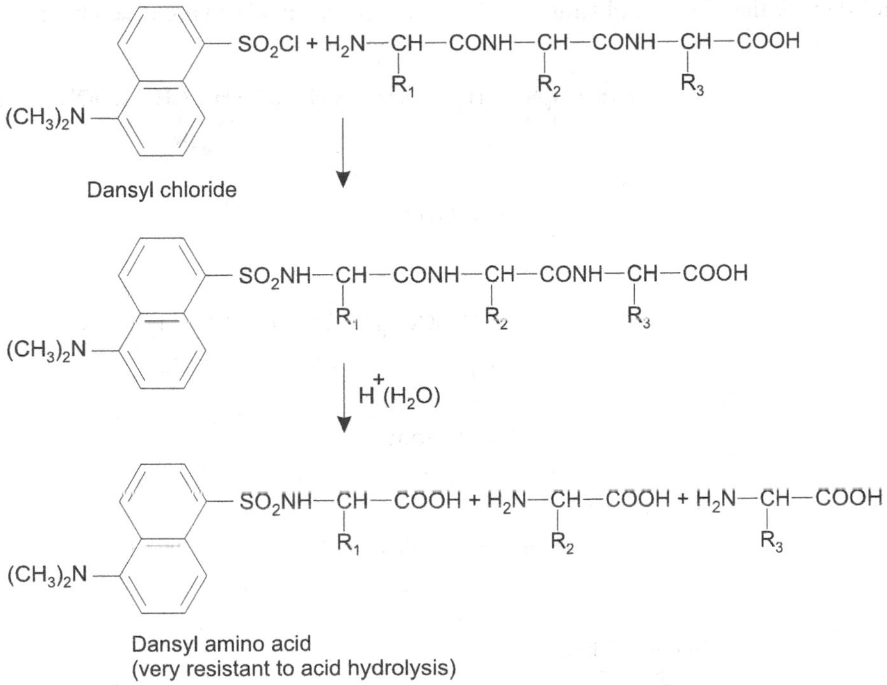

Dansyl chloride

Dansyl amino acid
(very resistant to acid hydrolysis)

The DNP- and 'dansyl' groups in fact serve as labels to identify the N-terminal residue after total hydrolysis although partial hydrolysis may lead to the elucidation of short sequences near the labelled group.

D. The Edman Method

This method comes under the second and more powerful type of methods for identifying the N-terminal residue of a polypeptide in which we attach a suitable substituent on the α-amino group followed by selective cleavage of the modified N-terminal residue. Repetition of the reaction then provides a method for the stepwise degradation of the peptide and progressive elucidation of its sequence.

The Edman method involves the reaction between phenyl isothiocyanate and the peptide at approximately pH 8. Reaction occurs with the unprotonated α-amino group. The solution is kept at constant pH by the addition of a base from pH-stat. The resultant N-phenylthiocarbamoyl derivative is freed from an excess of phenylisothiocyanate by extraction with an organic solvent. The labelled peptide is then treated with anhydrous HCl in an organic solvent. These conditions do not hydrolyse the amide linkages; the labelled amino acid undergoes a cyclisation reaction giving rise to the formation of 3-phenyl-2-thiohydantoin and the peptide containing one residue less than the original. The hydantoin so produced can be identified by comparing it with standard amino acid derivatives. The degraded peptide can be isolated and subjected to another cycle of the Edman degradation to identify the *N*-terminal amino acid. The reactions involved are given below:

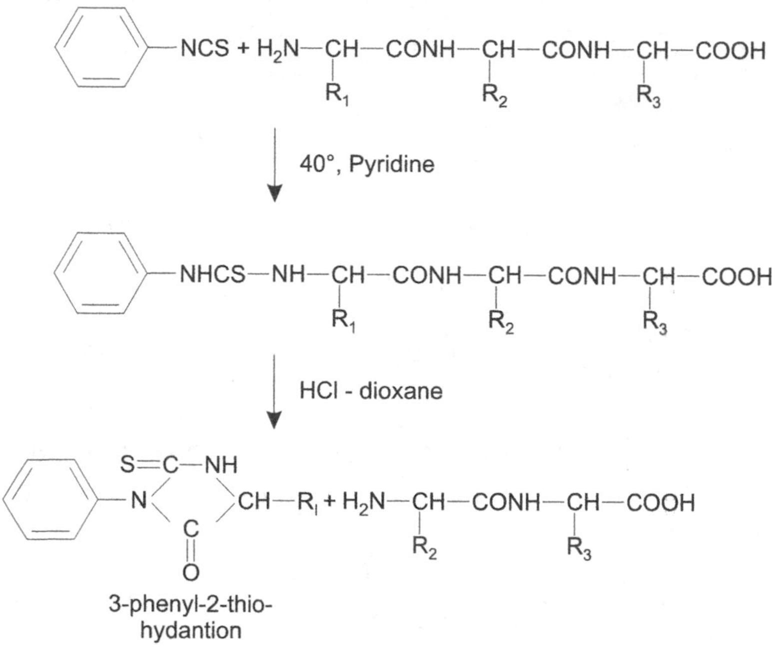

40°, Pyridine

HCl - dioxane

3-phenyl-2-thio-
hydantion

The process has been automated by Edman and Begg. They have described an apparatus which will carry out automatically most of the steps involved in the *N*-terminal sequence determination. The product from the *N*-terminal residue for each cycle is collected separately in a fraction collector and all are identified chromatographically at the end of the degradation. Fifteen cycles can be completed in 24 hours and the yields are about 98 per cent for each cycle. Edman was able to identify 60 amino acids in a protein containing 153 amino acids in its chain.

A number of other acid catalysts have been used for this degradation, but perhaps the best is trifluoroacetic acid. With this reagent cyclisation and degradation proceed rapidly at room temperature with little risk to the residual peptide. Use of dilute hydrochloric acid sometimes above room temperature, can cause hydrolysis of other peptide bonds so that subsequent steps liberate more than one 3-phenyl-2-thiohydantoin.

E. Condensation with Isocyanate

A recent method of stepwise degradation, which is related to the Edman procedure, involves carbamoylation of the terminal α-amino group with cyanate at pH-7 to 8. The reaction probably occurs between a cyanic acid molecule and the unprotonated α-amino group. The *N*-carbamoylpeptide on treatment with acid gives a hydantoin corresponding to the *N*-terminal residue and a peptide containing one residue less than the original. The hydantoin after isolation can be hydrolysed to the corresponding amino acid which is identified by chromatography.

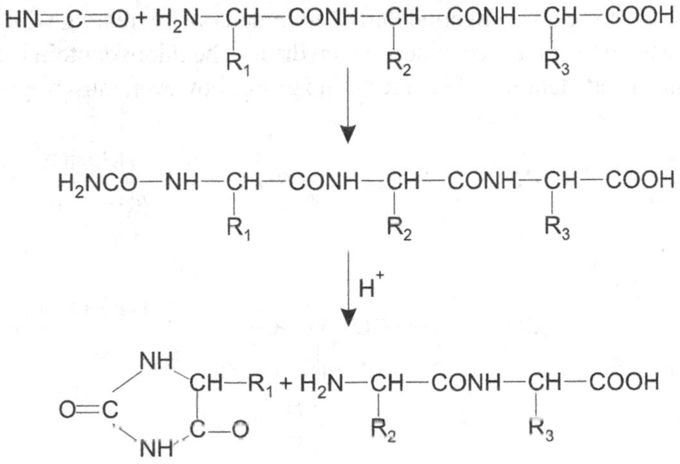

$$HN{=}C{=}O + H_2N{-}\underset{R_1}{CH}{-}CONH{-}\underset{R_2}{CH}{-}CONH{-}\underset{R_3}{CH}{-}COOH$$

$$\downarrow$$

$$H_2NCO{-}NH{-}\underset{R_1}{CH}{-}CONH{-}\underset{R_2}{CH}{-}CONH{-}\underset{R_3}{CH}{-}COOH$$

$$\downarrow H^+$$

$$O{=}C\underset{NH}{\overset{NH}{\diagup}}\underset{C{-}O}{\overset{CH{-}R_1}{\diagdown}} + H_2N{-}\underset{R_2}{CH}{-}CONH{-}\underset{R_3}{CH}{-}COOH$$

2.21   IDENTIFICATION OF THE C-TERMINAL AMINO ACID

Several chemical methods have been investigated for identification of C-terminal groups, but only a few of them have found extensive use.

A. Reaction with Benzylamine

Benzylamine has been used on the C-terminus in many cases. The peptide is converted to an ester and heated in a suspension of benzylamine to yield a benzylamide. On hydrolysis the C-terminus amino acid can be easily characterised as its benzylamide as given below:

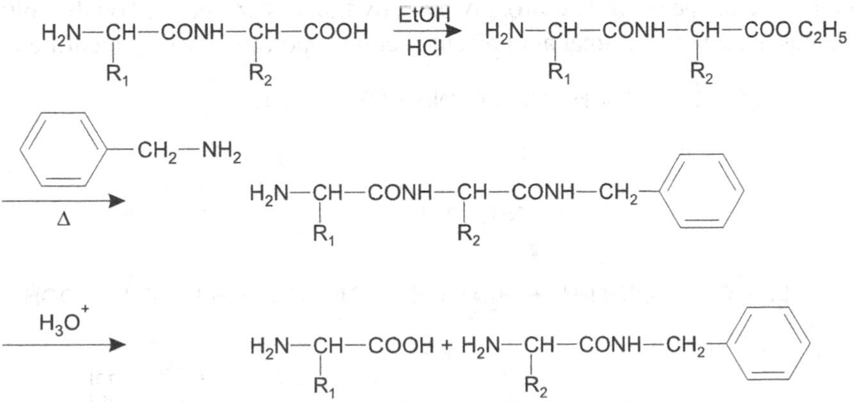

$$H_2N{-}\underset{R_1}{CH}{-}CONH{-}\underset{R_2}{CH}{-}COOH \xrightarrow[\text{HCl}]{\text{EtOH}} H_2N{-}\underset{R_1}{CH}{-}CONH{-}\underset{R_2}{CH}{-}COO\,C_2H_5$$

$$\xrightarrow{\Delta} H_2N{-}\underset{R_1}{CH}{-}CONH{-}\underset{R_2}{CH}{-}CONH{-}CH_2{-}\phi$$

$$\xrightarrow{H_3O^+} H_2N{-}\underset{R_1}{CH}{-}COOH + H_2N{-}\underset{R_2}{CH}{-}CONH{-}CH_2{-}\phi$$

Benzylamide of C-terminal
amino acid

B. Thiohydantoin Formation

This method is analogous to the Edman's method for *N*-terminus amino acids. The *C*-terminal amino acid is converted into a thiohydantoin by treating with ammonium isothiocyanate in the presence of acetic anhydride. The thiohydantoin is liberated by free cold alkali and identified. The free amino group, however, must be protected first.

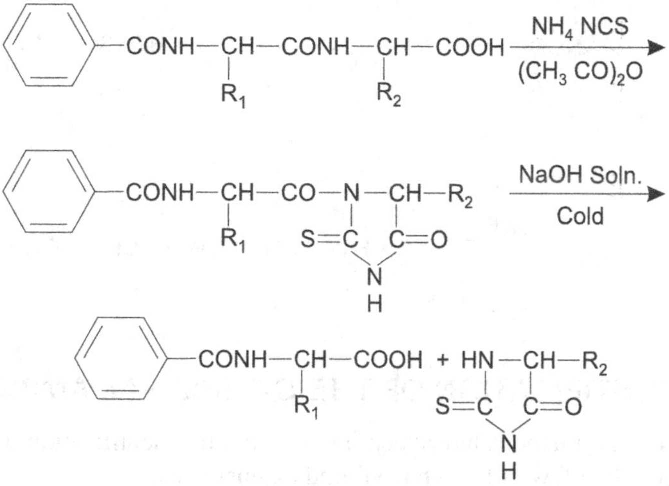

C. Reaction with Hydrazine (Hydrazinolysis)

This is the most widely used method for the identification of *C*-terminal groups. The peptide is heated with anhydrous hydrazine at 100°C. This cleaves the peptide bonds and forms the hydrazide of all the amino acids except the *C*-terminal residue. The mixture of products is subjected to column chromatography using a cation-exchange resin. The strongly basic hydrazides are retained on the column, whereas the free *C*-terminal amino acid is eluted and can be easily identified.

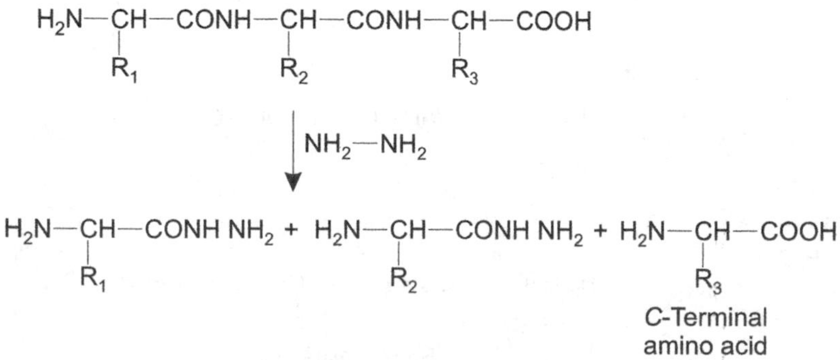

2.22   USE OF ENZYMES

The N-terminal residue may be determined enzymatically by treating the peptide chain or a protein with aminopeptidase. Aminopeptidases hydrolyse specifically the N-terminal amino acids from peptides and proteins. For example, the tripeptide,

$$H_2N—Gly—Val—Cys—COOH$$

on attack by an aminopeptidase will liberate Gly first. So, glycine would be the N-terminal amino acid. The process can go further and valine is released next. Use of aminopeptidases is of considerable importance for determining the N-terminal amino acids of short peptides. Under normal conditions the first three or four amino acids can be identified in this way.

The aminopeptidase from kidney has been thoroughly investigated. It preferentially cleaves amino acids with long alkyl side chains (*e.g.*, leucine); aromatic amino acids are cleaved rather slowly and peptides terminating in glycine, serine and proline are rather resistant. Another type of aminopeptidase has been discovered recently which is almost specific for cleavage of N-terminal lysine and arginine.

The C-terminal amino acid may be identified enzymatically by treating the peptide with carboxy-peptidase, which specifically catalyses the hydrolysis of the peptides from the C-terminus only. The reaction is shown below:

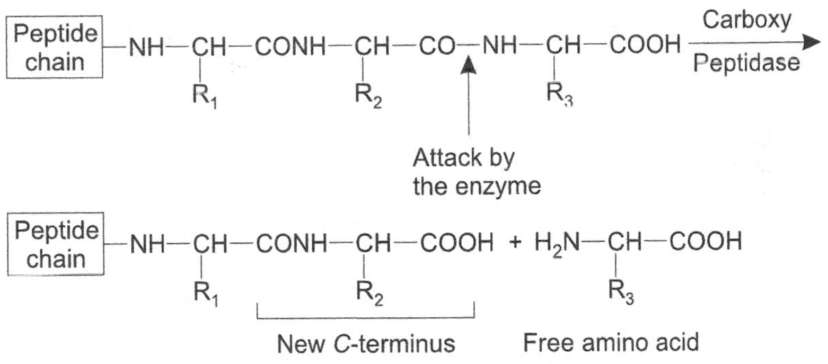

Thus, when the peptide or protein is treated with carboxypeptidase, the first amino acid to appear in solution is the one that occupies the C-terminal position. However, after the release of the first amino acid the enzyme continues to function and attack on the next residue and so on. Therefore, under favourable conditions the first three or four amino acid residues can be identified.

The use of two pancreatic enzymes, carboxypeptidases A and B, has been quite successful. These enzymes differ in their reactivity. Carboxypeptidase A has a fairly broad specificity although it preferentially liberates aromatic amino acids. C-terminal lysine, arginine and proline are not hydrolysed. Carboxypeptidase B is, however, much more selective and liberates only the basic amino acids, lysine and arginine when these are C-terminal.

2.23   TIME COURSE ANALYSIS

The sequence of amino acids is determined by following the increase in concentration of the amino acids liberated at regular time intervals after treating the peptide with aminopeptidase or carboxypeptidase. This method is called **Time Course Analysis**. For example, when the peptide is incubated with carboxy-peptidase the C-terminal amino acid would be the first to be released. If we plot a graph between the concentration of the amino acid liberated with time, we find that concentration of the C-terminus amino acid will be more than the next amino acid at any interval of time, *i.e.*, the C-terminus amino acid will be liberated first. For example, the time-course analysis for the tripeptide,

$$H_2N—Gly—Lys—Phe—COOH$$

with carboxypeptidase gives the following information

$$Phe > Lys = Gly$$

The tripeptide on treatment with carboxypeptidase liberates first phenylalanine indicating that it is the C-terminal amino acid. Simultaneously, the dipeptide $H_2N—Gly—Lys—COOH$ is generated. As time goes on, the dipeptide is hydrolysed by carboxypeptidase giving equal amounts of lysine and glycine. Therefore, concentration of Phe will always be greater than Lys at any interval of time (Fig. 2.11).

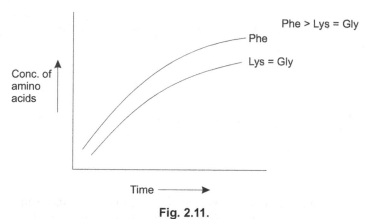

Fig. 2.11.

Thus, conclusion of the time-course analysis of the above tripeptide with carboxy-peptidase is that phenylalanine is the C-terminal amino acid. However, it does not give us any information about the sequence of the other two amino acid residues.

Similarly, time-course analysis of the same peptide with aminopeptidase gives us the following results:

$$Gly > Lys = Phe$$

Glycine is the first amino acid to be released. Its concentration will be more than lysine at any interval of time. It concludes that glycine is the N-terminal amino acid.

(Fig. 2.12) Again, the time-course analysis with aminopeptidase does not specify the position of the other two residues. However, if the results of the two time-course analyses are combined, we arrive at the complete sequence.

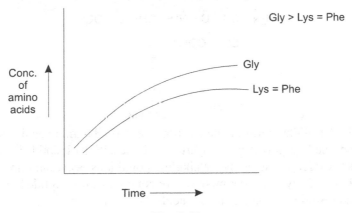

Gly > Lys = Phe

Fig. 2.12.

Results of the time-course analysis can be interpreted more clearly when the amino acids are released at the same rate. If the amino acids are released at different rates, we would have difficulty even in deciding which amino acid was the *C*-terminal amino acid.

For peptides having a number of amino acids the time-course analysis is useful for deciding the sequence of the two acids from both the *N*-terminus and *C*-terminus. For example, a heptapeptide gives the following results in the two time-course analyses:

1. Carboxypeptidase gives

Ala > Val

2. Aminopeptidase gives

Asp > Leu

From these results, we can interpret that alanine is the *C*-terminal amino acid, followed by valine and aspartic acid is the *N*-terminal residue followed by leucine. Thus, the sequence of two acids on each of the *N*-terminus and *C*-terminus is decided.

$$H_2N — Asp — Leu ----------- Val — Ala — COOH$$

N-Terminus   *C*-Terminus

2.24 SOME INTERESTING PEPTIDES

A. Aspartame, the Sweet Peptide

The synthetic sweetener, aspartame is the methyl ester of the dipeptide of the natural amino acids L-aspartic acid and L-phenylalanine.

Aspartame is of considerable commercial importance. It is about 200 times sweeter than sugar and is marketed as a sugar substitute under the trade name Nutrasweet.

$$\overset{+}{H_3N}-CH-\overset{\overset{\displaystyle O}{\|}}{C}-NH-CH-\overset{\overset{\displaystyle O}{\|}}{C}-OCH_3$$

with CH_2-COO^- and CH_2 attached to the respective CH groups, and a phenyl ring below.

Aspartame

The ethyl ester of the same dipeptide is not sweet. If a D-amino acid is substituted for either of the L-amino acids of aspartame, the resulting dipeptide is bitter rather than sweet. Under strong acidic or alkaline conditions, aspartame may generate methanol by hydrolysis. Under more severe conditions, the peptide bonds are also hydrolysed resulting in the free amino acids.

Aspartame is an artificial sweetener without the high energy value of sugar. It has a calorific value of about 4 kilocalories per gram. However, the quantity of aspartame needed to produce a sweet taste is so small that its contribution to calorific value is negligible, which makes it a popular sweetener for those trying to avoid calories from sugar. Many people want to curtail their sugar intake in the interest of fighting obesity. Others must limit their sugar intake because of diabetes. One of the most common ways of doing so is by drinking diet soft drinks. The soft-drink industry is one of the largest markets for aspartame.

The taste of aspartame is not identical to that of sugar. The sweetness of aspartame has a slower onset and longer duration than that of sugar.

Like many other peptides, aspartame may hydrolyse into its constituent amino acids under conditions of elevated **temperature** or high pH. This makes aspartame undesirable as a baking sweetener because it is prone to degradation in products hosting a high pH as required for a long shelf-life. Most soft drinks have a pH between 3 and 5, where aspartame is reasonably stable. In products that may require a longer shelf-life such as syrups for fountain beverages, aspartame is sometimes blended with a more stable sweetner such as saccharin.

B. Glutathione

Glutathione is a tripeptide of glutamate, cysteine and glycine. It is found in nearly all cells of plants, animals and microorganisms. Glutathione contains an unusual peptide linkage between the amino group of crysteine and the carboxyl group of the glutamate side chain.

The structure of glutathione is γ-glutamylcysteinylglycine. The γ-carboxyl group, *i.e.,* the side chain carboxyl group of the glutamic acid, is involved in the peptide bond; the amino group of the cysteine is bonded to it. The carboxyl group of the cysteine is then bonded to the amino group of glycine. The carbonyl group of glycine forms the *C*-terminal end. Its structure is shown below: it is the reduced form of glutathione (GSH).

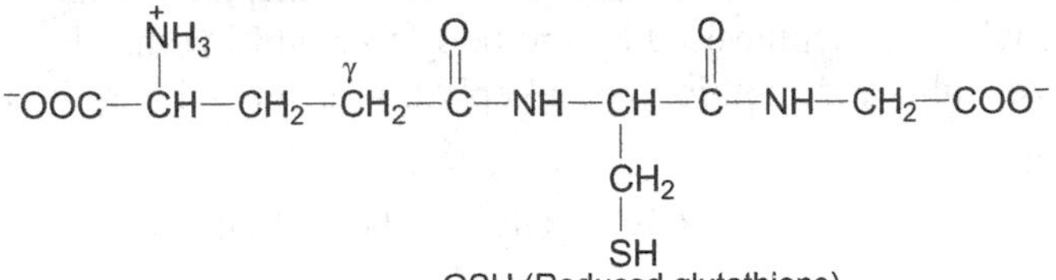

$$\overset{\overset{+}{N}H_3}{\underset{|}{^-OOC-CH-CH_2-\overset{\gamma}{CH_2}-\overset{\overset{O}{\parallel}}{C}-NH-CH-\overset{\overset{O}{\parallel}}{C}-NH-CH_2-COO^-}}$$

with CH₂—SH branch:

```
              +NH3                    O              O
               |                      ||             ||
    -OOC—CH—CH2—CH2—C—NH—CH—C—NH—CH2—COO-
                                      |
                                     CH2
                                      |
                                     SH
```

GSH (Reduced glutathione)

It is abbreviated as

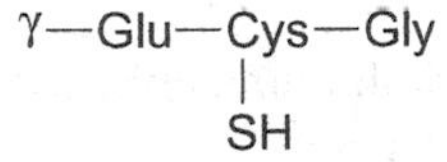

$$\gamma-\text{Glu}-\underset{\underset{\text{SH}}{|}}{\text{Cys}}-\text{Gly}$$

Glutathione acts as an antioxidant. It protects cells from toxins such as free-radicals. The most significant function of glutathione is to destroy harmful oxidising agents in the body. Oxidising agents are thought to be responsible for some of the effects of ageing and believed to play a role in cancer. Glutathione removes oxidising agents by reducing them: Consequently, glutathione is oxidised forming a disulphide bond between two glutathione molecules. An enzyme then reduces the disulphide bond, allowing glutathione to react with more oxidising agents. It is abbreviated as follows:

$$2\text{GSH} \underset{\text{Reduction}}{\overset{\text{Oxidation}}{\rightleftharpoons}} 2\text{GSSG}$$

The oxidised form of glutathione has the following structure:—

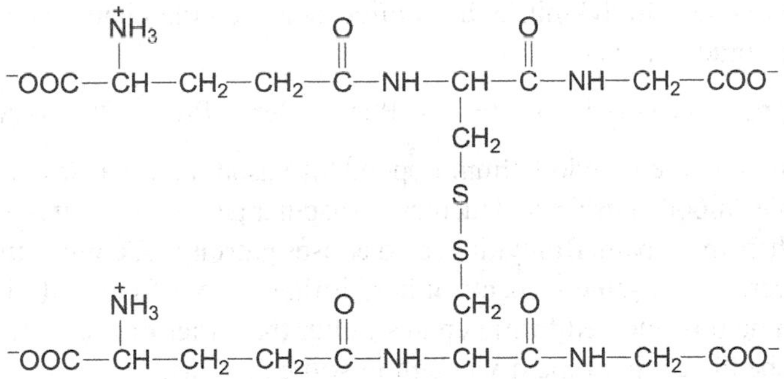

```
              +NH3                    O              O
               |                      ||             ||
    -OOC—CH—CH2—CH2—C—NH—CH—C—NH—CH2—COO-
                                      |
                                     CH2
                                      |
                                      S
                                      |
                                      S
                                      |
              +NH3                    O   CH2 O
               |                      ||   |  ||
    -OOC—CH—CH2—CH2—C—NH—CH—C—NH—CH2—COO-
```

GSSG (Oxidised glutathione)

Glutathione is found almost exclusively in its reduced form, since the enzyme that reverts it from its oxidised form, glutathione reductase, is quite active and inducible upon oxidative stress. In the reduced form, the thiol group of cysteine is able to donate a reducing equivalent ($H^+ +e$) to other unstable molecules such as reactive oxygen species. In donating an election, glutathione itself becomes reactive but readily reacts with another glutathione to form glutathione disulphide (GSSG). GSH can be regenerated from GSSG by the enzyme glutathione reductase.

In healthy cells and tissues more than 90% of the total glutathione is in the reduced form (GSH) and less than 10% exists in the disulphide form (GSSG). An increase in the GSSG-GSH ratio is considered indicative of oxidative stress.

C. Enkephalins

Two pentapeptides found in the brain are known as enkephalins. They are naturally occurring analgesics, synthesised by the body to control pain. The two peptides, leucine enkephalin and methionine enkephalin, differ only in their C-terminal amino acids.

Tyr — Gly — Gly — Phe — Leu

Leucine enkephalin

Tyr — Gly — Gly — Phe — Met

Methionine enkephalin

It is thought that the aromatic chains of tyrosine and phenylalanine in these peptides play a role in their physiological activities. It is believed that there are similarities between the three dimensional structures of opiates, such as morphine and those of the enkephalins. They decrease the sensitivity to pain by binding to the receptors in certain brain cells.

D. Bradykinin

Bradykinin is a nonapeptide that causes blood vessels to enlarge and therefore, causes blood pressure to lower.

Bradykinin is a physiologically and pharmacologically active peptide of the kinin group of proteins. Bradykinin is a 9-amino acid peptide chain. The amino acid sequence of bradykinin is

Arg — Pro —- Pro — Gly — Phe — Ser— Pro — Phe — Arg

Bradykinin is a potent endothelium dependent vasodilator, causes contraction of non-vascular smooth muscle, and increases vascular permeability. It is also involved in the mechanism of pain. Bradykinin also causes natriuresis, contributing to a drop in blood pressure. In some respects, it has similar actions to that of histamine and like histamine it is released from venules rather than arterioles. Bradykinin is also thought to be the cause of the dry cough in some patients.

E. Oxytocin

Oxytocin is a mammalian harmone. It induces labour pain in pregnant women and stimulates milk production in nursing mothers.

Oxytocin is synthesised in hypothalamic neurons and is stored in the posterior pituitary. It is then released into the blood from the posterior lobe of the pituitary gland.

Oxytocin is a peptide of nine amino acids; *i.e.*, a nonapeptide. The sequence is:

cysteine - tyrosine - isoleucine - glutamine -

asparagine - cysteine - proline - leucine - glycine

The cysteine residues form a sulphur bridge. Thus, oxytocin has got a cyclic structure. The sulphide bond is responsible for this cyclic structure. Oxytocin has an amide group rather than a free —COOH group at the C-terminal end. The abbreviated structure is represented below:

Oxytocin

Oxytocin is best known for its role in female reproduction. It controls contraction of uterine muscle. During pregnancy, the number of receptors for oxytocin in the uterine wall increases. The number of receptors for oxytocin becomes great enough to cause contraction of the smooth muscle of the uterus in the presence of small amounts of oxytocin produced by the body toward the end of pregnancy. The fetus moves toward the cervix of the uterus because of the frequency of the uterine contractions. The cervix stretches sending nerve impulses to the hypothalamus. The result is that more oxytocin is released by the posterior pituitary gland. The presence of more oxytocin leads to stronger contractions of the uterus so that the fetus is forced through the cervix and the baby is born.

Oxytocin also plays a role in stimulating the flow of milk in a nursing mother. The process of suckling sends nerve signals to the hypothalamus of the mother's brain. Oxytocin is released and carried by the blood to the mammary glands to contract forcing out the milk that is in them. As suckling continues, more hormone is released producing still more milk.

Recent studies have begun to investigate oxytocin's role in various behaviours including orgasm, social recognition, pair bonding, anxiety, trust and love.

F. Vasopressin

Arginine vasopressin (AVP), also known as vasopressin, is a peptide hormone, found in most mammals including humans. It is derived from a prehormone precursor that is synthesised in the hypothalamus and stored in vesicles at the posterior pituitary. Most of it is stored in the posterior pituitary to be released into the blood stream. However, some of it is also released directly into the brain.

Vasopressin is also a nonapeptide. Its structure is similar to oxytocin. The difference between the two is that oxytocin has an isoleucine residue at position 3 and a leucine residue at position 8 and vasopressin has a phenylalanine residue at position 3 and an arginine residue at position 8. Its structure is given on the next page.

$$
\begin{array}{c}
\overset{1}{Cys}\!-\!\overset{2}{Tyr}\!-\!\overset{3}{Phe} \\
| \qquad\qquad | \; 4 \\
S \qquad\qquad Gin \\
| \qquad\qquad | \; 5 \\
S \;\; 6 \qquad Asn \\
\;\;\;\searrow Cys \qquad\qquad O \\
7| \qquad\qquad\qquad || \\
\overset{}{Pro}\!-\!\overset{8}{Arg}\!-\!\overset{9}{Gly}\!-\!C\!-\!NH_2
\end{array}
$$

Vasopressin

Vasopressin and oxytocin differ only by two amino acids but their physiological effects are very different. Both oxytocin and vasopressin are peptide hormones. They were isolated and synthesised by Vincent du Vigneaud in 1953, for which he received Nobel Prize in chemistry in 1955. Oxytocin and vasopressin are the only known hormones released by the human posterior pituitary gland to act at a distance; *i.e.,* in spite of the similarity of their amino acid sequences, the two peptides have quite different physiological effects. Oxytocin occurs only in the female of a species and stimulates uterine contractions during child-birth, Vasopressin occurs in males and females both. Vasopressin plays a role in the control of blood pressure by regulating contractions of smooth muscle. Its major function in the body is as an antidiuretic. Physiologists often refer to vasopressin as an antidiuretic hormone (ADH). Its role is to regulate the body's retention of water. It is released when the body is dehydrated and causes the kidneys to conserve water thus concentrating the urine and reducing urine volume. In high concentrations, it also raises blood pressure by inducing moderate vascoconstriction.

G. Gramicidin S and Tyrocidine A

Gramicidin S and Tyrocidine A are peptide antibiotics produced by the *bacterium* Bacillus. Both of these are cyclic decapeptides. The cyclic structure is formed by the peptide bonds themselves. They contain D-amino acids as well as the more usual L-amino acids. In addition, both contain the amino acid ornithine (Orn) which does not occur in proteins. However, it plays a role as a metabolic intermediate in several common pathways. Ornithine resembles lysine but has one fewer methylene group in its side chain. Its structure is given below:

$$
\overset{+}{N}H_3\!-\!CH_2\!-\!CH_2\!-\!CH_2\!-\!\underset{\underset{\overset{+}{N}H_3}{|}}{CH}\!-\!COO^-
$$

Orrithine

Gramicidin S was isolated in 1944. Its cyclic structure contains two identical pentapeptides *viz-;* D-Phe-L-Pro-L-Val-D-Orn-L-Leu. In gramicidin S these two pentapeptides are condensed head to tail with each other. Its structure is as follows:

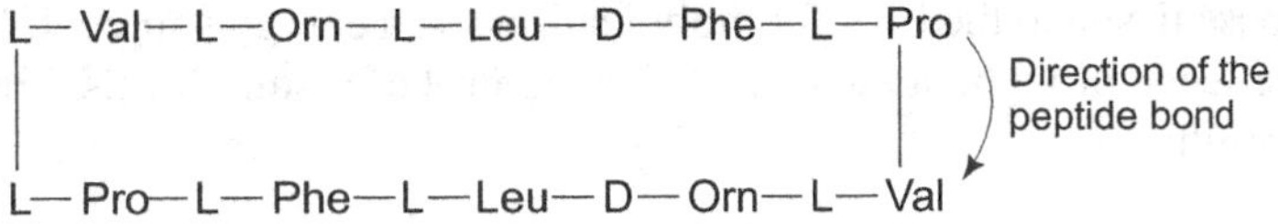

L—Val—L—Orn—L—Leu—D—Phe—L—Pro

Direction of the
peptide bond

L—Pro—L—Phe—L—Leu—D—Orn—L—Val

Gramicidin S

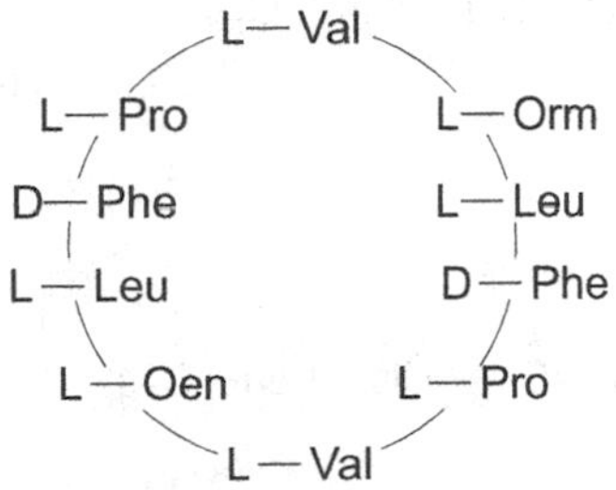

L—Val

L—Pro   L—Orm

D—Phe   L—Leu

L—Leu   D—Phe

L—Oen   L—Pro

L—Val

The structure of tyrocidine A is given below:

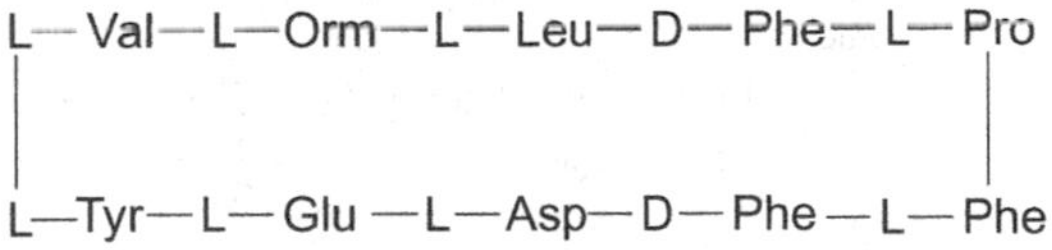

L—Val—L—Orm—L—Leu—D—Phe—L—Pro

L—Tyr—L—Glu—L—Asp—D—Phe—L—Phe

Tyrocidine A

Both gramicidin S and tyrocidine A exhibit broad spectrum antibiotic activities. Gramicidin is active against Gram-positive bacteria and against select gram-negative organisms. However, its use is limited due to its low bioactivity.

H. Insulin

Insulin is a peptide hormone secreted by the pancreas. It is produced in the islets of Langerhans (an endocrine gland) in the pancreas. It has extensive effects on metabolism and other body functions. It regulates glucose metabolism.

Insulin was the first protein whose amino acid sequence was determined by **F. Sanger** in 1953 after 10 years of work.

It has a total of 51 amino acid residues in two peptide chains called the A and B chains. Chain A has 21 amino acids and chain B has 30 amino acids. These chains are joined together by two disulphide linkages. The chain A contains an additional disulphide linkage between cysteine residues at positions 6 and 11 (Fig. 2.13).

Insulin is produced in the pancreas and released when any of several stimuli is detected. The stimuli include ingested protein and glucose in the blood produced from digested food.

Insulin causes cells in the liver, muscle and fat tissue to take up glucose from the blood, storing it as glycogen in the liver muscle and stopping use of fat as an energy source. When insulin is absent or low, glucose is not taken up by body cells and the body begins to use fat as an energy source, for example, by transfer of lipids from adipose tissue to the liver for mobilization as an energy source. It has several anabolic effects throughout the body. When control of insulin levels fails, *diabetes mellitus* results.

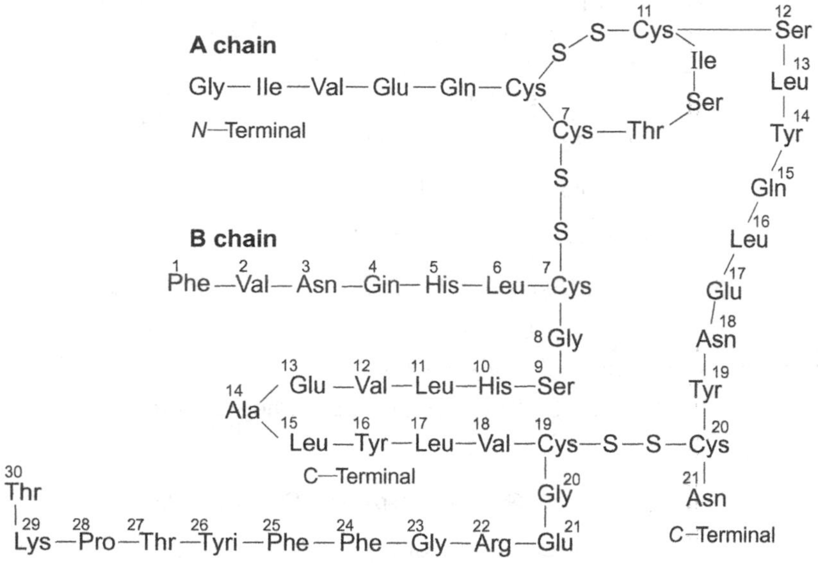

Fig. 2.13. The amino acid sequence of insulin.

Insulin is used medically to treat some forms of *diabetes mellitus*. Patients with Type-I *diabetes mellitus* depend on external insulin, (most commonly injected subcutaneously) for their survival because the hormone is no longer produced internally. Patients with Type-2 *diabetes mellitus* are insulin resistant and they may suffer from a relative insulin deficiency. Some patients with Type-2 diabetes may also require insulin when other medications fail to control blood glucose levels adequately.

Insulin is produced and stored in the body as a hexamer (a unit of six insulin molecules), while the active form is the monomer. The hexamer is an inactive form with long-term stability which serves as a way to keep the highly reactive insulin protected. The hexamer- monomer conversion is the basis of insulin formulations for injection. The hexamer is far more stable than the monomer, which is desirable for practical reasons, however the monomer is a much faster reacting drug because diffusion rate is inversely related to particle size.

2.25 IMPORTANCE AND BIOLOGICAL FUNCTIONS OF PROTEINS AND POLYPEPTIDES

Proteins are the primary constituents of all living matter as they are found in all living cells. They are the most abundant organic molecules in animals. Proteins perform several important biological functions. Some important functions are listed below:

1. Proteins serve as structural material

- One important function of the proteins is structural. We have already mentioned α-keratin, which is the important structural component of skin, hair, feathers, fur, horns and nails.

- Collagen is the material that forms the basis of the connective tissues—tendon, bone and cartilage.

- Fibroin is the silk of spider webs and of cocoons.

- Muscle tissue contains two important structural proteins—myosin is the stationary component and actin is the contractile component.

2. Proteins act as enzymes: These are biocatalysts which regulate almost all life processes, and catalyse the chemical reactions that occur in living organisms. We know these bring about digestion of food and carry out the biosynthesis of various cell components. We have already encountered carboxypeptidase, pepsin, trypsin and chymoptrypsin-enzymes that catalyse the hydrolysis of peptide chains.

3. A few proteins function as hormones: Insulin, secreted by the pancreas, is a polypeptide which regulates glucose metabolism. The pituitary gland secretes a growth hormone, which is a protein. β-Endorphin, a polypeptide with morphine like activity that appears to be a natural pain reliever.

4. Proteins are responsible for many physiological functions such as the transport and storage of oxygen in the body. As haemoglobins they transfer oxygen to remote corners and the protein, myoglobin, stores oxygen in the muscles. Proteins are also responsible for contraction of muscles. Cytochrome C and cytochrome P-450 also have a transport function. These compounds carry electrons in the oxidative phosphorylation cycle. Lipoproteins of the blood plasma transport lipids between the intestine, liver and adipose.

5. Protective proteins act as antibodies giving protection against disease: The blood proteins, thrombin and fibrinogen are resopnsible for blood clotting and thus prevent loss of blood at the time of injury. These materials form insoluble complexes with foreign substances that invade the blood-stream.

6. Proteins are also associated with hereditary characters: Nucleoproteins form chromatin material. During cell division, chromatin condenses to form chromosomes. These are carriers of hereditary characters.

7. Some proteins have other protective functions: Snake venoms and plant toxins protect their owners from other species. In addition to snake venom, the toxic proteins of cotton seed and caster bean are highly poisonous to vertebrates. Melittin is a 26-amino acid peptide that is responsible for the toxic property of bee venom. It has several biological functions including the ability to puncture the cell membrane of red blood cells. The primary structure of melittin is given here.

$H_2N - Gly - Ile - Gly - Ala - Val - Leu - Lys - Val - Leu - Thr - Thr$

$Gly - Leu - Pro - Ala - Leu - Ile - Ser - Trp - Ile - Lys - Arg - Lys - Arg - Gln - Gln - CONH_2$

8. Cell secretions like mucus are glycoproteins: These provide slippery texture and help in protection.

9. Spiders and silkworm secrete a thick solution of protein fibroin, which solidifies into a thread of high tensile strength and is used for weaving a web or cocoon. We get silk from the cocoon.

10. Proteins are stored as reserve food: Some proteins are used for storage. Ovalbumin is employed by nature as a food reservoir in egg white. Casein plays a similar role in mammalian milk. Similarly, proteins are stored as reserve food in various seeds, pulses, rice, maize and peas.

2.26 CLASSIFICATION OF PROTEINS

We have seen that the proteins are the most important, most varied and most abundant organic compounds in the body of all living organisms.They show such a wide range of structural and functional diversity that it is difficult to classify them on the basis of one single criterion. They may be classified according to their chemical composition or their shape and structure.

A. Classification of Proteins on the Basis of their Shape

On the basis of their shape, proteins are divided into two broad classes. These are

- Fibrous proteins
- Globular proteins

i. Fibrous Proteins

These are tough and usually insoluble in water and aqueous solutions of acids and bases. Fibrous proteins have long and thread-like structure and tend to lie side by side to form fibres. Fibrous proteins serve as the chief structural material of animal tissues.

Examples of simple fibrous proteins are:

1. Keratins: α-Keratins are found in outer layers of skin, in hair and wool. They consist of α-helices that are twisted together like the strands of a rope.

β-Keratins are found in hard tissues such as nails, horns, hooves and bird feathers. They have a β-pleated sheet structure.

2. Collagens: Collagen proteins are found in white fibrous connective tissue, which constitute tendons. These are also the main components of bone, teeth and cartilage.

3. Elastins: They are found in yellow elastin tissue like ligaments and blood vessels.

4. Fibroin: This protein is present in silk fibres.

ii. Globular Proteins

Globular proteins are soluble in water or aqueous solutions of acids, bases or salts. These solutions are colloidal in nature due to the large size of protein molecules. Globular proteins are folded into roughly spherical shapes.

Globular proteins serve a variety of functions which are related to maintenance and regulation of life processes. They do all such functions that require mobility and hence solubility. They usually function as enzymes, hormones or transport proteins. Examples of globular proteins are insulin, haemoglobin, ribonuclease and fibrinogen, which is converted into the insoluble fibrous protein, fibrin, which causes the clotting of blood.

B. Classification of Proteins on the Basis of Structure

Based on structure, proteins may be classified as simple, conjugated and derived.

i. Simple Proteins

Simple proteins on hydrolysis yield only amino acids. They do not contain any non-protein part attached to them. Most of the simple proteins are globular in nature. As mentioned above some fibrous proteins are also simple proteins.

Major groups of simple globular proteins are:

1. Albumins: These are soluble in water and aqueous solutions of acids, bases and salts. They can be precipitated by saturating the solution either with a neutral salt such as sodium sulphate in slightly acidic solution or with acidic salts such as ammonium sulphate. On heating, albumins get coagulated. Albumins are widely distributed in nature. Examples are egg white or ovalbumin, blood-serum albumin, soyabean albumin, gliadin (wheat protein), casein in milk, legumelin (protein of pulses), *etc*. These are stored as food reserves.

2. Globulins: These proteins are insoluble in water but readily dissolve in dilute neutral salts such as NaCl. They can be precipitated by half saturation with ammonium sulphate. These coagulate on heating and heat coagulation is enhanced by the addition of dilute acids. They are found both in animal and plant tissues. Examples are ovoglobulin (eggs), lactoglobulin (milk), serum globulin in blood plasma, myosin in muscles; these are animal globulins. Vegetable globulins include legumin (in peas), tuberin (potatoes), and edestin (wheat and hemp seeds).

3. Protamines: They are basic proteins, highly soluble in water and dilute aqueous solutions of acids, bases and ammonia. They form crystalline salts with mineral acids and are not coagulated by heat. These are the simplest of all naturally occurring proteins and have the lowest molecular weights. Protamines occur almost entirely in animals and are main components of the sperm cells of certain fishes.

4. Histones: They are also basic proteins of high molecular weight. They are soluble in water and dilute acids but not in ammonium hydroxide. They occur as part of nucleoproteins. Histones are not easily coagulated by heat.

5. Prolamines: These proteins are insoluble in water but are soluble in dilute aqueous solutions of acids and bases. They are found in plants only and are not coagulated by heat. Gliadin from wheat, zein from maize and hordein from barley are common examples of prolamins.

6. Glutelins: They are insoluble in water, alcohol and neutral salt solutions but are soluble in dilute aqueous solutions of acids and bases. They can be easily coagulated by heat. They are found exclusively in seeds of grains.

ii. Conjugated Proteins

These proteins are composed of a simple globular protein combined with some non-protein substance. This non-protein substance is known as a prosthetic group. Depending upon the nature of the prosthetic group, the conjugated proteins are divided into the following classes:

1. Glycoproteins: In glycoproteins, the simple proteins combine with carbohydrates. For example, mucin of saliva, heparin of bile juice, immunoglobulins of plasma, mucopolysaccharides of cartilage and tendon are glycoproteins.

2. Chromoproteins: In chromoproteins, simple proteins are combined with coloured pigments like flavins, cartotenoids, porphyrins, *etc*. Haemoglobin, myoglobin and cytochromes are some examples of chromoproteins.

3. Phosphoproteins: These proteins contain phosphoric acid as the prosthetic group. These are soluble in dilute alkalis and are precipitated by the addition of acids. Casein, the milk protein and ovovitellin of eggs are examples of phosphoproteins.

4. Metalloproteins: In these proteins, the protein molecule is bound to some metal ion like iron, copper or zinc. Siderophilin, an important plasma protein has a great affinity for iron.

5. Lipoproteins: In these, simple proteins are combined with lipids and have a variable composition. These proteins are present in the brain, blood plasma, milk and egg yolk *etc*.

6. Flavoproteins: These are enzymes. In these proteins, the protein moiety (apoenzyme) is permanently attached to flavin compound. Flavoproteins are enzymes of Kreb's cycle and participate in the electron transport system of the respiratory chain.

7. Nucleoproteins: In nucleoproteins, the protein molecules are combined with nucleic acid. These proteins are protamines or histones. The chromatin material of the nuclei is composed of nucleoproteins.

iii. Derived Proteins

These are the degradation products obtained from native proteins either by hydrolysis or by coagulation by heat. These are further classified according to the following types:

1. Coagulated Proteins: These are coagulated or denatured proteins obtained when ordinary proteins are heated. In these proteins, the molecular mass does not change; only the properties like solubility, precipitation, *etc.*, are altered, *e.g.*, coagulated egg white.

2. Meta Proteins: They are obtained by the hydrolysis of complex proteins by the action of digestive enzymes, acids or alkalis. The initial products are called proteoses. These are insoluble in water but soluble in dilute acids and alkalis. These can be salted out by half saturation with ammonium sulphate. Further hydrolysis gives peptones, which are soluble in water and are not coagulated by heat. They cannot be salted out by ammonium sulphate. Finally, hydrolysis gives polypeptides and simple peptides, which also behave like peptones.

2.27   MOLECULAR SHAPE OF PROTEINS

Proteins serve several important biological functions. They also serve as structural material. The structural proteins tend to be fibrous in nature. In these, the long polypeptide chains are lined up more or less parallel to each other and are bonded to one another by hydrogen bonds. A variety of structural forms may result depending on the actual three dimensional structure of the individual protein molecule and its interaction with other similar molecules, Thus, we get protective tissues such as hair, skin, nails and claws, which are known as α- and β-keratins. Other examples of simple fibrous proteins are collagens, and elastins. Collagens are found in connective tissues such as tendons and elastins are present in ligaments and blood vessels. Fibrous proteins are insoluble in water.

Proteins also have important roles as biological catalysts and regulators. They are responsible for increasing and regulating the speed of biochemical reactions and for the transport of various materials throughout an organism. Both the catalytic proteins, which we call enzymes and the transport proteins tend to be *globular* in nature. In globular proteins, the polypeptide chain is folded around itself in such a way as to give the molecule a rounded shape. Each globular protein has its own characteristic geometry, which is due to the interactions between different sites on the chain. The intrachain interaction may be of four types– disulphide bridging, hydrogen bonding, dipolar interactions or van der Waals attraction.

Each molecule of a globular protein may consist of a single long polypeptide chain twisted about and folded back upon itself or the molecule is composed of several sub-units. Each sub-unit is a single polypeptide chain that has its own unique three dimensional structure. The sub-units are bonded together by secondary forces like hydrogen bonding and van der Waals attraction. Thus globular proteins have specific shapes. In certain cases the surface of the three dimensional structure of the molecule contains a high percentage of amino acids having polar groups. In such a case, the globular protein is water soluble and exists in the cytoplasm or some other aqueous environment. If the surfaces of the globular proteins are covered with amino acids having non-polar side chains, such globular proteins exist embedded in intercellular membrane structures.

Globular proteins often carry a non-protein molecule as a part of their structure. Such a molecule is called a prosthetic group. The prosthetic group may be covalently bonded to the polypeptide chain or it may be held by other forces. They help globular proteins in carrying out various biological functions.

2.28   FACTORS THAT INFLUENCE MOLECULAR SHAPE

The linear amino acid sequence of a protein or a peptide is referred to as its **primary structure.** The backbone of the protein chain is the repeating unit

$$-NH-\underset{\underset{R}{|}}{CH}-\overset{\overset{O}{||}}{C}-$$

Thus, rigidity is imparted to the protein chain by the restricted rotation about the amide bond. However, the three-dimensional structure of the protein molecule is determined by two other factors, inter- or intrachain bonding and van der Waals interactions.

There are two kinds of secondary bonds that may exist between two different polypeptide chains or between different regions of the same chain. The first type is the disulphide bridges between cysteine units. If they are formed between cysteine units in separate chains, they result in cross-linking of the two chains. For example, in insulin, A and B chains are bonded together by two disulphide links. When the two cysteine units are in the same chain, disulphide bridging results in the formation of loops in the chain as we see in oxytocin.

Hydrogen bonding is a second type of secondary bonding that may occur between two different chains or between different regions of the same chain. Although hydrogen bonds are weak a polypeptide chain contains many >C=O and —NH groups. The total amount of bonding is substantial because it results from many small interactions. Thus, hydrogen bonding plays an important role in the actual shape or conformation of the molecule. Hydrogen bonding between different chains results in binding them together. Intrachain hydrogen bonding causes the chain to fold back on itself in some specific fashion.

The combined effect of disulphide bridges and hydrogen bonds is to give the protein a specific conformation that we refer as its **secondary structure.**

Another important factor that determines the final molecular shape of a protein is the polar or non-polar nature of the "side-chain" groups of the amino acids that make up the protein molecule. Some of the side-chain groups that project from the polypeptide backbone are nonpolar or hydrophobic. In globular proteins, these nonpolar groups are found to be equally distributed between the interior and the surface of the molecule.

If the side chains are polar, they can form hydrogen bonds with water molecules. Since the globular proteins exist mainly in aqueous solutions, the polar side chains are found mainly on the outer surface of the molecule.

The combined effect of the small interactions of the various amino acid side chains with each other and with the external environment result in giving a preferred conformation of the molecule. The molecular structure that is formed by these interactions is referred to as the *tertiary structure* of the protein molecule.

2.29   SOME FEATURES OF FIBROUS PROTEINS

The most important type of conformation found in fibrous proteins is the α-helix. In this structure, the polypeptide chain coils about itself in a spiral manner. The helix is held together by intrachain hydrogen bonding. The α-helix is right-handed and has a pitch of 5.4 Å or 3.6 amino acid units. A right-handed α-helix can form from either D- or L-amino acids but not from DL-amino acids. The right-handed conformation is more stable with the natural L-amino acids. However, not all polypeptide chains can form a stable α-helix. The stability of the α-helix depends on the nature of the side-chain groups and their sequence along the chain. For example, polyalanine where the side chains are small and uncharged forms a stable α-helix. However, polyserine does not form a stable α-helix.

pH plays an important role in stability of the α-helix arrangement. At pH 7 the terminal amino groups in the lysine side chains are all protonated. Electrostatic repulsion between the neighbouring ammonium groups disrupts the regular coil and forces poly-lysine to adopt a random coil conformation. However, at pH 12, the lysine groups are uncharged and the material spontaneously adopts the α-helical structure. In a similar way polyglutamic acid exists as a random coil at pH 7 and as an α-helix at pH 2 where the terminal carboxy groups are uncharged.

The case of proline is quite interesting. The α-amino group in proline is a part of a five-membered ring. Rotation about the carbon-nitrogen bond is impossible. The amide nitrogen bond in polyproline has no hydrogens and intrachain hydrogen bonding is not possible. Wherever proline occurs in a polypeptide chain, the α-helix is disrupted and a kink or bend results.

In some cases, such as keratins of hair and wool, several α helices coil about one another to produce a super helix. In other cases, the helices are lined up parallel to one another and are held together by intercoil hydrogen bonding.

Another type of conformation found in the fibrous proteins is the β- or pleated sheet structure of β-keratin as is present in silk. In this structure, the polypeptide chains are extended in a linear or zigzag arrangement. Neighbouring chains are bonded together by interchain hydrogen bonding. The result is a structure resembling a pleated sheet. Such a structure results in the side-chain groups being fairly close together.

Therefore, side chains that are bulky or have like charges disrupt the arrangement. In the β-keratin of silk fibroin 86% of the amino acid residues are glycine, alanine and serine, all of which have small side chains.

2.30   SOME FEATURES OF GLOBULAR PROTEINS

Globular proteins perform various metabolic functions. They are either soluble in the aqueous body fluids or present in the intercellular membrane structure. They have a unique structure that creates an active site where the catalytic or transport function of the protein is carried out. The various forces interact to give the specific coiling that produces the proper geometry of the protein molecule. Some folding is stabilised by disulphide bridging. In globular proteins, the non-polar side chains lie inside the bulk of the structure where they attract each other by van der Waals forces. The polar side chains tend to be on the surface of the molecule where they can form hydrogen bonds with the solvent molecules, and thus, they become water soluble. Further coiling of the structure results from intrachain hydrogen bonds between the amide linkages inside the bulk of the molecule.

Some segments of the polypeptide chain may have the typical α-helical structure and others may be random coil. In some other cases the chains may fold back on themselves in the β-or pleated sheet fashion.

If the protein contains a prosthetic group, that group will be embedded within the three-dimensional structure of the protein. This will be either covalently bonded to the polypeptide chain or simply held by secondary forces. For example, the prosthetic group heme is found both in haemoglobin and myoglobin. In both these proteins, the function of the prosthetic group is to bind an oxygen molecule.

Both of these are oxygen carriers—myoglobin in muscle and haemoglobin in the blood stream. In both cases, the polypeptide chain folds in such a way as to leave a hydrophobic site into which the heme just fits.

The stereo representation of the myoglobin shows only the backbone of the polypeptide chain and the heme. There are several α-helical regions in the chain.

The delicate three-dimensional structure of globular proteins may be disrupted under active conditions. This process is called denaturation. Denaturation commonly occurs when this protein is subjected to extremes in temperature or pH. Denaturation results in dramatic decrease in the water solubility of the protein. An example is the hardening of the white .and yolk of an egg upon heating. In many cases denaturation is a reversible process. The reverse process is called renaturation. Many cases are now known in which a soluble denatured protein reverts to its natural geometry when the pH and temperature are adjusted back to the point where the natural protein is stable. Thus, the three-dimensional structure of a protein seems to be a natural consequence of its primary structure. The unique conformation of each protein is simply the most stable structure that molecule can have under biological conditions.

2.31 STRUCTURE OF PROTEINS

A. Primary Structure

We have already discussed the primary structure of proteins. The primary structure of a protein refers to the linear sequence of amino acids in its polypeptide chain and location of disulphide bridges, if any. It represents the number, nature and sequence of amino acid molecules in the polypeptide chain. All the properties of a protein molecule are determined, directly or indirectly, by the primary structure. Any folding or catalytic activity of the protein molecule depends on the proper primary structure (Fig. 2.14).

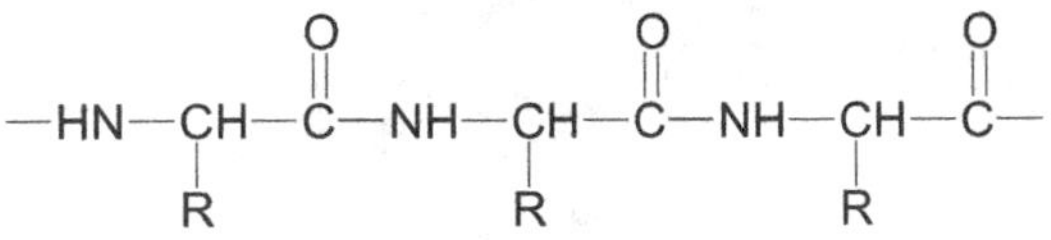

$$-HN-\underset{\underset{R}{|}}{CH}-\underset{\overset{\displaystyle O}{\|}}{C}-NH-\underset{\underset{R}{|}}{CH}-\underset{\overset{\displaystyle O}{\|}}{C}-NH-\underset{\underset{R}{|}}{CH}-\underset{\overset{\displaystyle O}{\|}}{C}-$$

Fig. 2.14.   A polypeptide chain showing primary structure.

B. Secondary Structure

The folding of linear polypeptide chains into a specific arrangement represents the secondary structure of a protein molecule. It results by hydrogen bonding between the carboxyl oxygen and amide hydrogen atoms of the component amino acids of the peptide chain. These hydrogen bonded arrangements, if present, are called the secondary structure of the protein. These bonds can occur either within the molecules of one polypeptide chain forming an α-helix, or between different polypeptide chains of protein to form a pleated sheet.

1. The α-Helix: The α-helix structure of proteins was first proposed by Linus Pauling and Robert Corey using X-ray analysis of proteins. They were able to understand the secondary structure of proteins for the first time. In the α-helix (a helix that looks like the thread on a right handed screw) conformation, the polypeptide chain assumes the form of a spiral staircase with 3.6 amino acid residues per turn (Fig. 2.15). Although an α-helix may turn in either direction, the α-helices of almost all natural proteins are right handed which means that they clockwise as they move away from a viewer at either end. The coils of the helix are held together by hydrogen bonds lying parallel to the main axis and all side chains

extend outward from the axis of the spiral. Such a structure is flexible and elastic but it gives stability to the peptide chain. The stability of this structure is attributed to a series of regularly spaced hydrogen bonds and van der Waals forces between atoms in the main chains (Fig. 2.16).

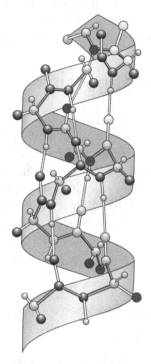

Fig. 2.15.   A representation of the α-helical structure of a polypeptide.

The α-helical structure is found in many proteins. It is the predominant structure of the polypeptide chains of fibrous proteins such as myosin, the protein of muscle, and of α-keratin, the protein of hair, unstretched wool and nails.

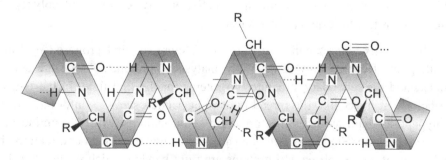

Fig. 2.16.   The α-helical arrangement.

2. The β-Pleated Sheet: The second type of secondary structure is the β-pleated sheet. In a β-pleated sheet structure hydrogen bonding occurs between two or more peptide chains and these bonds are at right angles to the main chain. In this arrangement, each carboxyl oxygen on one chain forms a hydrogen bond with an N-H hydrogen on an adjacent chain. The adjacent hydrogen-bonded peptide chains can run in the same direction or in the opposite direction. In a parallel β-pleated sheet the adjacent chains run in the same direction, whereas in an antiparallel β-pleated sheet, the adjacent chains run in the opposite direction (Fig. 2.17).

As the substituents (R) on the α-carbons of the amino acids on adjacent chains are close to each other, we can have maximum hydrogen-bonding interactions only if the substituents are small.

For example, silk has large segments of β-pleated sheets because it has a large number of small amino acids.

β-keratin found in feathers and claws is another example of this type of structure. Fibres containing β-pleated structures are not elastic because the polypeptide chains are fully extended. The number of side-by-side strands in a β-pleated sheet ranges from 2 to 15 in a globular protein.

A protein may or may not have the same secondary structure throughout the length. Some parts may be curled into an α-helix, while other parts are lined up in a pleated sheet. Parts of the chain may have no secondary structure at all. Such a structureless part is called a random coil. Most globular proteins contain segments of α-helix, or a β-pleated sheet separated by kinks of random coil allowing the molecule to fold into its globular shape.

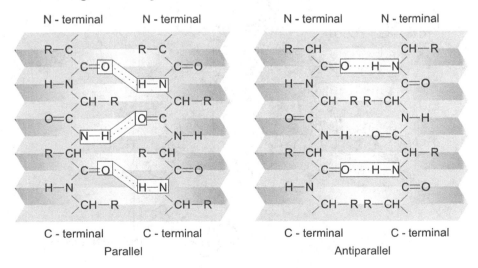

Fig. 2.17. The β-pleated structure.

C. Tertiary Structure

The tertiary structure of a protein is the final three-dimensional shape that arises due to further foldings of its polypeptide chains. These foldings do not occur randomly. They occur in a particular way, a way that is characteristic of a particular protein. Proteins fold in solution in order to maximize their stability. The stabilizing interactions that occur in folding are covalent bonds, hydrogen bonds, electrostatic attractions and van der Waals or hydrophobic interactions. So, a protein tends to fold in a way that maximizes the number of stabilizing interactions.

Disulphide bonds are the only covalent bonds that can form when a protein molecule folds. Other interactions that occur in folding are much weaker but there are so many of them that they become the most important interactions in determining how a protein folds. An important feature of most proteins is that the folding takes place in such a way as to expose the maximum number of polar hydrophilic groups to the aqueous environment and enclose a maximum number of non-polar hydrophobic groups within its interior.

The soluble globular proteins are highly folded than fibrous proteins. However, the fibrous proteins also have a tertiary structure. For example, α-keratin and fibrinogen are fibrous proteins and have tertiary structure. Albumin, myoglobin and plasma globulins are globular proteins possessing tertiary structure (Fig. 2.18).

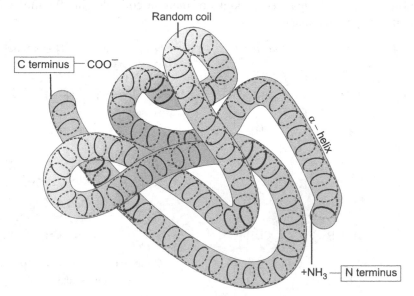

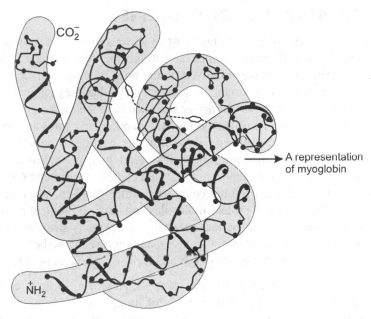

CO_2^-

A representation of myoglobin

$\overset{+}{N}H_2$

Fig. 2.18.   Tertiary structure of a protein.

D. Quaternary Structure

The quaternary structure involves the non-covalent association of two or more peptide chains in the protein molecule. Proteins that have more than one peptide chain are called oligomers. The individual chains are called sub-units. In a quaternary structure, these subunits are held together by the same kinds of interactions that hold the individual protein chains in a particular three dimensional conformation. They are hydrophobic interactions, hydrogen bonding and electrostatic interactions. The quaternary structure of a protein describes the way the sub-units are arranged in space. For example, haemoglobin consists of four peptide chains fitted together to form a globular protein (Fig. 2.19).

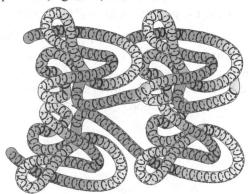

Fig. 2.19.   A representation of quaternary structure.

2.32 DENATURATION OF PROTEINS

We have seen earlier that proteins take part in various biological reactions. For a protein to be biologically active, it must have the correct structure at all levels. The sequence of amino acids in the primary structure must be right. The secondary and tertiary structures are very important. Proteins act only when they retain their natural conformations, *e.g.*, in enzymes the active site must have the right conformation. In primary structure, we have covalent bonds. But all other structures are maintained by weak solvation, non-covalent interactions or hydrogen bonding. Protein molecules may undergo disruption of their overall structure when subjected to heat or small changes in the environment. This change in structure (chemical or conformational) is called denaturation of protein molecules. Secondary and tertiary structures are disrupted and protein molecules lose their biological activity. The process of denaturation, however, does not change the primary structure of proteins. The amino acid sequence remains the same. Denaturation can also be brought about by strong acids or bases, X-rays, UV light, high pressure and heavy metals, etc.

Denaturation is generally an irreversible process. However, in some cases it can be made a reversible process. The cooking of an egg is an example of protein denaturation at high temperature. Egg white contains soluble globular proteins called albumins. When egg white is heated, the albumins unfold and coagulate to produce a solid mass. When cooked egg white is cooled it does not become uncooked. This is irreversible denaturation. We cannot get back the original egg white.

Similarly, when a protein is subjected to an acidic pH, some of the side-chain carboxyl groups are protonated and lose their ionic charge. Conformational changes result leading to denaturation. In basic solutions, amino groups become deprotonated losing their ionic charge which again causes conformation changes and denaturation. The process of curding is an example of irreversible denaturation. Milk turns sour because of bacterial conversion of carbohydrates to lactic acid. When pH increases, the soluble proteins in milk are denatured and precipitate. This process is called curding. However, some proteins are more resistant to acidic and basic conditions. For example, most digestive enzymes such as amylase and trypsin remain active under acidic conditions in the stomach even at pH of about 1.

Denaturation may be reversible, however, if the protein has undergone only mild denaturation. For example, a protein can be salted out of solution by a high salt concentration, which denatures and precipitates the protein. When the precipitated protein is redissolved in a solution with a lower salt concentration, it usually regains its activity and its natural conformation.

3

Enzymes

3.1 INTRODUCTION

We consume various food ingredients - carbohydrates, proteins and fats - any time we take our meals. These are digested and give us energy for performing the various metabolic functions in our body. The secret ingredient in living organisms is catalysis. The catalysts in living cells which facilitate the metabolic reactions are called **enzymes**. The miracle of life is that chemical reactions in the cell occur with great accuracy and at astonishing speed. Without the proper **enzymes** to process the food we eat, it might take us years to digest our breakfast or lunch.

With the exceptions of some RNAs that have catalytic activity, all enzymes are proteins or their derivatives and vary in molecular weight from 10,000 up to 5,00,000. Of all the functions of proteins, catalysis is probably the most important. In the absence of catalysis, most reactions in biological systems would take place too slowly to provide products at an adequate speed for a metabolising organism. A number of enzymes have been isolated and obtained in a crystalline form. Other enzymes are derivatives of proteins formed by combination with some other group such as a metal. Enzymes containing a wide range of metals including iron, copper, zinc, manganese and magnesium are known. In other cases, the protein is combined with a non-proteinous organic molecule. We shall study these under cofactors.

3.2 NOMENCLATURE AND CLASSIFICATION OF ENZYMES

The first enzymes to be discovered were named according to their source or mode of activity. Ptylin, which hydrolyses starches, is found in saliva (from the Greek word *ptyalon*, means 'spittle'). The enzyme pepsin is present in the digestive juices of the stomach (derived from the Greek, *pepsis*, means digestion). The first active enzymes were those present in yeast. Zymases (originated from the Greek *zyme* which means 'a leaven') are found in yeast. They catalyse the fermentation of glucose to ethyl alcohol and carbon dioxide. Many other enzymes have been given uninformative names *e.g.*, trypsin, chymotrypsin (present in pancreatic juices). Papain is present in the fruit papaya.

As more enzymes were discovered, a more systematic nomenclature developed. The trivial or common names of the enzymes are derived by adding the suffix–*ase* to the names of the substrate on which they act or the reaction they catalyse. A substrate is the molecule on which the enzyme exerts its catalytic action. For example, **urease** catalyses the hydrolysis of urea to ammonia and carbon dioxide, **arginase** catalyses the hydrolysis of arginine to ornithine and urea. The enzyme which catalyses the hydrolysis of the disaccharide, maltose, is named **maltase**. Similarly, sucrase acts on sucrose and **lipases** act on lipids. Hydrolases catalyse hydrolysis reactions and transferases facilitate the transfer of groups from one molecule to another.

However, the above nomenclature has not always been practical. As the number of newly discovered enzymes is increasing rapidly, a more systematic nomenclature has been developed by the International Commission on Enzymes which was set up by the International Union of Biochemistry (1961). The commission recommended two types of nomenclature, one systematic and the other working or trivial.

According to the new system of nomenclature, enzymes are divided into six major classes, which are further divided into subclasses according to the type of reaction catalysed.

Each enzyme is assigned a recommended name, usually short and appropriate for everyday use, a systematic name which identifies the reaction it catalyses and a classification number. The classification number is used where accurate identification of an enzyme is required. The coding system is mainly used by research workers and it is beyond the scope of this book.

A. Systematic Names

The systematic names of the enzyme consist of two parts. The first part contains the name of the substrate, while the second part indicates the nature of the reaction and it uses –ase as the suffix, for example, alcohol dehydrogenase in which alcohol is the substrate and it is oxidised by the enzyme. However, in the case of a bimolecular reaction the first part contains the names of both the substrates separated by a colon. For example, for the biomolecular reaction

$$ATP + Creatine \rightleftharpoons ADP + Phosphocreatine$$

the systematic name for the enzyme is ATP : Creatine phosphotransferase. The enzyme catalyses the transfer of a phosphate group from ATP to creatine. The recommended name of the enzyme, that normally used, is creatine kinase. Although, systematic names are quite logical, they are inconvenient to use. Therefore, the International Commission on Enzymes later decided to give more importance to the trivial names.

B. Major Classes of Enzymes

i. **Oxido-reductases:** They catalyse oxidation-reduction reactions. Transfer of electrons takes place from one molecule to another.

ii. **Transferases:** Transferases catalyse the transfer of functional groups from one substrate to another

$$AB + C \rightleftharpoons A + BC$$

iii. **Hydrolases:** They catalyse hydrolytic reactions, in which a substrate is hydrolysed into two simpler products. During hydrolysis, hydrogen atoms from water enter one of the products while the hydroxyl groups end up in the other product. The reaction is shown in a simplified form as follows:

$$AB + H_2O \rightleftharpoons AH + B.OH$$

iv. **Lyases:** They catalyse the reactions involving the removal of a group leaving a double bond or the addition of a group to a double bond. Decarboxylases which regulate the release of CO_2 from respiratory substrates are examples of lyase enzymes. Deaminases catalyse the release of ammonia from amino acids.

v. **Isomerases:** Isomerases control the conversion of one isomer of a compound to another isomer of the same compound, *e.g.,*

$$ABC \rightleftharpoons ACB$$

The interconversion of sugar isomers in glycolysis is catalysed by isomerase enzymes *e.g.,* glucose-6-phosphate is converted to fructose-6-phosphate by hexosephosphate isomerase.

$$\text{Glucose-6-phosphate} \rightleftharpoons \text{Fructose-6-phosphate}$$

vi. **Ligases:** This group of enzymes catalyses reactions in which new chemical bonds are formed. ATP provides the energy to make the new chemical bonds. For example, DNA and RNA ligases control the synthesis of macromolecules such as nucleic acids.

Let us examine some of these in detail.

C. Oxido-reductases

These enzymes are of two kinds:

i. **Oxidases:** Carbohydrates, fats and proteins, the main sources of energy, do not get oxidised easily in the presence of atmospheric oxygen. But in the body they are readily oxidised. The enzymes which catalyse these reactions by making use of molecular oxygen are called oxidases. Oxidases catalyse the transfer of hydrogen to molecules of oxygen as shown below:

$$\underset{\text{Substrate}}{AH_2} + \frac{1}{2} O_2 \rightleftharpoons \underset{\substack{\text{Oxidised} \\ \text{substrate}}}{A} + H_2O$$

Cytochrome oxidase is one of the most important of this group. It catalyses the oxidation of reduced cytochrome.

$$\underset{\substack{\text{Reduced} \\ \text{cytochrome}}}{Cyt. \ H_2} + \frac{1}{2} O_2 \rightleftharpoons \text{Cytochrome} + H_2O$$

Catalase acts on H_2O_2, converting it to H_2O and O_2. It is present in most living cells and its function appears to be removal of H_2O_2 which is toxic to living cells.

Flavin enzymes have prosthetic groups, which are mono or dinucleotides containing riboflavin. This functions by accepting hydrogen atoms and is subsequently oxidised by molecular oxygen. The enzyme xanthine oxidase, found in liver and milk contains flavin adenine dinucleotide, iron and molybdenum. It catalyses the oxidation of hypoxanthine to xanthine and the latter to uric acid.

The **phenol oxidases** are copper containing proteins which catalyse the oxidation of phenol derivatives to quinones.

Ascorbic acid oxidase works with ascorbic acid as the coenzyme.

ii. **Dehydrogenases:** Dehydrogenases catalyse the oxidation of substrates by transferring hydrogen to coenzymes such NAD^+. These dehydrogenases are highly specific in their action. The overall process for dehydrogenation may be represented as follows:

$$AH_2 \; + \; 2NAD^+ \; \rightleftharpoons \; A \quad + \; 2NADH$$

substrate   coenzyme   oxidised   reduced
                            substrate   coenzyme

For example, alcohol dehydrogenase controls the rate at which ethanol is oxidised to ethanal.

$$CH_3CH_2OH + 2NAD^+ \; \rightleftharpoons \; CH_3CHO + 2NADH$$

Ethanol   Ethanal

In reactions catalysed by oxido-reductases substrates are oxidised whilst oxygen or coenzymes are reduced.

D. Hydrolases

The earliest enzymes to be discovered were all hydrolases. These are especially important because many of them are associated with the digestion of foods in our body. The food materials we take in are composed of large molecules. They must be broken down into simpler components so that they can be absorbed and utilised by the body. The breakdown of large molecules in the digestive tract is carried out by hydrolytic enzymes or hydrolases. They catalyse introduction of the elements of water at a specific bond of the substrate.

Hydrolases are also found universally in plant and animal cells and in microorganisms. Because the concentration of the water is so high *in vivo*, hydrolytic reactions are generally irreversible. In addition to the solvents, hydrolases have only a single substrate. They are easy to detect and study. For this reason we know more about them than about other enzymes.

Hydrolases have been further divided into the following groups :

(i) **Esterases – Cleaving Ester Linkages:** These enzymes catalyse hydrolysis of organic esters to form alcohol and acids. Simple esterases such as liver esterase catalyse the hydrolysis of esters from lower alcohols and fatty acids. **Lipases** catalyse the hydrolysis of fats *i.e.*, of triglycerides to form glycerol and fatty acids. **Pancreatic lipase** (steapsin) is responsible for the hydrolysis of most of the fat that we take in.

Phosphatases hydrolyse esters of phosphoric acid. This is a large and complex group of enzymes, some of which appear to be highly specific. These are important factors in carbohydrate and lipid metabolism.

(ii) **Carbohydrases:** These enzymes hydrolyse the glycosidic linkage of simple glycosides, oligosaccharides and polysaccharides. Those which hydrolyse starch and glycogen are called amylases. There are two amylases, α- and β-amylase. α-Amylase hydrolyses starch and glycogen to dextrins, β-amylase hydrolyses starch to maltose. These enzymes are widely distributed in plants. Ptyalin, a salivary amylase (found in saliva) and amylosin, the pancreatic amylase, are thought to be mixtures of α- and β-amylase.

The disaccharide-splitting enzymes maltase, sucrase and lactase are important enzymes of the digestive tract as they hydrolyse common disaccharides. Maltase hydrolyses α-glycosides and emulsin attacks on β-glycosides.

(iii) **Nucleases:** Nucleases cause hydrolysis of nucleic acids liberating smaller nucleotides. Two specific nucleases have been obtained in crystalline form. Ribonuclease hydrolyses nucleic acids of ribose type. Deoxyribonuclease causes depolymerisation of DNA.

Nucleosidases hydrolyse nucleosides to sugars and purines or pyrimidines. As nucleotidases remove phosphoric acid from nucleotides to give nucleosides, they are phosphatases and often classified as such.

(iv) **Amidases:** Amidases hydrolyse amides. Two amidases are quite important – arginase and urease. Arginase is found in the liver and is an important factor in the synthesis of urea. Urease, the enzyme which hydrolyses urea to carbon dioxide and ammonia, is of special interest to the biochemist because it is used in the estimation of urea in blood and urine.

(v) **Proteases:** The proteases hydrolyse proteins to simpler molecules. They do not completely hydrolyse proteins to amino acids. The enzymes of the digestive tract are obvious examples. Pepsin found in the gastric juice and trypsin and chymotrypsin present in the pancreatic juice initiate the digestion of protein foods in the body. They attack only at specific peptide bonds.

Rennin, found in the gastric juice of calves, is a proteolytic enzyme, which acts on casein of milk. It is used commercially in the manufacture of cheese.

(vi) **Plant Proteinases:** Papain is obtained from the unripe fruit of papaya. Bromelin is found in pineapples and ficin is present in the milky sap of the fig tree. Bromelin is of special interest to the housewife. In making fruit gelatin desserts, fresh pineapples cannot be used because bromelin present will hydrolyse the gelatin forming products that will not gel. Cooked pineapple may be used because cooking ultimately destroys the enzyme.

(vii) **Peptidases:** They act on peptide bonds adjacent to a free amino or carboxyl group. If the amino acid containing a free amino group is split off, the enzyme acting is known as aminopeptidase, whereas a carboxypeptidase splits the amino acid having a free carboxyl group.

E. Transferases

These enzymes catalyse the transfer of a group from one substance to another. We consider some examples :

(i) **Transglycosidase:** Phosphorylase is an example of this type of enzyme which can split the 1,4- glucosidic bond in starch with the help of phosporic acid

$$\text{Starch} + PO_4^{3-} \xrightarrow{\text{Phosphorylase}} \alpha\text{-D-glucose-1-phosphate}$$

(ii) **Transphosphorylase:** The formation of D-glucose-6-phosphate from D-glucose with the help of ATP is catalysed by the enzyme phosphotransferase. One phosphate unit is transferred from ATP to glucose and ADP is obtained.

$$\text{D-Glucose} + \text{ATP} \rightarrow \text{D-Glucose-6-phosphate} + \text{ADP}$$

This enzyme is commonly known as glucokinase.

Phosphoglucomutase catalyses the conversion of glucose-1-phosphate to glucose-6-phosphate

(iii) **Transacetylase:** Acetyl-CoA is formed from coenzyme A. The transacetylation is performed by the coenzyme molecule. Acetyl-Co A can transfer the acetyl group to choline forming acetyl-choline and CoA.

(iv) **Transaminases:** These enzymes transfer an amino group of an amino acid to an α-keto acid and promote the formation of a new amino acid. An example is the reaction between glutamic acid and oxaloacetic acid to produce α-keto glutaric acid and aspartic acid. The transaminases are important enzymes associated with protein metabolism.

(v) **Transmethylase:** With these enzymes the transfer of methyl group takes place from one compound to another.

3.3   CHARACTERISTICS OF ENZYMES

Certain characteristics of enzymes are different from inorganic catalysts. Some of these are :

A. Catalytic Power

Enzymes are the most efficient catalysts known. They can increase the rate of a reaction by a factor of up to 10^{20} over an uncatalysed reaction. However, non-enzymatic catalysts enhance the rate of reaction by factors of 10^2 to 10^6 only.

We know that catalysts are effective in small amounts. When analysing the catalytic power of an enzyme, we should know the amount of starting material or substrate that is converted to product in a unit time by a given quantity of the enzyme. This quantity is called the **turnover number** and is defined as the number of moles of substrate converted into product per minute by one mole of enzyme. The turnover numbers of enzymes vary greatly ranging from one hundred to over three million. The turnover numbers of some enzymes is given in Table 3.1.

TABLE 3.1. Turnover Numbers of Some Enzymes

Enzymes	Turnover Numbers
Catalase	6×10^7
Carbonic anhydrase	1.0×10^6
Chymotrypsin	1.9×10^2
Lysozyme	0.5
Lactatedehydrogenase	1.0×10^3

Catalysts ordinarily have no effect on the equilibrium of a reversible chemical reaction. They merely speed up the reaction until it reaches equilibrium. Enzymes act in a similar way; they hasten the process in either direction.

Catalysts are usually unchanged in the reaction: This property of ideal catalysis does not hold true for enzymes, as proteins are easily inactivated or denatured by high temperature or very alkaline or acidic conditions. So, enzymes work under optimum conditions of temperature and pH. We shall discuss this in detail at a later stage.

B. Control of Enzyme Action

Biochemical pathways usually involve a number of linked reactions. Each reaction is catalysed by a specific enzyme. For example,

$$A \xrightleftharpoons{\text{Enzyme X}} B \xrightleftharpoons{\text{Enzyme Y}} C \xrightleftharpoons{\text{Enzyme Z}} D$$

In this there are three specific enzymes for each of the three steps involved in the conversion of A to D. In some instances accumulation of one of the products formed near the end of the chain inhibits the action of an enzyme in one of the earlier reactions. In the example shown above, the presence of a high concentration of the product D may slow down the rate at which enzyme X converts A to B. This is called negative feedback inhibition. Negative feedback ensures that reactants are used efficiently and prevents the excess manufacture of end products. Thus, control is exercised on the pathway of the above reactions. Such a control of enzyme action helps to maintain a stable environment in living oganisms.

Pathways can be shut down if an organism has no immediate need for their products, which saves energy for the organism.

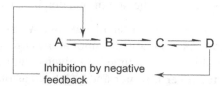

In glycolysis, three reactions are control points. The first is the conversion by glucose to glucose-6-phosphate, which is inhibited by glucose-6-phosphate itself. The second is the production of fructose-1, 6-bisphosphate, inhibited by ATP and lastly reaction of phosphoenolpyruvate to pyruvate is also inhibited by ATP.

It is frequently observed that control is exercised near the start and end of a pathway as well as at points involving the key intermediates as exemplified above.

C. Enzyme Specificity

Enzymes are highly specific in their reactivities. They catalyse only one reaction or a group of closely related reactions. An enzyme may be specific with regard to the type of reaction it catalyses or with regard to the kind of substrate on which it acts. That means some enzymes are reaction specific and others possess substrate specificity. For example, hydrolases carry out the hydrolysis reactions and oxidases catalyse only oxidations. A carbohydrase will attack only on carbohydrates and nucleases attack on nucleic acids. Maltase catalyses the hydrolysis of the disaccharide, maltose.

Urease, however, is a unique enzyme. It catalyses only the hydrolysis of urea. It has no effect on other compounds, even closely related ones such as amides. Such absolute specificity is rather rare among enzymes characterised to date

$$H_2N-\underset{\underset{O}{\|}}{C}-NH_2 + H_2O \xrightarrow{\text{Urease}} 2\,NH_3 + CO_2$$

Among reaction specific enzymes group specificity or substrate specificity is observed side by side. For example, α-glucosidases will hydrolyse all compounds containing an α-linked glucose. Invertase catalyses both hydrolysis of sucrose and of raffinose which have the same linkage, *i.e.*, the enzyme is β-fructofuranosidase. On the other hand lipases hydrolyse fat molecules irrespective of the nature of fatty acids present in the fat. Peptidases, however, are specific with regard to a particular linkage, *i.e.*, the peptide bond but they also require a special grouping in its vicinity. Carboxypeptidase, for example, cleaves only those terminal amino acids in a peptide chain having a free carboxyl group. Aminopeptidases on the other hand cleave the *N*-terminal amino acid *i.e.*, an amino acid having a free amino group in its vicinity.

Aminopeptidase   Carboxypeptidase

$$H_2N-\underset{\underset{R}{|}}{CH}-\underset{\underset{O}{\|}}{C}-NH \,-\text{- - - - - - - - -}\, \underset{\underset{O}{\|}}{C}-NH-\underset{\underset{R}{|}}{CH}-COOH$$

Substrate specificity of peptidases is given ahead.

Pepsin : It attacks at the amino positions of aromatic amino acids, phenylalanine, tyrosine and tryptophan. It does not attack ester bonds.

Chymotrypsin : It attacks on the carboxyl end of the above amino acids. It also hydrolyses esters.

Trypsin : It splits only those peptide bonds on the carboxyl sides of the basic amino acids, lysine and arginine.

Thrombin is even more specific. It cleaves only those bonds between arginine and glycine. It has no effect on the hydrolysis of other peptide linkages. It would not even attack the bonds between glycine and arginine.

In an oxidation-reduction reaction, we observe enzyme specificity with regard to both the reagents *i.e.*, both the substance oxidised and that reduced. Certain dehydrogenases will transfer hydrogen from substrates for which they are specific to coenzyme-I and not coenzyme-II. Some other dehydrogenases are specific to coenzyme-II and will not transfer hydrogen to coenzyme-I.

Enzyme specificity is very important in chemical reactions taking place in the cell. It ensures that proper reaction occurs in the proper place at the proper time. Enzyme specificity, therefore, plays a crucial role in the metabolism of the cell.

Enzymes permit reactions to take place in the living organisms which otherwise would not take place at an appreciable rate at ordinary temperature and may thereby exert a directive effect on the metabolism of the cell. For example, by accelerating certain pathways to a greater extent than others, they may virtually direct the mechanism along one of the possible routes.

D. Stereospecificity

Many enzymes show stereospecificity, *i.e.*, some enzymes act on only one of a pair of optical isomers. In addition some enzymes are involved in asymmetric syntheses producing only one of the two possible optical isomers. The following examples show stereospecificity among enzymes:

(i) Arginase catalyses the hydrolysis of L-arginine to ornithine and urea.

Arginase has, however, no effect on the rate of hydrolysis of D-arginine.

(ii) Pyruvic acid is converted to lactic acid by two different enzymes, each producing only one of the optical isomers.

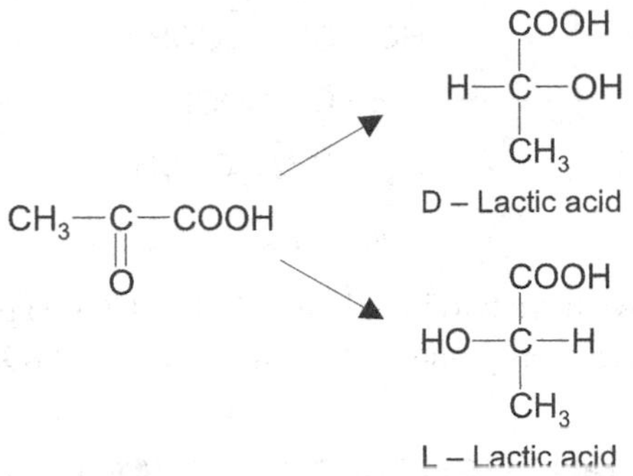

$$CH_3-\overset{\underset{\parallel}{O}}{C}-COOH$$

D – Lactic acid

$$\begin{array}{c} COOH \\ | \\ H-C-OH \\ | \\ CH_3 \end{array}$$

L – Lactic acid

$$\begin{array}{c} COOH \\ | \\ HO-C-H \\ | \\ CH_3 \end{array}$$

(iii) Fumarase catalyses addition of water to fumarate to give malate whereas maleate remains unaffected.

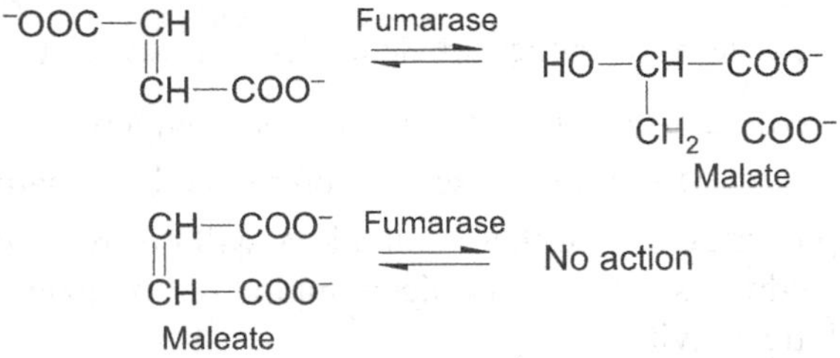

$$\begin{array}{c} ^-OOC-CH \\ \parallel \\ CH-COO^- \end{array} \quad \xrightarrow{\text{Fumarase}} \quad \begin{array}{c} HO-CH-COO^- \\ | \\ CH_2 \quad COO^- \end{array}$$

Malate

$$\begin{array}{c} CH-COO^- \\ \parallel \\ CH-COO^- \end{array} \quad \xrightarrow{\text{Fumarase}} \quad \text{No action}$$

Maleate

(iv) Glycerol is converted to L-phosphoglycerol by the enzyme glycerolkinase. Phosphate group comes from the ATP molecule.

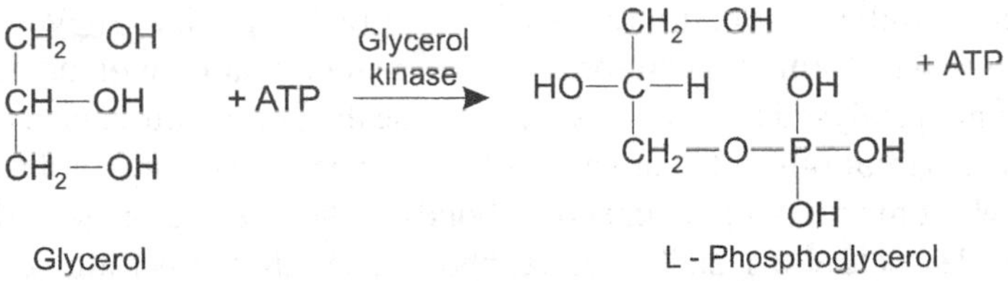

$$\begin{array}{c} CH_2 \quad OH \\ | \\ CH-OH \\ | \\ CH_2-OH \end{array} \quad + ATP \quad \xrightarrow[\text{kinase}]{\text{Glycerol}} \quad \begin{array}{c} CH_2-OH \\ | \\ HO-C-H \quad OH \\ | \quad \quad | \\ CH_2-O-P-OH \\ | \\ OH \end{array} \quad + ATP$$

Glycerol   L - Phosphoglycerol

(v) During citric acid metabolism, a series of enzymatic reactions occur in which we observe the selective action of an enzyme on only one of two chemically identical groups in a compound. For example, oxaloacetic acid is converted to citric acid by the condensing enzyme. If the carboxyl group adjacent to the —CH_2 group is labelled, the citric acid obtained will have a labelled position only in one of its terminal carboxyl groups.

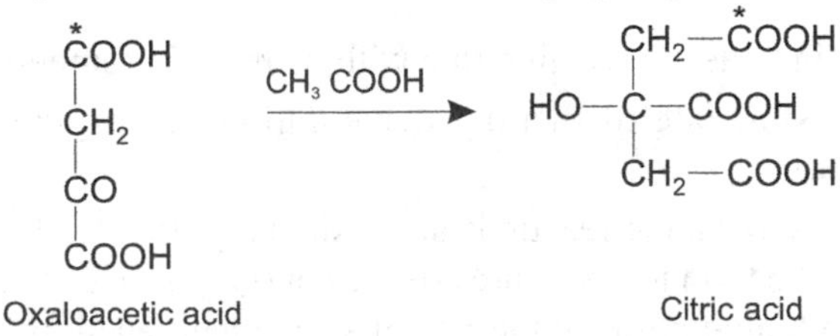

$$\begin{array}{c} \overset{*}{C}OOH \\ | \\ CH_2 \\ | \\ CO \\ | \\ COOH \end{array} \quad \xrightarrow{CH_3\,COOH} \quad \begin{array}{c} CH_2-\overset{*}{C}OOH \\ | \\ HO-C-COOH \\ | \\ CH_2-COOH \end{array}$$

Oxaloacetic acid   Citric acid

This citric acid can be converted to α-ketoglutaric acid (*via* isocitric acid) which contains C^{14} in the α-carboxyl group.

$$CH_2-\overset{*}{C}OOH$$
$$HO-\overset{|}{C}-COOH$$
$$\overset{|}{C}H_2-COOH$$

Citric acid

$$HO-CH-\overset{*}{C}OOH$$
$$H-\overset{|}{C}-COOH$$
$$\overset{|}{C}H_2-COOH$$

Isocitric acid

$$CO-\overset{*}{C}OOH$$
$$\overset{|}{C}H_2$$
$$\overset{|}{C}H_2-COOH$$

α-Ketoglutaric acid

In citric acid, we have two identical —CH_2—COOH groups, but the enzyme selectively attacks only one particular —CH_2—COOH group.

3.4 FACTORS INFLUENCING ENZYME ACTIVITY

A. Effect of Temperature on Enzyme Activity

The catalytic properties of enzymes are dependent on two features :

1. that the enzyme is able to form an intermediate complex with the substrate and
2. that the protein part of the molecule is preserved in its native state.

Factors which prevent either of these conditions will destroy catalysis. Thus, all substances or conditions which cause denaturation of an enzyme will inevitably destroy its catalytic activity.

We know that heat supplies kinetic energy to reacting molecules causing them to move more rapidly. The chances of molecular collision taking place are thus increased at higher temperatures, so it is more likely that enzyme-substrate complexes will be formed. Therefore, increase in temperature will increase the rate of an enzyme-catalysed reaction. As we increase the temperature, heat energy also increases the vibrations of the atoms which make up the enzyme molecules. If the vibrations become too violent, chemical bonds in the enzyme break and the three-dimensional structure is lost. Therefore, above a certain temperature denaturation will take place and the enzyme loses its activity.

Enzyme-catalysed reactions often appear to have **optimum temperature** at which the reaction proceeds most rapidly. If the temperature is raised beyond this point, the reaction rate decreases due to denaturation of the enzyme (Fig. 3.1). This temperature is not necessarily that at which the enzyme is most stable. It is the resultant of the two contrary processes:

1. the usual increase in reaction rate with increase in temperature and
2. the increasing rate of thermal denaturation of enzyme above a critical temperature.

The rate of most enzymatic reactions approximately doubles for every 10°C rise in temperature. The term temperature coefficient (Q_{10}) is used to express the effect of a 10°C rise in temperature on the rate of a chemical reaction.

$$Q_{10} = \frac{\text{rate of reaction at } (t+10)°C}{\text{rate of reaction at } t°C}$$

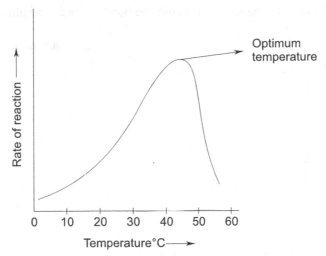

Fig. 3.1 Effect of temperature on enzyme catalysed reaction.

Optimum temperature for enzymes in the human body is 37°C; the Q_{10} for enzyme-catalysed reactions is 2.

Although most enzymes are inactivated at temperatures above 55°C, some are quite stable and retain activity at much higher temperatures; *e.g.*, enzymes of various species of bacteria inhabiting hot springs are active at temperatures exceeding 85°C. Some enzymes, such as ribonuclease lose activity on heating but quickly regain it on cooling indicating that their unfolded polypeptide chain quickly reverts back into its natural conformation.

B. Effect of pH on Enzyme Activity

The symbol pH refers to the concentration of hydrogen ions in solution. The activity of an enzyme varies with the pH of the medium. Most enzymes have a characteristic pH at which their activity is maximum. Above or below this pH the activity declines. When we plot activity versus pH most enzymes yield a bell-shaped curve with a more or less sharply defined maximum. That means enzymes have maximum activity at **optimum pH.** Figure 3.2 illustrates the relationship between the pH and activity of the enzyme invertase.

We give below in Table 3.2, the optimum pH for some typical enzymes. The optimum pH of an enzyme is not ncessarily identical with the pH of its normal intracellular surroundings.

Even small changes in pH can have a great effect on enzyme activity. Small changes in pH mean relatively large changes in [H$^+$]. A change of 1 on the pH scale involves a ten-fold increase or decrease in [H$^+$], while a change in pH of 2 represents a hundred-fold change in [H$^+$]. The concentration of [H$^+$] affects the stability of the electrovalent bonds which help to maintain the *tertiary* structure of protein molecules. Extremes of pH cause the bonds to break resulting in enzyme denaturation.

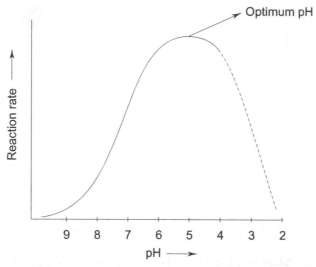

Fig. 3.2 Relationship between the pH and activity of the enzyme.

The affinity of an enzyme for its substrate may be altered by variations in pH. Changes in $[H^+]$ can alter the ionisation of the amino acid side chains at the active centres of enzymes. Ionisation of the substrate molecules can also be affected. The formation of enzyme-substrate complexes depends on the active centres and substrate molecules having opposite electrostatic charges. If the charges are altered by changes in pH, some enzymes fail to function.

TABLE 3.2. Optimum pH for Some Enzymes

Enzyme	Optimum pH
Lipase (pancreas)	8.0
Lipase (stomach)	6 – 5
Lipase (castor oil)	6.7
Pepsin	1.5 – 1.6
Trypsin	7.8 – 8.7
Urease	7.0
Invertase	6.5
Maltase	6.1 – 6.8
Amylase (pancreas)	6.7 – 7.0
Amylase	6.6 – 5.2

It is clear from Table 3.2, that for every enzyme there is an optimum pH at which the reaction it catalyses proceeds most rapidly. Most enzymes work within a pH range of 5 – 9 and catalyse reactions most efficiently at pH 7. There are, however, some exceptions. For example, pepsin and rennin secreted in the mammalian stomach work best at pH 1.5 – 2.5. Alkaline phosphatase in the kidneys has an optimum pH of 10 (Fig. 3.3).

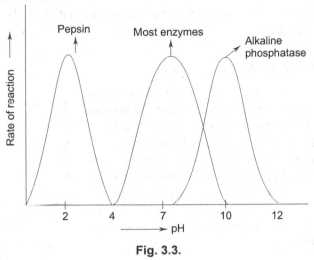

Fig. 3.3.

C. Effect of Substrate Concentration

If the concentration of the enzyme is kept constant and the substrate concentration is varied, we get the reaction profile as shown in Fig. 3.4. At low concentration of the substrate there is a linear relationship between the reaction rate and substrate concentration *i.e.*, reaction rate increases with increase in substrate concentration.

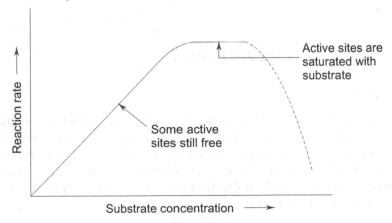

Fig. 3.4.

In these conditions, the ratio of the enzyme to substrate molecules is high. Some active sites are always free for the substrate molecules to bind with the enzyme. With increasing substrate concentration a point is reached when a further increase in substrate concentration does not cause the reaction to go any faster *i.e.*, the reaction velocity remains unaffected. The enzyme to substrate ratio is then lower and there are more substrate molecules present than there are free active centres with which to bind. Adding more substrate will not make the reaction go more quickly. The third phase comes when the substrate concentration becomes very high, the enzyme activity is inhibited by high concentration of the substrate and the reaction rate declines.

D. Effect of Enzyme Concentration

Enzymes catalyse reactions rapidly at very low enzyme concentrations. If the substrate concentration is maintained constant, the reaction rate will increase as concentration of the enzyme is increased. This is because enzyme molecules form complexes with substrates only very briefly. The products of the reaction are quickly released and the enzyme is then available for further activity (Fig. 3.5). This relationship holds good over a wide range of enzyme concentrations.

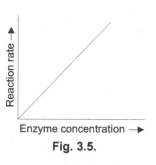

Fig. 3.5.

The rate at which enzymes use substrates is described as the turnover number. For some enzymes the turnover number is very high. A molecule of catalase for example can break down 60,000 molecules of hydrogen peroxide into water and oxygen every second.

The larger the number of enzyme molecules present the greater the amount of substrate used in a given period of time provided that there is an excess of substrate available.

3.5 COFACTORS–COENZYMES

A. Definition

Some enzymes in order to exhibit catalytic reactivity require additional chemical compounds called cofactors. Cofactors are molecules that attach to an enzyme during chemical reactions. In general, all compounds that help enzymes in their catalytic reactivity are called cofactors. A cofactor is any non-protein component in an enzyme. It is an organic molecule or metal ion which the enzyme requires in order to catalyse a reaction.

We have seen that most enzymes are simple globular proteins. Some others are conjugated proteins which have a non-protein fraction called the **prosthetic** group. A prosthetic group is an essential cofactor attached to the protein part of a conjugated enzyme. That means cofactors which are bound tightly to an enzyme are termed as prosthetic groups. These can be organic vitamins, sugars, lipids etc.

An enzyme without a cofactor is called an **apoenzyme** and the enzyme-cofactor complex is called a **holoenzyme**. Apoenzyme is enzymatically an inactive protein. Cofactors can be divided into two groups.

- Organic cofactors, which are called coenzymes.
- Inorganic cofactors – essential metal ions.

B. Coenzymes

Organic cofactors are known as coenzymes. A coenzyme is an organic non-protein compound that binds with an enzyme to catalyse a reaction.

A coenzyme cannot function alone but can be reused several times when paired with an enzyme. Coenzymes are heat stable, low molecular weight organic compounds required for the activity of enzymes. Coenzymes act as group transfer reagents. These are reusable non-protein molecules that contain carbon. They bind loosely to an enzyme at the active site to help catalyse reactions. They are linked to enzymes by non-covalent forces. Most coenzymes are vitamins, vitamins derivatives or derived from nucleotides.

C. Cofactors

Unlike coenzymes true cofactors are reusable non-protein molecules that do not contain carbon *i.e.*, they are inorganic cofactors. Usually they are metal ions such as iron, zinc, cobalt, copper, magnesium etc. that loosely bind to an enzyme's active site to carry out the enzymatic reactions.

Metal ions are Lewis acids *i.e.*, electron pair acceptors. Therefore, they can act as Lewis acid base cataysts. We know that they can form coordination compounds by behaving as Lewis acids. These coordination compounds are an important part of metal ions in biological systems. These coordination compounds formed by metal ions tend to have specific geometries, which help in positioning the group involved in a reaction for optiomum catalysis.

Without coenzymes or cofactors enzymes cannot catalyse reactions effectively. In fact, the enzyme may not function at all. If reactions cannot occur at the normal catalysed rate, then an organism will have difficulty in sustaining life.

D. Functions of Coenzymes

We have seen that when an enzyme gains a coenzyme it becomes a holoenzyme or active enzyme. Active enzymes convert substrates into products an organism requires to carry out the essential functions whether chemical or physiological. Coenzymes like enzymes can be reused and recycled without changing reaction rate. When an enzyme is denatured by extreme temperature or pH, the coenzyme can no longer attach to the active site.

The coenzymes perform two functions. One is to bind to the enzyme creating an active site so that the substrate can bind and the other is to provide some functional groups in order to help in the catalytic activity of the enzyme.

For enzymatic reactions a coenzyme binds to an apoenzyme to give holoenzyme or active enzyme. It is the holoenzyme that binds to the substrate to perform the reaction. That means the role of the coenzyme is in the formation of a stable complex between the substrate and enzyme as shown in Fig. 3.6.

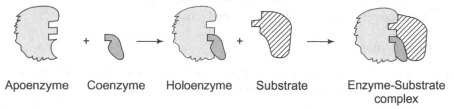

| Apoenzyme | Coenzyme | Holoenzyme | Substrate | Enzyme-Substrate complex |

Fig. 3.6   Role of coenzyme in the formation of stable complex.

E. Coenzyme Classification

There are two types of coenzymes – vitamin derived coenzymes and metabolite coenzymes or non-vitamins

i. Vitamin Derived Coenzymes

Many coenzymes are vitamins or derived from vitamins. Some vitamins directly act as coenzymes but vitamins help the body to produce coenzymes. These are generally obtained from nutrients. If vitamin intake is too low, then an organism will not have the coenzymes needed to catalyse various biological reactions. Water soluble vitamins which include B-complex vitamins and vitamin C lead to the production of coenzymes.

Two of the most important and widespread vitamin derived coenzymes are nicotinamide adenine dinucleotide (NAD) and coenzyme A. NAD is derived from nicotinamide, a B-group vitamin (B_3). Coenzyme A, T also known as acetyl-CoA is derived from vitamin B_5. Similarly, riboflavin (vitamin B_2) forms flavin-adenine dinucleotide (FAD).

Many of the coenzymes are involved in the oxidation reduction reactions which provide energy for the organism. Others serve as group transfer agents in metabolic processes.

Some coenzymes with their reactions and vitamin precursors are given in Table 3.3

Table 3.3 Some Enzymes with their Reactions and Processor

Coenzymes	Reactions Type	Vitamin Processor
NAD & NADH	Oxidation reduction	Niacin
Thiamine Pyrophate (TPP)	Aldehyale transfer	Thiamine (Vitamin B_1)
FMN and FAD	Oxidation – reduction	Riboflavin (Vitamin B_2)
Pyridoxal-5-phosphate	Transamination	Pyridoxal (Vitamin B_6)
Coenzyme A	Acyl transfer	Panthothenate (Vitamin B_5)
Biotin	–	Biotin
Lipoic acid	Acyl transfer	

ii. Metabolite Coenzymes-Non-vitamins

These are synthesised by micro-organisms and plants and can be produced from nucleotides such as adenosine. Non-vitamin coenzymes aid in the transfer of groups for enzymes. They ensure physiological functions like blood clotting and metabolsm which occur in an organism. Adenosine triphosphate (ATP) is the most important example of a non-protein coenzyme. It is the most widely distributed coenzyme in the human body. It transports substances and supplies energy needed for necessary chemical reactions.

F. Inorganic Cofactors

We have seen some essential metal ions act as cofactors. They are further classified as – activations (loosely bound) and metalloenzymes (tightly bound). Activator ions are Ca^{2+}, k^+, Mg^{2+} and Mn^{2+} whereas zinc, copper, cobalt and iron are metalloenzymes. Metalloenzymes retain their metal atom under all conditions because they have high affinity for the metal ion.

Table 3.4 given below gives some metal ions which are used as cofactors with certain enzymes.

Table 3.4 Some Metal Cofactors

Enzyme	Metal ion	Function Type.
Cytochrome oxidase	Cu^{2+}	Oxidation-reduction
Alcohol dehydrogenase	Zn^{2+}	Ulsed with NAD^+
Urease	Ni^{2+}	Part of catalytic site
Ascorbic acid oxidase	Cu^{2+}	Oxidation–reduction
Pyruvate dehydrogenase complex	Mg^{2+}	Pyruvate to CO_2 and acetyl CoA
Acotinase	Fe^{2+}/Fe^{3+}	Citrate to isocitrate
Arginase	Mn^{2+}	Removes electrons
Carboxy peptidare	Zn^{2+}	Used with NAD^+

3.6   STRUCTURE AND FUNCTIONS OF VARIOUS COENZYMES

A. NAD^+ and $NADP^+$

We have seen NAD^+ is the most important and widespread vitamin-derived coenzyme. Both nicotinamide adenine dinucleotide (NAD^+) and nicotinamide adenine dinucleotide phosphate ($NADP^+$) are derived from vitamin B_3, niacin or nicotinic acid.

NAD^+ is a dinucleotide and contains nicotinamide, adenine, D-ribose and phosphoric acid whereas $NADP^+$ has an additional phosphate group attached to the 3′-position of D-ribose attached to adenine. The structures of these coenzymes are given in Fig. 3.7.

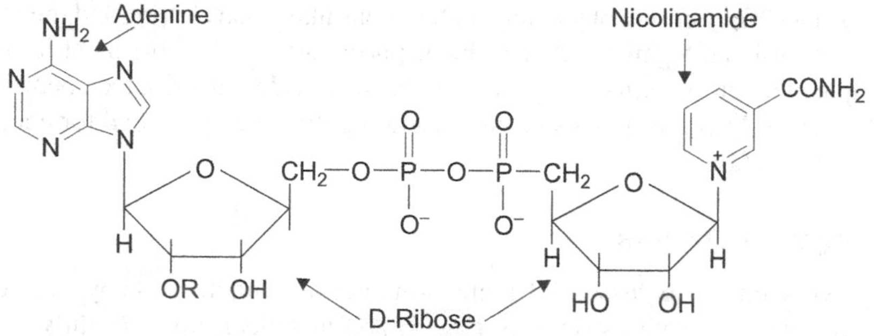

Fig. 3.7   Structures of NAD^+ (R = H) and $NADP^+$ R = PO_3^{2-}.

NAD^+ and $NaDP^+$ act as coenzymes for many dehydrogenases where they are involved in the transfer of hydrogen atoms and electrons causing either oxidation or reduction of the substrates. The reactive site of both of these coenzymes is the nicotinamide moiety. The para-position (C4) of nicotinamide is the site of hydrogen transfer. The oxidised form of these coenzymes accepts a pair of electrons whereas the reduced form donates a pair of electrons as shown in Fig. 3.8.

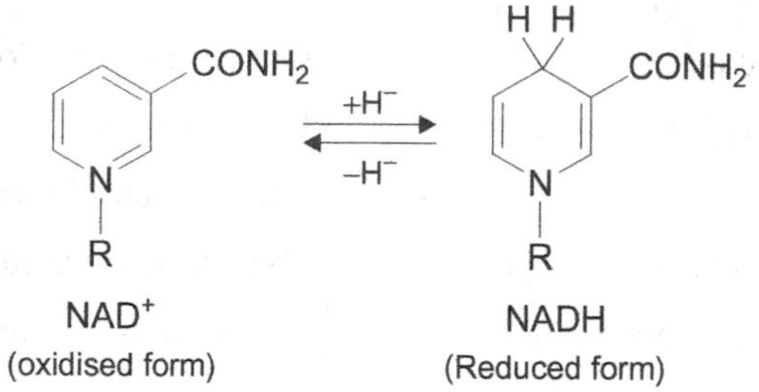

Fig. 3.8   Mode of action of NAD^+.

NAD^+ primarily transfers electrons needed for redox reactions specially those involved in parts of the citric acid cycle. When NAD^+ gains electrons NADH is formed. NADH, called coenzyme 1, has many functions. In fact, it is considered the number one coenzyme in the human body because this coenzyme carries electrons for the reactions and produces energy from food. A lack of NADH causes energy deficiency resulting in fatigue. Also this coenzyme is recognised as the powerful antioxidant for protecting cells against harmful substances.

NAD^+ acts as a coenzyme in many enzymatic reactions involving dehydrogenases. Some NAD^+ containing dehydrogenases are lactate dehydrogenases, alcohol dehydrogenase, malate dehydrogenase, glyceraldehyde phosphate dehydrogenase *etc*. Two examples are given below – one of reduction and the other of oxidation

1. Lactic acid is oxidised to pyruvic acid where NAD^+ acts as H-acceptor

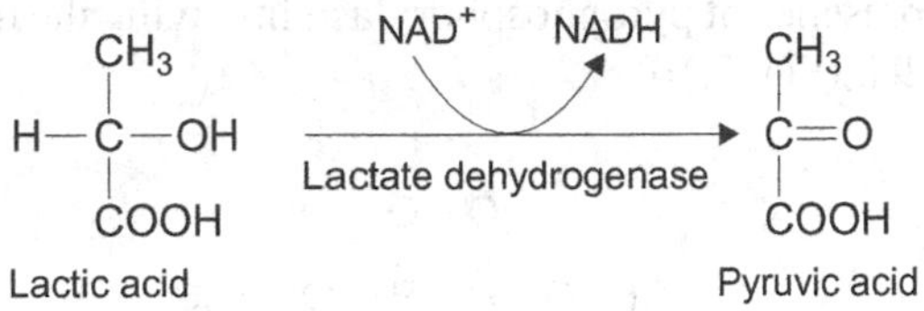

Lactic acid → Pyruvic acid (Lactate dehydrogenase, NAD^+ → NADH)

2. In another example, acetaldehyde is reduced to ethanol where NADH acts as H-donor

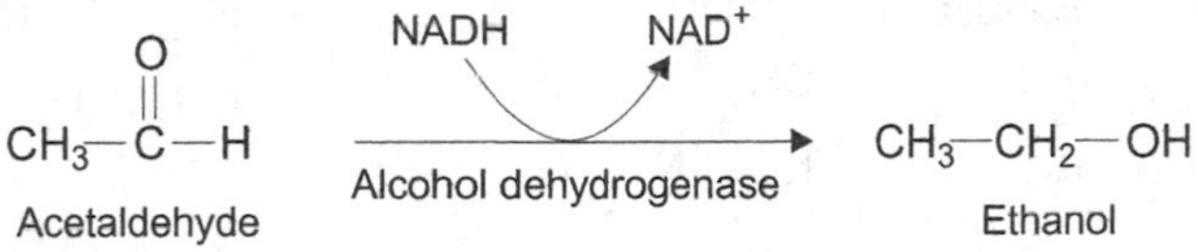

Acetaldehyde → Ethanol (Alcohol dehydrogenase, NADH → NAD^+)

Similarly, isocitric acid is oxidised to oxalosuccinic acid in the presence of isocitrate dehydrogenase where $NADP^+$ acts as a coenzyme.

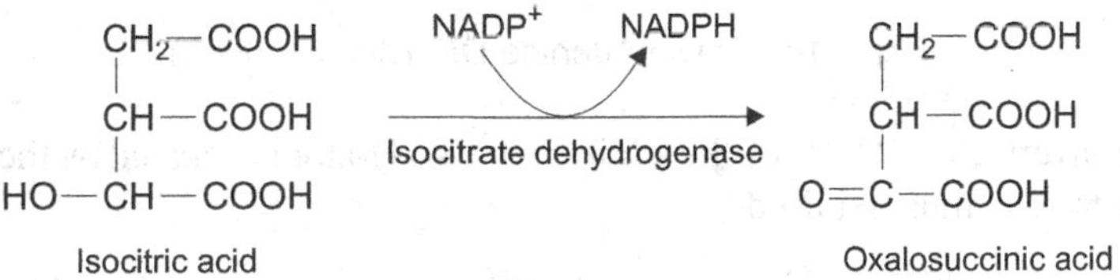

Isocitric acid → Oxalosuccinic acid (Isocitrate dehydrogenase, $NADP^+$ → NADPH)

B. Flavin Mononucleotide (FMN) and Flavin Adenine Dinucleotide (FAD)

FMN and FAD are both derived from riboflavin (Vitamin B_2). They act as coenzymes in redox reactions catalysed by various enzymes. The enzymes requiring FMN and FAD as cofactors are called flavoproteins. Generally, these coenzymes function as prosthetic groups as they are more tightly bound to the apoenzyme. As a result they cannot be separated by dialysis.

The coenzyme, FMN is obtained by phosphorylation of riboflavin using ATP and the enzyme flavokinase (Fig. 3.9).

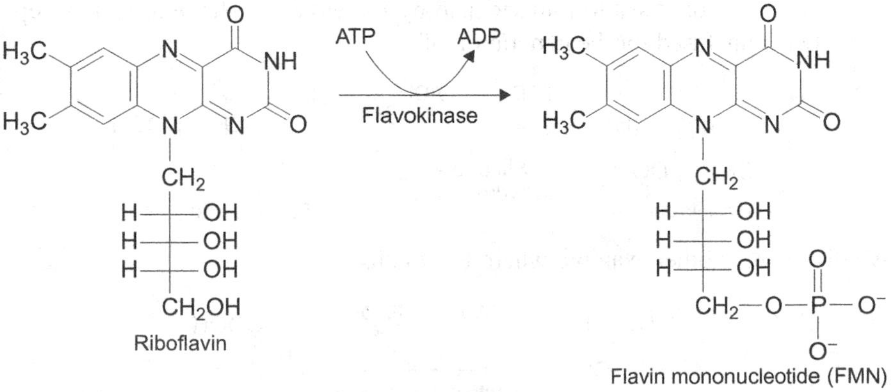

Riboflavin → Flavin mononucleotide (FMN) (Flavokinase, ATP → ADP)

Fig. 3.9   Synthesis of FMN.

However, FAD is synthesised by the reaction of FMN with ATP. The reaction is carried out in the presence of pyrophosphorylase involving the transfer of the AMP unit of ATP to FMN. (Fig. 3.10)

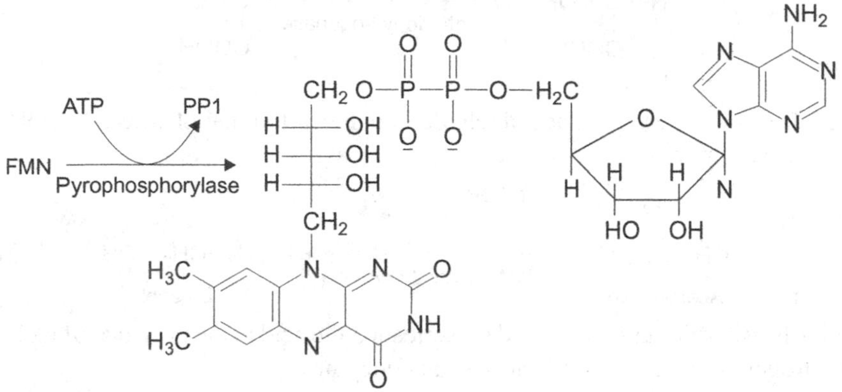

Fig. 3.10 Flavin Adenine Dinucleotide (FAD).

FAD is converted to $FADH_2$ by the addition of two H-atoms donated by the substrate. The substrate is thus oxidised

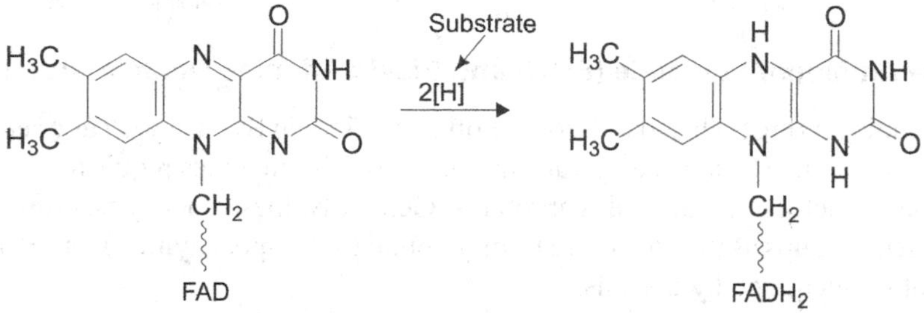

Both FMN and FAD are used with a variety of substrates. An example of an FAD containing enzyme is succinate dehydrogenase occurring in the Krebs' cycle. Succinic acid is oxidised to fumaric acid by the enzyme. The hydrogen accepted by FAD is transfused for the generation of ATP.

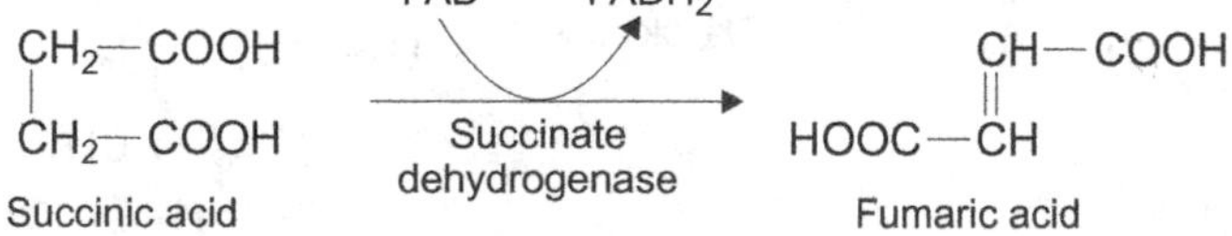

We give here another example where FAD is used

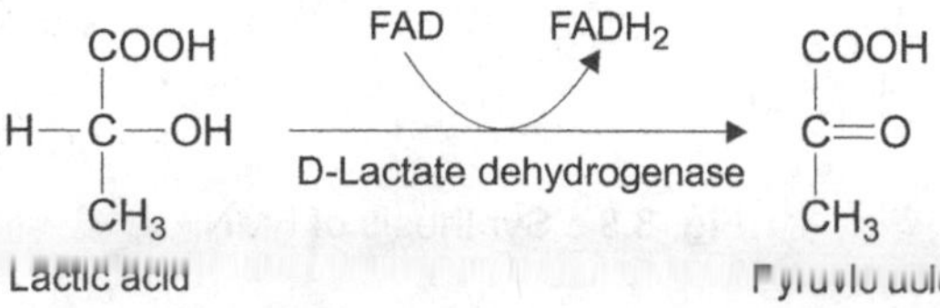

C. Thiamine Pyrophosphate (TPP)

TPP is a coenzyme which is involved in the oxidative decarboxylation reactions and transketolase reaction. It also catalyses several other biological reactions.

TPP is derived from thiamine (vitamin B_1) and is produced in the brain and liver cells by the phosphorylation of thiamine in the presence of TPP synthase. The reaction involves the transfer of the pyrophosphate group from ATP to vitamin and formation of AMP as a by-product (Fig. 3.11)

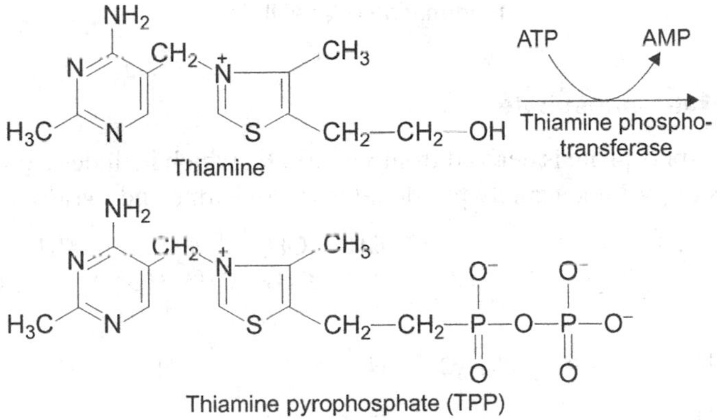

Fig. 3.11   Synthesis of TPP.

TPP is involved in the transfer of aldehyde groups like acetaldehyde and glycol aldehyde. The thiazole group of the coenzyme molecule accepts the aldehyde group and transfers it to an acceptor *via* other coenzymes like lipoic acid and coenzyme A.

In TPP, the active moiety is the thiozolium cation which can lose a proton from the carbon between the N and S atoms of the ring in a base catalysed mechanism (Fig. 3.12)

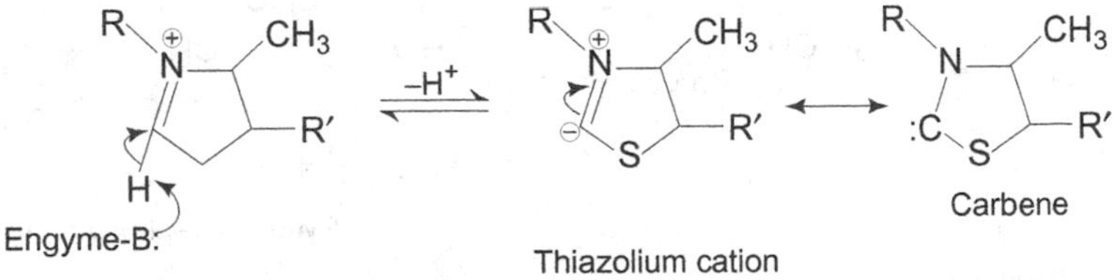

Fig. 3.12

The active site is responsible for the transfer of the acetyl group. If we use acetaldengde it adds to TPP to give the active acetal as shown below:

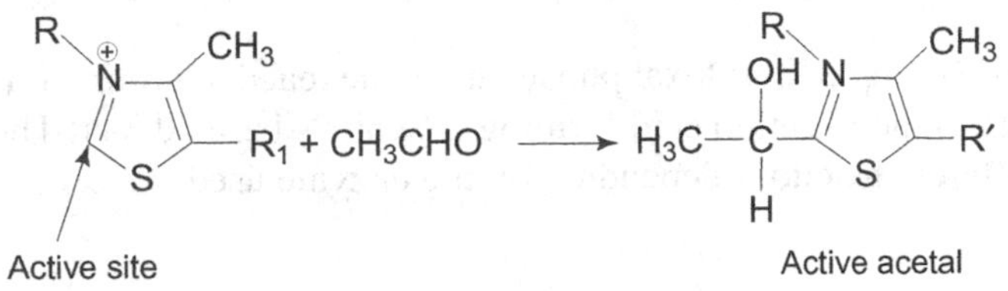

An example of an enzyme complex involving TPP, lipoic acid and coenzyme A is pyruvate decarboxylase. The reaction is shown in a simplied way (Fig. 3.13)

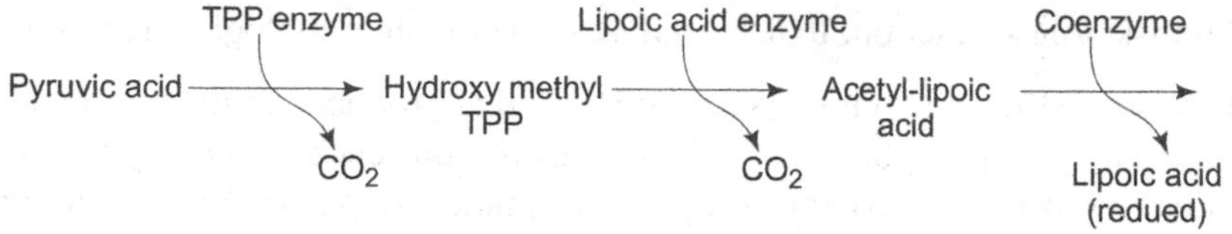

Fig. 3.13   Oxidative decarboxylation of pyruvic acid leading to formation of acetyl CoA.

D. Pyridoxal-5-phosphate

Pyridoxal-5-phosphate is derived from vitamin B_6 which includes a group of three derivatives of pyridine namely pyridoxal, pyridoxamine and pyridoxine.

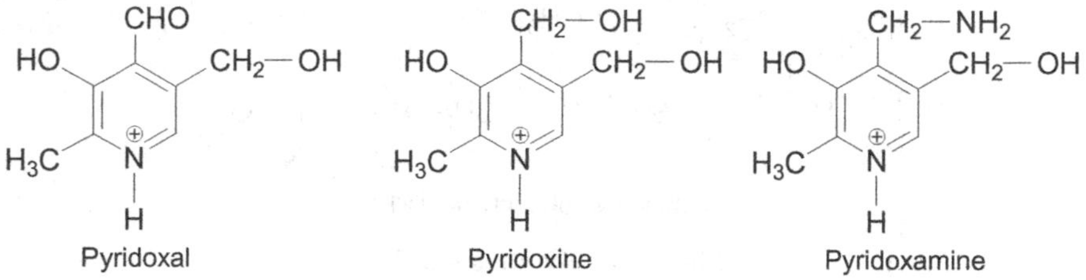

In our food these are present as pyridoxine, pyridoxal-5-phosphate and pyridomine-5-phosphate. Out of these pyridoxal-5-phosphate is the active coenzyme. During digestion some hydrolysis of the phosphate group may take place but these are phosphorylated back by the enzyme pyridoxalkinase present in the cells with the help of ATP.

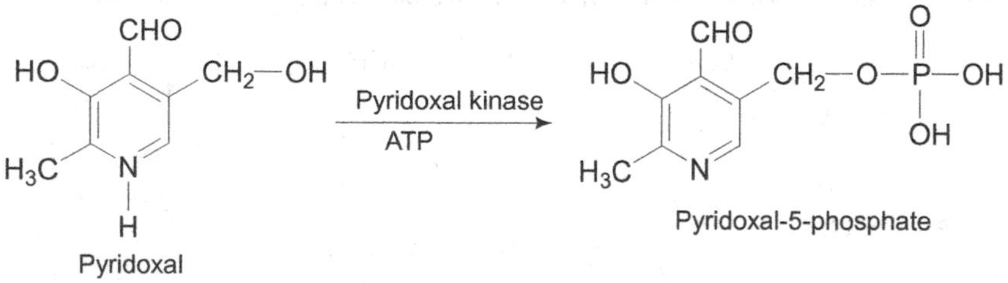

Pyridoxal-5-phosphate functions in a variety of reactions for the synthesis and catabolism of α-amino acids. These include isomerisation, transamination, racemisation, decarboxylation, $α − β$, $β − γ$, elimination reactions involving amino acids.

The aldehyde group of pyridoxal phosphate is the reactive group of the coenzyme which binds to the α-amino acid forming a Schiff's base adduct. The adduct can undergo different reactions depending on the enzyme used.

Pyridoxal-5-phosphate has a special role in transamination reactions. The coenzyme is associated with transaminases which catalyse transfer of the amino group of amino acids to keto acids. In this transfer process phyridoxal-5-phosphate (PLP) acts as the acceptor of the amino group and is converted to pyridoxamine phosphate (PEP).

Similarly, PEP can react with a keto acid to produce an amino acid. PLP and PEP remain bound to the protein part of the transaminase enzyme during the transfer of the amino group. The reactions catalysed by transaminases can be represented in the following way:

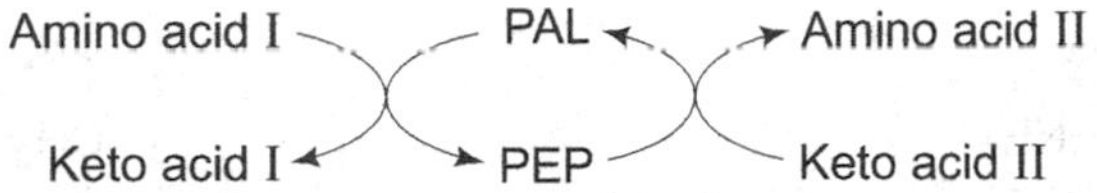

We can show the transamination of amino and keto acid by the following scheme:

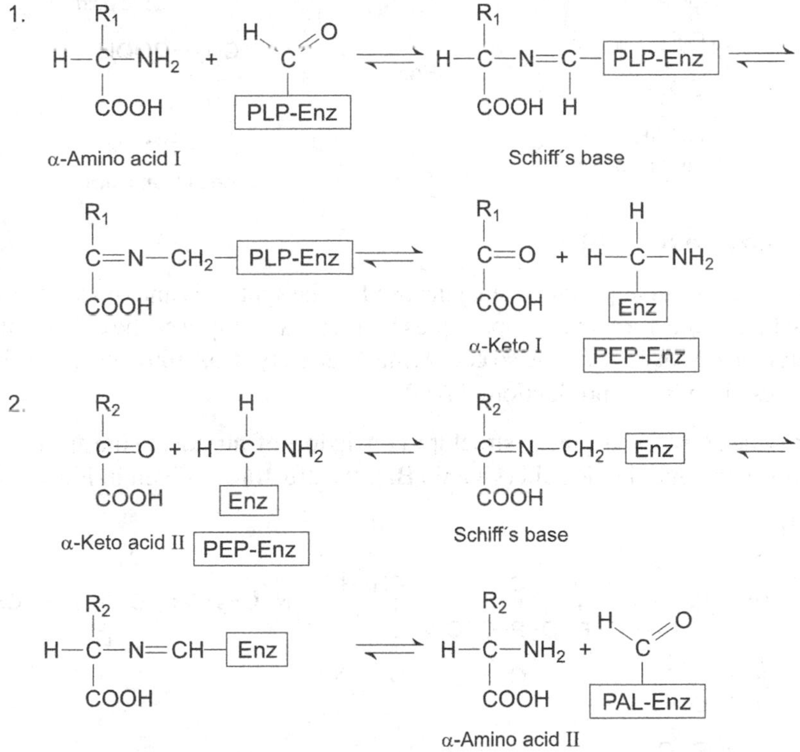

As an example we have here the transfer of the α-amino group of aspartic acid to α-keto glutaric acid giving oxaloacetic acid and glutamic acid. This reaction is catalysed by the enzyme asparate aminotransferase.

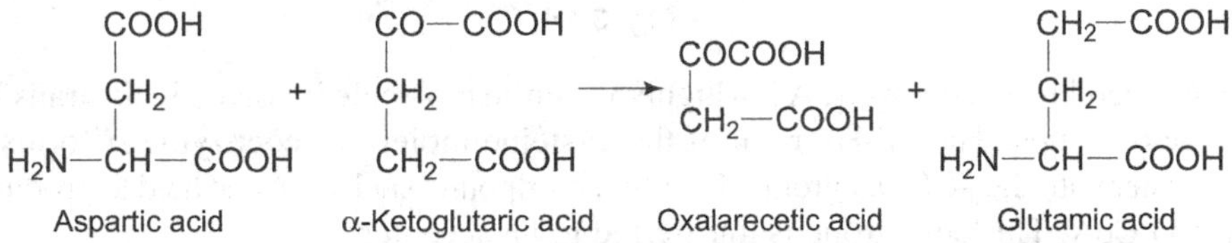

In this reaction, the —NH$_2$ group of aspartic acid is transferred to α-ketoglutaric acid giving glutamic acid and oxaloactic acid.

We can also represent the above reaction in the following way.

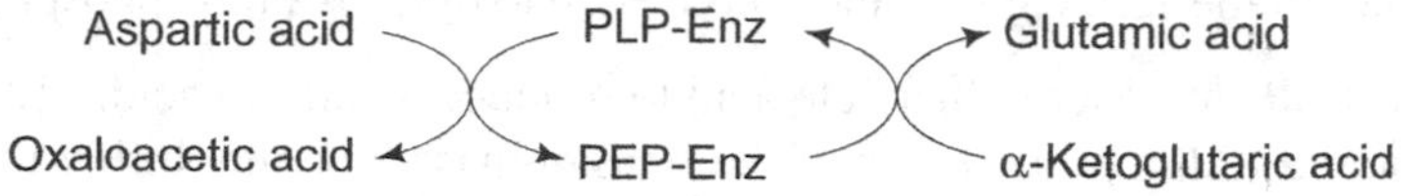

Aspartic acid $\diagdown$ PLP-Enz $\diagup$ Glutamic acid

Oxaloacetic acid $\diagup$ PEP-Enz $\diagdown$ α-Ketoglutaric acid

In another example the —NH$_2$ group of alanine is transferred to α-ketoglutaric acid; thereby it gets converted to pyruvic acid and glutamic acid.

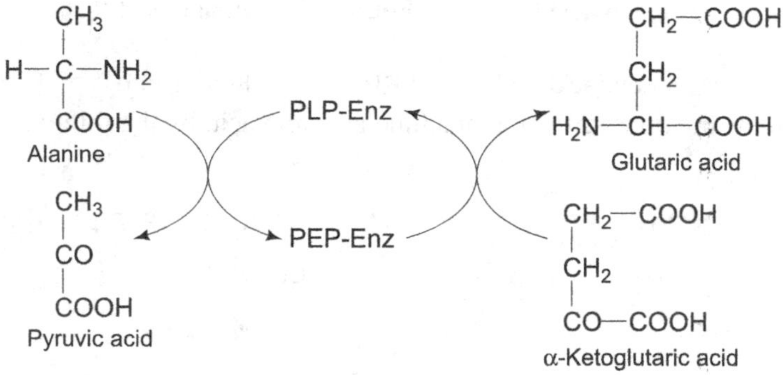

$$
\begin{array}{c}
CH_3 \\
| \\
H-C-NH_2 \\
| \\
COOH \\
\text{Alanine}
\end{array}
\qquad
\text{PLP-Enz}
\qquad
\begin{array}{c}
CH_2-COOH \\
| \\
CH_2 \\
| \\
H_2N-CH-COOH \\
\text{Glutaric acid}
\end{array}
$$

$$
\begin{array}{c}
CH_3 \\
| \\
CO \\
| \\
COOH \\
\text{Pyruvic acid}
\end{array}
\qquad
\text{PEP-Enz}
\qquad
\begin{array}{c}
CH_2-COOH \\
| \\
CH_2 \\
| \\
CO-COOH \\
\text{α-Ketoglutaric acid}
\end{array}
$$

E. Coenzyme A (CoA)

Coenzyme A is an important coenzyme used in the synthesis and oxidation of fatty acids. Fatty acids form the phospholipids bilayer that comprises the cell membrane, a feature necessary for life. Also coenzyme A as acetyl CoA initiates the citric acid cycle resulting in the production of ATP.

Coenzyme A has a complex structure consisting of an adenosine triphosphate, cysteine and pentothenic acid (vitamin B$_5$). Its structure is given in Fig. 3.14.

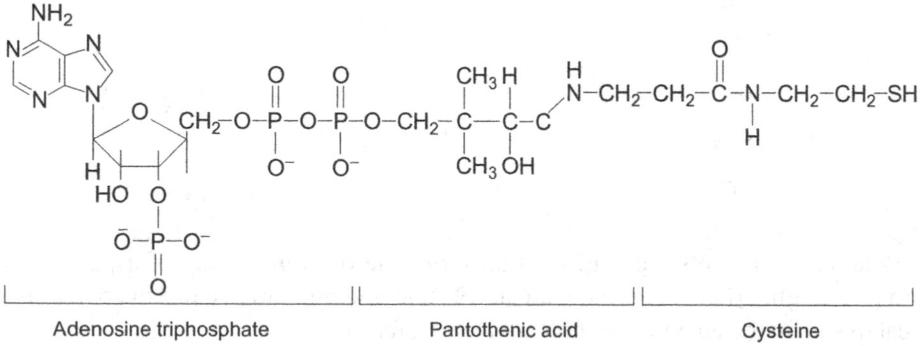

Fig. 3.14

In its acetyl form coenzyme A is a highly versatile molecule involved in the transfer of acid groups. The – SH group of the cysteine moiety of coenzyme A forms a thioester with the –COOH group of the acyl compound such as acetic acid to produce acetyl CoA. This acyl group is transferred to an acceptor.

We give below (Fig. 3.15) the transfer of acetyl group of oxaloacetic acid to form citric acid in the presence of condensing enzyme.

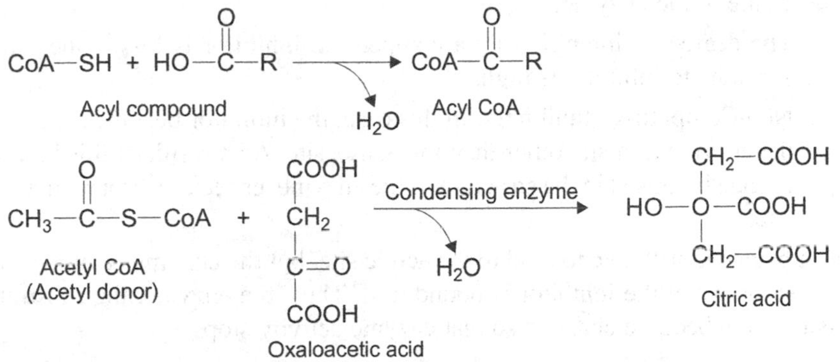

Fig. 3.15   CoA catalysed reaction

3.7 ENZYME INHIBITORS

An inhibitor, as the name implies, is a substance that interferes with the action of an enzyme and slows the rate of a reaction. Many substances inhibit the activity of enzymes. Enzymes are sensitive to a variety of chemical reagents that react with groups at or near the surface of the protein, producing partial or complete inhibition of chemical activity. Inhibitors fall into two categories – reversible and non-reversible.

A. Reversible Inhibitors

Reversible inhibitors are substances which prevent enzymes from combining with substrates. Activity of the enzyme is restored when the inhibitor is removed. A reversible inhibitor can bind to the enzyme and subsequently be released, leaving the enzyme in its original condition.

There are two major classes of reversible inhibitors –

1. Competitive Inhibitors: These compete with the normal substrate molecule for combining with the active site on the enzyme. They can bind to the active site and block the substrate from binding to the enzyme. Thus, they affect enzyme action by becoming attached to the active centres stopping the substrate from binding to the enzyme. Such inhibitors are very similar in structure to the substrate.

   A well-known example of this behaviour is inhibition of the enzyme succinate dehydrogenase by malonic acid. Succinate dehydrogenase catalyses the oxidation of succinate to fumarate in the Kreb's cycle. In the presence of malonate the reaction rate is slowed down. Malonate has a molecular structure very similar to that of succinate. Inhibition occurs because the active centre of some of the enzyme molecules become occupied by malonate rather than by succinate. In fact, both malonate and succinate are competing for the active centre of the enzyme.

The degree of inhibition by a competitive inhibitor is less if the ratio of substrate to inhibitor is high.

2. Non-Competitive Inhibitors: In this case, the inhibitor becomes attached to the enzyme at a site other than the active site. As a result of this binding a change is caused in the structure of the enzyme, especially around the active site.

The substrate is still able to bind to the active site, but the enzyme cannot catalyse the reaction when the inhibitor is bound to it. Thus, the enzyme, the substrate or possibly both become changed so that enzyme activity stops.

Disulphide bridges are important in maintaining the tertiary structure of enzyme molecules. If the disulphide bridges are broken the three-dimensional shape of the enzyme changes. The ions of heavy metals such as mercury, silver and copper affect enzymes in this way. Hg^{2+}, Ag^+ and Cu^{2+} ions combine with thiol (—SH) groups in enzymes. They denature the enzyme molecule and thus inhibit enzyme activity. Cyanide is another non-competitive inihibitor. It blocks the action of some enzymes by combining with iron which may be present in a prosthetic group or which may be required as an enzyme activator. For this reason, salts of heavy metals and cyanide are potent poisons to living organisms. However, non-competitive inhibitors do not bind strongly to enzymes and can be removed by dialysis. Enzyme activity is then restored.

B. Non-Reversible Inhibitors

Non-reversible or irreversible inhibitors react with the enzyme to produce a protein that is not enzymatically active and from which the original enzyme cannot be regenerated. The enzymes undergo irreversible inactivation when they are treated with agents capable of permanently modifying a functional group required for catalysis, making the enzyme molecule inactive.

Organophosphorus insecticides such as malathion are good examples of non-reversible inhibitors. They become firmly bound to active centres so that substrate molecules cannot bind to enzymes and activity of the enzyme is permanently stopped. Insecticides of this type inactivate the enzyme cholinesterase, which is essential for the functioning of the nervous system.

3.8   MECHANISM OF ENZYME CATALYSIS

Enzymes are soluble in water and work in aqueous solution in living cells. They are sometimes described as organic catalysts. A catalyst is a substance which accelerates the rate of a chemical reaction. Many of the biochemical reactions catalysed by enzymes are reversible reactions. An example of a reversible reaction is given below:

$$A + B \rightleftharpoons C$$
reactants   product

At equilibrium the rate at which A and B are converted to C may or may not be equal to the rate at which C is converted back to A and B. The position of equilibrium depends on the energy difference betwen the reactants and product (Fig. 3.16).

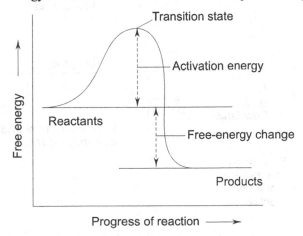

Fig. 3.16 Activation energy profile for a typical reaction.

Enzymes, like all catalysts, speed up reactions but they cannot alter the equilibrium constant or the free energy change. The reaction rate depends on the free energy of activation or activation energy *i.e.*, the energy required to initiate the reaction. The activation energy for an uncatalysed reaction is higher than that for a catalysed reaction. That means an uncatalysed reaction requires more energy to get started than that required for a catalysed reaction (Fig. 3.17).

Enzymes do not alter the direction of a reaction; they speed up the rate at which equilibrium is reached. In doing so, they can catalyse reversible reactions in either direction provided it is energetically feasible.

Biochemical reactions involve the formation or the destruction of chemical bonds. When two or more reactants are joined, chemical bonds are formed. When a complex molecule is split into simpler components, chemical bonds are destroyed. In both the cases, energy is required to bring about the changes. That energy is the activation energy. Heat can be used as a source of activation energy. Many chemical reactions do not proceed quickly unless the reactants are raised to relatively high temperatures. In living cells, however, reactions take place rapidly at relatively low temperatures. Enzymes lower the amount of activation energy needed, making it possible for reactions to occur at temperatures which are otherwise energetically unfavourable.

Enzymes catalyse reactions by lowering the energy of activation for a chemical reaction. Thus, they act in a similar way to an increase in temperature *i.e.*, by increasing the proportion of colliding molecules which react. Like inorganic catalysts, enzymes accelerate reactions but do not affect the position of equilibrium.

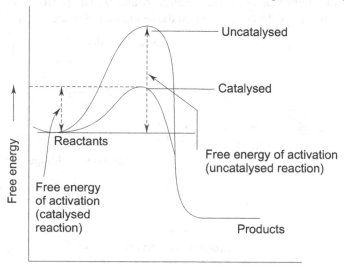

Fig. 3.17   A comparison of activation energy for catalysed and uncatalysed reactions.

It is not always possible to recover an enzyme from the reaction mixture unchanged at the end of the reaction, but the essential feature is that it should not affect the final equilibrium mixture; *i.e.*, it does not affect the free energy of the reaction.

Michaelis and **Menten** first postulated that enzymes act like inorganic catalysts and follow the same principle as **Oswald's law of chemical catalysis**. They suggested that the energy 'hump', which must be overcome for reactions to take place, was lower for the enzyme-substrate complex than for the free substrate in the absence of enzyme. The enzyme-substrate complex then breaks down to give the products and free enzyme as shown below:

Enzyme + substrate ⇌ Enzyme – substrate
                                                    complex

⇅

products + enzyme

As shown above reactions are reversible. The enzyme catalyses both the forward and reverse reactions. The enzyme does not change the extent of reaction. It only changes the rate at which the reaction comes to an equilibrium. An equilibrium may be shifted towards the left or right by the usual factors such as temperature, concentration of the reactants and products or removal of a reactant or product.

The enzyme-catalysed reaction may be represented in the following way:

$$E + S \; \underset{K_{-1}}{\overset{K_{+1}}{\rightleftharpoons}} \; ES \; \xrightarrow{K_2} \; E + products$$

This is the Michaelis-Menten rate equation for reactions catalysed by enzymes where K_{+1} is the rate constant for the formation of the enzyme-substrate complex, ES

K_{-1} is the rate constant for the reverse reaction.

K_2 is the rate constant for the formation of the product P.

3.9   HOW DO SUBSTRATES BIND TO ENZYMES ?

We have seen that in an enzyme catalysed reaction, the enzyme binds to the substrate to form a complex. The formation of the complex leads to the formation of the transition-state species, which then forms the product. A substrate binds, usually by non-covalent interactions, to a small portion of the enzyme called the **active-site**. This active site is frequently situated in a cleft or crevice in the protein which consists of certain amino acids that are essential for enzymatic activity.

The catalysed reaction takes place at the active site usually in several steps:

The **first** is the binding of the substrate to the enzyme at the active site.

After the substrate is bound and the transition state is subsequently formed, catalysis can occur. This means that bonds must be rearranged.

In the **third** step, bonds are broken and new bonds are formed, the substrate is converted into the product.

Then the product is released from the enzyme. The enzyme can now catalyse the reaction of more substrate to form more product.

Two important models have been developed to describe the binding process.

A. The Lock and Key Model

This model was proposed by **Emil Fischer** in 1896 to explain how enzymes and substrates interact when mixed. Fischer suggested that enzyme and substrate molecules combine to form an enzyme-substrate complex before the products of the reaction are released. In doing so the substrate must fit a portion of the enzyme *i.e.*, active site in a similar way just as a key must fit a lock in order to open it. As enzyme and substrate molecules are three-dimensional structures, the formation of an enzyme-substrate complex requires that shapes of the reactants and the active centres of enzymes are complementary. Otherwise, the enzyme and substrate cannot unite (Fig. 3.18). It is comparable to the way in which a lock can only be opened by a key of a specific shape.

The specificity of enzymes supports the lock and key model. As different locks have specific keys, similarly specific enzymes act on different substrates.

(*a*) Bond Formation

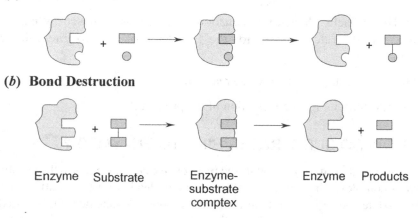

(*b*) Bond Destruction

Enzyme   Substrate   Enzyme-   Enzyme   Products
substrate
comptex

Fig. 3.18 The lock and key model of enzyme action.

B. Induced Fit Model

The above lock and key model is now largely of historical interest because it does not take into account an important property of the proteins *i.e.*, conformational flexibility. This mechanism was proposed by **Koshland**. The induced fit model takes into account the three dimensional flexibility of proteins. According to this model, the shape of the active centre is changed when a substrate molecule binds to such enzymes. The binding of the substrate induces a conformational change in the enzyme that results in a complementary fit after the substrate is bound (Fig. 3.19).

The binding site has a different three-dimensional shape before the substrate is bound. The shape changes after the enzyme-substrate complex is formed. This model gives more satisfactory explanation for the lowered activation energy that occurs with an enzyme catalysed reaction.

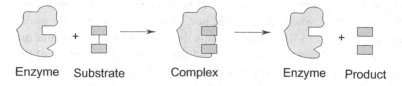

Enzyme   Substrate   Complex   Enzyme   Product

Fig. 3.19 The induced-fit mechanism.

3.10   ALCHOLIC FERMENTATION

A. Historical Background

Although alcoholic fermentation was known to mankind since prehistoric times, nothing was known until 1860. During this period, the French Chemist, **GayLussac** had described the production of alcohol from sugar by the following equation :

$$C_6H_{12}O_6 \rightarrow 2CO_2 + 2CH_3-CH_2-OH$$

In the middle of the nineteenth century two schools of controversy developed – one group considered the process to be a chemical one, while the other group considered it to be a process due to the activities of living organisms. At that time **Louis Pasteur** was studying how lactic acid is produced in milk and how acetic acid is formed in certain wines. At that time there was no clear concept of enzymes, so he worked only with whole organisms, namely yeast cells. From his experiments he concluded that fermentation takes place anaerobically only in the presence of certain microorganisms. When these organisms were excluded, no fermentation occurred. Thus, fermentation is a physiological process that is closely bound up with the life of cells.

Earlier **Lavoisier** had concluded that all living things need oxygen for respiration. This was contradicted by Pasteur in that fermentation is the result of life acti-vities being carried on in the absence of air. He was also able to show that aerobic metabolism was a much more efficient way to degrade sugars than anaerobic metabolism. We will see under aerobic metabolism that we get much more energy from glucose if oxygen is available.

In 1897, **Buchner** was interested in making protoplasm extracts from yeast to be injected into animals. He wanted to preserve this extract. Ordinary antiseptics could not be used, so he tried the usual kitchen chemistry of adding large amounts of sugar. Sucrose was rapidly fermented by the yeast juice. So, he observed fermentation in the presence of living cells for the first time. Yeast was later found to be capable of fermenting a large number of sugars, such as glucose, fructose, mannose, sucrose and maltose. Glucose was converted by the juice into ethyl alcohol and carbon dioxide according to the Gay-Lussac equation.

In 1905, **Harden** and **Young** studied the activity of yeast juice. When they added fresh yeast juice to a solution of glucose at pH 5, fermentation began almost at once. The rate of carbon dioxide soon fell off, so they restored it by adding inorganic phosphate. Again the rate of fermentation dropped as the percentage of phosphate declined. So, they suspected the formation of an organic phosphate. Soon, they were able to isolate fructose diphosphate, which was fermented very actively by fresh yeast juice. It was concluded by them that this substance was probably an intermediate in alcoholic fermentation.

Robison was able to isolate glucose-6-phosphate and fructose-6-phosphate. Since both of these sugar phosphates could be fermented it became clear that all these sugar phosphates are intermediates in alcoholic fermentation. Later studies proved that fermentation of glucose by yeast juice is very similar to the anaerobic breakdown of glycogen by muscle extracts. Further, Harden and Young found that yeast juice loses its activity when it is dialysed *i.e.*, when the juice is placed in a selective permeable bag surrounded with water. The dialysed juice did not ferment but the activity could be restored by adding the material which had diffused into water. Material outside the bag was found to be quite stable, whereas the contents of the bag could be easily destroyed by heat. The large molecules inside the bag, which we now know, are enzymes and the dialysable factors, that move away from the catalysts inside the bag, are called coenzymes. Thus, yeast juice was thought to contain zymase, which is composed of non-dialysable thermolabile enzyme and a coenzyme which is dialysable and thermostable. Zymase is today known to be a complex mixture of many enzymes.

B. The Overall Pathway in Alcoholic Fermentation

If glucose and inorganic phosphate are added to dialysed juice from yeast no fermentation occurs and no sugar phosphate esters are formed showing that one of the coenzymes must play a role in the reaction. If ATP is added to the dialysed juice, phosphorylation of the sugar is begun and hexose diphosphate can be isolated. In alcoholic fermentation one molecule of glucose is converted to fructose-1, 6-diphosphate, which eventually gives rise to two molecules of pyruvate. Pyruvate loses carbon dioxide producing acetaldehyde which in turn is reduced to give ethanol.

All these reactions can be divided into three phases.

C. Conversion of Glucose to Glycoraldehyde-3-phosphate

It involves the first five steps :

Step 1. Glucose is phosphorylated to give glucose-6-phosphate

The enzyme that catalyses this reaction is hexokinase. This reaction uses ATP and gets converted to ADP. The term kinase is applied to the class of ATP-dependent enzymes that transfer a phosphate group from ATP to a substrate. The substrate of hexokinase is not necessarily glucose; rather it can be any one of a number of hexoses such as glucose, fructose and mannose. Mg^{2+} is the cofactor.

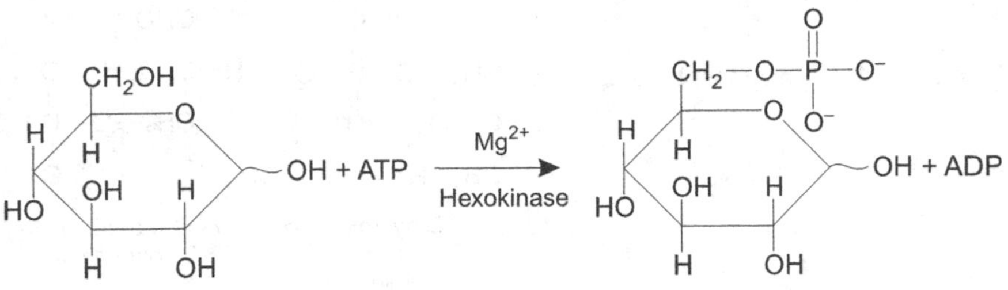

Glucose -6-phosphate

Step 2. Glucose-6-phosphate isomerises to give fructose-6-phosphate

Glucosephosphate isomerase is the enzyme that catalyses this reaction. The C–1 aldehyde group of glucose-6-phosphate is reduced to a hydroxyl group and C–2 hydroxyl group is oxidised to give the ketone group of the fructose-6-phosphate.

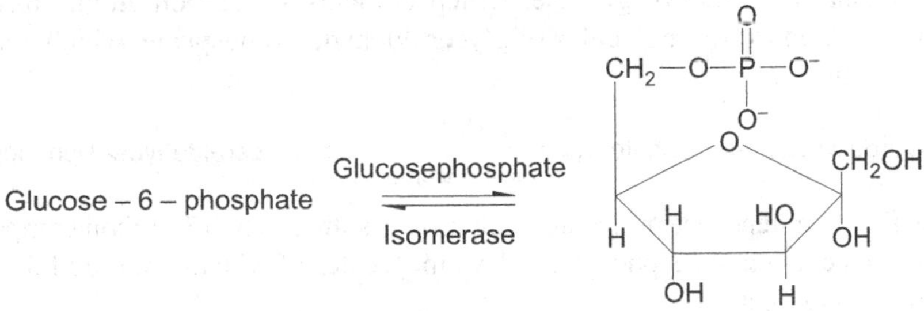

Glucose – 6 – phosphate

Fructose – 6 – phosphate

Step 3. Fructose-6-phosphate is further phosphorylated producing fructose-1, 6-bisphosphate

The reaction takes place in the presence of ATP. Phosphofructokinase is the enzyme and Mg^{2+} acts as cofactor. After fructose-1,6-bisphosphate is formed from the original sugar, no other pathways are available and the molecule must undergo the rest of the reactions of alcholic fermentation. This reaction is highly exothermic and irreversible.

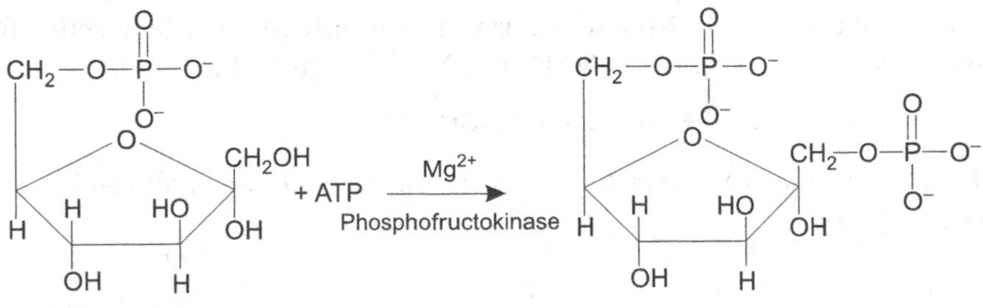

Fructose – 6 – phosphate

Fructose – 1,6-bisphosphate

Step 4. Fructose-1,6-bisphosphate is split into two 3-carbon fragments

The cleavage reaction here is the reverse of an aldol condensation. The enzyme that catalyses it is called aldolase.

Fructose – 1, 6 – bisphosphate $\rightleftharpoons$

$$\underset{\text{Dihydroxy - acetone - phosphate}}{\begin{array}{c} \quad\quad O \\ \quad\quad \| \\ CH_2-O-P-O^- \\ | \quad\quad\quad | \\ C=O \quad\quad O^- \\ | \\ CH_2OH \end{array}} \quad + \quad \underset{\text{D - Glyceraldehyde-3 - phosphate}}{\begin{array}{c} CHO \\ | \\ H-C-OH \quad\quad O \\ | \quad\quad\quad\quad \| \\ CH_2-O-P-O^- \\ | \\ O^- \end{array}}$$

Step 5. The dihydroxyacetone phosphate is converted to glyceraldehyde-3-phosphate

The enzyme that catalyses the reaction is triosephosphate isomerase. One molecule of glyceraldehyde has already been produced by the aldolase reaction. We now have a second molecule of glyceraldehyde-3-phosphate produced in this reaction. The original molecule of glucose, which contains six carbon atoms, has now been converted to two molecules of glyceraldehyde-3-phosphate, which contains 3-carbon atoms.

$$\text{Dihydrooxycetone-phosphate} \underset{\text{phosphateisomerase}}{\overset{\text{Triose}}{\rightleftharpoons}} \text{Glyceraldehyde-3-phosphate}$$

In the first five steps, one molecule of glucose is split into two 3-carbon compounds i.e., D-glyceraldehyde-3-phosphate. Two molecules of ATP are required for these reactions (Fig. 3.20).

D. The Conversion of Glyceraldehyde-3-Phosphate to Pyruvate

Uptil now glucose has been converted to two molecules of glycraldehyde-3-phosphate. We have not seen any oxidation reaction yet, but now we shall encounter them. In this phase of the pathway, ATP is produced instead of being used.

Step 6. The oxidation of glyceraldehyde-3-phosphate to 1,3-bisphosphoglycerate

Here NAD^+ is the coenzyme. The reaction involves the addition of a phosphate group to glyceraldehyde-3-phosphate as well as an electron-transfer reaction from glyceraldehyde-3-phosphate to NAD^+ i.e., NAD^+ is reduced to NADH

$$NAD^+ + 2H^+ + 2e \rightarrow NADH + H^+$$

The enzyme that catalyses this conversion is glyceraldehyde-3-phosphate dehydrogenase.

$$\underset{\substack{\text{Glyceraldehyde 3 -}\\ \text{phosphate}}}{\begin{array}{c} CHO \\ | \\ H-C-OH \quad\quad O \\ | \quad\quad\quad\quad \| \\ CH_2-O-P-O^- \\ | \\ O^- \end{array}} + NAD^+ + \underset{}{\begin{array}{c} O \\ \| \\ HO-P-O^- \\ | \\ O^- \end{array}} \underset{\substack{\text{3-phosphate}\\ \text{dehydrogenase}}}{\overset{\text{Glyceraldehyde -}}{\rightleftharpoons}} \underset{\substack{\text{1,3 - Bisphospho-}\\ \text{glycerate}}}{\begin{array}{c} O \quad\quad O \\ \| \quad\quad \| \\ C-O-P-O^- \\ | \quad\quad\quad | \\ H-C-OH \; O^- \; O \\ | \\ CH_2-O-P-O^- \\ | \\ O^- \end{array}}$$

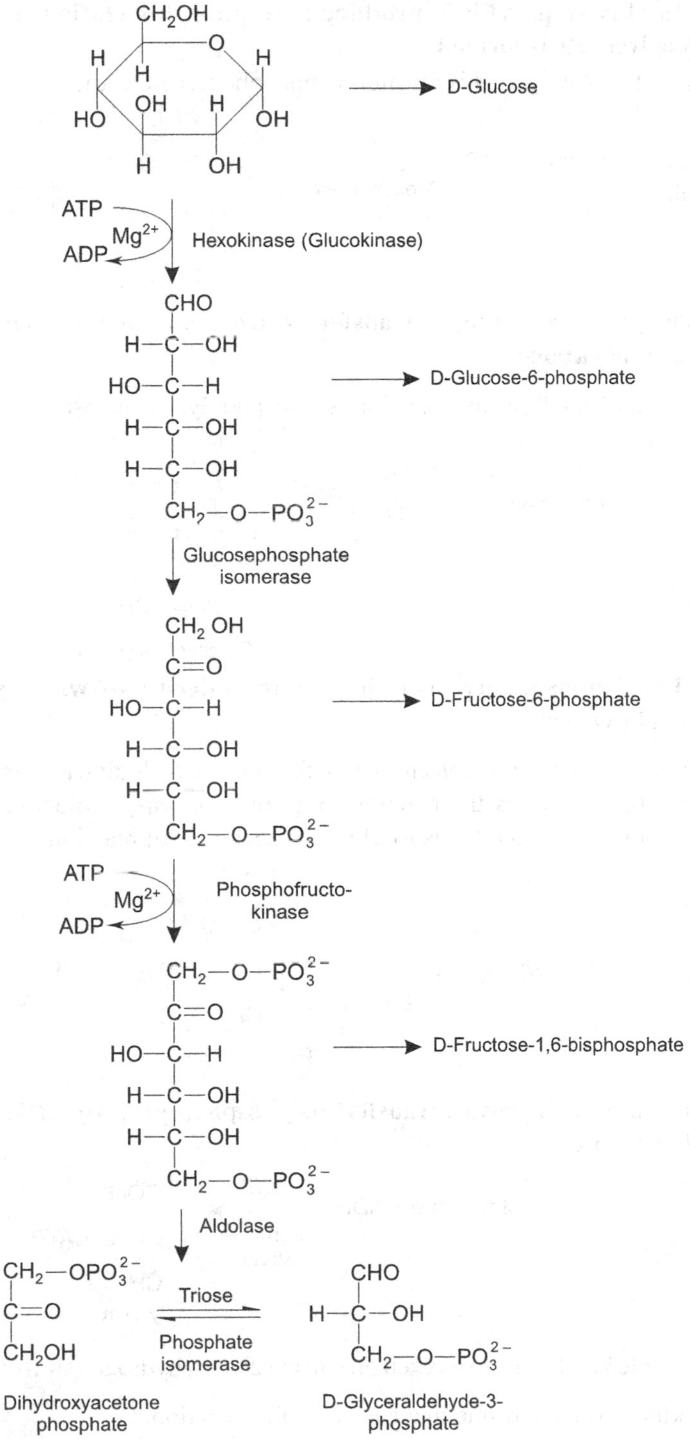

Fig. 3.20.

Step 7. In this step, ATP is produced by phosphorylation of ADP and 3-phosphoglycerate is formed

The enzyme that catalyses this reaction is phosphoglycerate kinase.

$$
\text{1,3-Bisphospho- + ADP} \underset{\substack{\text{Phosphoglycerate} \\ \text{kinase}}}{\overset{\text{Mg}^{2+}}{\rightleftharpoons}}
$$

glycerate

$$
\begin{array}{c}
\text{COO}^- \\
| \\
\text{H}-\text{C}-\text{OH} \quad \text{O} \\
| \qquad\qquad || \\
\text{H}_2\text{C}-\text{O}-\overset{\displaystyle |}{\underset{\displaystyle \text{O}^-}{\text{P}}}-\text{O}^- + \text{ATP}
\end{array}
$$

3-Phosphoglycerate

Step 8. The phosphate group is transferred from carbon 3 to carbon 2 of the glyceric acid backbone

The enzyme that catalyses this reaction is phosphoglyceromutase.

3-Phospho-
glycerate

$$
\underset{\substack{\text{Phosphoglycero-} \\ \text{mutase}}}{\overset{\text{Mg}^{2+}}{\rightleftharpoons}}
$$

$$
\begin{array}{c}
\text{O} \\
|| \\
\text{C}-\text{O}^- \quad \text{O} \\
| \qquad\qquad || \\
\text{H}-\text{C}-\text{O}-\overset{\displaystyle |}{\underset{\displaystyle \text{O}^-}{\text{P}}}-\text{O}^- \\
| \\
\text{CH}_2-\text{OH}
\end{array}
$$

2-Phosphoglycerate

Step 9. The 2-phosphoglycerate loses one molecule of water producing phosphoenol pyruvate

This reaction does not involve electron transfer; it is a dehydration reaction. Enolase, the enzyme that catalyses the reaction, requires Mg^{2+} as a cofactor. The water molecule that is eliminated binds to Mg^{2+} in the course of reaction.

2-Phosphoglycerate

$$
\underset{\text{Enolase}}{\overset{\text{Mg}^{2+}}{\rightleftharpoons}}
$$

$$
\begin{array}{c}
\text{O} \\
|| \\
\text{C}-\text{O}^- \quad \text{O} \\
| \qquad\qquad || \\
\text{C}-\text{O}-\overset{\displaystyle |}{\underset{\displaystyle \text{O}^-}{\text{P}}}-\text{O}^- + \text{H}_2\text{O} \\
|| \\
\text{CH}_2
\end{array}
$$

Phosphoenolpyruvate

Step 10. Phosphoenolpyruvate transfers its phosphate group to ADP, producing ATP and pyruvate

Phosphoenolpyruvate + ADP

$$
\underset{\substack{\text{Pyruvate} \\ \text{kinase}}}{\overset{\text{Mg}^{2+}}{\longrightarrow}}
$$

$$
\begin{array}{c}
\text{COO}^- \\
| \\
\text{C}=\text{O} + \text{ATP} \\
| \\
\text{CH}_3
\end{array}
$$

Pyruvate

Here the double bond shifts to oxygen on carbon 2 and a hydrogen shifts to carbon 3.

Pyruvate kinase is the enzyme that catalyses this reaction.

Thus, in the second phase of alcoholic fermentation glyceraldehyde-3-phosphate is converted to pyruvate. The reactions are summarised in Fig. 3.21. These reactions yield four molecules of ATP, two for each molecule of pyruvate produced.

E. The Conversion of Pyruvate to Acetaldehyde and then to Ethanol

Pyruvate is decarboxylated to produce acetaldehyde. The enzyme that catalyses this reaction is pyruvate decarboxylase. This enzyme requires Mg^{2+} and a cofactor, thiamine pyrophosphate (TPP) (Thiamine is vitamin B_1). Carbon dioxide splits off, leaving a two carbon fragment covalently bonded to TPP. Thus, acetaldehyde is produced. This step is irreversible.

$$
\begin{array}{c}
COO^- \\
| \\
C=O \\
| \\
CH_3
\end{array}
\quad
\xrightarrow[\substack{\text{decarboxylase} \\ \text{TPP acts as} \\ \text{cofactor}}]{\text{Pyruvate}}
\quad
CH_3-CHO + CO_2
$$

Pyruvate   Acetaldehyde

The carbon dioxide produced is responsible for the bubbles in beer and in sparkling wines.

Acetaldehyde is then reduced to produce ethanol and at the same time one molecule of NADH is oxidised to NAD^+, for each molecule of ethanol produced.

$$CH_3CHO + NADH \xrightarrow[\text{dehydrogenase}]{\text{alcohol}} CH_3CH_2OH + NAD^+$$

This step provides recycling of NAD^+ and thus allows further anaerobic reactions.

The net reaction for alcoholic fermentation is

$$\text{Glucose} + 2ADP + 2Pi + 2H^+ \rightarrow 2 \text{ Ethanol} + 2 ATP + 2CO_2 + 2H_2O.$$

NAD^+ and NADH do not appear in the net equation. It is essential that recycling of NADH to NAD^+ takes place at this step so that further anaerobic oxidation takes place. Alcohol dehydrogenase is the enzyme that catalyses the conversion of acetaledehyde to ethanol. It is a NADH-linked dehydrogenase.

3.11   GLYCOLYSIS

A. Introduction

We have seen earlier that carbohydrates are the major energy sources. In plants energy is stored in the form of starches, while glycogen is the storage polysaccharide present in animals. Glycogen is found in animal cells in granules similar to the starch granules in plant cells. Glycogen granules are observed in well-fed liver and muscle cells.

Glycogen is a branched-chain polymer of α-D-glucose and in this respect it is similar to the amylopectin fraction of starch. Like amylopectin, glycogen consists of α (1 → 4) linkages with α (1 → 6) linkages at the branch points. However,

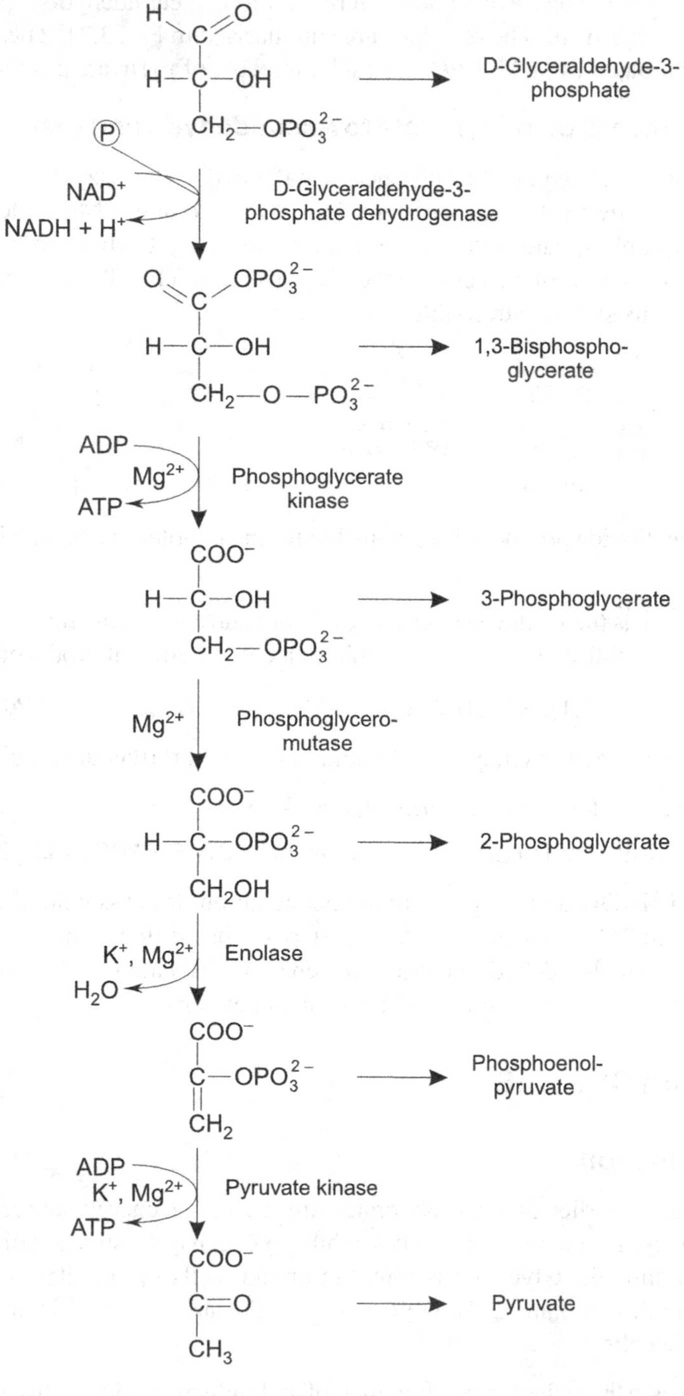

The pathway begins with a molecule:

$$\begin{array}{c} H \diagdown \\ C = O \\ H - C - OH \\ CH_2 - OPO_3^{2-} \end{array} \quad \longrightarrow \quad \text{D-Glyceraldehyde-3-phosphate}$$

NAD$^+$ / NADH + H$^+$ — D-Glyceraldehyde-3-phosphate dehydrogenase

$$\begin{array}{c} O = C \diagup OPO_3^{2-} \\ H - C - OH \\ CH_2 - O - PO_3^{2-} \end{array} \quad \longrightarrow \quad \text{1,3-Bisphospho-glycerate}$$

ADP, Mg^{2+} / ATP — Phosphoglycerate kinase

$$\begin{array}{c} COO^- \\ H - C - OH \\ CH_2 - OPO_3^{2-} \end{array} \quad \longrightarrow \quad \text{3-Phosphoglycerate}$$

Mg^{2+} — Phosphoglycero-mutase

$$\begin{array}{c} COO^- \\ H - C - OPO_3^{2-} \\ CH_2OH \end{array} \quad \longrightarrow \quad \text{2-Phosphoglycerate}$$

K$^+$, Mg^{2+} / H$_2$O — Enolase

$$\begin{array}{c} COO^- \\ C - OPO_3^{2-} \\ \parallel \\ CH_2 \end{array} \quad \longrightarrow \quad \text{Phosphoenol-pyruvate}$$

ADP, K$^+$, Mg^{2+} / ATP — Pyruvate kinase

$$\begin{array}{c} COO^- \\ C = O \\ CH_3 \end{array} \quad \longrightarrow \quad \text{Pyruvate}$$

Fig. 3.21.

glycogen is more highly branched than amylopectin. Branch points occur about every 10 residues in glycogen and every 25 residues in amylopectin. In glycogen the average chain length is 13 glucose residues and there are 12 layers of branching.

When we digest a meal high in carbohydrates, we have a supply of glucose that exceeds our immediate needs. Therefore, excess glucose is stored in our body in the form of glycogen primarily in liver and muscle. When we need energy, various degradative enzymes break down glycogen to smaller polysaccharides or free glucose units. In muscle glucose-6-phosphate, obtained from glycogen breakdown, enters the glycolytic pathway directly rather than being hydrolysed to glucose first. However, liver glycogen breaks down to glucose-6-phosphate which is hydrolysed to give glucose. The release of glucose from the liver by this breakdown of glycogen replenishes the supply of glucose in the blood.

The conversion of glycogen to glucose-6 phosphate takes place in three steps.

In the **first reaction** each glucose residue cleaved from glycogen reacts with inorganic phosphate to give glucose-6-phosphate. This cleavage reaction is phosphorolysis and not hydrolysis. The enzyme that catalyses this reaction is known as glycogen phosphorylase, which cleaves one glucose unit at a time from the non-reducing end of a branch of glycogen to produce glucose-1-phosphate and the remainder of the glycogen molecule. This then enters the metabolic pathways of carbohydrate breakdown

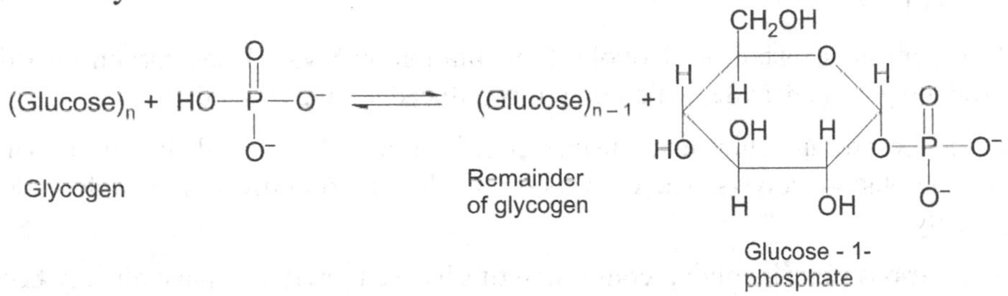

In the **second reaction**, glucose-1-phosphate isomerises to give glucose-6-phosphate by the enzyme phosphoglucomutase.

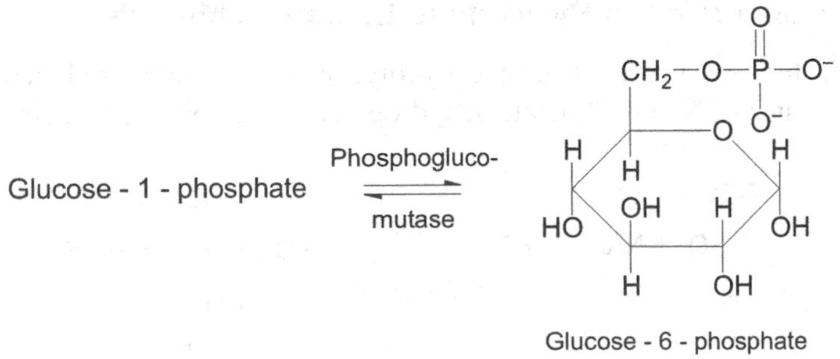

Complete breakdown of glycogen also requires debranching enzymes that degrade the α $(1 \rightarrow 6)$ linkages and thereby expose additional $1 \rightarrow 4$ glycosidic linkages to be attacked by phosphorylase.

When an organism needs energy quickly, glycogen breakdown is important. Muscle tissue can mobilise glycogen more easily than fat and can do so anaerobically. When we do low-density exercise such as jogging, fat is the preferred food, but as the intensity increases, muscle and liver glycogen becomes more important.

Thus, glycolysis is the first stage of glucose metabolism. It is an anaerobic process and gives only a small amount of energy in the form of only two molecules of ATP. We will see that the complete aerobic oxidation of glucose to carbon dioxide and water *via* the citric acid cycle yields the high energy equivalent of 30 – 32 molecules of ATP.

B. Overall Pathway in Glycolysis

We have seen the various steps in alcoholic fermentation. After the formation of glucose-6-phosphate, glycolysis and alcoholic fermentation follow a common pathway under anaerobic conditions until pyruvate is formed. In glycolysis, pyruvate is converted to lactic acid, which will eventually be exported from the muscle to the liver.

When pyruvate is formed, it can have one of several fates:

Under aerobic conditions, pyruvate loses carbon dioxide and the remaining two carbon atoms become linked to coenzyme A as an acetyl group to form acetyl-CoA which then enters the citric acid cycle. There are two fates for pyruvate in anaerobic metabolism.

In organisms capable of alcoholic fermentation, pyruvate loses carbon dioxide producing acetaldehyde, which in turn is reduced to give ethanol.

In the second anaerobic metabolism, lactate is produced, especially in muscle. This is anaerobic glycolysis. The conversion of glucose to pyruvate is simply called glycolysis.

The various steps involving conversion of glucose to pyruvate have already been described under alcoholic fermentation. The same reactions take place in anaerobic glycolysis. We give below the conversion of pyruvate to lactate in muscle.

C. The Conversion of Pyruvate to Lactate in Muscle

This is the final reaction of anaerobic glycolysis; *i.e.*, pyruvate is reduced to lactate in the presence of NADH. Lactate dehydrogense is the enzyme that catalyses this reaction.

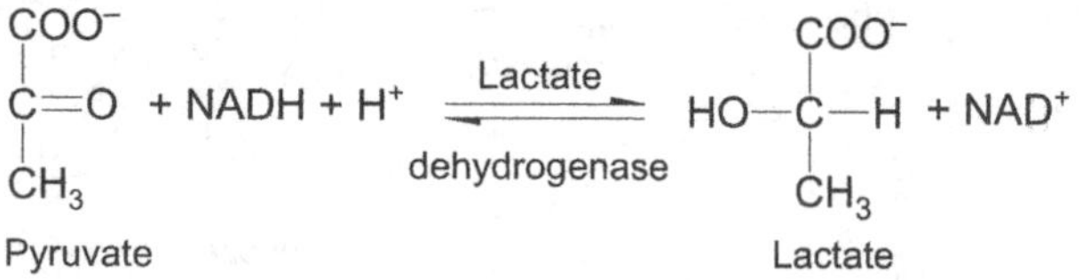

$$
\underset{\text{Pyruvate}}{\overset{\displaystyle COO^-}{\underset{\displaystyle CH_3}{|}}\, C{=}O} + NADH + H^+ \;\underset{\text{dehydrogenase}}{\overset{\text{Lactate}}{\rightleftharpoons}}\; \underset{\text{Lactate}}{HO{-}\overset{\displaystyle COO^-}{\underset{\displaystyle CH_3}{|}}C{-}H} + NAD^+
$$

The NADH produced from NAD^+ by the earlier oxidation of glyceraldehyde-3-phosphate is used up with no net change in the relative amounts of NADH and

NAD^+ in the cell. This regeneration is required under anaerobic conditions in the cell so that NAD^+ will be present for further glycolysis to take place.

D. Control Points in Glycolysis

Pathways can be shut down if an organism has no immediate need for their products. This saves energy for the organism.

In glycolysis, there are three control points. The first is the reaction of glucose to glucose-6-phosphate which is catalysed by hexokinase. This conversion is inhibited by glucose-6-phosphate itself.

The second control point is the production of fructose-1,6-bisphosphate. This step is catalysed by phosphofructokinase and is inhibited by ATP.

The last point is the conversion of phosphoenol pyruvate to pyruvate, catalysed by pyruvate kinase. This step is also inhibited by ATP (Fig. 3.22).

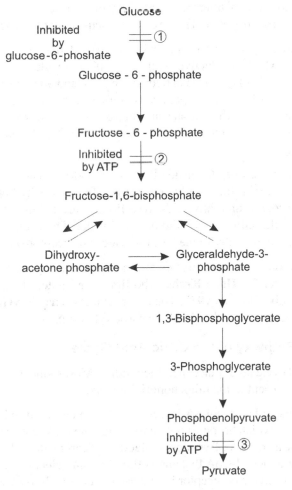

Fig. 3.22 Control point in clycolysis.

3.12 CITRIC ACID CYCLE

A. Introduction

We have seen that glycolysis releases only a very small amount of chemical energy which is available in the glucose molecule. Much more energy is released when the glucose molecule is oxidised completely to CO_2 and water. This is **aerobic metabolism**. The evolution of aerobic metabolism by which nutrients are oxidised to carbon dioxide and water was an important step in the history of life on earth. Organisms can obtain far more energy from nutrients by aerobic oxidation than by anaerobic oxidation. We have seen earlier that glycolysis produces only two molecules of ATP for each molecule of glucose metabolised. However, in complete aerobic oxidation each molecule of glucose produces about 30 to 32 molecules of ATP.

Three processes play roles in aerobic metabolism: the citric acid cycle, electron transport and oxidative phosphorylation. Here we shall discuss only the citric acid cycle.

Metabolism consists of catabolism, which is the oxidative breakdown of nutrients and anabolism, which is reductive synthesis of biomolecules. The citric acid cycle is amphibolic; *i.e.*, it plays a role in both catabolism and anabolism. The citric acid cycle, as the hub of metabolic pathways, serves to connect the breakdown and synthesis of proteins, carbohydrates and lipids. Most of the metabolites of major nutrients can be fed into the citric acid cycle as acetyl-CoA and then oxidised to produce energy.

The citric acid cycle operates inside the mitochondrion which acts as its engine room. Here, metabolic fuels – glucose derived from carbohydrates, amino acids derived from proteins and fatty acids from lipids – are fed into the cycle and will ultimately be oxidised to carbon dioxide and water. The energy is transferred to electron carriers and finally to the terminal electron acceptor-oxygen.

There are two other common names for the citric acid cycle. One is the **Krebs cycle,** named after **Sir Hans Krebs**, who first investigated the pathway and was awarded the Nobel Prize in 1953. The other name is **tricarboxylic acid cycle** (or TCA cycle) because of acids with three carboxyl groups.

B. Overall Pathway of the Citric Acid Cycle

The citric acid cycle takes place in mitochondria. Most of the enzymes of the citric acid cycle are present in the mitochondrial matrix.

Pyruvate produced by glycolysis is oxidised to one carbon dioxide molecule and one acetyl group which becomes linked to coenzyme A to give acetyl-CoA. It then enters the citric acid cycle. Two more molecules of carbon dioxide are produced for each molecule of acetyl-CoA that enters the cycle and electrons are transferred in the process. The electron acceptor in all cases but one is NAD^+, which is reduced to NADH. In one case the electron acceptor is FAD (flavin adenine dinucleotide),

This takes up two electrons and two hydrogen ions to produce $FADH_2$. The electrons are passed from NADH and $FADH_2$ through several stages and the final acceptor is oxygen with water as the product.

There are eight steps in the citric acid cycle (Fig. 3.23), each catalysed by a different enzyme. Four of the eight steps – steps 3, 4, 6 and 8 – are oxidation steps.

The oxidising agent is NAD^+ in all except step 6, in which it is FAD. In step 5, a molecule of GDP (guanosine diphosphate) is phosphorylated to produce GTP.

In the first reaction of the cycle, the two carbon acetyl group condenses with four-carbon oxaloacetate to produce six-carbon citrate ion. In the next few steps citrate isomerises and then oxidative decarboxylation takes place to give α-keto-glutarate, which again is oxidatively decarboxylated to produce the four-carbon compound succinate. The cycle is completed by regeneration of oxaloacetate from succinate *via* the fumarate in several steps.

C. Conversion of Pyruvate to Acetyl-Co A

We have seen earlier that pyruvate is generated from glycolysis under anaerobic conditions. It moves from the cytosol into the mitochondrion *via* a specific transporter. Here pyruvate is converted to carbon dioxide and the acetyl portion of acetyl-Co A. An oxidation reaction precedes the transfer of the acetyl group to the Co A.

The whole process involves several enzymes, all of which are part of the pyruvate dehydrogenase complex. The overall reaction is

Pyruvate + Co A—SH + NAD^+ → Acetyl-CoA + CO_2 + H^+ + NaDH.

There is a —SH group at one end of the CoA; it is frequently shown in equations as CoA—SH. The reaction is catalysed by four cofactors, TPP, FAD, Mg^{2+} and lipoic acid.

D. The Individual Reactions of the Citric Acid Cycle

There are eight steps in the citric acid cycle. These reactions and the enzymes that catalyse them are listed in Table 3.5. Let us study these reactions one by one.

Step 1. Formation of Citrate: This is the first step in the citric acid cycle. It involves the reaction of acetyl-CoA and oxaloacetate to form citrate and CoA-SH. It is a condensation reaction because a new carbon-carbon bond is formed. The reaction is catalysed by the enzyme citrate synthase

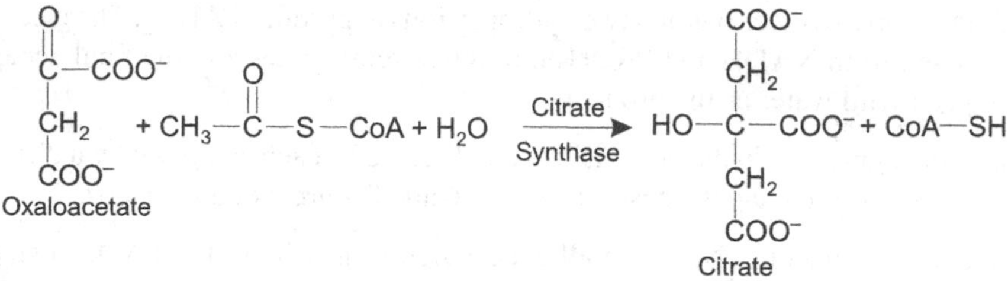

Step 2. Isomerisation of Citrate to Isocitrate: This reaction is catalysed by aconitase. The enzyme requires Fe^{2+} as cofactor. The enzyme is able to select one end of the citrate molecule in preference to the other.

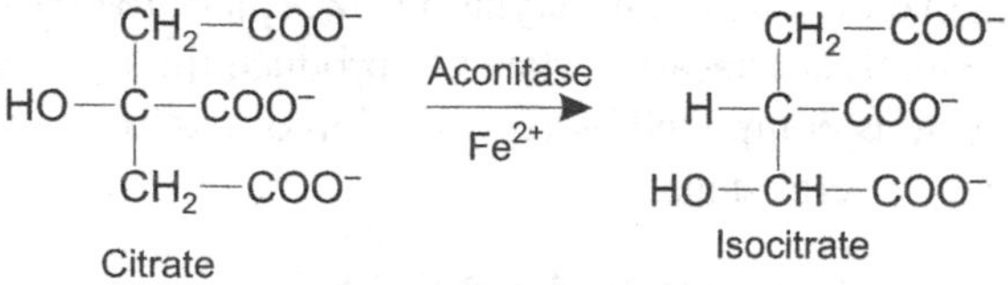

TABLE 3.5 The Reactions of the Citric Acid Cycle

Step	Reaction	Enzyme
1.	Acetyl-CoA + Oxaloacetate + H_2O → Citrate + CoA-SH	Citrate synthase
2.	Citrate → Isocitrate	Aconitase
3.	Isocitrate + NAD^+ → α-Ketoglutarate + NADH + CO2 + H^+	Isocitrate dehydrogenase
4.	α-Ketoglutarate + NAD^+ + CoA-SH → Succinyl-CoA + NADH + CO_2 + H^+	α-Keto-glutarate dehydrogenase
5.	Succinyl-CoA + GDP + P → succinate + GTP + CoA-SH	Succinyl-CoA synthetase
6.	Succinate + FAD → Fumarate + $FADH_2$	Succinate dehydrogenase
7.	Fumarate + H_2O → L-Malate	Fumarase
8.	L-Malate + NAD^+ → Oxaloacetate + NADH + H^+	Malale-dehydrogenase

Step 3. Formation of α-ketoglutarate and CO_2 (First Oxidation Step)

This is the oxidative decarboxylation of isocitrate to α-ketoglutarate and carbon dioxide. The enzyme that catalyses it is isocitrate dehydrogenase. This is the first reaction in which NADH is produced.

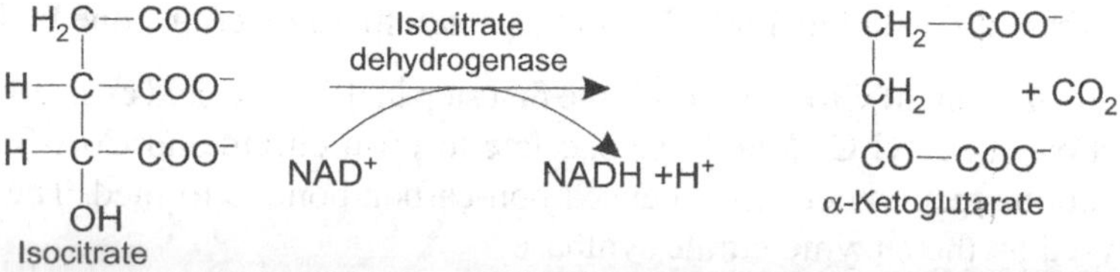

Step 4. Formation of Succinyl-CoA and CO_2 (Second Oxidation Step)

This reaction occurs in several stages and is catalysed by an enzyme system called α-ketoglutarate dehydrogenase complex. The cofactors used are TPP, FAD, Mg^{2+} and lipoic acid.

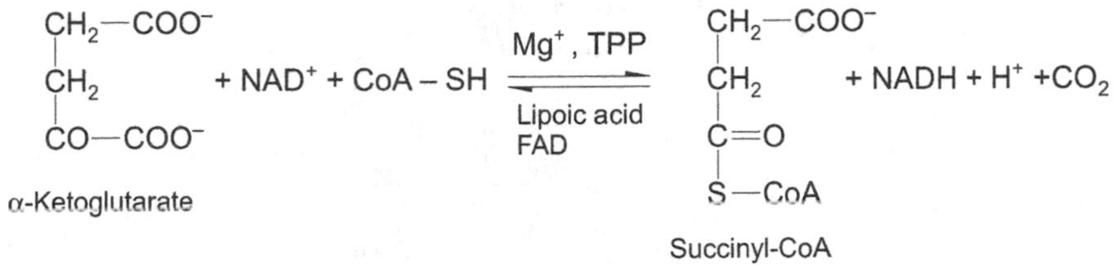

$$\begin{array}{c} CH_2-COO^- \\ | \\ CH_2 \\ | \\ CO-COO^- \end{array} + NAD^+ + CoA-SH \underset{\substack{\text{Lipoic acid}\\ \text{FAD}}}{\overset{Mg^+,\, TPP}{\rightleftharpoons}} \begin{array}{c} CH_2-COO^- \\ | \\ CH_2 \\ | \\ C=O \\ | \\ S-CoA \end{array} + NADH + H^+ + CO_2$$

α-Ketoglutarate   Succinyl-CoA

Step 5. Formation of Succinate

In this step, succinyl-CoA is hydrolysed to produce succinate and CoA-SH. The reaction is catalysed by the enzyme succinyl-CoA synthetase. The accompanying reaction is the phosphorylation of GDP to GTP. The energy required for the phosphorylation of GDP to GTP is provided by the hydrolysis of succinyl-CoA to produce succinate and CoA.

$$\begin{array}{c} COO^- \\ | \\ CH_2 \\ | \\ CH_2 \\ | \\ C=O \\ | \\ S-CoA \end{array} + GDP + P \xrightarrow[\text{CoA synthetase}]{\text{Succinyl}} \begin{array}{c} COO^- \\ | \\ CH_2 \\ | \\ CH_2 \\ | \\ COO^- \end{array} + GTP + CoA-SH$$

Succinyl-CoA   Succinate

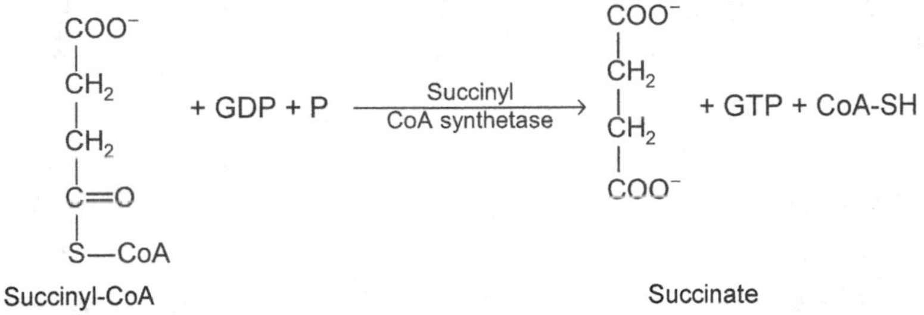

Step 6. Formation of Fumarate (Oxidation-Step)

Succinate is oxidised to fumarate, a reaction that is catalysed by the enzyme succinate dehydrogenase. In this reaction, FAD is the coenzyme which is reduced to $FADH_2$.

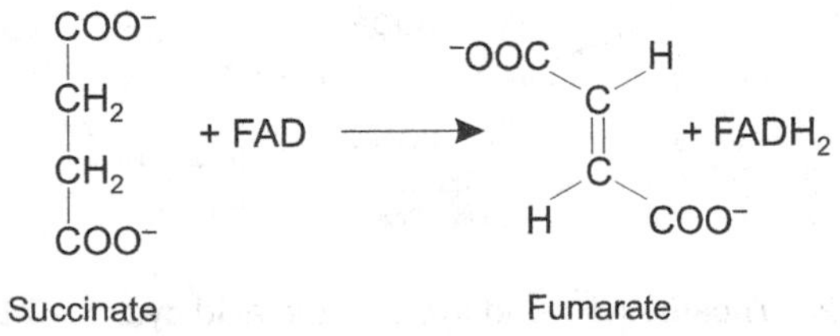

$$\begin{array}{c} COO^- \\ | \\ CH_2 \\ | \\ CH_2 \\ | \\ COO^- \end{array} + FAD \longrightarrow \begin{array}{c} {}^-OOC \quad\quad H \\ \diagdown \;\; / \\ C \\ \| \\ C \\ / \;\; \diagdown \\ H \quad\quad COO^- \end{array} + FADH_2$$

Succinate   Fumarate

Step 7. Formation of L-Malate

This reaction is catalysed by the enzyme fumarase. Water is added across the double bond of fumarate in a hydration reaction to give malate. Here the reaction is stereospecific. It gives only L-malate.

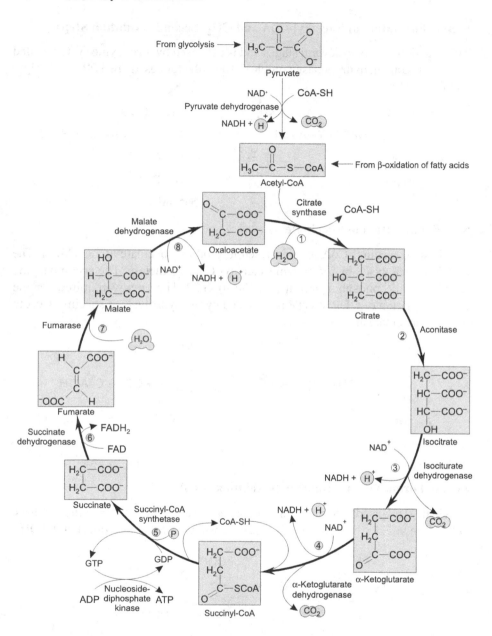

Fig. 3.23 Tricarboxylic acid cycle (Citric acid cycle) Krebs cycle.

$$\underset{\text{Fumarate}}{\overset{\text{-OOC}}{\underset{\text{H}}{\overset{\text{H}}{\underset{\text{COO}^-}{\overset{\text{C}}{\shortparallel}}}}}} + H_2O \longrightarrow \underset{\text{L - Malate}}{\overset{COO^-}{\underset{COO^-}{\overset{HO-C-H}{\overset{|}{\underset{CH_2}{\overset{|}{}}}}}}}$$

Step 8. Regeneration of Oxaloacetate (Final Oxidation Step)

This is the final step in which malate is oxidised to oxaloacetate. The reaction is catalysed by malate dehydrogenase. Another molecule of NAD^+ is reduced to NADH. The oxaloacetate produced can then react with another molecule of acetyl-CoA to start another round of the cycle.

$$\overset{COO^-}{\underset{COO^-}{\overset{|}{\underset{CH_2}{\overset{HO-C-H}{\overset{|}{}}}}}} + NAD^+ \longrightarrow \overset{COO^-}{\underset{COO^-}{\overset{|}{\underset{CH_2}{\overset{C=O}{\overset{|}{}}}}}} + NADH + H^+$$

The oxidation of pyruvate by the pyruvate dehydrogenase complex and the citric acid results in the production of three molecules of CO_2. During this period one molecule of GDP is phosphorylated to GTP, one molecule of FAD is reduced to $FADH_2$ and four molecules of NAD^+ are reduced to NADH. Three molecules of NADH come from the citric acid cycle and one comes from the reaction of the pyruvate dehydrogenase complex.

From the pyruvate dehydrogenase complex

$$\text{Pyruvate} + \text{CoA-SH} + NAD^+ \rightarrow \text{Acetyl-CoA} + \text{NaDH} + CO_2 + H^+$$

From the citric acid cycle

$$\text{Acetyl-CoA} + 3NAD^+ + FAD + GDP + P + 2H_2O \rightarrow 2CO_2 + \text{CoA-SH} + 3\text{NaDH} + 3H^+ + FADH_2 + GTP$$

3.13 BIOCATALYSIS–IMPORTANCE IN GREEN CHEMISTRY AND THE CHEMICAL INDUSTRY

A. Introduction to Green Chemistry

Organic chemistry deals with a large number of compounds, both synthetic as well as natural. The study and synthesis of these compounds involves carrying out reactions in the laboratory as well as on the industrial scale. This may lead to environmental problems. In recent years, the influence of the chemical industry on the environment has been in focus. Now the term "Green Chemistry" is used for the technology that reduces or eliminates the use or generation of hazardous substances in the design, manufacture and application of chemical products, Biocatalytic processes will certainly contribute in this respect.

With the development of "Green Chemistry" the chemists all over the world are motivated to use such synthetic methods that reduce the use and generation of hazardous substances or byproducts. Thus, Green Chemistry focuses on developing such processes for the synthesis of new products but also green synthesis of existing chemicals. This has been possible in many cases to avoid the use of hazardous organic solvents or by the replacement of organic solvents by water.

In recent years, the manufacture of chemicals has seen large advances in "Green Chemistry". The term "Green Chemistry" has come to embrace some well-defined concepts in the manufacture of chemicals. As enunciated by **Paul Anastas**, considered to be the "father of green chemistry", a chemical process is greener when it meets certain well-defined criteria. Among those characteristics that enable a process to be called greener, many are common sense objectives. The basic aim is to devise pathways for the prevention of pollution. We discuss below some of the basic principles of green chemistry.

B. Principles of Green Chemistry

i. Use of Solvents

The use of solvents and separating agents should be avoided wherever possible. A number of solvents are used in a large number of reactions due to their excellent solvent properties. The halogenated solvents like methylene chloride, chloroform, carbon tetrachloride, benzene and other aromatic hydrocarbons have been proved to be carcinogenic in nature. A greener chemical process avoids the use of such chlorinated hydrocarbons that are slow to break down or difficult to dispose of. Reactions using these solvents will not be considered green.

The solvent selected for a particular reaction should not cause any environmental pollution and health hazard. Use of solvents that are more biodegradable is preferred in green chemistry. Green reactions, if possible, should be carried out in the aqueous phase or without the use of a solvent in the solid phase. Moreover, the pathway for a reaction should be such that there is no need for separation or purification.

ii. Minimum Byproducts

The selection of pathway for a green reaction should be such that formation of waste or byproducts should be minimum or absent. The waste that is discharged in the atmosphere, sea or land causes pollution and requires expenditure for cleaning up. It is better to prevent waste byproducts than to treat and remove the byproducts after they are formed.

iii. Maximum Utilisation of all the Materials Used in the Process

This is the basic principle of green chemistry. The reaction is considered to be green if there is maximum utilisation of the starting materials or reagents in the final product *i.e.,* the percentage atom utilisation should be maximum. Final products must contain all the materials used. The reactions should go to 100 percent completion and steps involved should be minimum.

The greenness of a process is summarised by a term called the "**E-factor**" which measures the total amount of solvents, reagents and consumables used in kilograms and divides it by the kilograms of the final product produced.

iv. Use of Catalytic Reagents

The use of catalytic reagents in carrying out organic reactions is very important in developing green synthesis. We know a catalyst facilitates the reaction without being consumed or without being incorporated into the final product. Catalysts are also selective in their action *i.e.,* the degree of the reaction that takes place is controlled. By the use of catalysts we can enhance the utilisation of the starting materials and reduce the formation of byproducts. The activation energy of the reaction is reduced and also the temperature necessary for the reaction is lowered. Use of phase transfer catalysts and crown ethers is very well-known in developing green synthesis.

v. Energy Requirements

We know that the time required for completion of the reaction should be minimum, so that the least amount of energy is required.

The use of a catalyst has the advantage of lowering the requirement of energy of a reaction. The process should be designed in such a way that there is no need of separation or purification. If it is possible, the energy to a reaction can be supplied photochemically by using microwave or sonication. Use of microwaves and sonication saves a lot of energy and gives much better yields.

vi. Synthesis in Solid State

Organic synthesis carried out in the solid state *i.e.,* solvent-free synthesis, is mostly green. A large number of reactions have now been performed without any solvent.

vii. Selection of Starting Materials

Starting materials should be obtainable from renewable sources. They should not cause any harm to the person handling them.

Substances like carbon dioxide (generated from natural sources) and methane gas (obtained from natural gas etc.) are available in abundance, so these are considered as renewable starting materials.

viii. Use of Protecting Groups

Unnecessary use of protecting groups should be avoided wherever possible. Their use results in lot of wastage.

ix. Products Should be Biodegradable

It is very important that the products that are synthesised by green reactions should be biodegradable. They should not be persistant chemicals, such chemicals which are non biodegradable remain in the same form in the environment or are taken up by the various plants and animals and accumulate in their systems. This problem is generally associated with pesticides and plastics and a large number of other organic molecules.

x. Biocatalysis in Green Chemistry

Biocatalysis, *i.e.,* the use of enzymes as catalysts belongs to the fundamental strategies of "Green Chemistry". Under green reactions a large number of methods have been developed where enzymes have been used for carrying out the reactions. We know enzymes possess good catalytic properties and can be used for carrying out simple organic reactions like oxidations, reductions, hydrolysis, hydroxylations, etc. In addition to their selectivity, enzymes are also stereoselective. We require milder reaction conditions and reactions are free of undesirable products. Biocatalytic processes will certainly contribute in controlling the environmental pollution by eliminating the generation of hazardous substances.

C. Biocatalysis in Chemical Industry

The use of growing cells in order to catalyse chemical reactions is not new. We have already studied the manufacture of ethyl alcohol from molasses and conversion of lactose into lactic acid. These fermentation processes have been known to mankind since prehistoric times. However, the use of pure enzymes or cells, in aqueous or organic media in order to catalyse specific organic reactions, or biotransformations has emerged in recent years. This field of research is at present undergoing rapid development. There is both general and industrial interest to develop biocatalysis for organic chemistry as well as for the chemical industry. A collaboration of organic chemists, biochemists and molecular geneticists is needed for success. The development of biocatalytic methods undoubtedly involves a strong cooperation of these researchers for extending the scope of enzyme activities and reactions to be used in chemical industry. The use of biocatalysts in the chemical industry is clearly an activity within the field of biotechnology. Many conventional chemical enterprises have already made use of biotechnology to improve traditional chemical and engineering processes. The use of biocatalysis is one of the typical applications.

D. Why Biocatalysis?

Compared to non-enzymatic chemical catalysts biocatalysts (enzymes or whole cell microorganisms) are known to present some interesting and advantageous features. These are discussed below:

i. High Efficiency

Enzymes possess good catalytic properties. Enzymatic reactions are highly specific. They are free of undesirable reactions and give maximum yield of the products. Enzymatic reactions are much faster. Reactions not possible otherwise can be achieved by enzymes.

ii. Mild Reaction Conditions

All kinds of reactions of chemical products are conducted under so many severe conditions such as high temperatures, high pressures and specific catalysts. In contrast to this various complex reactions within organisms are conducted by specific enzyme catalysts at moderate temperatures and pressures. That is enzymatic reactions are carried out under mild environmentally friendly operation conditions. The reactions can be carried out in aqueous media at optimum pH, temperature and pressure.

iii. Easy Availability

Enzymes used in biocatalysis are easily available. The sources of enzymes include animals, plants and microorganisms. Among these three categories, microorganisms are the primary source, which is decided by many reasons. Microorganisms have quite a large number of varieties with rich resources and they are easy to cultivate. Moreover, under manual control it is suitable for microorganisms to be applied in large scale industrialised production, which can accelerate the application of biotechnology in the chemical industry. Thus, isolation of enzymes from microorganisms made it possible to use them on the industrial scale.

iv. Versatility

Enzymatic reactions are highly versatile. All types of organic reactions can be carried out by the use of specific enzymes. The most important application of enzymes in organic synthesis involves enzymatic oxidations, hydroxylations, hydrolysis, reductions and isomerisations.

v. High Selectivity

We have already studied the high selectivity of enzymatic reactions. Enzymes are highly specific in their reactivities and possess stereoselectivity, *i.e.,* they act on only one of a pair of optical isomers. Enzymatic reactions are stereospecific and stereoselective.

The selectivity and particularly the stererochemical properties observed when biocatalysts act on their natural substrates, if extended to the usual objects of organic synthesis, may be of outstanding interest for the preparation of the enantiopure compounds now required for use as drugs and agrochemicals.

We know enzymes possess poor stability. Their isolation sometimes requires complex purification procedures. All these factors affect the application of enzymes in the chemical industry. In order to solve such problems, scientists combine an enzyme with carrier materials containing numerous gaps such as agar and cellulose by various means of physical adsorption, chemical bonding or biological integration. This kind of immobilisation technique of the enzyme can result in a continuous process of catalytic reactivity of the enzyme.

In order to maintain catalytic activity of a biological enzyme in the process of extracting, purifying and immobilising people usually make use of cell immobilisation or microorganism immobilisation, which doesn't need to separate enzymes with its cells and organisms but to immobilise the whole cell or the organism.

This kind of immobilisation is useful for synthetic products which are produced through multistep reactions catalysed by various enzymes in organisms or their cells. The multistep reactions within organisms can be finished in one step by microorganisms.

❑❑❑

4

Pyrimidines, Purines, and Nucleic Acids

> ## Learning Objectives

In this chapter we will study

- Importance of Nucleic Acids
- Pyrimidines — Aromatic Character, Chemical Properties and Synthetic Methods
- Structure, Reactions and Synthesis of Uracil, Thymine and Cytosine
- Purines — Structure, Chemical Reactions and Synthesis
- Structure, Reactions and Synthesis of Adenine and Guanine
- Ureides
- Uric Acid — Structure and Synthesis
- Caffeine — Isolation, Structure and Synthesis
- Nucleosides and Nucleotides
- Nucleic Acids — Deoxyribonucleic Acid and Ribonucleic Acid
- Structure of DNA – Conformational Variations in DNA
- Denaturation and Renaturation of DNA, Replication of DNA
- Principal Kinds of RNA
- The Genetic Code, Transcription, Translation and Gene Therapy

4.1 INTRODUCTION AND BIOLOGICAL IMPORTANCE OF NUCLEIC ACIDS

Nucleic acids play an important role in all biological systems. They help in storing the genetic information and in protein biosynthesis. Nucleic acids preserve hereditary information. They transcribe and translate it in a way that allows the synthesis of all the proteins of the cell.

There are two types of nucleic acids: ribonucleic acid (RNA) and deoxyribonucleic acid (DNA). All livings things have some form of nucleic acids, which allow the translation of proteins in order to carry out important functions of the cell. DNA is the hereditary material in all living organisms. In plant viruses, RNA forms the genetic material where DNA is absent. Genetic information is passed from cell to cell by a process called DNA replication. We shall study replication separately at a later stage.

Both DNA and RNA are biopolymers in which the repeating monomer units are called **nucleotides**. A nucleotide is made up of one unit each of phosphate, a sugar and a heterocyclic base. The heterocyclic base is either a pyrimidine or a purine. In RNA the sugar is the pentose D-**ribose** and the bases present are **uracil, cytosine, adenine** and **guanine**.

In DNA, the sugar is 2-**deoxy-D-ribose** and the heterocyclic bases are the same except that **thymine** replaces uracil.

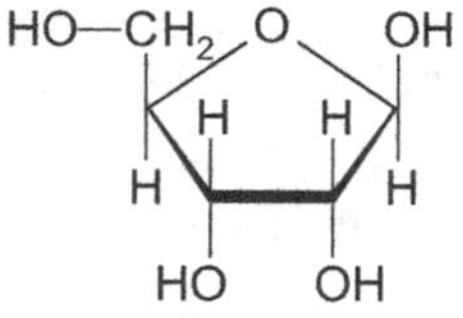

D-Ribose
(RNA)

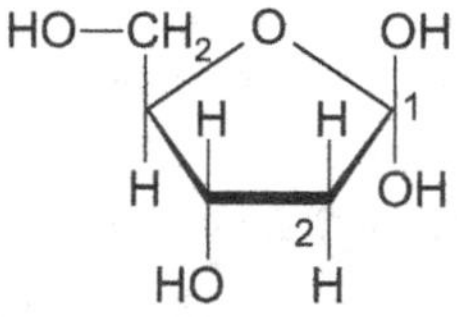

2-Deoxy-D-ribose
(DNA)

Pyrimidines

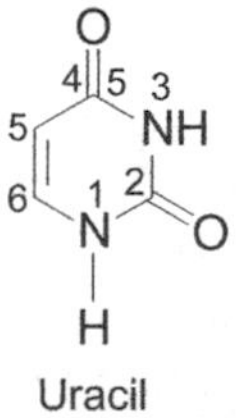

Uracil
(RNA)

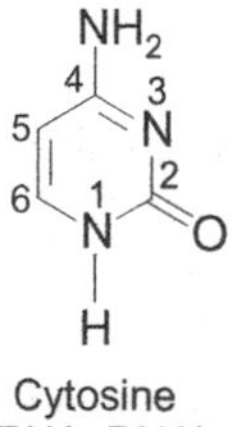

Cytosine
(RNA, DNA)

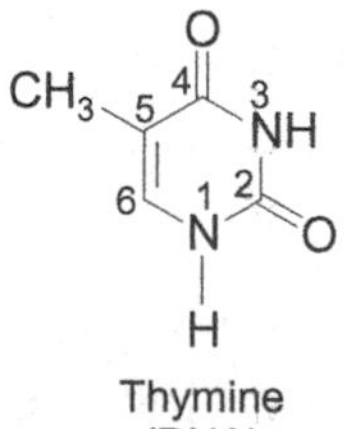

Thymine
(DNA)

Purines

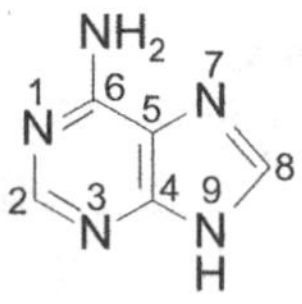

Adenine
(RNA, DNA)

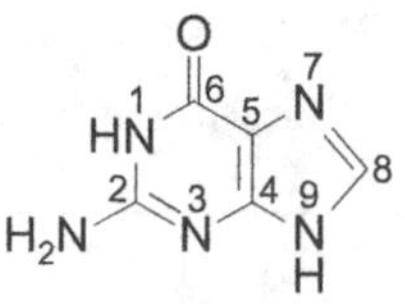

Guanine
(RNA, DNA)

Base-catalysed hydrolysis of a nucleotide removes the phosphate group and yields a **nucleoside**. A nucleoside is a glycoside formed from the pentose and the heterocyclic base. The nucleosides of RNA are **cytidine, uridine, adenosine** and **guanosine**.

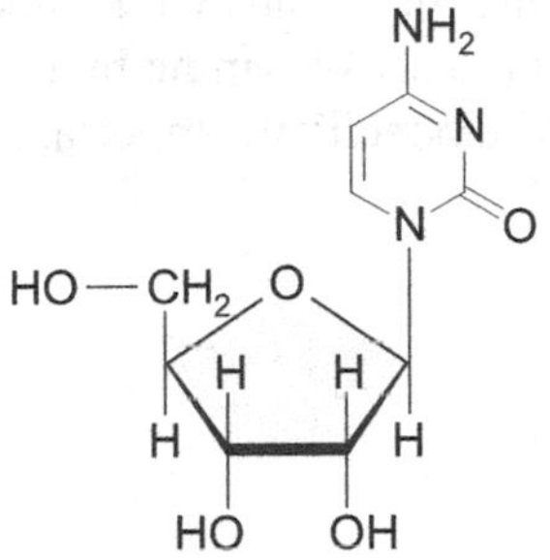

Cytidine

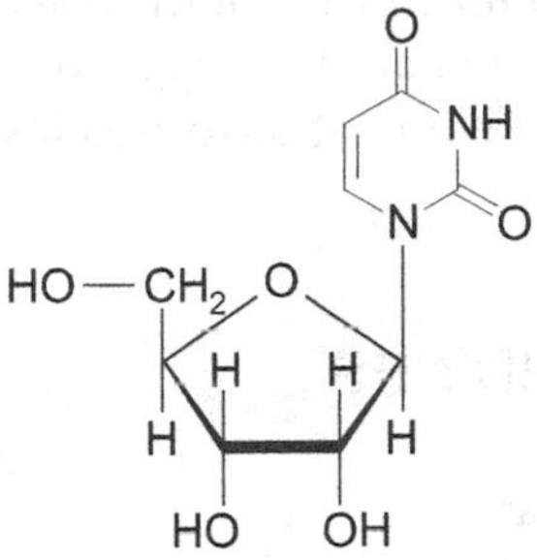

Uridine

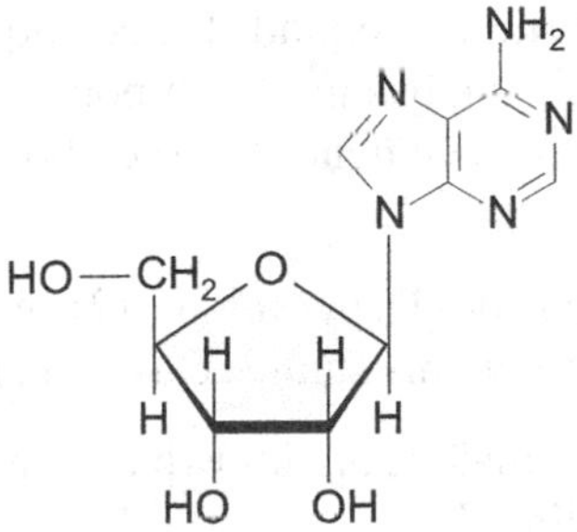

Adenosine

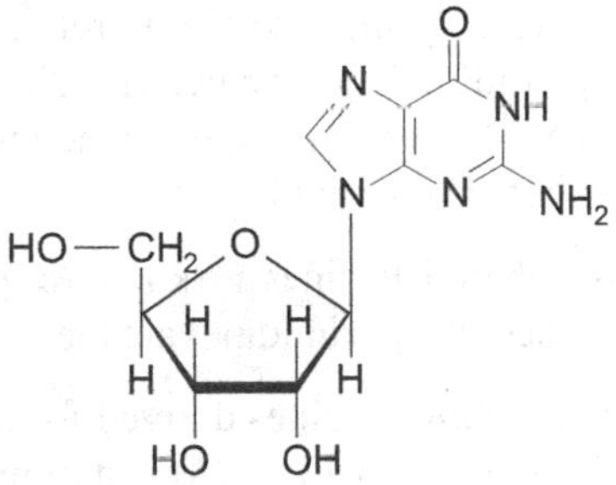

Guanosine

The nucleosides for the DNA are the corresponding 2-deoxy analogs with **2-deoxythymidine** replacing uridine. The others are **2-deoxycytidine**, **2-deoxyadenosine** and **2-deoxyguanosine**. Structure of these are given on page 288.

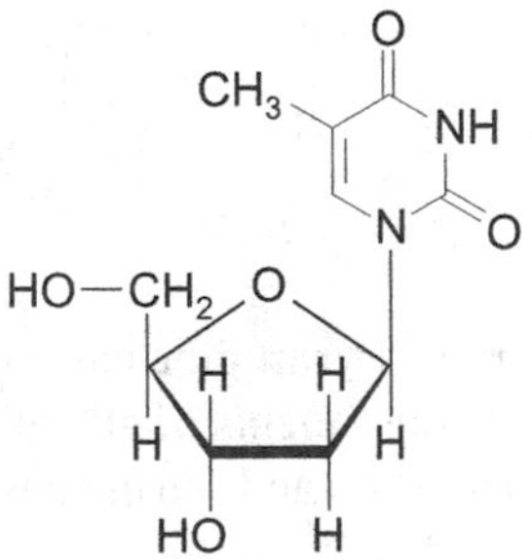

2-Deoxythymidine

We know that DNA occurs in the nuclei of all cells and is the molecule in which genetic information is stored. It acts as the master blueprint for the production of proteins, which act as essential catalysts for all cellular reactions. In a process mediated by RNA, the sequence of DNA bases specifies the sequence of amino acids in a single protein chain. This sequence of amino acids in a protein determines its structure and function. Thus, the base sequence of DNA ultimately determines the activities of proteins, the essential machinery of life.

We shall see at a later stage that DNA has a double helix structure. Its base pairs are adenine(**A**), thymine(**T**), guanine(**G**) and cytosine(**C**). A binds with T and G binds with C. RNA is similar in structure to DNA but replaces thymine by uracil(**U**). It translates genetic information to proteins, which help facilitate important functions of the cell.

4.2 PYRIMIDINES

A. Introduction

Pyrimidines are the compounds having the same parent heterocyclic nucleus – the pyrimidine (2). Pyrimidine is a purely synthetic compound. It does not occur in nature. Pyrimidine is a diazine. The diazines are a group of compounds formally derived from benzene and napthalene by the replacement of two carbon atoms of a six-membered ring by nitrogen.

The six-membered diazines are known as pyridazine (1), pyrimidine (2) and pyrazine (3). Out of these the pyrimidines are the most thoroughly studied class of compounds.

The corresponding diazines derived from naphthalene are known as cinnoline (4), phthalazine (5), quinazoline (6) and quinoxaline (7).

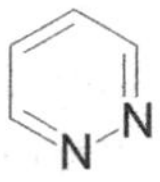

(1) 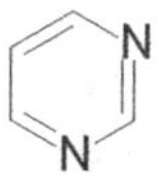   (2) 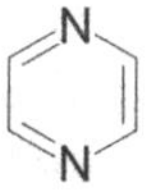   (3)

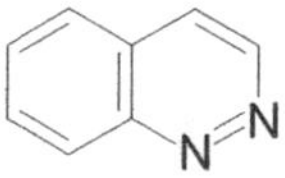

(4) 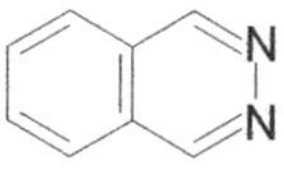   (5) 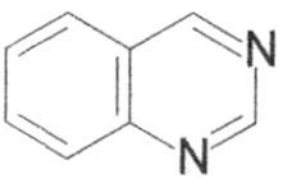   (6) 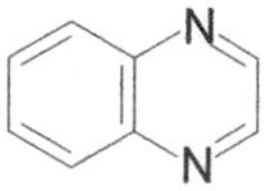   (7)

Pyridazines do not occur naturally. Pyrimidine natural products are quite important. The nucleic acids are essential constituents of all cells and thus of all living matter. The nucleic acids contain pyrimidine and purine bases.

The pyrimidine bases present in nucleic acids are uracil, thymine, cytosine. Synthetic derivatives of barbituric acid are used as hypnotics. An example is that of veronal. Vitamin B_1 also contains a pyrimidine ring.

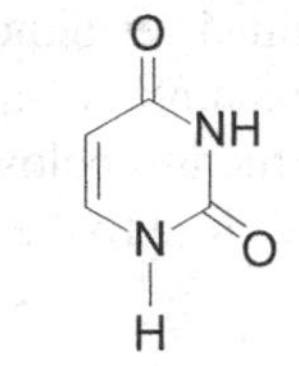

Uracil 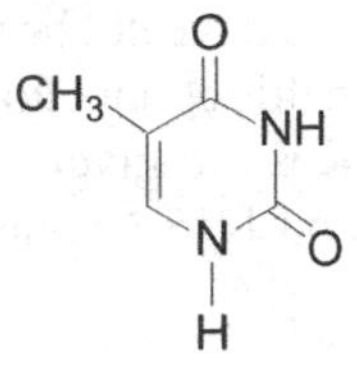   Thymine 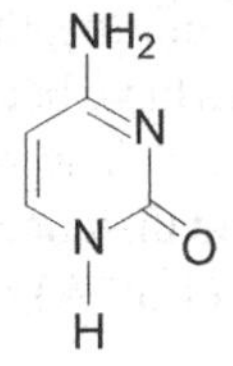   Cytosine

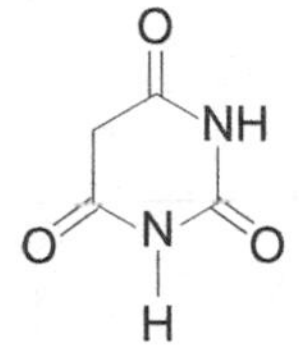

Barbituric acid 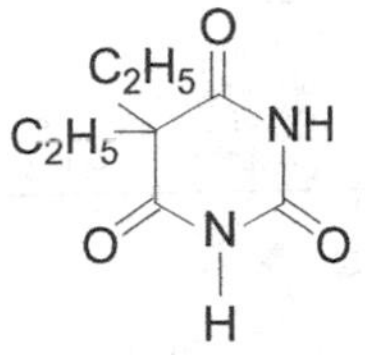   Veronal 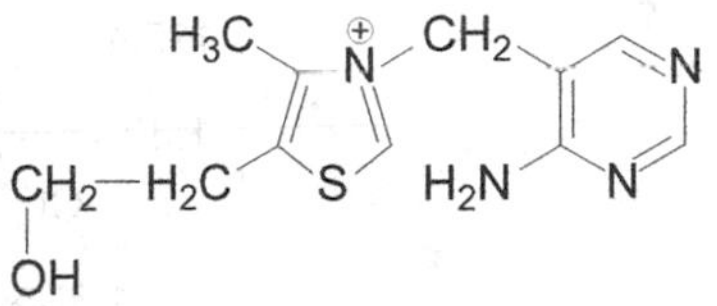   Vitamin B$_1$

Luminal is a hypnotic and is used in medicine

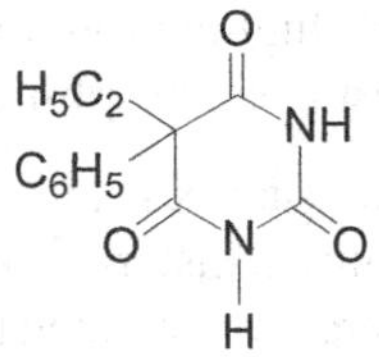

Luminal

The function and structure of a number of enzymes possessing the pyrimidine ring have been established. Purines can also be strictly considered as pyrimidine derivatives.

Deoxyribonucleic acid contains adenine and guanine.

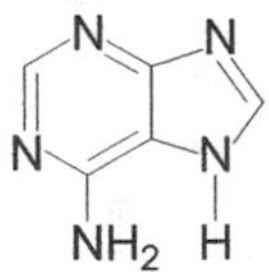

Adenine 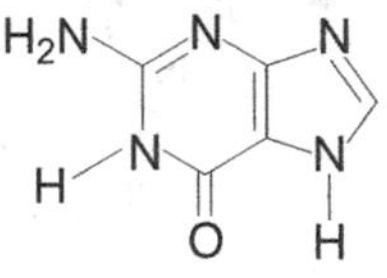   Guanine

B. Aromatic Character of Pyrimidine

All the parent diazine ring systems may be regarded as aromatic. This conclusion has been derived from the experimental observations of the C—C and C—N bond lengths which have been found to be intermediate in length between the values expected for the characteristic single and double bond in question.

In case of simple pyrimidine, the dimensions have not be determined yet, although a number of derivatives have been studied in this connection. An example is 4-amino-2,6-dichloropyrimidine. The bond distances are given in its structure below. Out of the two C—C bond distances (1.40 Å and 1.35 Å) one is very close to the C—C distance of benzene (1.39 Å).

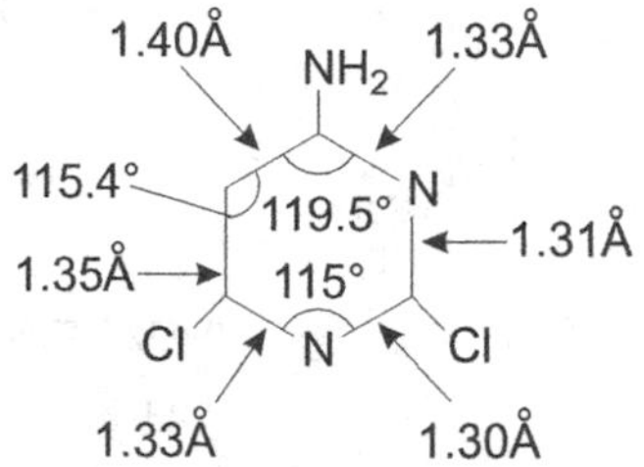

4-Amino-2,6-dichloropyrimidine

The general similarity of the size and shape of the pyrimidine ring to that of benzene and pyridine is consistent with its highly aromatic chemical characteristics. Thus, pyrimidine also undergoes the usual electrophilic substitution reactions.

The differences between the bond angles and bond distances between the ring systems suggest that there is maximum distortion in the case of pyrimidine. This is in agreement with the resonance energies calculated by molecular orbital methods. The values are benzene 36 K cal, pyridine = 31 K cal and pyrimidine = 26 K cal/ mole.

Pyrimidine is best considered as a resonance hybrid of eight structures, out of which only two are uncharged structures. The p-electron densities at the 2, 4 and 5-positions have been calculated as 0.776, 0.825 and 1.103 respectively.

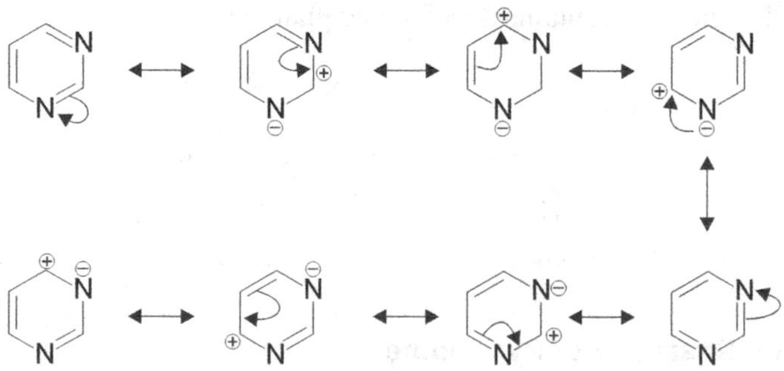

C. Chemical Properties of Pyrimidine

An enormous amount of work has been done on the derivatives of pyrimidines but the chemical reactions of pyrimidine itself have been investigated only recently. This is partly because pyrimidine was not readily available. However, it can now be obtained quite easily.

i. Basic Character

The diazines, in general are weakly basic because of the added electron deficiency of the second nitrogen which conveys to the heterocyclic system.

The basic strengths of the diazines are given below. It is to be noted that pyrimidine is particularly weaker than pyridazine, whereas pyrazine is the weakest of all these.

Compound	pK_a
Pyridazine	2.3
Pyrimidine	1.3
Pyrazine	0.6

Comparing the basic strenghts of pyrimidine with other heterocyclic ring systems, we find that pyrimidine is a much weaker base than pyridine (pK_a = 5.23) or imidazole (pK_a = 7.2)

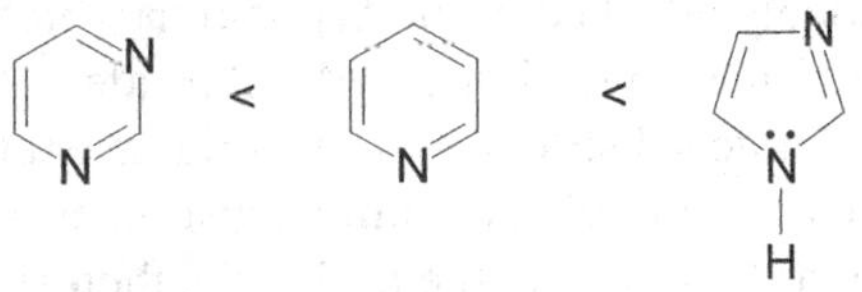

This is due to the fact that addition of a proton does not increase the possibilities of resonance.

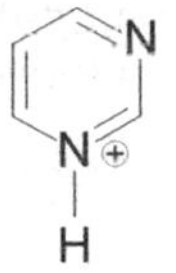

ii. Methylation-Formation of Quaternary Salts

The quaternization of these substances has been studied to a limited extent. Only one of the nitrogen atoms of pyrimidine is alkylated by alkylating agents such as dimethyl sulphate, e.g.,

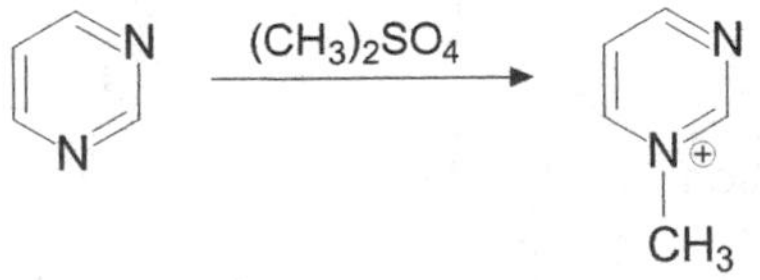

In case of substituted pyrimidines, the reaction with alkylating agents is subject to direction by the attached groups e.g., 4-amino-2-methylpyrimidine is methylated slowly at N − 1.

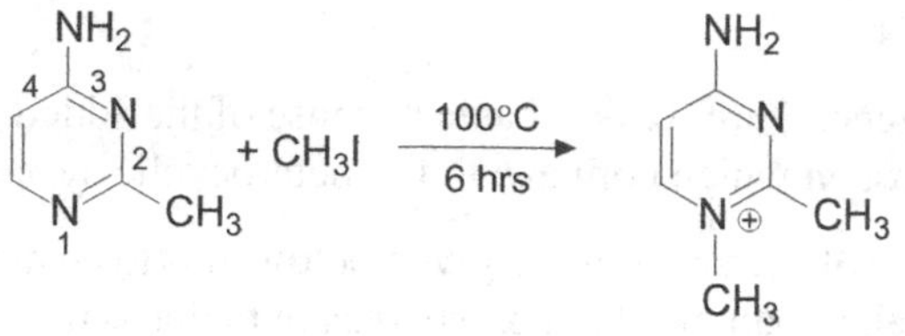

iii. Electrophilic Substitution

We know that pyridine is somewhat resistant to electrophilic substitution. The substitution, however, takes place only under drastic conditions. The presence of a second nitrogen atom in the monocyclic diazines would be expected to increase further the electron deficiency of these heterocycles and thus make these compounds pyridazine, pyrimidine and pyrazine extremely resistant to attack by positively charged entities. This unreactivity is quite pronounced and ring systems are often destroyed when vigorous drastic conditions are used.

Pyrimidines are invariably attacked at the 5-position which should correspond to position 3 of pyridine. Thus, pyrimidine hydrochloride is brominated at position 5, but no other electrophilic substitution of pyrimidine itself has been reported. If activating groups such as hydroxyl and amino substituents or even the N-oxide are present at other positions (2,4 or 6) in the molecule, then electrophilic substitution (nitration, nitrosation, diazocoupling *etc.*) usually occurs but only at position 5. Halogenation can also be effected when one or two activating groups are present. We can explain this on the basis of stabilities of the intermediate complex when attack is at 4, 5 or 6-positions. The intermediate carbocation, obtained when attack is at the 5-position, is more stable than those obtained from attack at the 4- or the 6- position (Fig. 4.1).

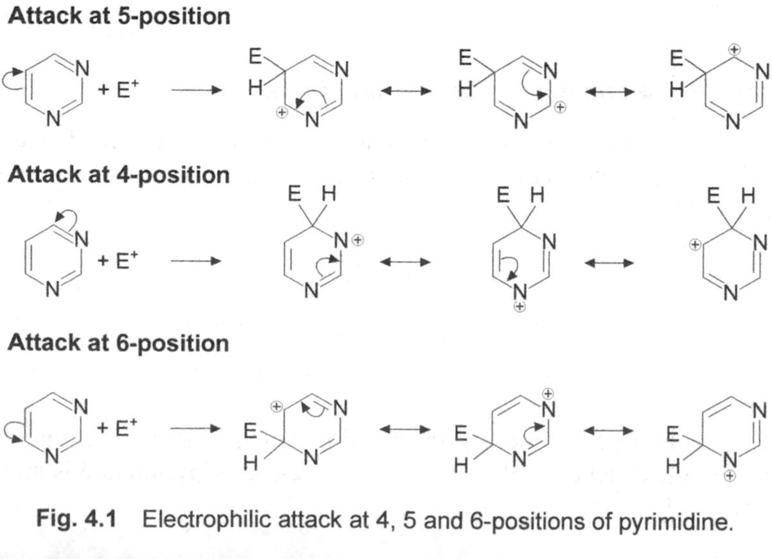

Attack at 5-position

Attack at 4-position

Attack at 6-position

Fig. 4.1 Electrophilic attack at 4, 5 and 6-positions of pyrimidine.

Some examples are:

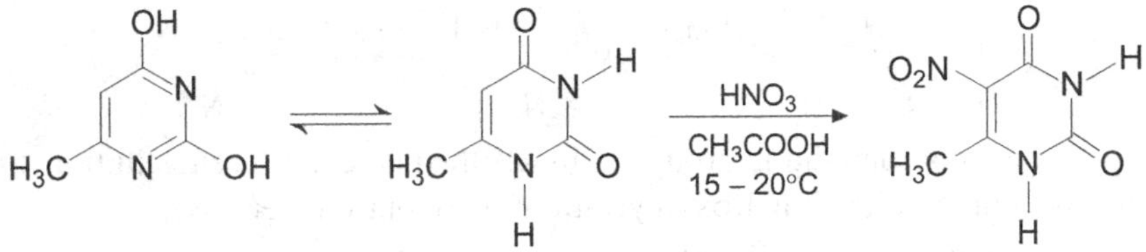

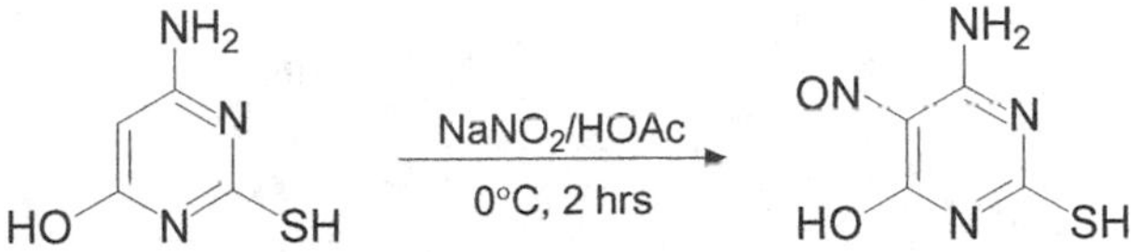

iv. *N*-Oxide Formation

Mono-*N*-oxides are known for all the diazines and several of them are known to be capable of di-*N*-oxide formation. Pyrimidine forms only a mono-*N*-oxide which, unlike pyridine-*N*-oxide, has not been nitrated successfully.

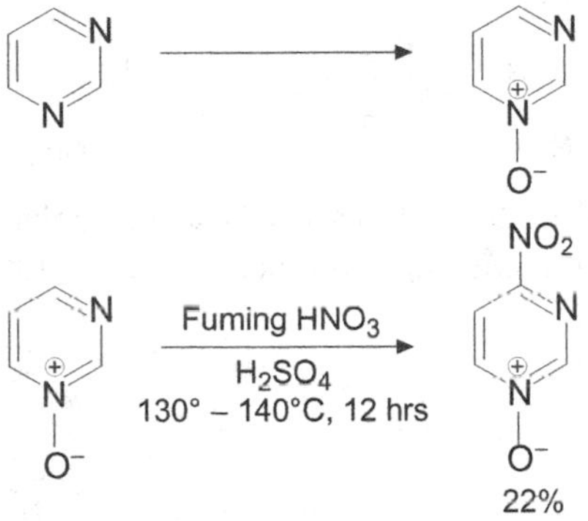

v. Nucleophilic Substitution

We know that in pyridine position 2 is attacked by nucleophilic reagents. Similarly, in diazine series positions α and γ to the nitrogen atoms are reactive towards nucleophilic reagents. This behaviour is similar to the behaviour of the analogous positions in pyridine, quinoline and isoquinoline.

In the few cases investigated, we find that pyrimidines are attacked by nucleophilic reagents such as sodamide and phenyl magnesium bromide. For example,

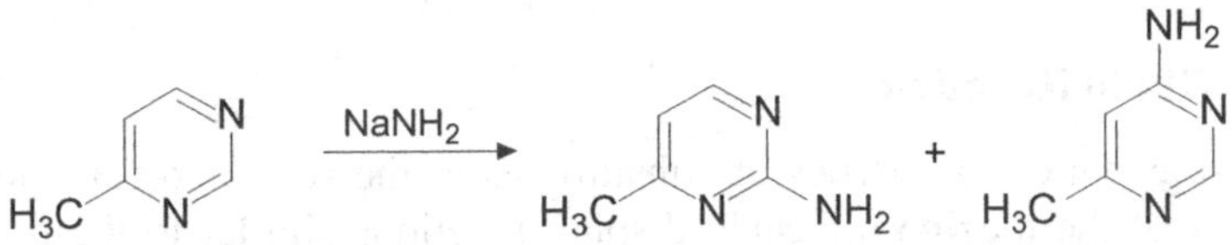

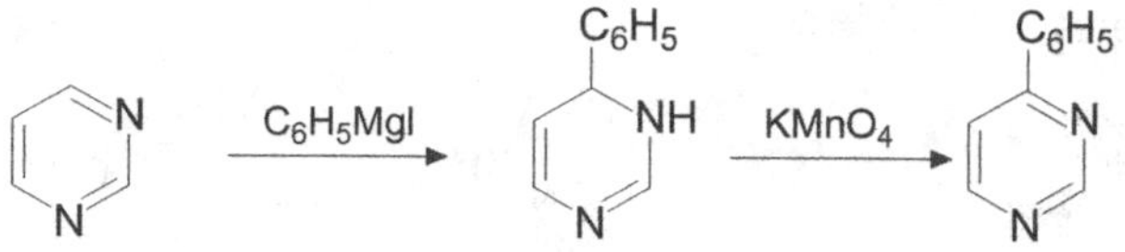

The presence of an activating group facilitates the nucleophilic substitution. Some reactions of uracil (2,4-dihydroxy pyrimidine) are outlined below:

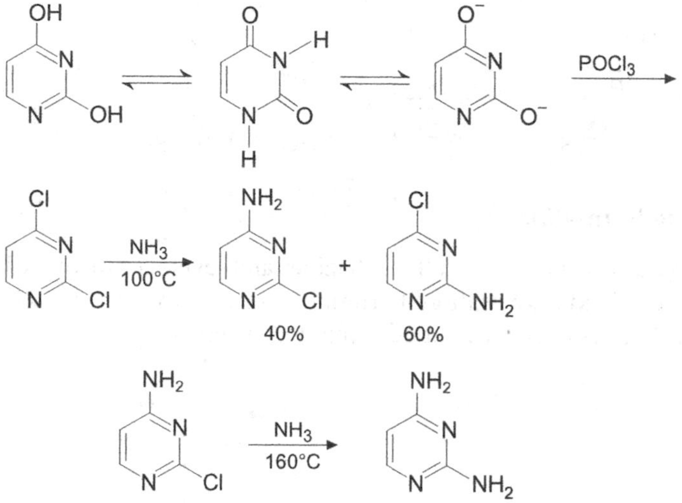

Halogen atoms at 2, 4 or 6-positions can also be replaced by —OH groups on hydrolysis with dilute mineral acid. Replacement by —OCH$_3$ group gives a 50:50 mixture.

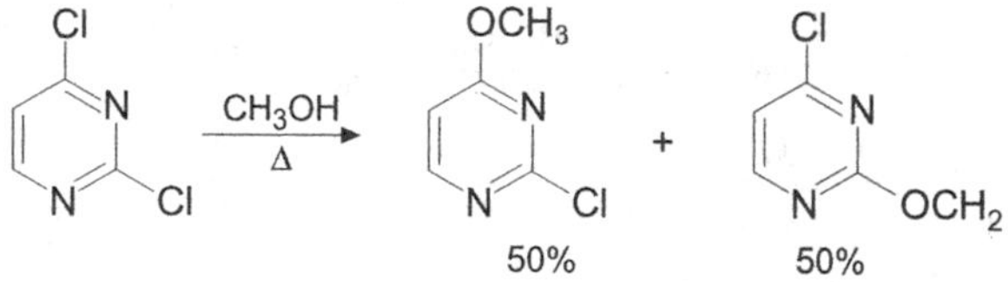

vi. Hydrogenation

Hydrogenation of pyrimidine and its derivatives under acid conditions gives 1, 4, 5, 6-tetrahydropyrimidine which has the properties of an aliphatic amidine.

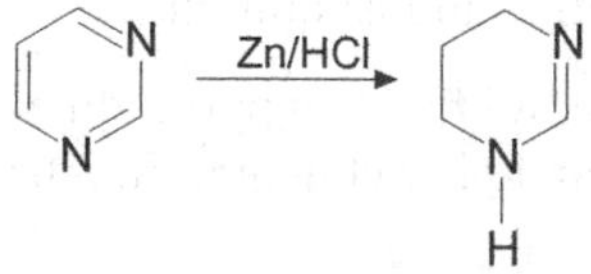

vii. Side Chain Reactivity

In general, chemical properties of substituents at the so called activated α- and γ-positions of the diazine molecule display reactions similar to the reactions of pyridine derivatives, for alkyl groups at these sites possess acidic hydrogens and undergo condensation reactions in the presence of appropriate bases or Lewis acids.

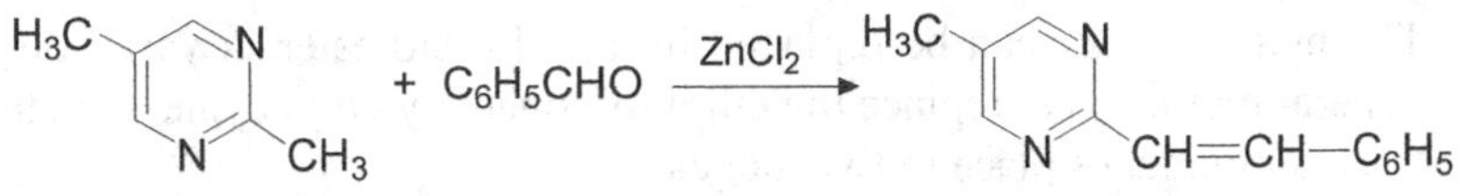

Hydroxy diazines generally exist in the keto forms. Aminodiazines, however, exist predominantly in the amino forms.

Carboxyl groups situated at the α- and γ-positions are prone to decarboxylation at elevated temperature

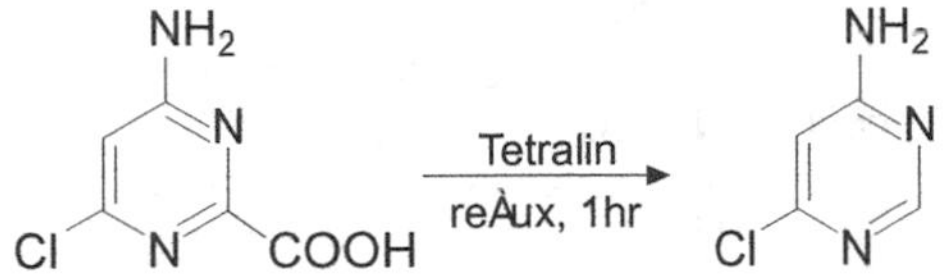

D. Synthetic Methods

i. The most common synthesis involves the condensation of a three carbon unit to a species having a N-C-N linkage. This general method is extremely versatile because of the large variety of molecules which undergo cyclisation. The 3-carbon unit may be a β-dialdehyde, β-ketoaldehyde, β-ketoester, malonic ester, or β-ketonitrile. The N-containing unit may be a thiourea, amidine, urea or guanidine. For example,

(1)

(2)

(3)

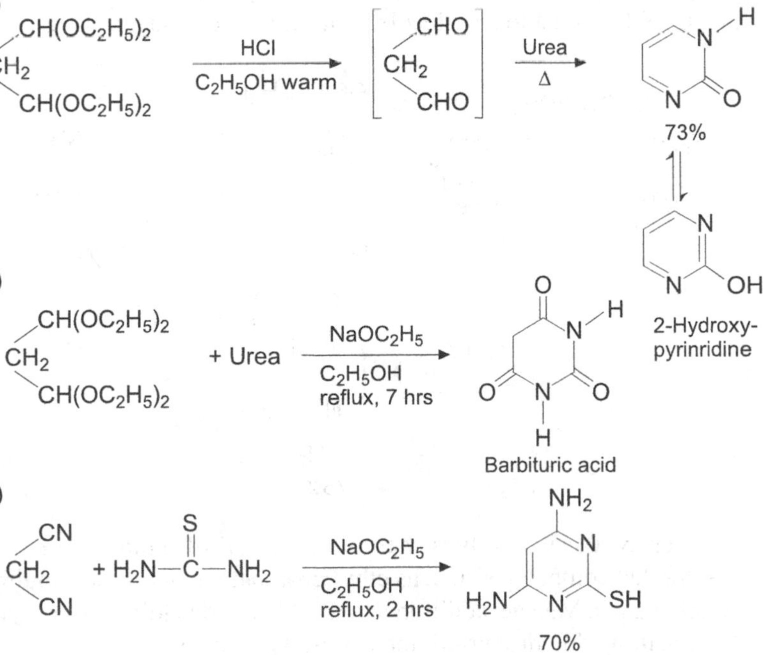

The malonic ester can be replaced by any β-keto ester. A variation of the general method is to replace the ethyl malonate by ethyl cyanoacetate when the reaction takes place in two stages.

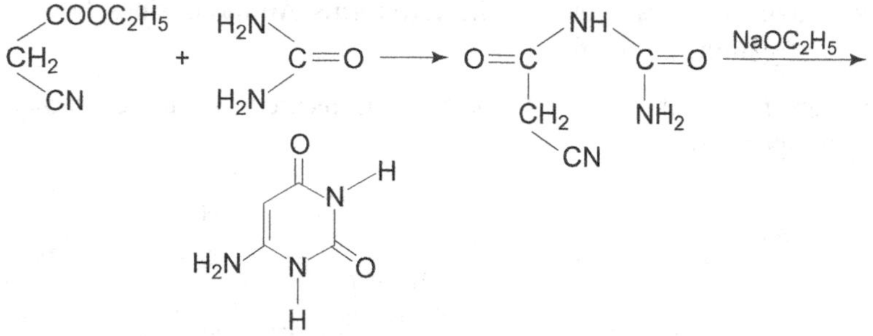

ii. A second pyrimidine synthesis, which is of considerable use, involves the condensation of formamide with β-dicarbonyl compounds or their precursors at elevated temperatures. This affords pyrimidines unsubstituted at the 2-position

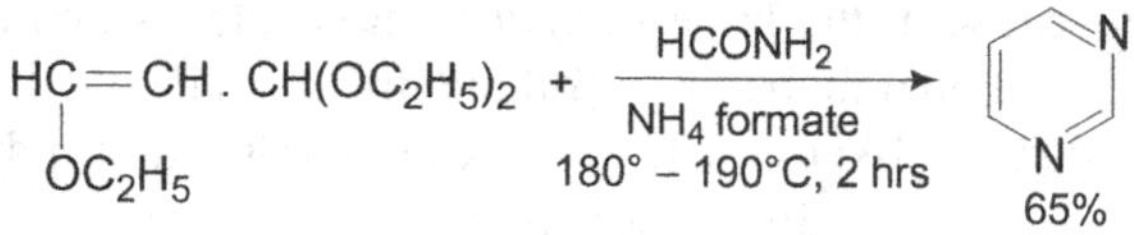

iii. The third important method of pyrimidine synthesis consists in the addition of a C–N fragment to a molecule containing the sequence C–C–CN.

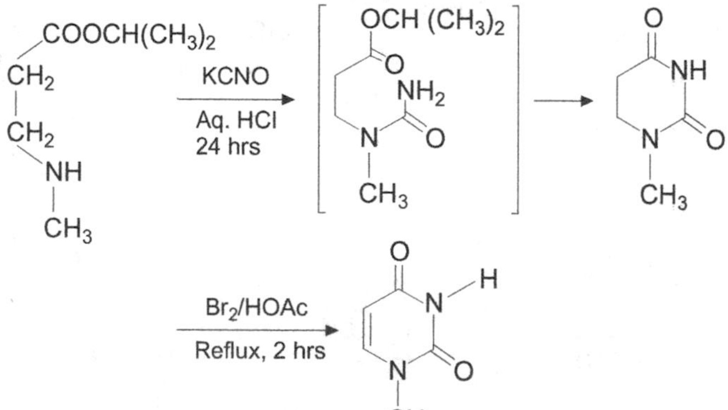

iv. Another synthesis involves the condensation of amidines or ureas with unsaturated compounds such as ethyl crotonate under basic conditions. It is possible that a Michael addition to a double bond is followed by cyclisation. The resulting dihydropyrimidine is readily oxidised.

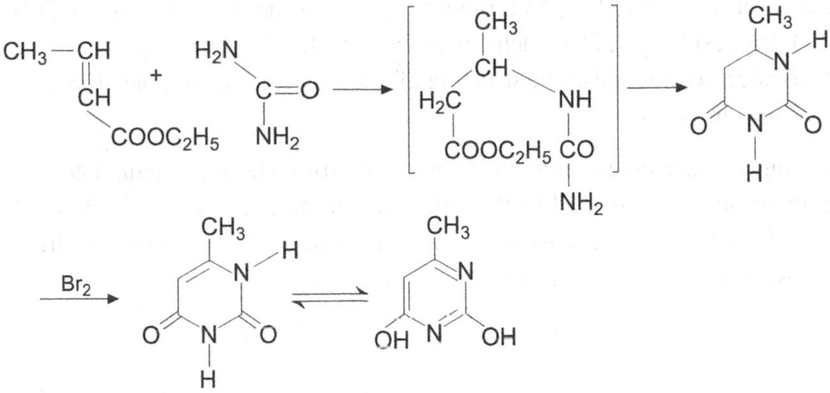

4.3 URACIL

A. General Features

Uracil is a common and naturally occurring pyrimidine derivative. It is 2,4-dioxypyrimidine. It is a planar, unsaturated compound that has the ability to absorb light.

We have already seen that uracil is one of the four nucleo bases present in RNA that are represented by the letters A, G, C and U. The others are adenine, cytosine and guanine. In RNA, uracil base-pairs with adenine and replaces thymine during DNA transcription.

B. Properties and Reactions of Uracil

Methylation of uracil produces thymine. In DNA the substitution of uracil for thymine increases the stability of DNA and improves the efficiency of DNA replication.

Uracil can base-pair with any of the bases, but readily pairs with adenine because the methyl group is repelled into a fixed position. Uracil pairs with adenine through two hydrogen bonds as shown below in Fig. 4.2.

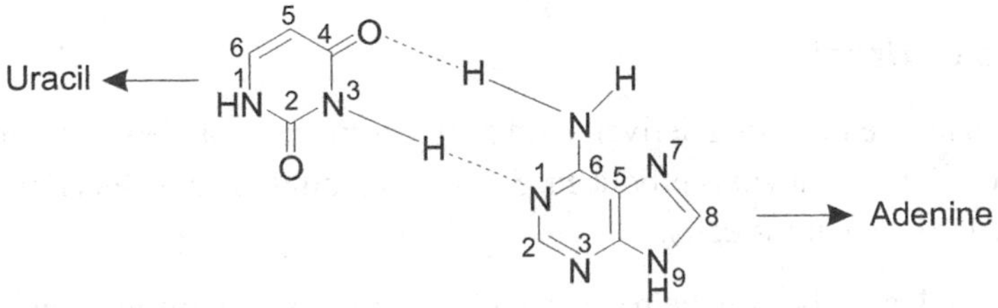

Fig. 4.2 Hydrogen bonding between uracil and adenine.

In RNA, uracil binds with the sugar D-ribose to give ribonucleoside, uridine. When a phosphate attaches to uridine, uridine 5'-monophosphate (UMP) is produced.

Uridine further adds phosphate units to give uridine disphosphate (UDP) and uridine triphosphate (UTP). Each one of these molecules is synthesised in the body and has specific functions. The structure of uridine 5′-monophophate is given under nucleotides in Fig. 4.7.

Uracil undergoes amide-imidic acid tautomeric shifts. The amide tautomer is referred to as the lactam structure, while the imidic acid tautomer is referred to as the lactim structure. These tautomeric forms are predominant at pH 7. The lactam structure is the most common form of uracil.

Lactam
(amide structure)

Lactim
(imide structure)

Degradation of uracil produces the substrates aspartate, carbon dioxide and ammonia whereas oxidative degradation of uracil produces urea and maleic acid in the presence of H_2O_2 and Fe^{2+}.

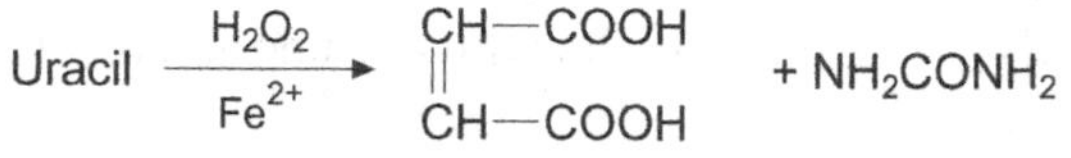

Uracil readily undergoes the simple reactions including oxidation, nitration and alkylation. Nitration and alkylation take place at position 5 of uracil. Methylation of uracil produces thymine.

Uracil also reacts with halogens because of the presence of more than one strongly electron withdrawing groups. When chlorine reacts with uracil 5-chlorouracil is produced.

C. Uses of Uracil

Uracil can be used for drug delivery and as a pharmaceutical. When fluorine reacts with uracil, 5-fluorouracil is produced. 5-Fluorouracil is an anticancer drug. It stops the growth of cancerous cells.

Uracil is used in the body to help carry out the synthesis of many enzymes necessary for cell function through bonding with riboses and phosphates.

Uracil serves as an allosteric regulator and coenzyme for reactions in the human body and in plants.

Uridine diphosphate-glucose regulates the conversion of glucose to galactose in the liver and other tissues in the process of carbohydrate metabolism. Uracil is also involved in the biosynthesis of polysaccharides and the transportation of sugars containing aldehydes.

D. Synthesis of Uracil (2,4-Dioxypyrimidine)

These are many laboratory syntheses of uracil available.

i. The first reaction is the simplest of syntheses by adding water to cytosine to produce uracil and ammonia.

$$\text{Cytosine} + H_2O \longrightarrow \text{Uracil} + NH_3\uparrow$$

ii. The most common way to synthesise uracil is by condensation of maleic acid with urea in fuming sulphuric acid.

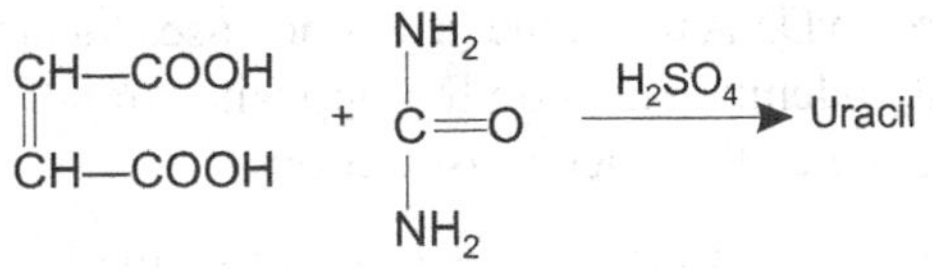

iii. **Fischer and Roder Synthesis**

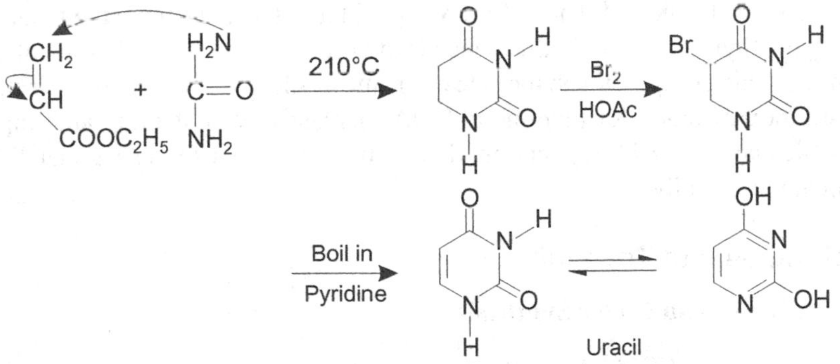

iv. **Wheeler and Liddle Synthesis**

Uracil can also be synthesised by a double decomposition of thiouracil in aqueous chloroacetic acid.

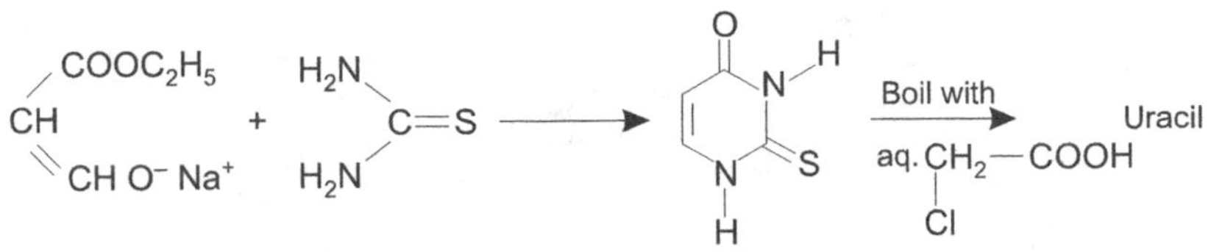

v. Davidson *et al.* Synthesis

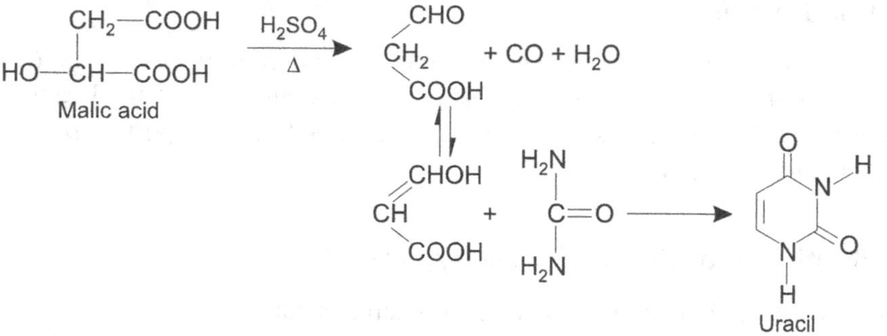

Malic acid

Uracil

4.4   THYMINE (5-Methyluracil)

A. Structure and General Features

Thymine is 5-methyluracil. Methylation of uracil produces thymine. We have learnt that thymine is one of the four bases in DNA, the other being adenine, cytosine and guanine. Within the DNA molecule, thymine bases located on one strand form hydrogen bonds with adenine bases on the opposite strand. The sequence of four DNA bases encodes the cell's genetic instructions.

We know both uracil and thymine bases pair with adenine. However, RNA contains uracil and DNA contains thymine. The question comes to our mind– why does DNA contain thymine and not uracil? It is now believed that RNA was the original hereditary molecule and that DNA developed later. If we compare the structure of uracil and thymine, the only difference is the presence of a methyl group at C-5 of thymine. This group is not on the side of the molecule involved in base pairing. The substitution of uracil for thymine in DNA increases DNA stability and improves the efficiency of DNA replication. Thus, thymine helps to ensure that DNA is replicated faithfully.

B. Synthesis of Thymine

i.   Fischer and Roder Synthesis

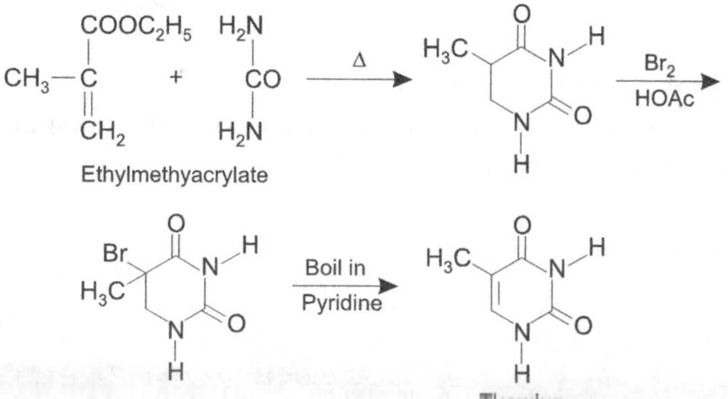

Ethylmethyacrylate

Boil in Pyridine

Thymine

ii. Wheeler and Liddle Synthesis

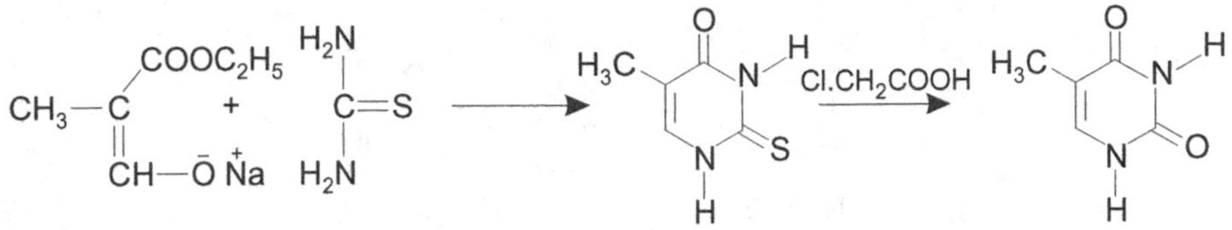

Thymine

iii. Bergmann *et al.* Synthesis

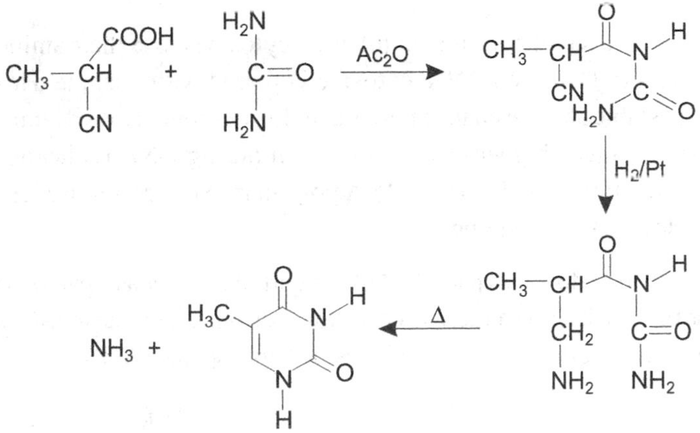

4.5 CYTOSINE

A. General Features

We know cytosine is a pyrimidine derivative with a heterocyclic aromatic ring and two substituents attached (an amino group at position 4 and a keto group at position 2). Cytosine is one of the four main bases found both in DNA and RNA. The nucleoside of cytosine is cytidine. In Watson-Crick base pairing, it forms three hydrogen bonds with guanine as given in Fig. 4.12.

B. Chemical Reactions

Methylation of cytosine occurs at carbon number 5. Cytosine can also be methylated by an enzyme called DNA methyltransferase or it can be methylated and hydroxylated to make 5-hydroxymethyl cytosine.

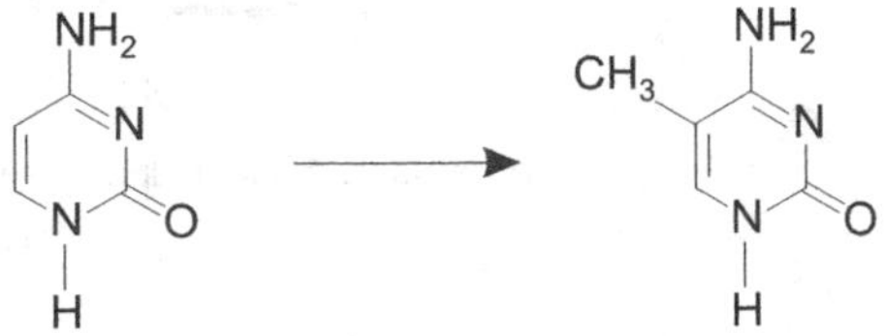

Cytosine   5-Methylcytosine

One of the most common spontaneous mutations of bases is the natural deamination of cytosine to uracil.

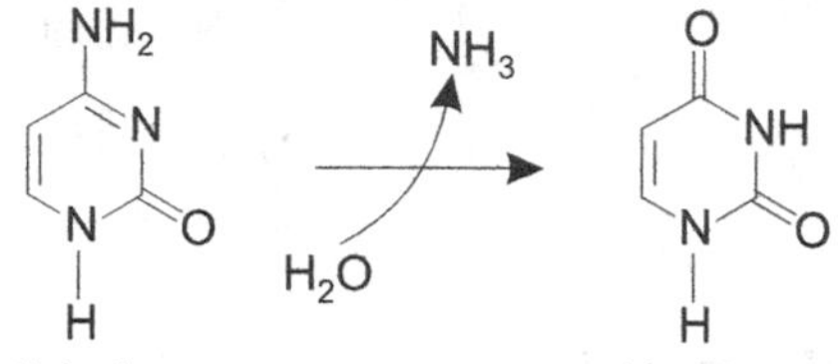

Cytosine   Uracil

At any moment, a small but finite number of cytosines lose their amino groups to become uracil. In DNA and RNA, cytosine is paired with guanine. However, it is inherently unstable and can change into uracil due to spontaneous deamination. This would lead to a much higher level of mutation during DNA replication. Because uracil is an unnatural base in DNA, DNA polymerase can recognise it as a mistake and can replace it with thymine.

We know cytosine is found as part of DNA, as part of RNA or as a part of a nucleotide. As cytidine triphosphate, it can act as a co-factor to enzymes and can transfer a phosphate to convert adenosine diphosphate to adenosine triphosphate (ATP).

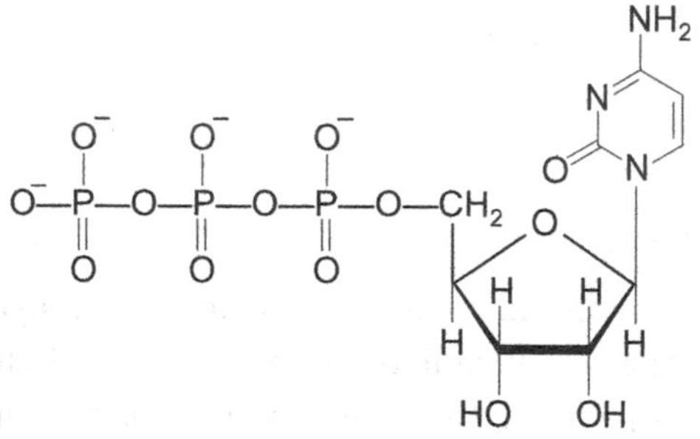

Cytidine triphosphate

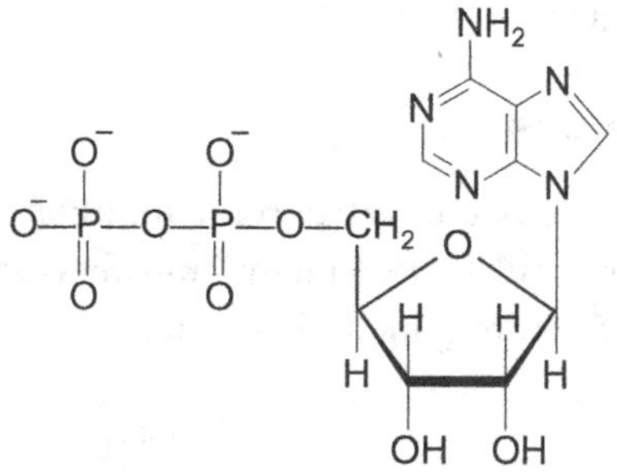

Adenosine 5′-diphosphate (ADP)

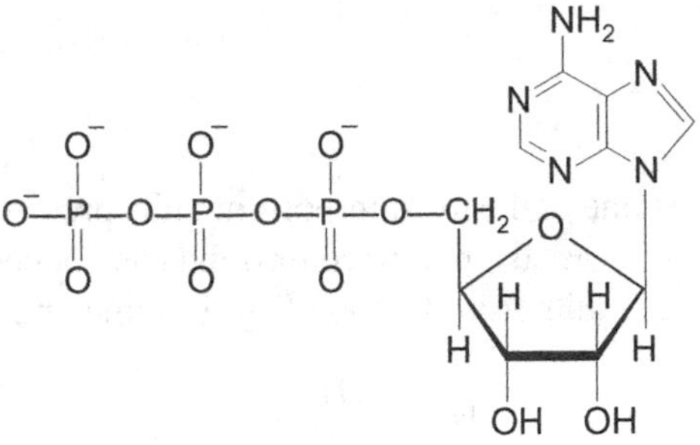

Adenosine 5′-triphosphate (ATP)

C. Synthesis of Cytosine

i. Wheeler and Johnson Synthesis

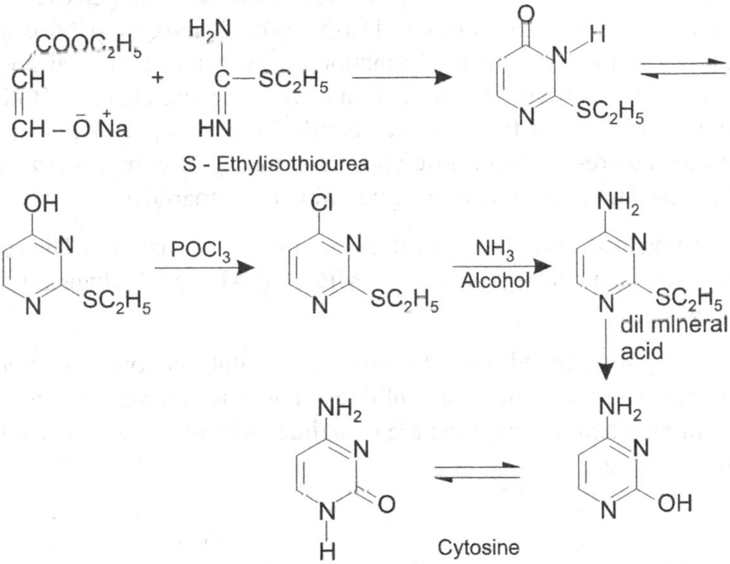

S - Ethylisothiourea

Cytosine

ii. Tariso *et al.* Synthesis

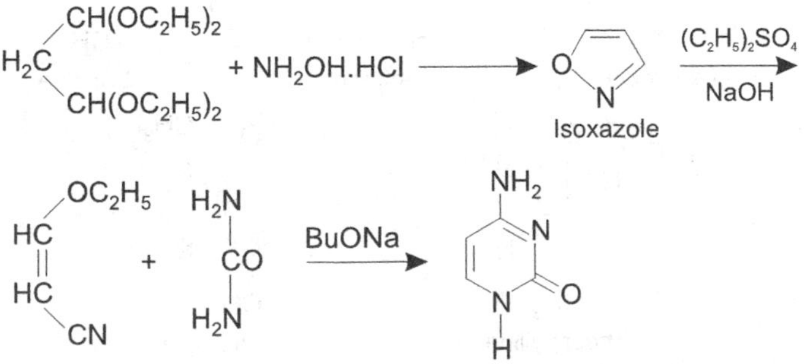

Isoxazole

4.6 PURINES

A. Introduction

The purine ring system is one of the most important units present in living systems, but as in the case of pyrimidines the parent compound has not been found in nature. Theoretically, the purine results from fusion of pyrimidine and imidazole nuclei.

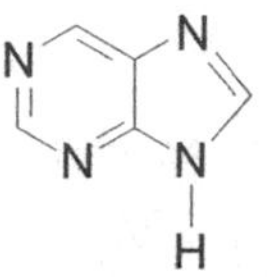

Purine

The purine unit is present in many natural products. We have already seen that adenine and guanine are the building blocks of both RNA and DNA. Attachment of either of the two sugar residues D-ribose or 2-deoxy-D-ribose at N-9 of adenine and guanine results in the formation of glycosides known as nucleosides. Purine metabolism involves the formation of adenine and guanine. Both adenine and guanine are derived from the nucleotide inosinemonophosphate, which is synthesised on a pre-existing ribose phosphate through a complex pathway using atoms from the amino acids glycine, glutamine and aspartic acid.

The purine ring system is also present in a number of coenzymes. The coenzymes adenosine mono-, di- and triphosphate (AMP, ADP, ATP) are biologically important nucleotides.

Several other purine based natural products constitute important components of several plants. Hypoxanthine and xanthine occur in tea extract and animal tissue, whereas caffeine and theophylline are constituents of tea leaves. Theobromine is found in cocoa beans.

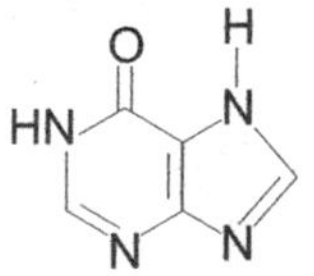

Hypoxanthine 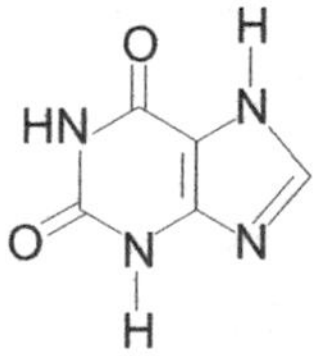   Xanthine 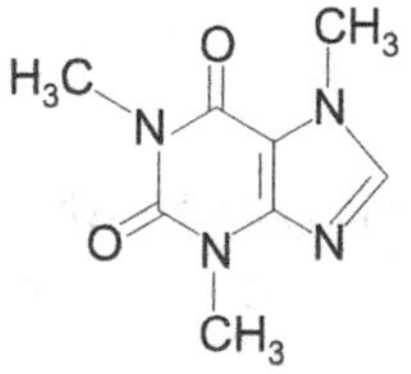   Caffeine

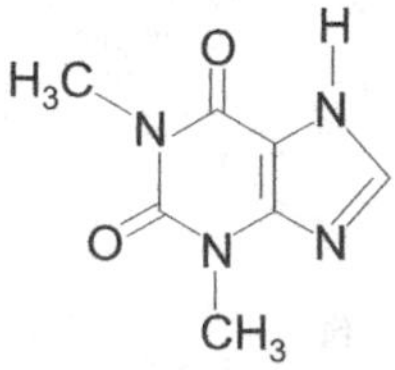

Theophylline 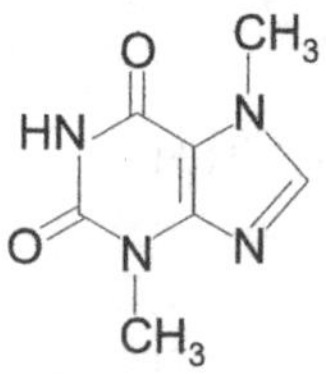   Theobromine

B. Structure and Chemical Reactions of Purine and its Simple Derivatives

Purine itself is a colourless compound, melting point 212°C, which is very soluble in water but sparingly soluble in organic solvents. The structure of purine has not been examined by physical methods but dimensions of 6-aminopurine hydrochloride (*i.e.*, adenine) have been calculated by X-ray diffraction techniques. The C–C and C–N distances indicate that purine is aromatic in nature.

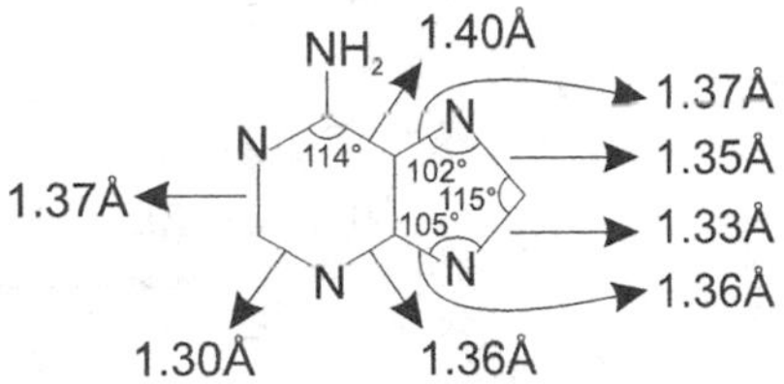

By analogy positions 2 and 6 of purine should be similar chemically to positions 2 and 4 of pyrimidine and position 8 to position 2 of imidazole.

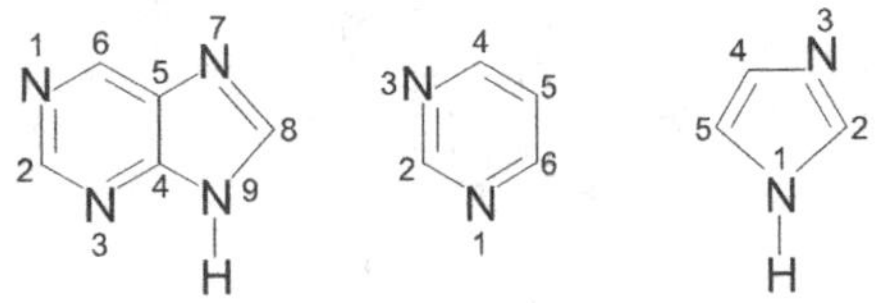

| Purine | Pyrimidine | Imidazole |

The results of molecular orbital calculations of π-electron densities at various positions of purine indicate that position 8 has the lowest electron density and therefore, would be expected to be least susceptible to electrophilic attack. However, in actual practice this is contrary to this fact.

Position	Electron density
2	0.902
6	0.907
8	0.895

Purine does not react with diazonium salts whereas 2, 6-dihydroxy purine couples at position 8 and 2, 8- and 6, 8-dihydroxypurines do not react at all.

Also some amino and hydroxy purines brominate at position 8. π-Electron densities alone can, therefore, be a misleading guide to the positions of attack of a reagent on a molecule.

This is best shown by considering the stabilities of the intermediates obtained after the electrophilic attack at various positions. We find that the intermediate carbocation formed is most stable when the attack is there at position 8 (Fig. 4.3).

Attack at Position 8

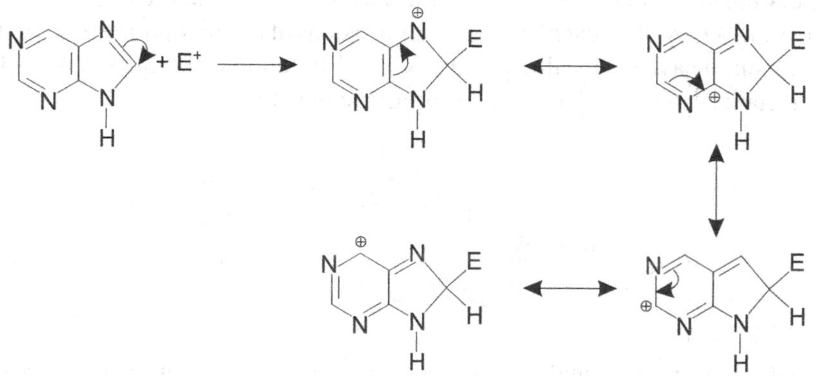

Attack at Position 2

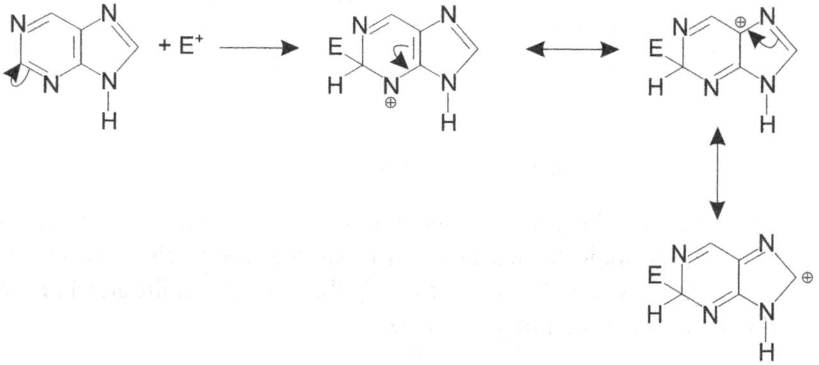

Attack at Position 6

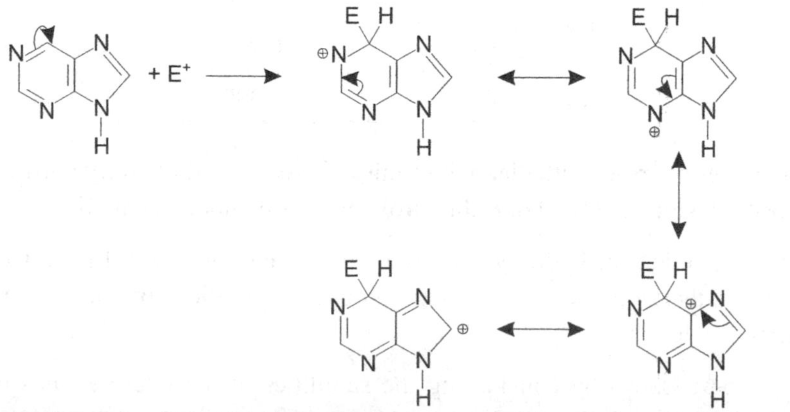

Fig. 4.0 Electrophilic attack at 8, 2 and 6-positions of purine.

Basic Character

Purine (pK_a 4.4) is a stronger base than pyrimidine (pK_a 1.3) but weaker than imidazole (pK_a 7.2).

Diazomethane or dimethyl sulphate in the presence of alkali methylates purine in the 9-position.

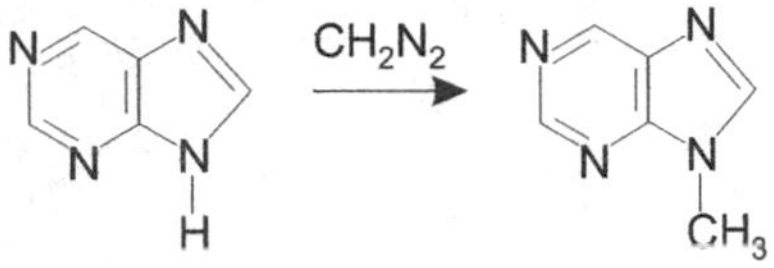

C. Synthesis of Purine

1. Purine may be synthesised from uric acid as follows:

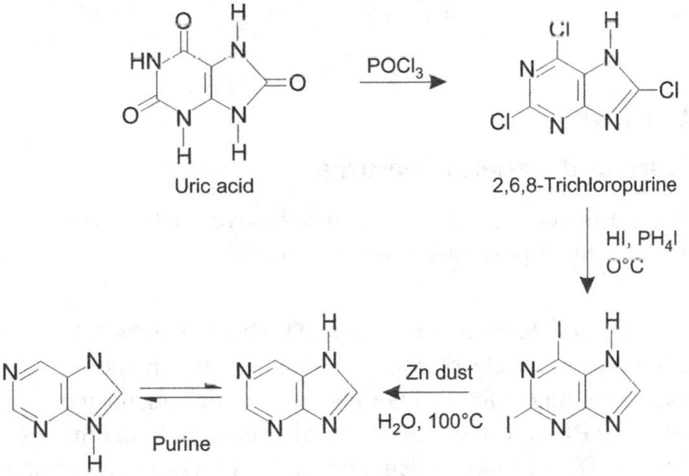

Uric acid

2,6,8-Trichloropurine

Purine

Catalytic reduction of trichloropurine (H_2-Pd in aqueous sodium acetate) also gives purine.

2. In general synthetic methods, the pyrimidine ring is synthesised first and then the imidazole ring is built up on this.

Traube's Method

In this 4,5-diaminopyrimidine is synthesised first which is then condensed with formic acid to produce the imidazole ring. When the formyl derivative or its sodium salt is heated, ring closure takes place to give purine.

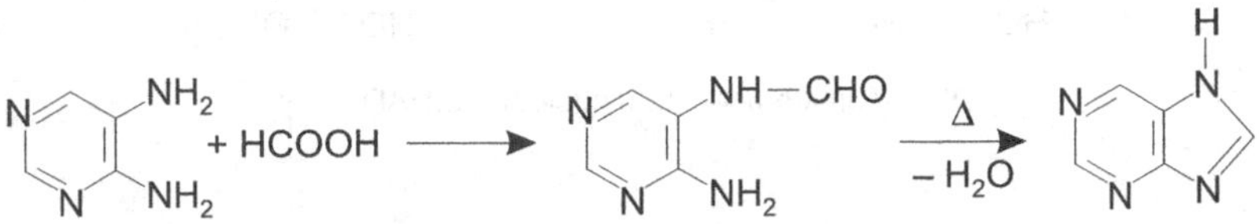

This synthesis leads to the preparation of purines that are unsubstituted in position 8. 8-Hydroxy purines may be prepared by using ethyl chloroformate instead of formic acid. Also, diamino pyrimidines may be fused with urea to produce 8-hydroxy purines.

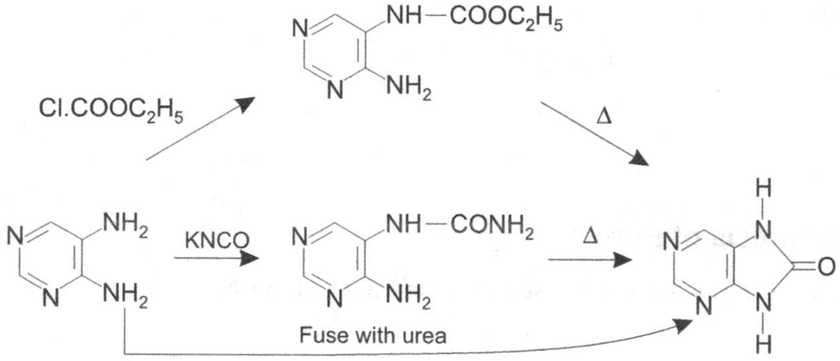

4.7 ADENINE

A. Structure and General Features

Adenine is a purine derivative. Its structure is given below with standard numbering of positions. It is 6-aminopurine.

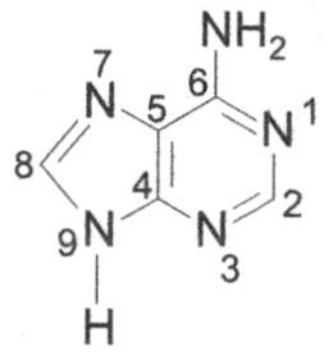

Adenine

We know adenine is a chemical component of both DNA and RNA. However, it plays a variety of roles in biochemistry including cellular respiration, in the form of both the energy rich adenosine triphosphate (ATP) and the cofactors nicotinamide adenine dinucleotide (NAD^+) and flavin adenine dinucleotide (FAD), and protein synthesis. The shape of adenine is complementary to either thymine in DNA or uracil in RNA.

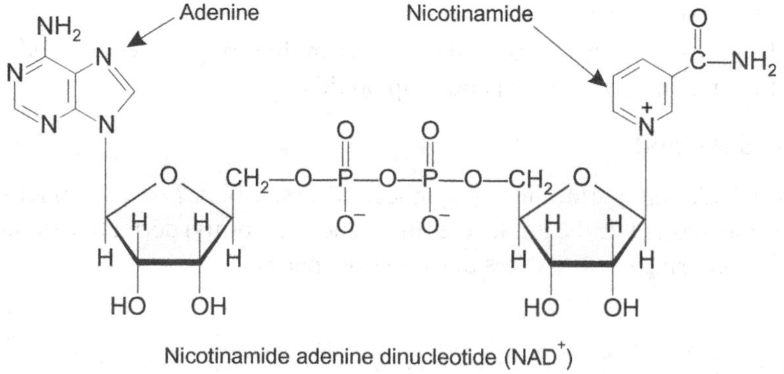

Nicotinamide adenine dinucleotide (NAD^+)

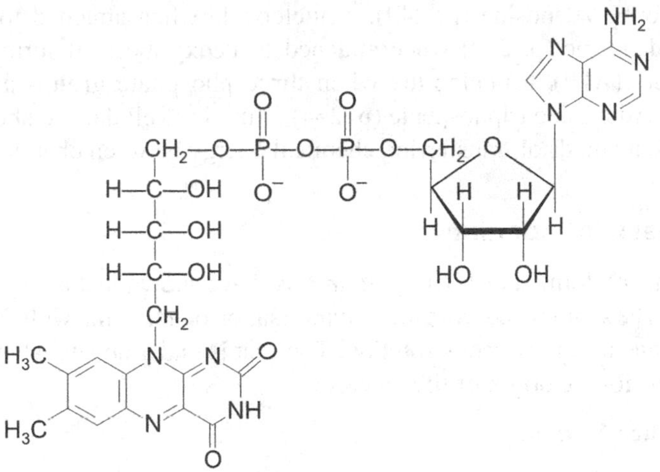

Flavin adenine dinucleotide (FAD)

In addition to nucleic acids, adenine occurs in the pancreas of cattle and in tea extracts. General reactions of adenine are similar to that of purine. On treatment with nitrous acid, it can be converted into hypoxanthine (an oxypurine). Its structure was established by synthesis.

B. Properties of Adenine

Adenine forms several tautomers. However, in an inert gas matrix the 9H-adenine tautomer is found.

In older literature, adenine was sometimes called vitamin B_4. It is no longer considered a true vitamin. However, two B vitamins niacin and riboflavin bind with adenine to form the essential cofactors NAD^+ and FAD as given above.

Adenine is one of the purine bases (the other being guanine) used in forming nucleotides of the nucleic acids. In DNA, adenine binds to thymine *via* two hydrogen bonds (Fig. 4.4) to assist in stabilizing the nucleic acid structures. In RNA, which is used for protein synthesis, adenine binds to uracil (Fig. 4.2, page 259.)

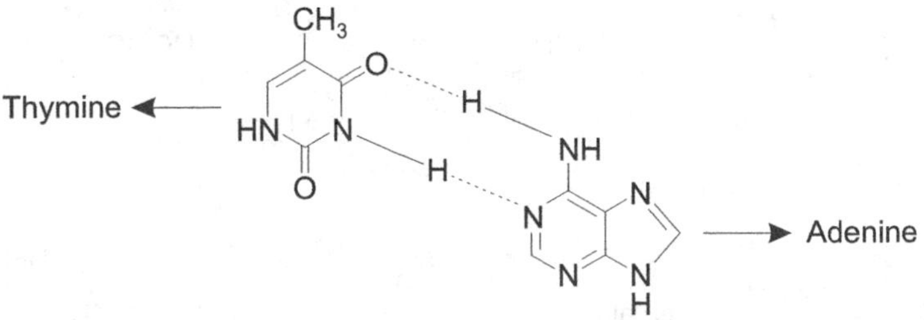

Fig. 4.4 Hydrogen bonding between thymine and adenine.

Adenine forms adenosine (p. 249), a nucleoside when attached to ribose and 2-deoxy-adenosine (p. 288) when attached to deoxyribose. It forms adenosine triphosphate (ATP), a nucleotide when three phosphate groups are added to adenosine. Adenosine triphosphate (p. 264) is used in cellular metabolism as one of the basic methods of transferring chemical energy between chemical reactions.

C. Synthesis of Adenine

Experiments performed in 1961 by **Joan Oro** have shown that a large quantity of adenine can be synthesised from the polymerisation of ammonia with five hydrogen cyanide molecules in aqueous solution. The fact is under debate whether this has implications for the origin of life on earth.

i. Fischer Method

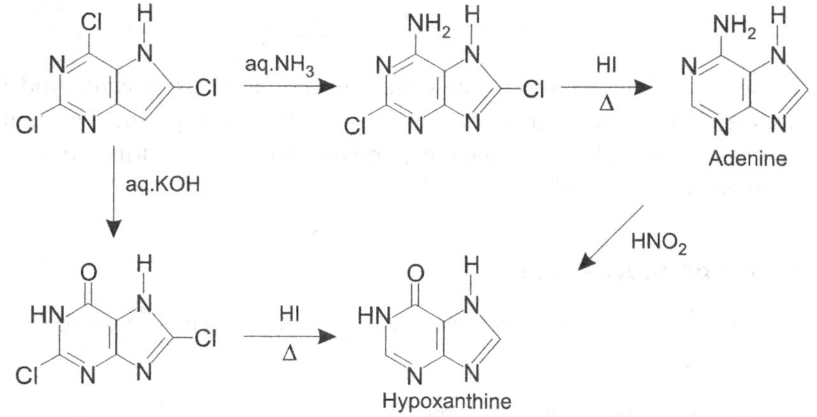

aq.NH$_3$

HI Δ

Adenine

aq.KOH

HNO$_2$

HI Δ

Hypoxanthine

ii. Traube Method

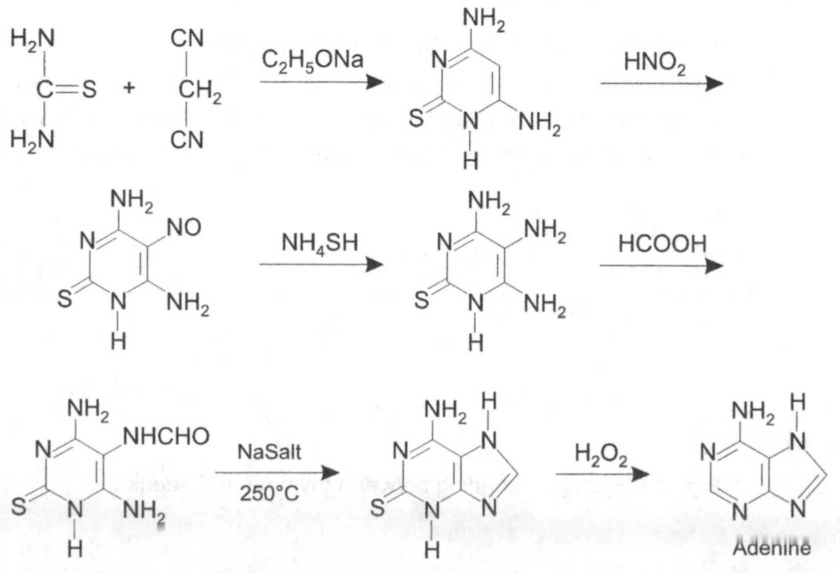

C$_2$H$_5$ONa

HNO$_2$

NH$_4$SH

HCOOH

NaSalt 250°C

H$_2$O$_2$

Adenine

iii. Todd *et al.* Method

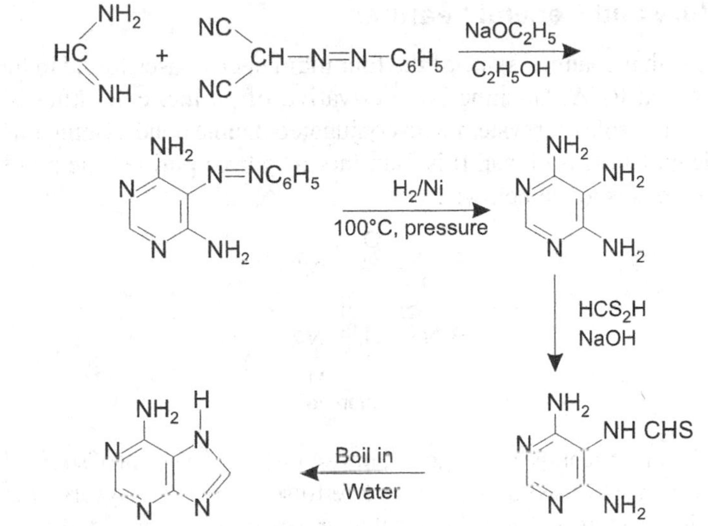

iv. Bredereck *et al.* Method

They modified the synthesis given by Traube

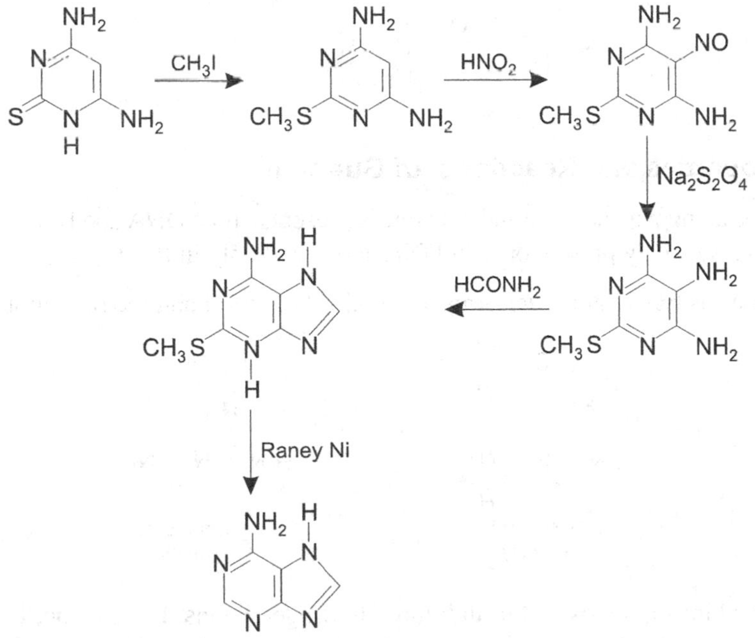

4.8 GUANINE

A. Structure and General Features

We have seen that guanine is one of the four main nucleobases found in the nucleic acids, DNA and RNA. Guanine is a derivative of purine, consisting of a fused pyrimidine-imidazole ring system with conjugated double bonds. Being unsaturated, the bicyclic molecule is planar. It is 2-amino-6-hydroxypurine. The numbering of various positions is given below:

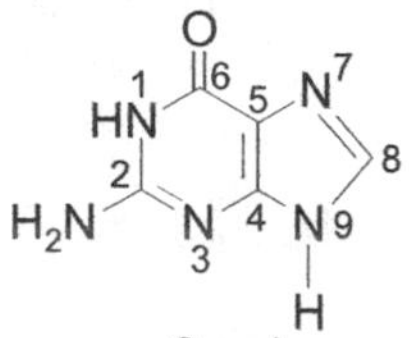

Guanine

The first isolation of guanine was reported in 1844 from the excreta of sea birds, known as **guano**, which was used as a source of fertiliser. About fifty years later, Fischer determined its structure and also showed that uric acid can be converted to guanine. In addition to nucleic acids, guanine occurs in the pancreas of cattle and in certain fish scales.

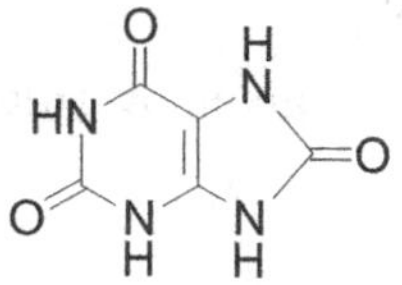

Uric acid

B. Properties and Reactions of Guanine

Guanine along with adenine and cytosine is present in both DNA and RNA, whereas thymine is usually present only in DNA and uracil only in RNA.

Guanine has two tautomeric forms, the major keto form and the rare enol form

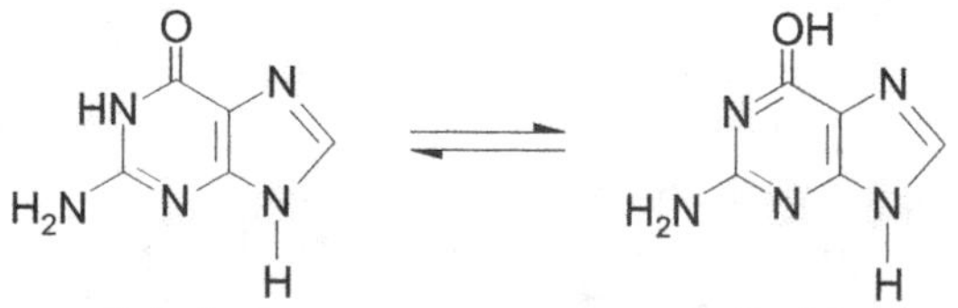

Keto form   Enol form
(major)   (minor)

Guanine binds to cytosine through three-hydrogen bonds. In cytosine, the amino group acts as the hydrogen donor and the C-2 carbonyl and the N-3 amine as the hydrogen bond acceptor. Guanine has a group at C-6 that acts as the hydrogen acceptor, while the group at N-1 and the amino at C-2 act as the hydrogen donors as shown in Fig. 4.5.

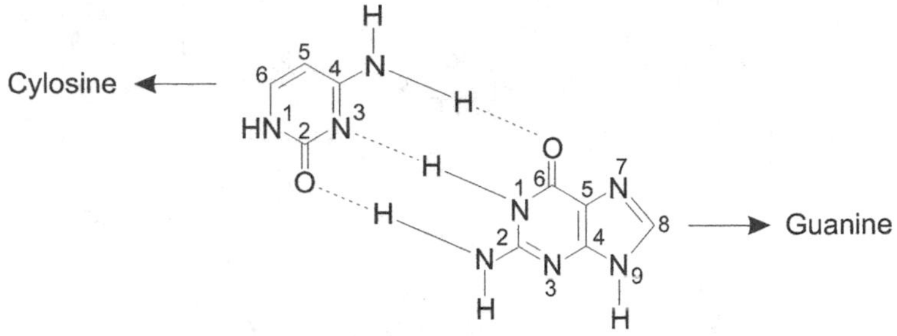

Fig. 4.5 Hydrogen bonds between cytosine and guanine.

Guanine can be hydrolysed with strong acid to glycine, ammonia, carbon dioxide and carbon monoxide. Guanine is first deaminated to xanthine. It gives xanthine on treatment with nitrous acid. The conversion is also effected by boiling with 25% hydrochloric acid. Guanine oxidises more readily than adenine.

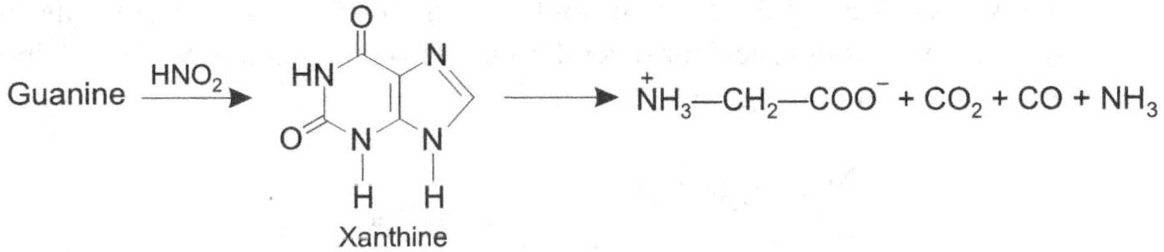

The high melting point of guanine (350°C) reflects the intermolecular hydrogen bonding between the oxo and amino groups in the molecules in the crystal form. Because of this intermolecular H-bonding, guanine is relatively insoluble in water, but it is soluble in dilute acids and bases.

C. Synthesis of Guanine

Trace amounts of guanine are formed by the polymerisation of ammonium cyanide. These results indicate guanine could arise in frozen regions of the primitive earth.

i. Fischer-Tropsch Synthesis

A Fischer-Tropsch synthesis can be used to form guanine along with uracil.

Heating an equimolar gas mixture of CO, H_2 and NH_3 to 700°C for 15 to 24 minutes, followed by quick cooling and then sustained heating to 100 to 200°C for 16 to 44 hours with an alumina catalyst yielded guanine and uracil.

$$10\ CO + H_2 + 10\ NH_3 \longrightarrow 2C_5H_8N_5O + 8H_2O$$
$$\text{Guanine}$$

ii. Fischer

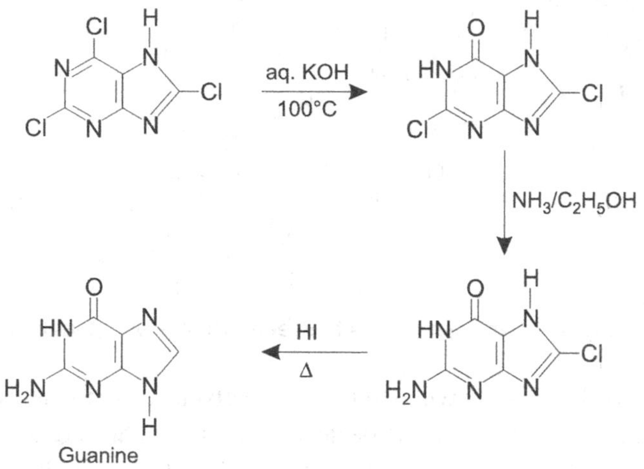

Guanine

iii. Traube's Method

It involves heating 2,4,5-triamino-1, 6-dihydro-6-oxypyrimidine (as the sulphate) with formic acid for several hours. This intermediate can be obtained easily by the following route

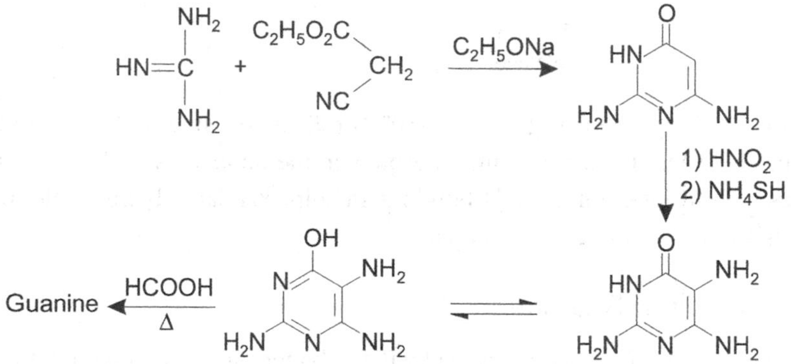

D. Uses of Guanine

In the cosmetics industry, crystalline guanine is used as an additive to various products (e.g., shampoos) where it provides a pearly iridescent effect. It is also used in metallic paints and simulated pearls and plastics. It provides lustre to eye shadow and nail polish.

Guanine crystals are rhombic platelets composed of multiple transparent layers but they have a high index of refraction that partially reflects and transmits light from layer to layer, thus producing a pearly lustre. It can be applied by spray, painting or dipping.

4.9 UREIDES

Acyl derivatives of urea are called ureides, *e.g.*, acetyl urea in which one hydrogen of urea is replaced by an acetyl group.

They may be classified as:

1. Simple ureides
2. Cyclic ureides

A. Simple Ureides

They are generally prepared by the action of acyl chlorides or acid anhydrides of mono-carboxylic acids on urea. For example,

$$CH_3\ COCl + H_2N-\overset{\overset{\displaystyle O}{\|}}{C}-NH_2 \longrightarrow CH_3\ CO-NH-\overset{\overset{\displaystyle O}{\|}}{C}-NH_2 + HCl$$
$$\text{Acetyl urea}$$

$$CH_3-CONH-\overset{\overset{\displaystyle O}{\|}}{C}-NH_2 + CH_3COCl \longrightarrow CH_3-CONH-\overset{\overset{\displaystyle O}{\|}}{C}-NH-COCH_3 + HCl$$
$$\text{Diacetylurea}$$

Simple ureides resemble amides *i.e.*, they are almost neutral in character and are hydrolysed by alkali into urea and the acid.

The halogen derivatives of some simple ureides have been used as hypnotics. An example is that of **bromural**, which is α-bromoisovaleryl urea.

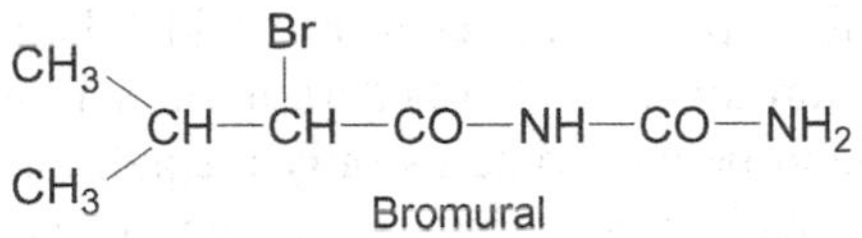

$$\begin{array}{c} CH_3 \\ \diagdown \\ \diagup \\ CH_3 \end{array}\!\!CH-\overset{\overset{\displaystyle Br}{|}}{CH}-CO-NH-CO-NH_2$$
$$\text{Bromural}$$

B. Cyclic Ureides

They are prepared by the action of dicarboxylic acid derivatives with urea. For example, we get parabanic acid, which is oxalylurea obtained by the action of urea on oxalyl chloride

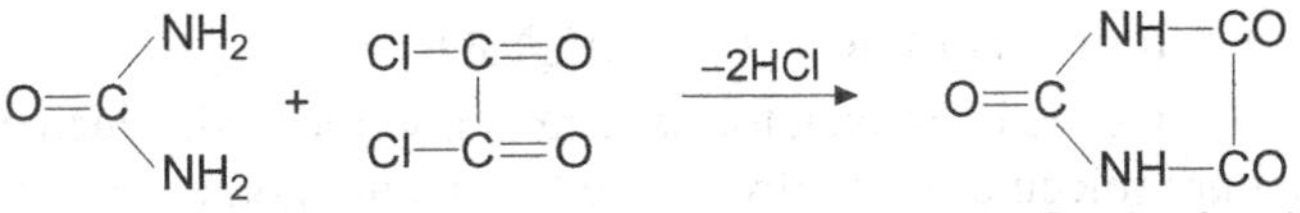

$$\begin{array}{c} \diagup NH_2 \\ O=C \\ \diagdown NH_2 \end{array} + \begin{array}{c} Cl-C=O \\ | \\ Cl-C=O \end{array} \xrightarrow{-2HCl} \begin{array}{c} \diagup NH-CO \\ O=C \qquad\qquad | \\ \diagdown NH-CO \end{array}$$
$$\text{Parabanic acid}$$

Barbituric Acid

It is an important pyrimidine derivative.

We get barbituric acid (malonyl urea) by the action of malonic ester with urea in ethanolic solution containing sodium ethoxide.

$$\begin{array}{c}
\diagup COOC_2H_5 \\
CH_2 \\
\diagdown COOC_2H_5
\end{array}
\quad + \quad
\begin{array}{c}
NH_2 \\
C=O \\
NH_2
\end{array}
\quad \xrightarrow[C_2H_5OH]{NaOC_2H_5} \quad
\begin{array}{c}
CO-NH \\
CH_2 \quad CO \\
CO-NH
\end{array}$$

Barbituric acid

Barbituric acid is a solid, m.p. 250°C. It is not very soluble in water.

Cyclic ureides, unlike simple ureides, are acidic in character and form salts with bases. They are hydrolysed by bases to give urea and the corresponding acid.

Barbituric acid and its derivatives, particularly 5, 5-disubstituted derivatives, are used as hypnotics and sedatives. Two important examples are of barbitone (5,5-diethyl barbituric acid) and phenobarbitone (5-phenyl, 5-ethyl barbituric acid).

$$\begin{array}{c}
CO-NH \\
C_2H_5 \diagdown \quad | \quad | \\
\quad\quad C \quad CO \\
C_2H_5 \diagup \quad | \quad | \\
CO-NH
\end{array}
\qquad\qquad
\begin{array}{c}
CO-NH \\
C_6H_5 \diagdown \quad | \quad | \\
\quad\quad C \quad CO \\
C_2H_5 \diagup \quad | \quad | \\
CO-NH
\end{array}$$

Barbitone   Phenobabitone

4.10   URIC ACID

A. Introduction

Uric acid is a product of animal metabolism. It is present in the excreta of birds and reptiles. Guano, the excrement of birds found on islands near the Western Coast of South America, contains up to 25 per cent uric acid. About 90 percent of snakes' excrement is ammonium urate. It also occurs in human urine in small amounts. Normal blood of human beings contains only a small amount (0.2 mg percent) of uric acid, but in the patients suffering from gout, the amount of uric acid rises considerably. Uric acid was first discovered by **Scheele** in urinary calculi.

B. Structure of Uric Acid

The structure of uric acid has been elucidated through the extensive research work of **Liebig, Wohler, Baeyer** and **Fischer**.

1. Molecular formula of uric acid is $C_5H_4N_4O_3$
2. On exhaustive methylation it forms tetramethyluric acid, which on hydrolysis gives four molecules of methylamine but not ammonia.

$$C_5H_4N_4O_3 \xrightarrow{CH_3I} C_9H_{12}N_4O_3 \xrightarrow{H_2O} 4CH_3NH_2$$
Uric acid   Tetramethyl   Methylamine
                   uric acid

This shows that uric acid contains four amino groups and each methyl group in tetramethyl uric acid is attached to a nitrogen atom.

3. Oxidation of uric acid with nitric acid gives alloxan and urea in equimolecular quantities

$$\text{Uric acid} + H_2O \xrightarrow{HNO_3} C_4H_2N_2O_4 + NH_2-\overset{\overset{\displaystyle O}{\|}}{C}-NH_2$$

Alloxan

4. Structure of alloxan

The structure of alloxan was established as (8) on the basis of hydrolytic experiments

(i) On hydrolysis with alkali alloxan gives one molecule of urea and one molecule of mesoxalic acid.

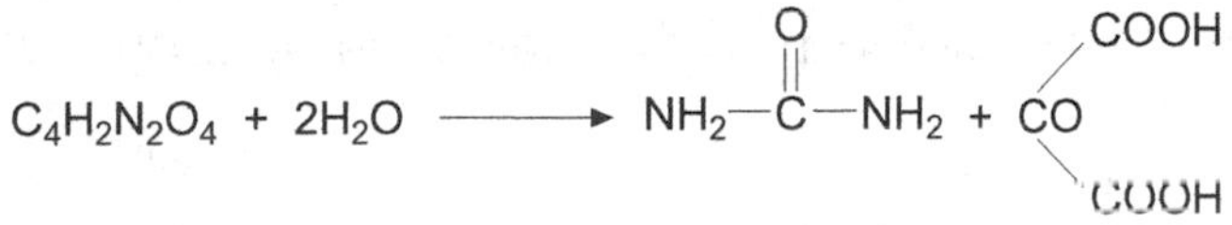

$$C_4H_2N_2O_4 + 2H_2O \longrightarrow NH_2-\overset{\overset{\displaystyle O}{\|}}{C}-NH_2 + \overset{\overset{\displaystyle \diagup COOH}{}}{\underset{\diagdown COOH}{CO}}$$

Alloxan does not contain any free amino or carboxyl groups. Alloxan is, therefore, a cyclic ureide *i.e.*, it is mesoxalyl urea.

The structure of alloxan has been confirmed as it can be obtained by direct condensation of urea and mesoxalic acid.

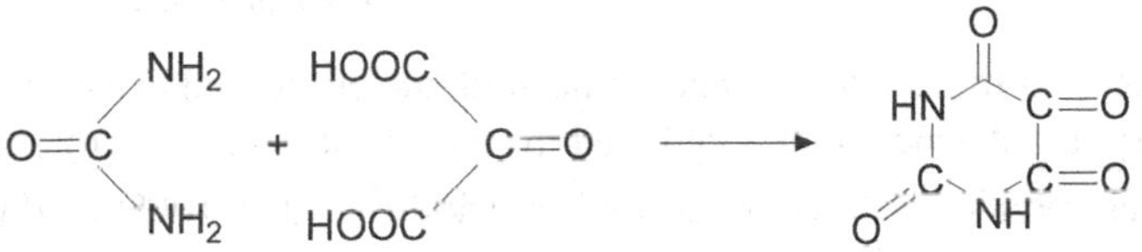

(8) Alloxan

(ii) Barbituric acid can also be converted to alloxan monohydrate as given below:

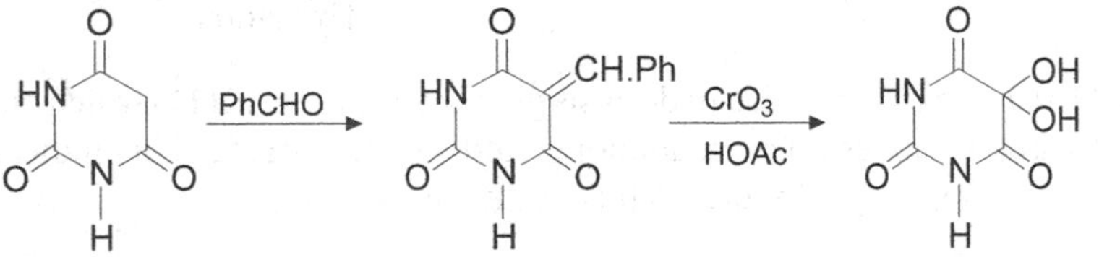

Barbituric acid   Alloxan monohydrate

5. Uric acid on oxidation with an aqueous suspension of lead dioxide gives allantoin (9) and carbon dioxide.

$$\text{Uric acid} \xrightarrow{(O)} C_4H_6N_4O_3 + CO_2$$

Allantoin

6. Structure of allantoin

(i) Allantoin on hydrolysis with alkali forms two molecules of urea and one molecule of glyoxylic acid.

$$C_4H_6N_4O_3 \xrightarrow{H_2O} 2NH_2CONH_2 + \underset{\underset{COOH}{|}}{CHO}$$

Thus, allantoin is the diureide of glyoxylic acid.

(ii) On oxidation with nitric acid, allantoin forms urea and parabanic acid in equal amounts.

$$\text{Allantoin} \xrightarrow{HNO_3} NH_2CONH_2 + \underset{\text{Parabanic acid}}{C_3H_2N_2O_3}$$

(iii) Parabanic acid is oxalyl urea as on hydrolysis it gives urea and oxalic acid.

$$\text{Parabanic acid} \xrightarrow{H_2O} NH_2CONH_2 + \underset{\underset{COOH}{|}}{COOH}$$

The structure of parabanic has been confirmed by synthesis

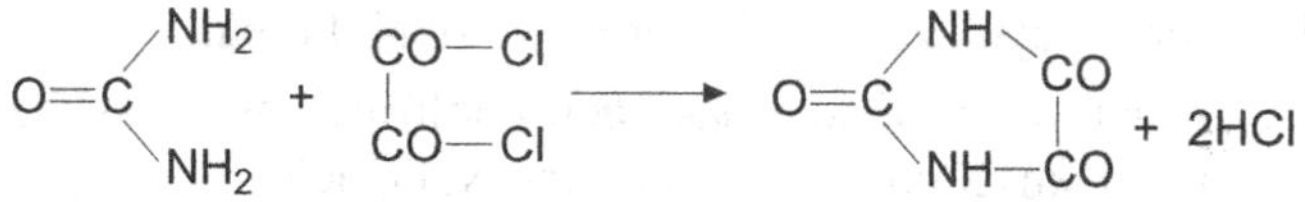

Parabanic acid

Thus, allantoin contains a parabanic acid nucleus joined to a molecule of urea. The point of attachment is deduced from the following experimental evidence.

(iv) When reduced with concentrated hydriodic acid at 100°C, allantoin forms urea and hydantoin

$$\text{Allantoin} \xrightarrow[100°C]{HI} NH_2CONH_2 + C_3H_4N_2O_2$$
$$(10)$$
$$\text{Hydantion}$$

(v) Hydantoin on controlled hydrolysis gives hydantoic acid (11) which on further hydrolysis gives glycine, ammonia and carbon dioxide. This indicates that hydantoin is glycollylurea. Thus, we have,

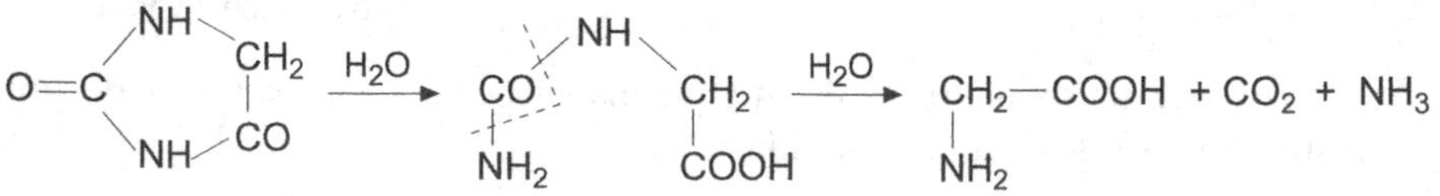

(10)   (11)

This structure (10) for hydantoin has been confirmed by synthesis

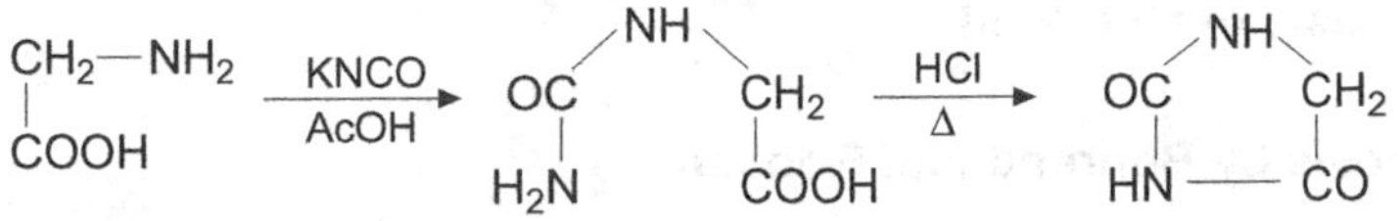

$$CH_2-NH_2 \; | \; COOH \quad \xrightarrow[AcOH]{KNCO} \quad OC \overset{NH}{\underset{H_2N}{\diagdown}} CH_2 \; | \; COOH \quad \xrightarrow[\Delta]{HCl} \quad OC \overset{NH}{\underset{HN}{\diagdown}} \overset{CH_2}{\underset{CO}{|}}$$

(10)

Thus, the structure of allantoin is (9). All the above reactions can be shown as below:

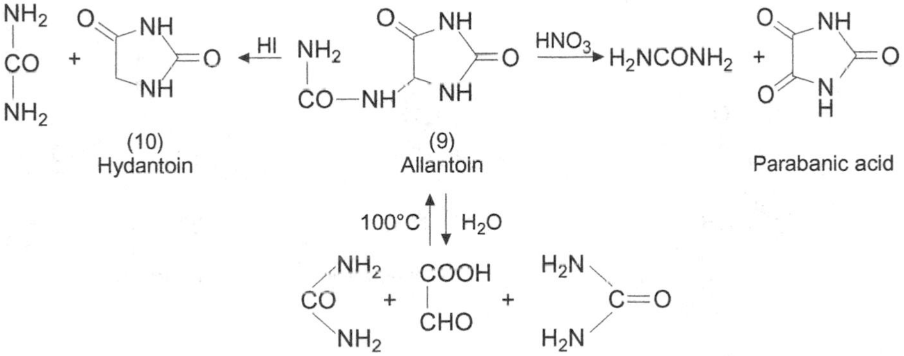

$$NH_2 \; | \; CO \; | \; NH_2 \quad + \quad \overset{O}{\diagdown} \overset{NH}{\diagdown} =O \quad \xleftarrow{HI} \quad NH_2 \; | \; CO-NH \overset{O}{\diagdown} \overset{NH}{\diagdown} =O \quad \xrightarrow{HNO_3} \quad H_2NCONH_2 \quad + \quad \overset{O}{\diagdown} \overset{NH}{\diagdown} =O$$

(10)
Hydantoin

(9)
Allantoin

Parabanic acid

$$100°C \uparrow \downarrow H_2O$$

$$\overset{NH_2}{\underset{NH_2}{\diagdown}} CO \quad + \quad \overset{COOH}{\underset{CHO}{|}} \quad + \quad \overset{H_2N}{\underset{H_2N}{\diagdown}} C=O$$

The structure for allantoin has been confirmed by synthesis. It can be obtained by heating urea with glyoxylic acid at 100°C. The structure can also be confirmed by the following synthesis:

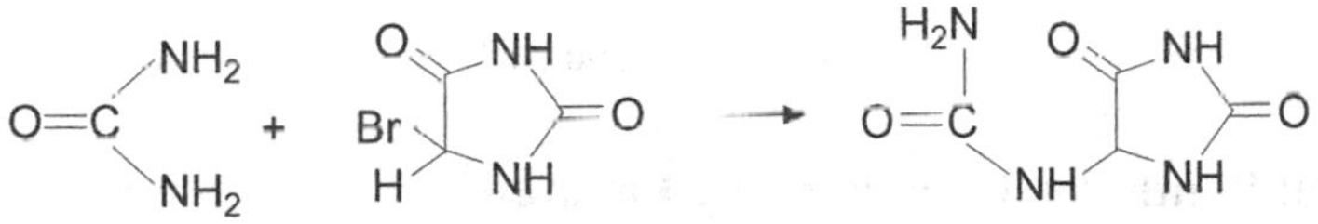

$$O=C \overset{NH_2}{\underset{NH_2}{\diagdown}} \quad + \quad Br \overset{O}{\underset{H}{\diagdown}} \overset{NH}{\underset{NH}{\diagdown}} =O \quad \rightarrow \quad O=C \overset{H_2N}{\underset{NH}{\diagdown}} \overset{O}{\underset{NH}{\diagdown}} \overset{NH}{\diagdown} =O$$

Allantoin

7. Now uric acid on oxidation loses one carbon atom and gives allantoin and at the same time it gives alloxan on oxidation with nitric acid. The formation of these two products can only be explained from the following structure for uric acid 12

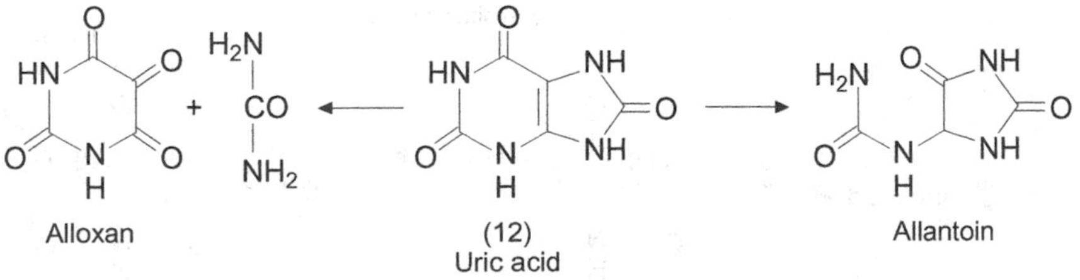

Alloxan

$$+ \quad \overset{H_2N}{\underset{NH_2}{|}} CO$$

(12)
Uric acid

Allantoin

8. The structure was confirmed by Fischer who prepared two isomeric mono methyl uric acids. One gave methyl alloxan and urea on oxidation with nitric acid and the other gave alloxan and methyl urea.

Four monomethyl derivatives of uric acid are possible, which were also prepared by Fischer.

C. Synthesis of Uric Acid

i. Synthesis by Behrend and Roosen

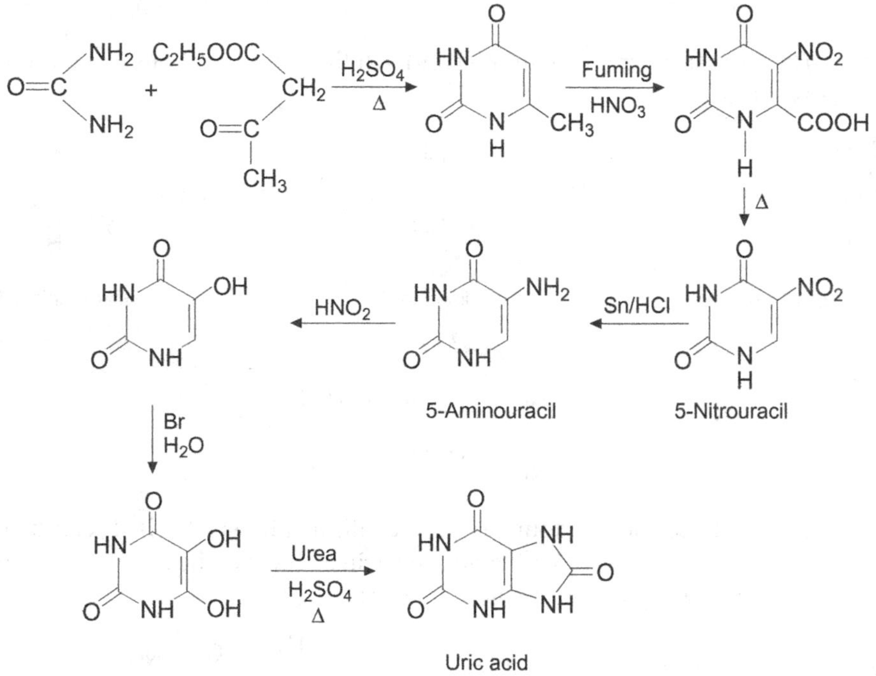

5-Aminouracil

5-Nitrouracil

Uric acid

ii. Baeyer Synthesis Completed by Fischer

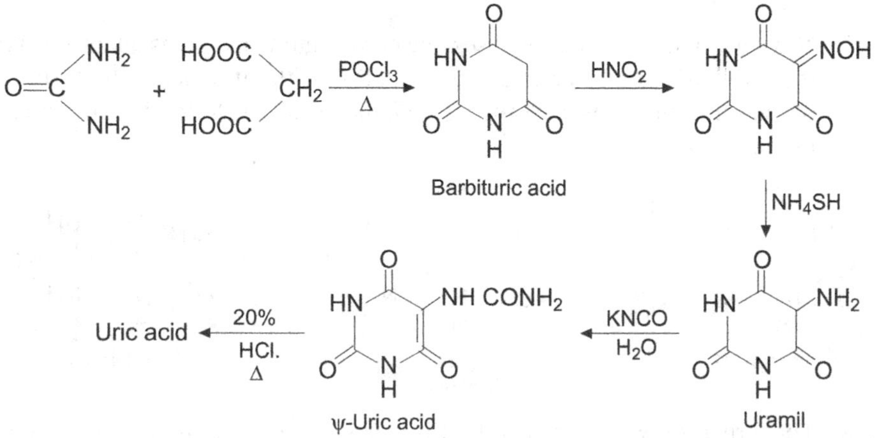

Barbituric acid

Uric acid

ψ-Uric acid

Uramil

iii. Traube's Synthesis

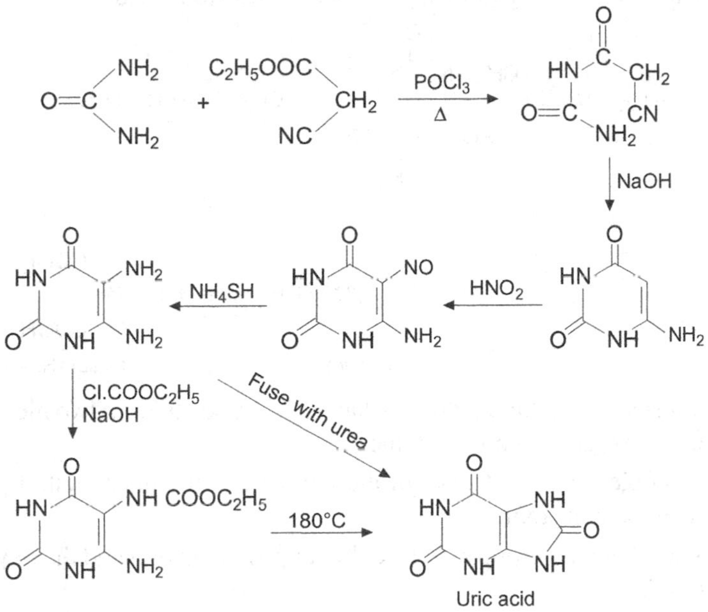

4.11   CAFFEINE

We all consume a large quantity of caffeine in the form of tea and coffee. It is an important methylated xanthine that occurs naturally in tea leaves and coffee beans. Caffeine has stimulant action on the central nervous system.

A. Isolation from Tea Leaves

The dried tea leaves are digested with boiling water. The extract is treated with an excess of lead acetate solution. Proteins and tannins get precipitated. They are filtered off and the filtrate is acidified with dilute sulphuric acid to remove an excess of lead acetate as lead sulphate. The filtrate is concentrated and extracted with chloroform. Chloroform is distilled off and the residue is crystallised from water to give silky needles of caffeine.

Caffeine is a colourless crystalline solid, m.p. 236°C. It is fairly soluble in alcohol and chloroform but sparingly soluble in water and ether. It has a bitter taste and sublimes on heating.

B. Structure of Caffeine

  i. The molecular formula of caffeine is $C_8H_{10}N_4O_2$.
 ii. Caffeine and uric acid have the same skeleton structure. This is shown by the following facts:
    1. On oxidation with potassium chlorate in hydrochloric acid, it gives di-methylalloxan and methyl urea in equimolar proportions

2. The structure of dimethyl alloxan is established as (13) because on hydrolysis it gives N, N'-dimethyl urea and mesoxalic acid.

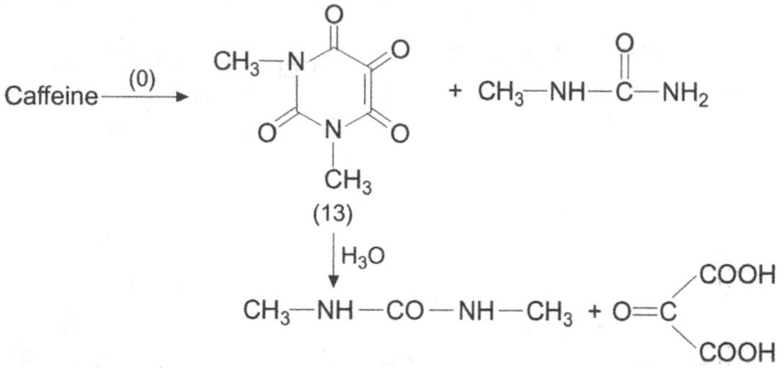

(13)

Dimethyl urea   Mesoxalic acid

The formation of dimethylurea establishes the positions of two methyl groups and one oxygen atom in caffeine.

The problem now is to assign the position of the third methyl group and second oxygen atom.

The third methyl group may be either at position 7 or 9 and the oxygen atom at 6 or 8.

iii. Position of the Third Methyl Group

We have seen the oxidation of caffeine to dimethyl alloxan and methyl urea. Fischer isolated another oxidation product which was shown to be N-methylhydantoin (14) because on hydrolysis it gives N-methylglycine, carbon dioxide and ammonia.

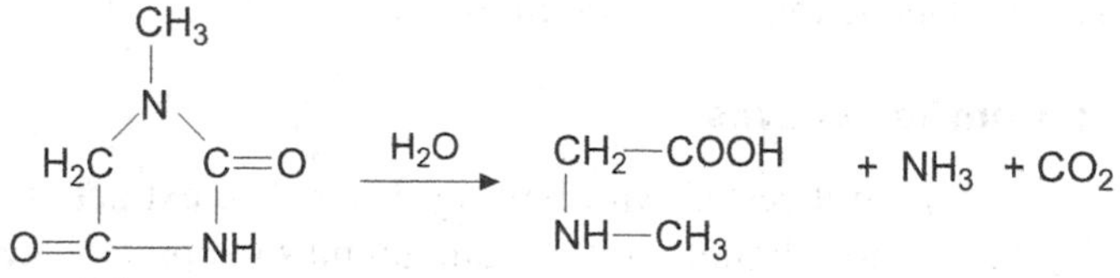

(14)

Therefore, two skeleton structures are possible (15 or 16) for caffeine; each could give the above oxidation products

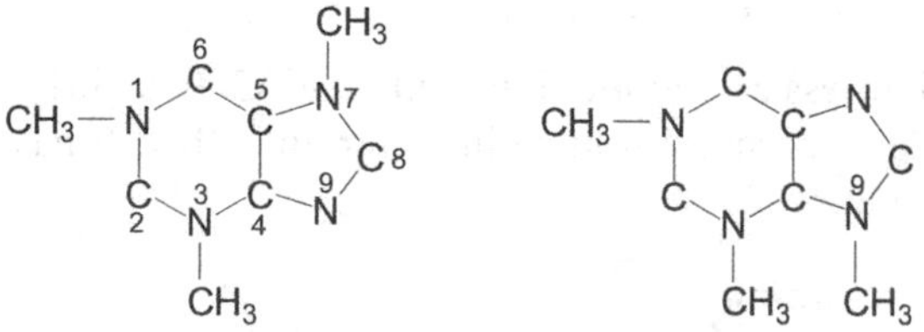

Methyl at 7 position   Methyl at 9 position
(15)   (16)

Fischer was able to isolate a fourth oxidation product which was found to be N, N'-dimethyloxamide. Examination of the structures showed that oxamide can be obtained from structure (15) and not from structure (16). So structure (15) is the skeleton of caffeine.

iv. Position of the Second Oxygen Atom

As mentioned above the second oxygen atom may be at either position 6 or 8. Therefore, two structures are possible for the caffeine molecule.

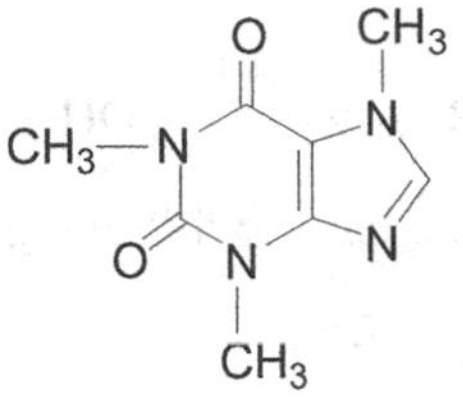

(17) 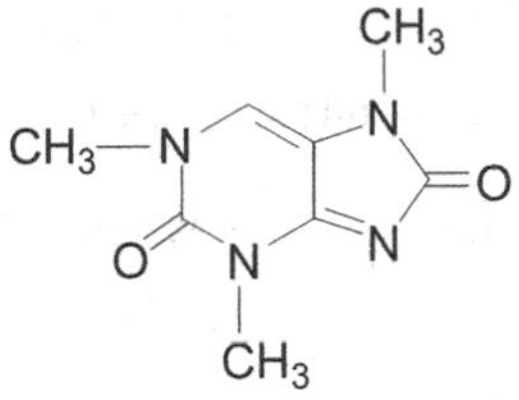   (18)

v. Final Structure of Caffeine

Structure (17) was shown to be caffeine by Fischer. He carried out the following reactions:

| Caffeine | Cl_2 | Chlorocaffeine | CH_3OH | Methoxy caffeine |
| $C_8H_{10}N_4O_2$ | | $C_8H_9ClN_4O_2$ | NaOH | $C_8H_9N_4O_2-OCH_3$ |

dil HCl
Δ

Oxycaffeine
$C_8H_{10}N_4O_3 + CH_3Cl$

Oxycaffeine was shown to be identical with trimethyl uric acid as it was converted into tetramethyl uric acid on methylation with methyl iodide in the presence of aqueous sodium hydroxide. Therefore, methoxycaffeine is either (19) or (20) and oxycaffeine (21) or (22).

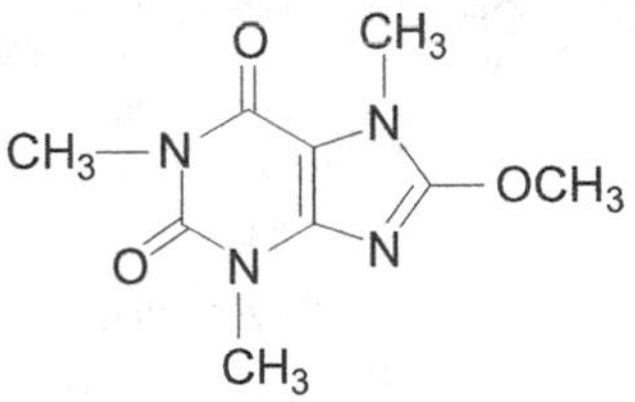

(19)   (20)

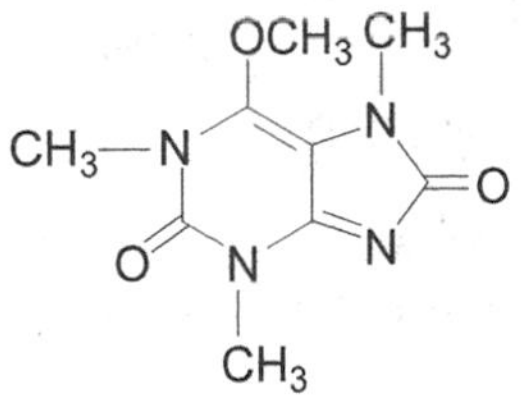

Methoxycaffeine

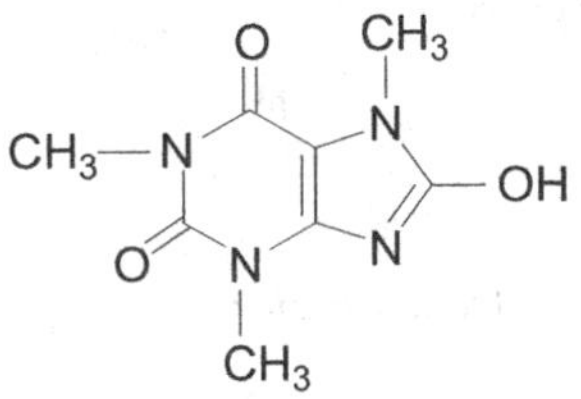

(21)   or   (22)

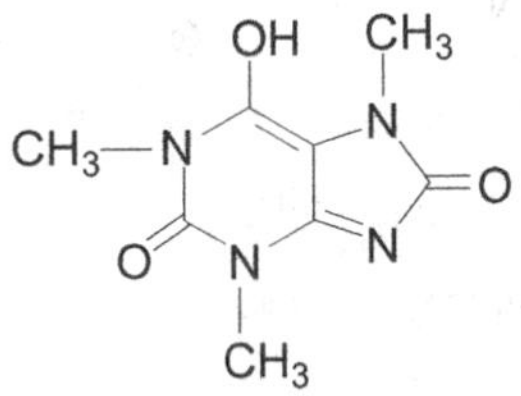

Oxycaffeine

Further, when oxycaffeine is heated with methyl iodide, it is converted into a mixture of tetramethyluric acid and methoxycaffeine. The formation of these two products simultaneously suggests that oxycaffeine is a tautomeric substance *i.e.,* it contains amido-imidol system

$$-NH-\overset{|}{C}=O \rightleftarrows -N=\overset{|}{C}-OH$$

Such a system is possible only when caffeine has the following structure.

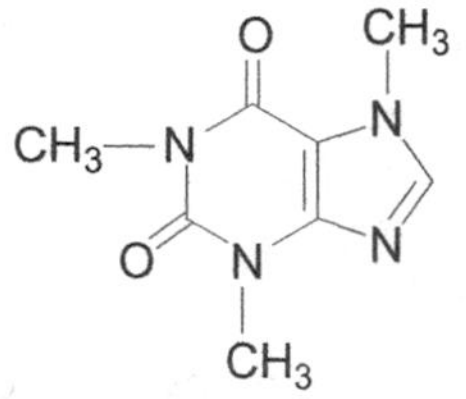

Caffeine

The formation of oxycaffeine from caffeine can then be shown in the following manner:

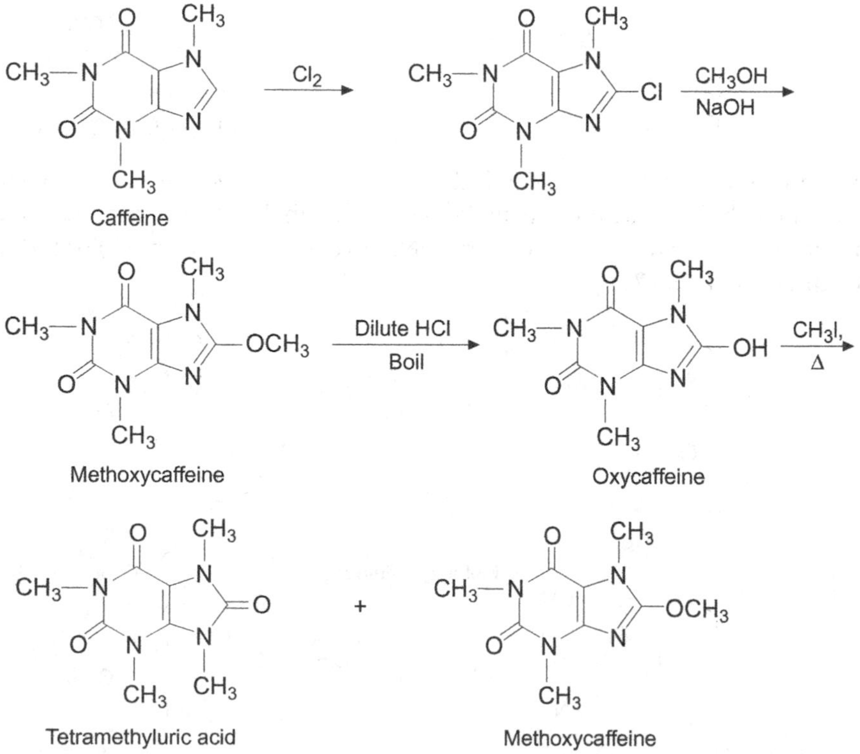

Caffeine

Methoxycaffeine

Oxycaffeine

Tetramethyluric acid

+

Methoxycaffeine

The above structure for caffeine has been confirmed by synthesis.

C. Synthesis of Caffeine

i. Fischer's Synthesis

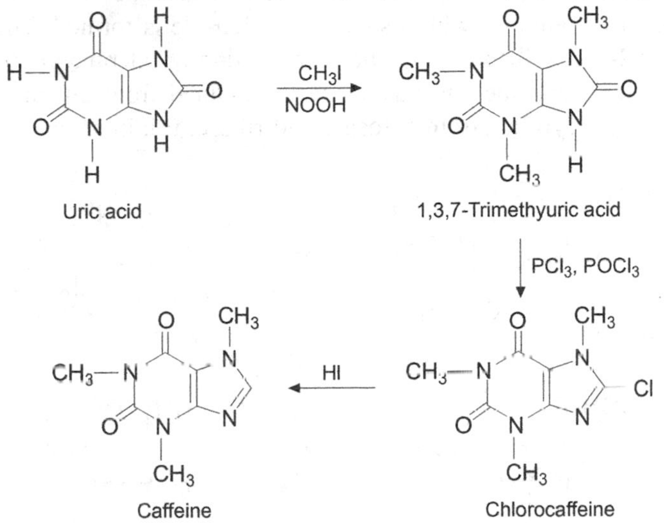

Uric acid

1,3,7-Trimethyuric acid

CH_3I
$NOOH$

PCl_3, $POCl_3$

Caffeine

Chlorocaffeine

HI

ii. Synthesis based on Traube's Method

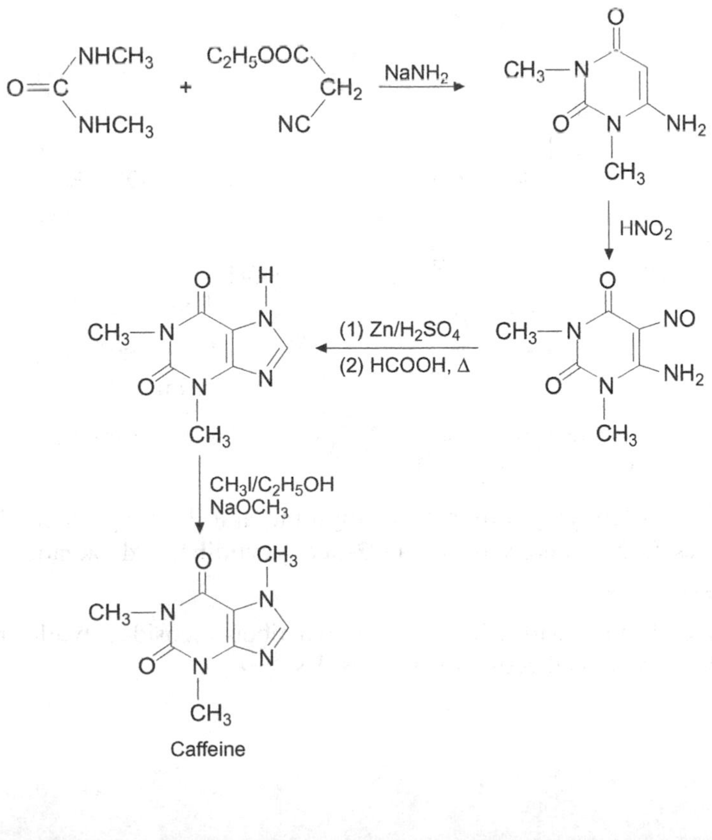

$NaNH_2$

HNO_2

(1) Zn/H_2SO_4
(2) HCOOH, Δ

CH_3I/C_2H_5OH
$NaOCH_3$

Caffeine

4.12 NUCLEOSIDES

We have already seen pyrimidine and purine bases are present in nucleic acids. Pyrimidine bases commonly found in nucleic acids are cytosine, uracil and thymine. The most impotant purine bases are adenine and guanine. When a purine or a pyrimidine base combines with a sugar, a nucleoside is formed. The sugar may be ribose or 2-deoxyribose. Thus, the combination of adenine and ribose forms the nucleoside, adenosine and guanosine is obtained from guanine and ribose. Similarly, we get cytidine from cytosine and ribose, uridine from uracil (p. 249)

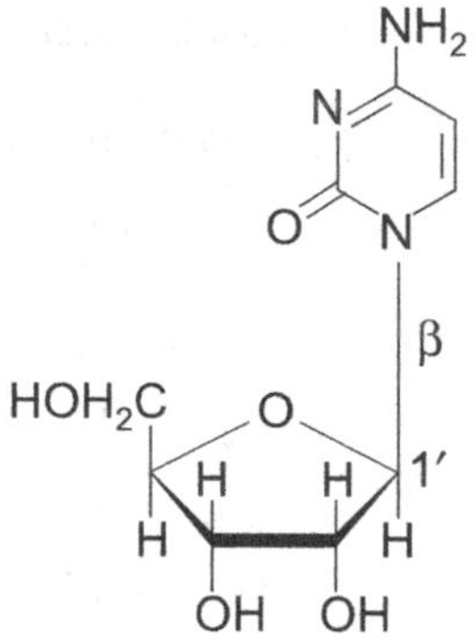

Cytidine
A ribonucleoside

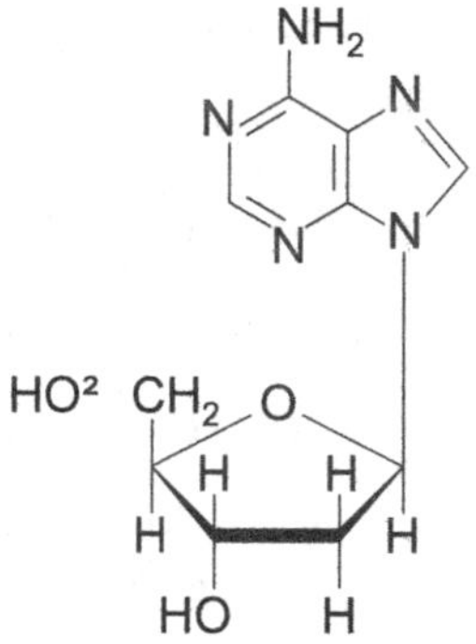

2-Deoxyadenosine

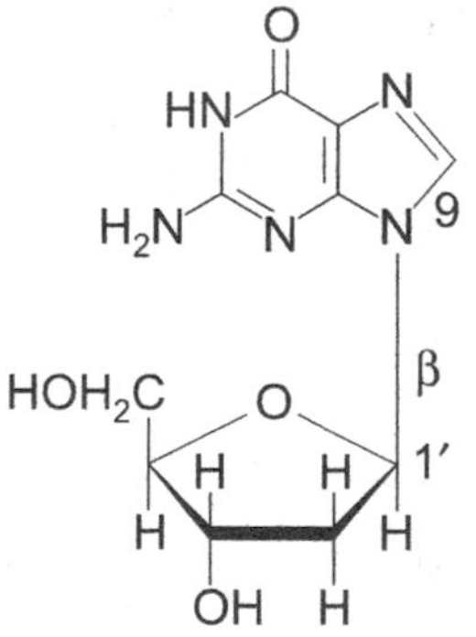

2-Deoxy guanosine
A deoxy ribonucleoside

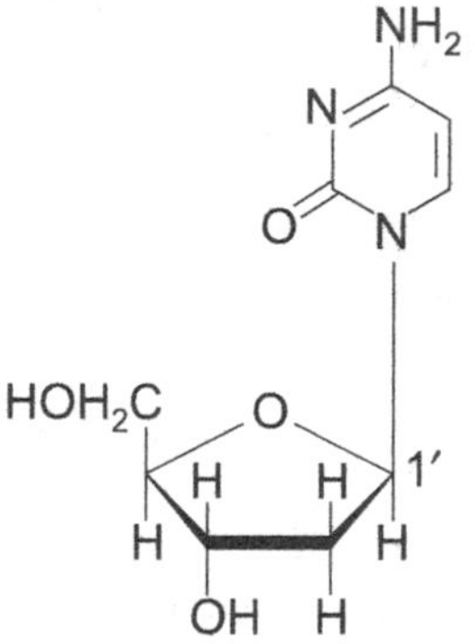

2-Deoxycytidine

and ribose; 2-deoxythymidine from thymine and 2-deoxyribose. The other nucleosides from 2-deoxyribose are 2-deoxycytidine, 2-deoxyadenosine and 2-deoxyguanosine.

Nucleosides formed with ribose are termed ribonucleosides, while those with deoxyribose are termed deoxyribonucleosides.

In a nucleoside, a base forms a glycosidic linkage with the sugar. Since nucleosides are non-reducing the aldehyde group of the sugar cannot be free. That means C–1 of the sugar is joined to the base. Now the problem is to decide which atom of the base is joined to C–1 of the sugar. This was decided by studying the reactions of cytidine and uridine.

- Cytidine on treatment with nitrous acid is converted into uridine. This means the sugar residue is linked in the same position in both of these nucleosides. The point of linkage cannot be 3 or 4 since cytidine has a free amino group at position 4 and there cannot be a hydrogen atom on N–3.

- When uridine is treated with an excess of bromine followed by the addition of phenylhydrazine, a uridine derivative is obtained which contains two phenylhydrazine groups at the positions 5 and 6. This type of compound can be obtained only if uracil is substituted in position 1 and positions 5 and 6 are free. Thus, the sugar is attached to N–1. In a similar way, it has been proved that the other pyrimidine nucleosides have the sugar residue linked at N–1. This linkage has also been confirmed by the X-ray analysis of cytidine.

In case of nucleosides derived from purine bases, it has been proved that C–1 of the sugar is attached to the N–9 position of the purine base. This has been established by the following facts:

Adenosine has a free amino group at position 6. Therefore, the sugar cannot be at C–6 or N–1. Similarly guanosine has a free amino group at position-2; the sugar cannot be at C–2 or N–3. Therefore, only positions 7, 8 and 9 are possible points of attachment. Position 8 was excluded because this point would involve a carbon-carbon bond which would be quite stable and not hydrolysed readily by dilute acids. Thus, positions 7 or 9 are free. **Todd** *et al.* had synthesised adenosine and guanosine in which the sugar is known to be in the 9-position. They showed that their synthetic compounds are identical with the natural products.

Nature of the ring in the sugars

Degradative experiments have shown that the sugar is present as the furanose form in the nucleosides. We have already studied that ring size of glucose and fructose can be determined by methylation studies. Here also methylation of a pyrimidine riboside followed by hydrolysis gives trimethylribose which on oxidation with nitric acid forms dimethylmesotartaric acid. The formation of this acid shows that the ribose ring is in the furanose form. Had it been pyranose, it would have given trimethoxyglutaric acid (Fig. 4.6). This is also confirmed by oxidation with periodic acid which consumes one molecule of the reagent with no loss of carbon and a dialdehyde is formed.

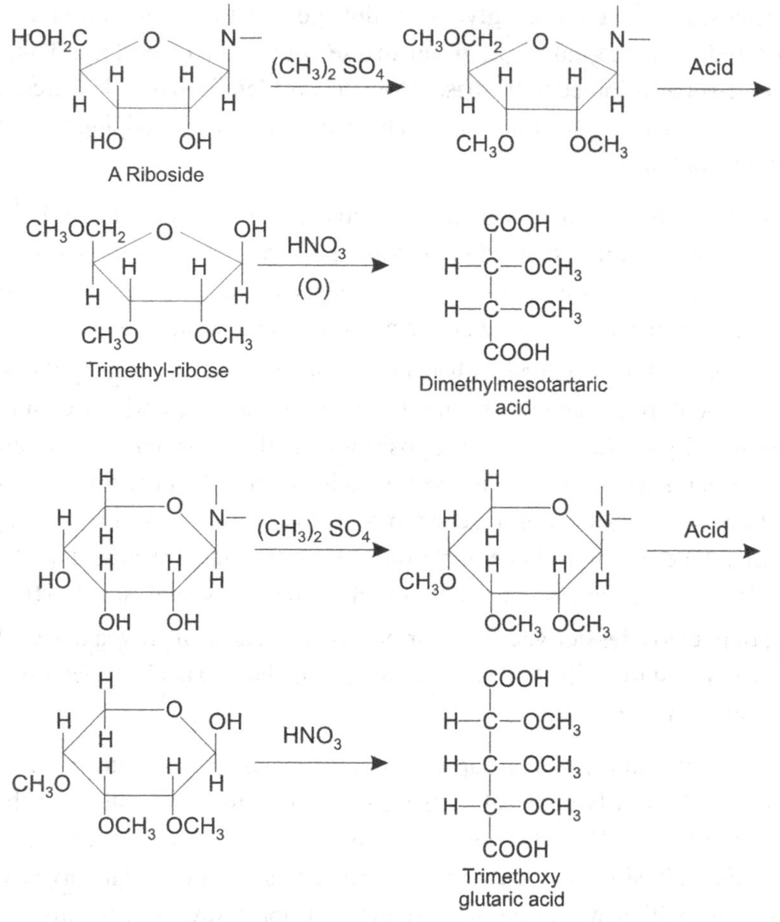

Fig. 4.6

Deoxyribose has also been shown to have a furanose structure. It is found that pyrimidine deoxyribosides are not degraded by periodic acid. This agrees with the 2-deoxyribofuranose structure since the molecule does not contain two adjacent hydroxyl groups.

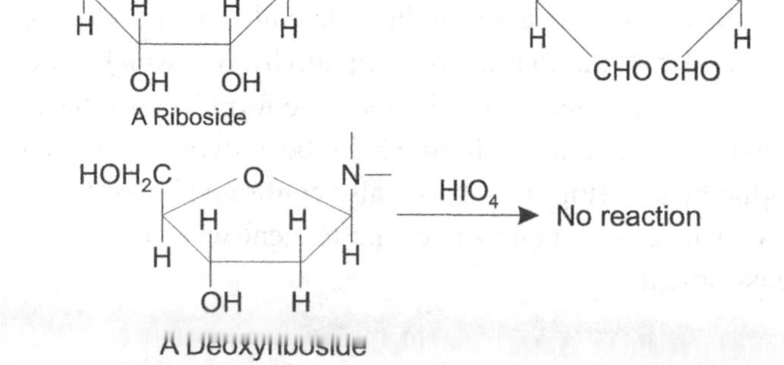

Lastly, the configuration of the furanoside link has been shown to be β- by different experiments.

Oxidation of adenosine with periodic acid was shown to be identical with the product obtained from the oxidation of 9-β-D-mannopyranosidyladenine. This proves that the sugar residue is at position 9, has the furanose structure and that the linkage is β. Similar experiments with other ribonucleosides suggest that all these compounds have a β-configuration.

In case of cytidine, it was shown by X-ray analysis that the sugar residue is attached to N–1 and is β-ribofuranoside. Since other ribonucleosides exhibit the same general pattern, it is inferred that all are furanosides with the β-configuration. Deoxyribonucleosides also exist in the β-configuration as shown by Manson *et al.* from absorption spectra measurements.

Nucleosides are usually stable towards alkaline hydrolysis but are readily hydrolysed by acids. Deoxynucleosides undergo acid hydrolysis more readily than ribonucleosides. In general, the order of hydrolysis is guanosine > adenosine > cytidine > uridine > thymidine. The reasons for this order are not clear. We already know that the furanose structure occurs only when the sugar is in the form of a glycoside. On hydrolysis, the furanose liberated immediately changes into the more stable pyranose form.

4.13 NUCLEOTIDES

When phosphoric acid is esterified to one of the hydroxyl groups of the sugar portion of the nucleoside, we get a nucleotide. Thus, a nucleotide contains a purine or a primidine base, ribose or 2-deoxy-ribose and a phosphoric acid unit. The point of attachment may be 2′, 3′ or 5′ in the ribose molecule and 3′ or 5′ in the deoxyribose molecule.

A nucleotide is named for the parent nucleoside with the suffix "monophosphate" or 'diphosphate' depending upon the number of phosphate units added. For example, adenosine can react with phosphoric acid to form the ribonucleotide, adenosine mono-phosphate (AMP). The position of the phosphate ester is specified by the number of the carbon at the hydroxyl group to which it is esterified. Thus, we have adenosine 3′-monophosphate or deoxycytidine 5′-monophosphate. The 5′-nucleotides are most commonly encountered in nature. Some examples are given in Fig. 4.7.

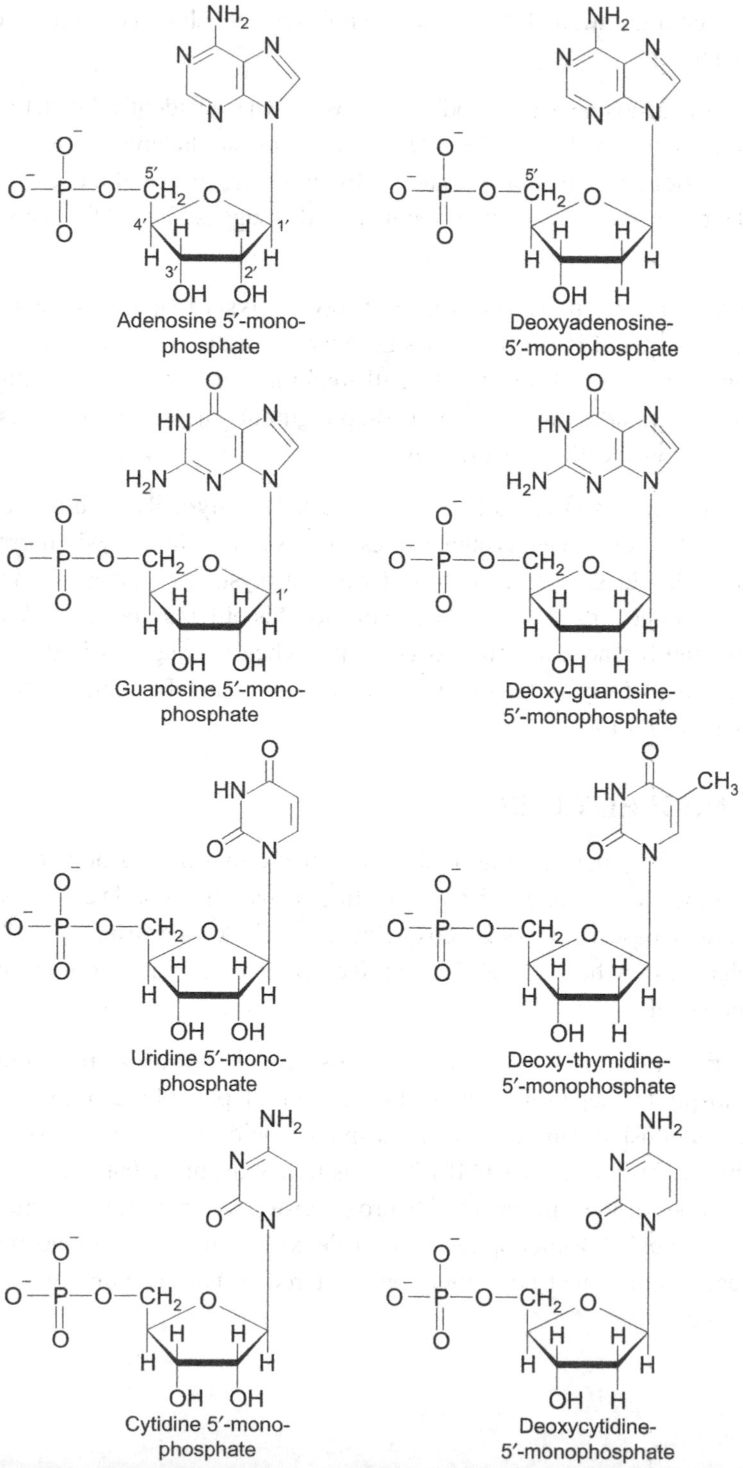

Adenosine 5'-mono-
phosphate

Deoxyadenosine-
5'-monophosphate

Guanosine 5'-mono-
phosphate

Deoxy-guanosine-
5'-monophosphate

Uridine 5'-mono-
phosphate

Deoxy-thymidine-
5'-monophosphate

Cytidine 5'-mono-
phosphate

Deoxycytidine-
5'-monophosphate

Fig. 4.7

If additional phosphate groups form anhydride linkages with the first phosphate, the corresponding nucleoside diphosphates and triphosphates are formed. For example, adenosine forms adenosine diphosphate (ADP) and adenosine triphosphate (ATP). Similarly, deoxyadenosine diphosphate (d ADP) and deoxyadenosine triphosphate (d ATP) are formed (Fig. 4.8). Nucleoside triphosphates serve directly as precursors for DNA and RNA synthesis.

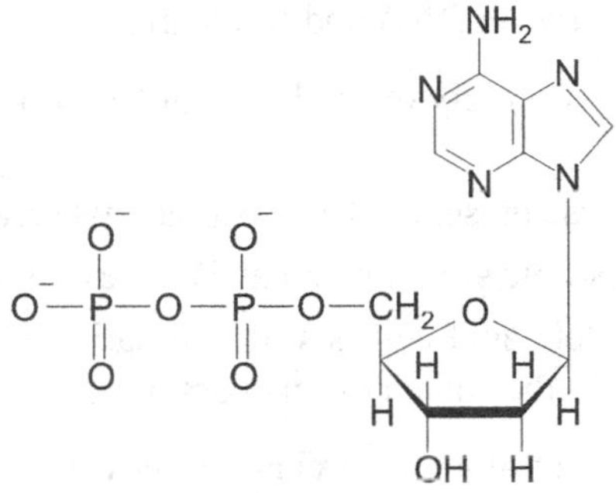

Deoxyadenosine 5′-diphosphate (d ADP)

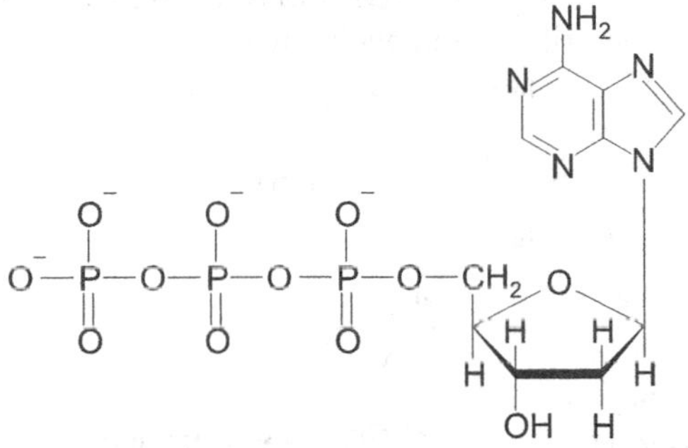

Deoxyadenosine 5′-triphosphate (d ATP)

Fig. 4.8

4.14 POLYNUCLEOTIDES

Nucleotides polymerise to form long chain molecules known as polynucleotides which actually constitute the nucleic acids. When two nucleotides are joined together the resultant molecule is called a dinucleotide while three form a trinucleotide. Oligonucleotides may contain nucleotides up to ten and if the number of nucleotides is more than ten they form a polynucleotide.

Thus, nucleic acids are obtained as a result of polymerisation of nucleotides. The linkage between monomers in nucleic acids involves formation of two ester bonds by phosphoric acid. The hydroxyls to which the phosphoric acid is esterified are those bonded to 3′ and 5′ carbons on adjacent residues. In other words, the polynucleotide chains are composed of monomer units linked by 3′,5′-phosphodiester bonds. The nucleotide residues of nucleic acids are numbered from the 5′-end, which normally carries a phosphate group, to the 3′-end which normally has a free hydroxyl group.

4.15   THE NUCLEIC ACIDS

We already know that there are two nucleic acids. In deoxyribonucleic acid (DNA), the monomers are deoxyribonucleotides and the chief bases are adenine (A), guanine (G), cytosine (C) and thymine (T). In RNA (ribonucleic acid) the monomers are ribonucleotides and the chief bases are A, G, C and uracil (U).

The basic differences between DNA and RNA are:

1. Ribose is the sugar moiety of the RNA molecule whereas it is deoxyribose in DNA.
2. The pyrimidine base present in RNA is uracil whereas it is thymine in DNA.
3. RNA exists as a single strand whereas DNA exists as a double strand.

The backbone of a nucleic acid consists of alternating phosphates and pentoses. The bases are attached to the sugars of this backbone.

The structure of a fragment of an RNA chain is shown in Fig. 4.9. In this repeated linkage is a 3′, 5′-phosphodiester bond. As mentioned earlier the nucleotide residues are numbered from the 5′ end. Thus, the bases in RNA are A, C, G and U. There is a β-glycosidic bond between ribose and each base.

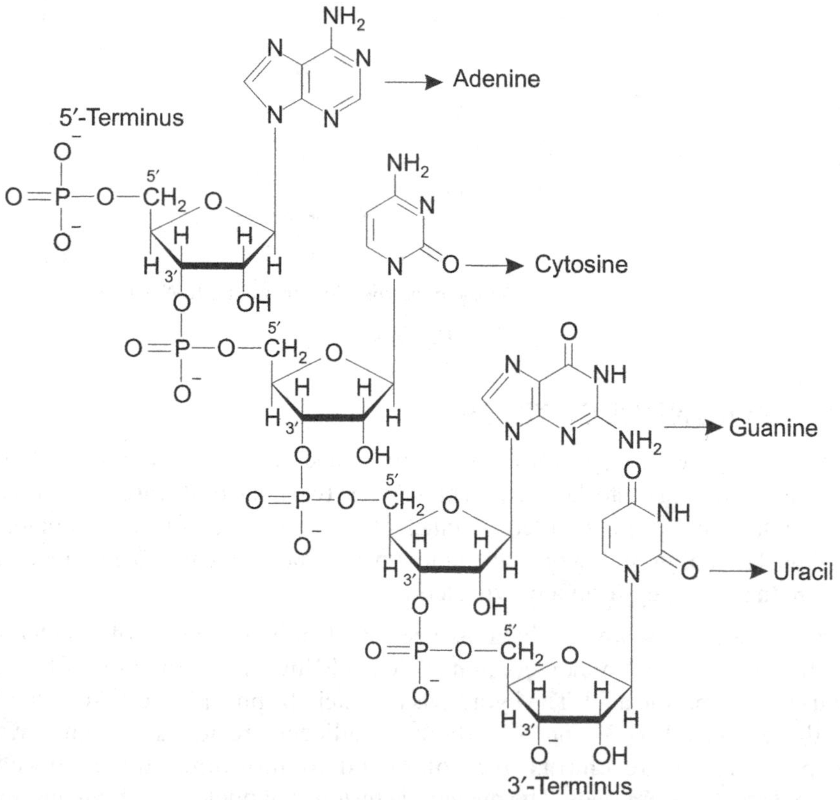

Fig. 4.9 A fragment of an RNA chain.

A portion of the DNA chain is shown in Fig. 4.10. It differs from the RNA chain only in the fact that here the sugar is 2′-deoxyribose rather than ribose. The bases from the 5′-terminus are thymine, guanine, cytosine and adenine. Each base forms a β-glycosidic bond with 2′-deoxyribose.

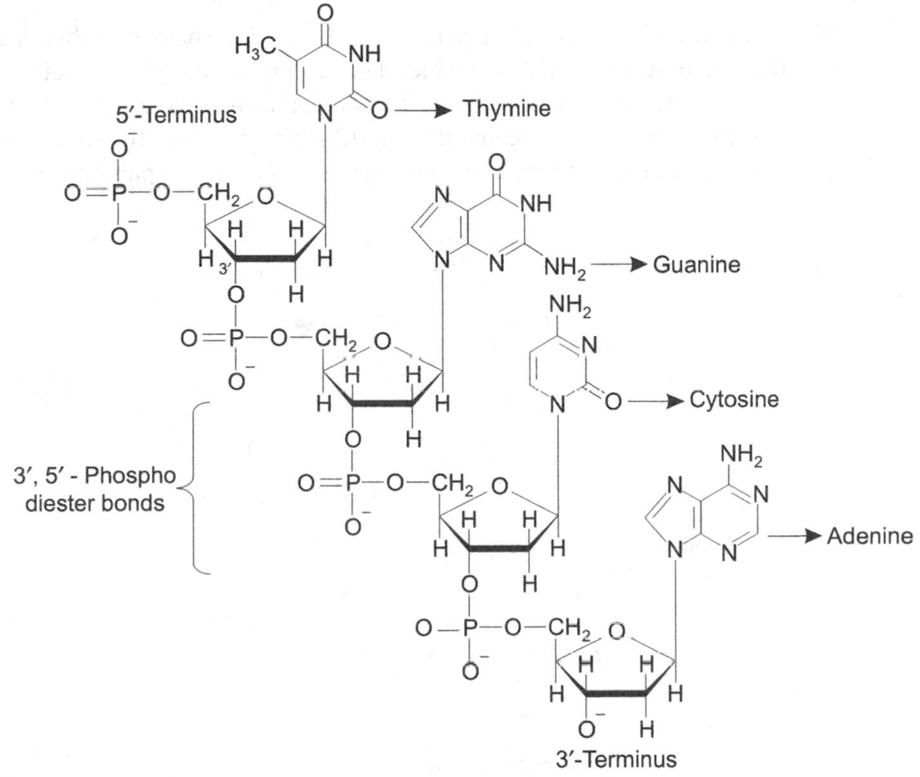

Fig 4.10 A portion of a DNA chain.

4.16   STRUCTURE OF DNA (THE DOUBLE HELIX)

Fig. 4.10 represents the primary structure of the DNA molecule. But it was suspected that most naturally occurring DNAs did not exist as single chains. Instead two or more chains appeared to interact with one another in some way. The resulting macromolecules were known to be long, thin and rigid such that they formed highly viscous solutions in water.

The double helix structure of DNA was proposed by **James Watson** and **Francis Crick** in 1953. It was based primarily on model building and X-ray diffraction patterns. Information from X-ray patterns was added to information from chemical analysis that showed that the amount of **A** was always the same as the amount of **T** and that the amount of **G** always equaled the amount of **C**. Both of these lines of evidence were used to conclude that DNA consists of polynucleotide chains wrapped around each other to form a helix. Hydrogen bonds between bases on opposite chains determine the alignment of the helix. The paired bases lie in planes perpendicular to the helix axis. The sugar-phosphate backbone is the outer part of the helix. The chains run in antiparallel directions one 3′ to 5′ and the other 5′ to 3′ (Fig. 4.11).

Other important features of the **Watson** and **Crick** Model are:

1. The base pairing is complementary, meaning that adenine pairs with thymine and that guanine pairs with cytosine. This complementary base pairing occurs along the entire double helix; the two chains are also referred to as complementary chains. An adenine-thymine (A-T) base pair has two hydrogen bonds between them whereas a guanine-cytosine (G-C) base pair has three.

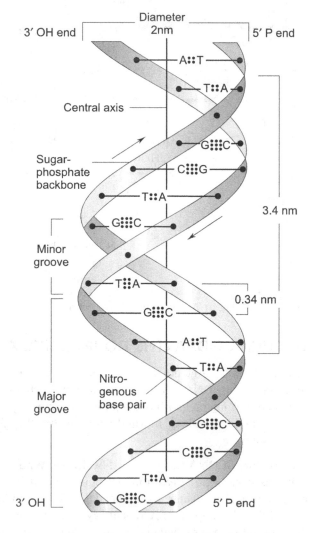

Fig. 4.11 Double helical structure of a part of DNA molecule.

2. The inside diameter of the sugar phosphate backbone of the double helix is about 11 Å (1.1 nm). The distance between the points of attachment of the bases to the two strands of the sugar-phosphate backbone is the same for the two base pairs (A-T and G-C), about 11 Å (1.1 nm). This gives DNA an extremely ordered structure.

3. The outside diameter of the helix is 20 Å (2 nm). The length of one complete turn of the helix along its axis is 34 Å (3.4 nm) and contains ten base pairs.

4. The double helix has empty spaces known as grooves. There is a major groove and a smaller minor groove in the double helix. Both can be sites at which drugs or polypeptides bind to DNA.

We can summarise the important features of Watson and Crick's double helical model of DNA as follows:

- The double helix comprises of two complementary polynucleotide chains.
- The two polynucleotide chains are wrapped clockwise around a central axis to form a circular stair-case type structure.
- The two strands of the double helix are antiparallel *i.e.*, they run in opposite directions. One chain is aligned in $5' \rightarrow 3'$ direction and the other is aligned in $3' \rightarrow 5'$ direction.
- Each polynucleotide chain has a sugar-phosphate backbone and nitrogenous bases are directed inside the helix *i.e.*, towards the central axis. The sugar-phosphate backbone is formed of covalently bonded deoxyribose and phosphate groups and therefore, occur on the outside of the helix.
- The phosphate groups give the molecule a negative charge.
- The nitrogenous bases are hydrophobic in nature. They avoid water and therefore, remain inside the helix.
- The diameter of the double-helix is 20 Å or 2 nm.
- The nitrogenous bases lie almost perpendicular to the axis of the molecule. Therefore, these are stacked one on top of another. The hydrophobic interactions between the bases provide stability to DNA.
- The nitrogenous bases of two anti-parallel polynucleotide strands are linked through hydrogen bonds. There are two-hydrogen bonds between A and T and three between G and C (Fig. 4.12). The hydrogen bonds are the only attractive force between the two polynucleotides of the double helix. These serve to hold the strands together.
- There are ten base pairs in one turn of the helix. Therefore, each turn of the helix is 34 Å or 3.4 nm.
- The double helix has two helical grooves – a major groove, which is deep and wide and a minor groove which is shallow and narrow.

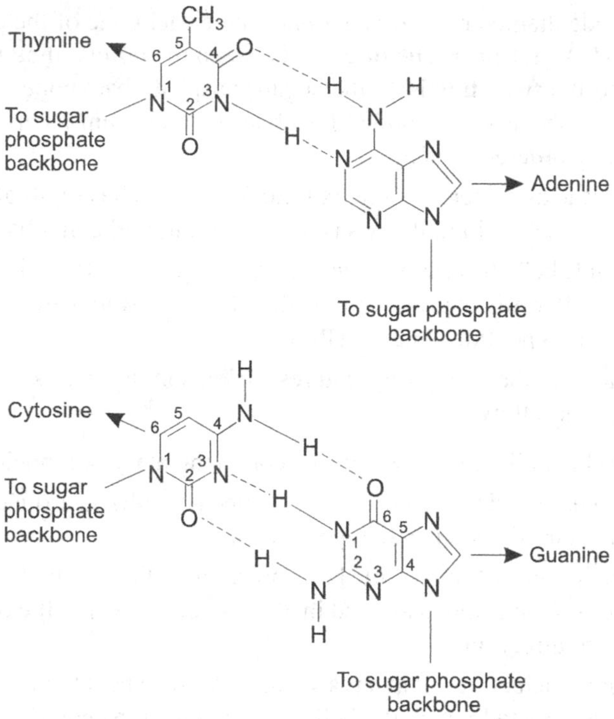

Fig. 4.12   Hydrogen bonds between A and T and between G and C.

4.17 CONFORMATIONAL VARIATIONS IN DNA (FORMS OF DNA)

We have discussed above the common form of DNA, which is called B-DNA. The double helical structure described by Watson and Crick was also of B-DNA. However, other secondary structures can occur depending on conditions such as the nature of the positive ion associated with DNA and the specific sequence of bases. Thus, we have three important forms of DNA. These are A, B and Z-forms. B and Z-conformations occur as cellular DNA.

1. B-DNA

This form of DNA occurs in all living beings under normal conditions *i.e.*, this is the naturally occuring form. It occurs under low salt concentration and high degree of hydration. Its features have been described above. As given above its each turn measures 34 Å and each turn has ten base pairs. Each base pair occupies 3.4 Å. Its two strands are wound in right-handed coils and each strand follows a clockwise path.

2. A-DNA

It is the dehydrated form of DNA. It occurs under conditions of decreased hydration and increased concentration of Na^+ ions. Its polynucleotide chains are coiled clockwise and have right-handed configuration. It has eleven base pairs per turn of the helix and a diameter of 23 Å. A-DNA is more compact and bulky as the distance between base pairs is 2.7 Å instead of 3.4 Å. The orientation of bases is also different. It is presumed that A-DNA is derived from B-DNA as a result of hydrophobic molecules under dry conditions.

3. Z-DNA

Another variant form of the double helix is Z-DNA, which is left-handed *i.e.*, it winds in the direction of the fingers of the left hand. It occurs in a small proportion in cellular DNA. It is called Z-DNA because the phosphodiester backbone of its polynucleotide strands follows a zig-zag course. Z-DNA was discovered by **Alexander Rich** in 1979. It has 12 base pairs per turn. One complete helix is 45 Å as against 34 Å in B-DNA. The diameter of Z-DNA is 18 Å. It has only one groove. Z-DNA is less stable than B-DNA. It is formed only when purines and pyrimidines are present alternatively in the chain. Z-DNA is more compact. The Z-conformation is stabilised by high salt concentration.

4.18   DENATURATION AND RENATURATION OF DNA

We have already seen that the hydrogen bonds between the base pairs are responsible for holding the double helix together. The amount of stabilizing energy associated with the hydrogen bonds is not so great, but the hydrogen bonds hold the two polynucleotide chains in the proper alignment. Moreover, the stacking of the bases contribute the largest part of the stabilization energy.

Energy must be added to a sample of DNA to break the hydrogen bonds and to disrupt the stacking interactions. If we heat a DNA sample in solution, the two strands of the DNA molecule separate. The phenomenon of separation of the two strands of the DNA molecule by heating is known as **denaturation or melting**. The denatured DNA becomes single-stranded.

If native DNA molecules are heated to nearly 100°C, the hydrogen bonds between bases of two antiparallel polynucleotide strands break and two polynucleotide chains separate. This heat denaturation of DNA can be monitored experimentally by observing the absorption of ultraviolet light. There is a sharp increase in absorbance of UV light of 260 nm.

The strands of a DNA molecule separate over a temperature range. The mid-point of this range is called **melting temperature** or **transition temperature** and is represented by *Tm*. It depends upon base composition. DNA rich in G-C pairs melts at a higher temperature. In addition to the effect of base pairs, G-C pairs are more hydrophobic than A-T pairs, so they stack better, which also affects the transition temperature (Fig. 4.13).

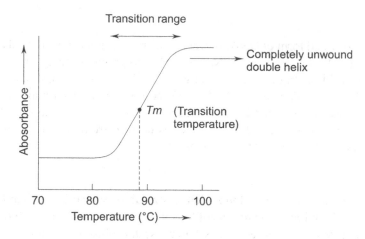

Fig. 4.13

Renaturation of denatured DNA is possible on slow cooling. When heated DNA solution is slowly cooled, the separated strands can recombine and form the same base pairs responsible for maintaining the double helix.

4.19 REPLICATION OF DNA

Whenever a cell divides to produce daughter cells, it is necessary to have duplication of DNA giving rise to a new DNA molecule with the same base sequence as the original. This duplication process is called **replication**. During DNA duplication, the two chains dissociate and each one serves as a template for the synthesis of a new complementary chain. In this way, two DNA molecules are produced, each having exactly the same molecular constitution. Approximately twenty or more enzymes and proteins are needed during replication. Replication occurs at the rate of between 50 nucleotides per second in mammals and 500 nucleotides per second in bacteria.

Watson and Crick suggested a simple mechanism of DNA replication.

The hydrogen bonds between the complementary bases on the two strands of the parent DNA molecule break so that the two polynucleotide strands of DNA separate. The strands thus separated are complementary to each other. Each separated strand acts as a template for the synthesis of a new complementary strand. Thus, two identical daughter DNA molecules are formed. As the base pairing is very specific, each nucleotide of the separated chains attracts its complementary nucleotide from the cell cytoplasm. After the nucleotides are attached by the hydrogen bonds, the sugar molecules join through their phosphate components, and the formation of the new polypeptide is completed.

Thus, each strand of the double helix DNA serves as a template or model on which its complementary strand is synthesised. This method of DNA replication is described as **semi-conservative**, because each new DNA molecule contains one strand from the original DNA and one newly synthesised strand (Fig. 4.14).

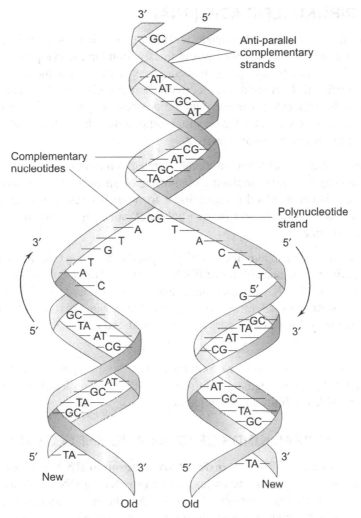

Fig. 4.14 Semi-conservative replication of DNA.

We can summarise the basic rules for DNA replication as below:

1. The base pairing is specific during DNA replication *i.e.*, adenine pairs with thymine and guanine with cytosine.

2. Nucleotide monomers are added one by one to the 3′ end of the growing strand by the enzyme DNA polymerase.

3. The sequence of bases in each daughter strand is complementary to the base sequence in the template strand.

4. 3′ Carbon of deoxyribose present on the 3′ end of the new polynucleotide chain of DNA has OH group and is free to bind to another nucleotide. 5′ Carbon of the deoxyribose on the 5′ end of the polynucleotide chain has a phosphate. Therefore, the new polynucleotide chain is always synthesised in 5′ → 3′ direction.

4.20 RIBONUCLEIC ACID (RNA)

We have already seen that ribonucleic acid (RNA) is a single-stranded nucleic acid which is found in all living cells. RNA is chiefly found in the cytoplasm but also in the nucleolus. Inside the cytoplasm it occurs freely as well as in the ribosomes. The different forms of RNA are associated with the transmission of information from the nucleus into the cytoplasm and with the synthesis of proteins for the regulation of cell activities. RNA is also the only macromolecule which also functions for the storage of information and also as a catalyst.

The primary structure of RNA has already been given in Fig. 4.9. More commonly RNA is a single-stranded molecule consisting of an unbranched polynucleotide chain, but it is often folded back on itself forming helices. However, DNA is a double stranded structure and its two polynucleotide chains are wound spirally around the main axis.

Like DNA, in RNA also we have 3′–5′ phosphodiester bonds. As the sugar found in RNA is ribose, the nucleotides of RNA are called ribonucleotides. We know the four bases found in RNA are adenine, cytosine, guanine and uracil. In addition to these bases, some unusual nitrogenous bases are also found in RNA. However, here the base composition of RNA does not agree with the ratio,

$$A :: U \simeq G :: C = 1.$$

In DNA, nucleotides of the two polynucleotide strands pair through hydrogen bonds. In RNA, intramolecular pairing between the nucleotides of the single strand of RNA provides stability to the molecule.

4.21 PRINCIPAL KINDS OF RNAs AND THEIR STRUCTURES

There are six kinds of RNA–**transfer RNA, ribosomal RNA, messenger RNA, small nuclear RNA, micro RNA** and **small interfering RNA**. All these play an important role in the life processes of cells. DNA is the hereditary material, whereas RNAs perform different functions during protein synthesis. The various kinds of RNAs participate in the syntehsis of proteins in a series of reactions ultimately directed by the base sequence of the cell's DNA. The base sequences of all types of RNA are determined by that of DNA. The process by which the order of bases is passed from DNA to RNA is called transcription. Transfer RNA, ribosomal RNA and messenger RNA are the three functional types of RNAs.

A. Transfer RNA (*t*-RNA)

It is the smallest of the three important kinds of RNA. Different types of *t*-RNA molecule can be found in every living cell because at least one *t*-RNA bonds specifically to each of the amino acids that commonly occur in proteins. A *t*-RNA is a single-stranded polynucleotide chain containing 75 to 93 nucleotides. Each *t*-RNA molecule has a 3′-OH terminus and a 5′-mono-phosphate terminus. Its polynucleotide chain undergoes secondary and tertiary foldings because of internal complementary base pairing. Therefore, a *t*-RNA molecule acquires a precise L-shaped three dimensional configuration. In 1965, **R. W. Holley** proposed a clover leaf model of a *t*-RNA molecule. For this, he got Nobel Prize with **Khorana** and **Nirenberg** in 1968.

The transfer RNA plays a key role in protein synthesis. It picks up a specific amino acid from the cytoplasm, carries it to the site of protein synthesis and attaches itself to a ribosome in accordance with the sequence specified by *m*-RNA. Finally, it transmits its amino acid to the polypeptide chain which is being synthesised.

B. Ribosomal RNA (*r*-RNA)

The ribosomal RNA constitutes the bulk of the cellular RNA up to 85% of the total RNA. In contrast with *t*-RNA, *r*-RNA molecules tend to be quite large and only a few types of *r*-RNA exist in a cell. Cells contain millions of ribosomes and each ribosome contains several molecules of *r*-RNA. The RNA portion of a ribosome accounts for 60%–65% of the total weight and the protein portion constitutes the remaining 35%–40% of the weight. Ribosomes are the structures where the proteins are synthesised.

All ribosomes are constructed from two subunits; the larger subunit is twice the size of the smaller one. Both subunits contain RNA and protein. Ribosomal RNA cannot be removed from the ribosomes without destroying them. All RNAs are single stranded; however many of these are hydrogen bonded.

r-RNA differs in base content from *t*-RNA and *m*-RNA. It is relatively rich in guanine and cytosine. The base components in **E. Coli** have a molar ratio of adenine 21 : uracil 19 : guanine 37 : cytosine 23.

C. Messenger RNA (*m*-RNA)

It is the least abundant of the main types of RNA. It constitutes about 5–10% of the total cellular RNA. Messenger RNA carries genetic information from chromosomal DNA to the cytoplasm, where it acts as a template for protein synthesis. It is complementary to DNA and carries the copy of the same base sequence as found in that part of DNA from which it is copied. The only difference is that thymine is substituted by uracil. The sequences of bases in *m*-RNA specify the order of amino acids in proteins. In rapidly growing cells, many different proteins are needed within a short time interval. Therefore, fast turnover in protein synthesis becomes essential. It is logical that *m*-RNA is formed when it is needed, directs the synthesis of proteins and then is degraded so that the nucleotides can be recycled. Of the main types of RNA, *m*-RNA is the one that usually turns over most rapidly in the cell. Both *t*-RNA and *r*-RNA can be recycled intact for many rounds of protein synthesis.

We can consider here some characteristic features of *m*-RNA:

- It is formed as a complementary strand to one of the two strands of DNA.
- As it carries the same sequence of base arrangement as found in that part of DNA from which it is copied, *m*-RNA, therefore, contains the same information as coded in that part of the DNA.
- The molecules of *m*-RNA are linear and longest of all the three types of RNAs.
- There is one *m*-RNA for each polypeptide chain. *m*-RNA acts as a template for protein synthesis.
- It has a high turnover. However, it has a short life span and withers away after a few translations.
- Messenger RNA molecules are heterogeneous in size as are the proteins whose sequences they specify.

4.22 THE GENETIC CODE

A. Introduction—Definition

We all know that there are only four nitrogenous bases in a DNA molecule, namely adenine (A), cytosine (C), guanine (G) and thymine (T). The sequence of these four bases on the DNA strand directs the proper amino acids to their proper places in protein synthesis. This takes place through the intervention of m-RNA.

The **genetic code** is the set of rules by which information encoded in genetic material (DNA or m-RNA sequences) is translated into proteins (amino acid sequences) by living cells. It gives us the relationship between the nucleotide sequence in nucleic acids and the amino acid sequence in proteins. As a result of this relationship, the information for the structure and functioning of all living things is passed from one generation to the next.

The theory for the existence of the genetic code which is widely accepted was proposed by **F.H.C. Crick**. According to this theory the smallest unit which codes for one amino acid is known as a **codon**. A codon (or code) is the nucleotide or nucleotide sequence in m-RNA which codes for a particular amino acid whereas the genetic code is the sequence of nitrogenous bases in the m-RNA molecule which has information for the synthesis of protein molecules.

The genetic code defines how sequences of three nucleotides, called codons, specify which amino acid will be added next during protein synthesis. With some exceptions, a three nucleotide codon in a nucleic acid sequence specifies a single amino acid. Because the vast majority of genes are encoded with exactly the same code, this particular code is often referred to as the standard genetic code or simply, the genetic code.

B. General Features of the Genetic Code

The various characteristics of the genetic code are as follows:

i. Genetic Code is a Triplet Code

George Gamow postulated that a three-letter code must be employed to encode the 20 standard amino acids used by living cells to encode proteins. With four different nucleotides a code of two nucleotides could only code for a maximum of 4^2 or 16 amino acids. A code of three nucleotides could code for a maximum of 4^3 or 64 amino acids, which are more than enough to code for twenty amino acids. That means each codon consists of three nitrogenous bases to specify one amino acid. Thus, the genetic code is a **triplet code**. Studies by **Nirenberg**, **Ochoa** and **Khorana** led to the complete deciphering of all the 64 codons in the m-RNA. These have been listed in Table 4.1.

ii. Genetic Code is Unambiguous

The genetic code is said to be unambiguous because a particular codon always codes for the same amino acid. The same amino acid may be coded by more than one codon but one codon never codes for two different amino acids.

In certain rare cases, the genetic code may be ambiguous *i.e.,* same codons coding for different amino acids under different conditions.

iii. Genetic Code is Universal

The genetic code appears to be essentially the same in all kinds of living organisms. The code is almost universal. However, there are a few differences in the genetic code in mitochondria, chloroplasts and some organisms.

iv. Genetic Code is Non-Overlapping

Evidences are available to show that translation of the genetic code is an m-RNA that begins at the correct point and there is no overlapping of codons *i.e.*, the genetic code is non-overlapping.

v. Genetic Code is Degenerate

Since several codons usually correspond to a single amino acid, the code is said to be highly degenerate. Since there are 64 possible codons for 20 amino acids, it was concluded that in most cases a single amino acid is coded for by two or more different codons. This multiple system of coding is known as a degenerate system or degenerate genetic code. Except for methionine and tryptophan which have only one codon each, all other amino acids have two or more codons.

vi. Start/Stop/Codons

The genetic code contains 64 codons. Out of these, there are stop codons or termination codons. They are UGA, UAA and UAG. The most common start codon is AUG, which is read as methionine or in bacteria as formyl-methionine. As start codon is also termed as an initiation codon.

The remaining sixty codons code for 20 amino acids.

Table 4.1 Triplet codons for amino acids

	Amino acid	Abbreviation	m-RNA codon
1.	Alanine	Ala	GCU, GCC, GCA, GCG
2.	Arginine	Arg	CGU, CGA, CGG CGC, AGA, AGG
3.	Asparagine	Asn	AAU, AAC
4.	Aspartic acid	Asp	GAU, GAC
5.	Cysteine	Cys	UGU, UGC
6.	Glutamine	Gln	CAA, CAG
7.	Glutamic acid	Glu	CAA, GAG
8.	Glycine	Gly	GGU, GGC, GGA, GGG
9.	Histidine	His	CAU, CAC
10.	Isoleucine	Ilu	AUA, AUU, AUC
11.	Leucine	Leu	UUA, UUG, CUU, CUC, CUA, CUG
12.	Lysine	Lys	AAA, AAG
13.	Methionine	Met	AUG
14.	Phenylalanine	Phe	UUU, UUC
15.	Proline	Pro	CCU, CCC, CCA, CCG
16.	Serine	Ser	UCU, UCC, UCA, UCG, AGU, AGC
17.	Threonine	Thr	ACU, ACC, ACA, ACG
18.	Tryptophan	Try	UGG
19.	Tyrosine	Tyr	UAU, UAC
20.	Valine	Val	GUU, GUC, GUA, GUG
	Terminating triplets	—	UAA, UAG, UGA

4.23   TRANSCRIPTION

A. Introduction

We have seen earlier that DNA is the primary genetic material and because of its molecular configuration, it is capable of storing information. It was also stated that DNA specifies amino acid sequences for proteins. DNA is located in the nucleoid or nucleus and protein synthesis occurs in the cytoplasm of the cell. Thus, the genetic information coded in the nucleotide sequences of DNA must be transferred to the sites of protein synthesis in the cytoplasm (*i.e.,* ribosomes). This transference of information from DNA is called **transcription**.

DNA does not move to the site of the protein synthesis to directly help the process. Instead it transfers its information to single-stranded m-RNA molecules which move to the ribosomes to direct protein synthesis. The process of formation of all the three families of RNA, *i.e.*, *m*-RNA, *t*-RNA and *r*-RNA from DNA templates is called transcription. Thus, it involves transferring the genetic message coded in DNA to the RNA molecule. It is the major control point of the expression of genes and the production of proteins.

We know that the RNA sequence differs from DNA in one respect only. The DNA base thymine (T) is replaced by the RNA base uracil (U). Of all the DNA in a cell only some are transcribed. Transcription produces all the types of RNA, *m*-RNA, *t*-RNA, *r*-RNA etc.

The transcribed *m*-RNA moves out of the nucleus to the ribosomes in the cytoplasm and directs protein synthesis.

B. Mechanism

The fundamental mechanism for the synthesis of all types of RNA is very similar to that of DNA. In both cases, the DNA double helix unwinds and the two strands separate due to the breakdown of hydrogen bonds, which hold the complementary strands together. There must exist some kind of control mechanism which dictates whether the separated DNA strands will synthesise a deoxyribonucleotide chain or ribonucleotide chain. At this stage, appropriate enzyme-DNA polymerase or RNA-polymerase plays a crucial role. However, these enzymes can line up the nucleotides only in the presence of the DNA template.

Thus DNA acts as a template for the transcription of RNA; *i.e.*, the RNA molecule is copied from the DNA code. Only one strand of DNA serves as a template for the formation of a single strand of m-RNA. RNA polymerase reads it from 3′ to 5′.

We give below the basic features of RNA synthesis:

1. Transcription is the process of synthesis of RNA using a DNA template. The enzyme that catalyses the process is DNA dependent RNA polymerase.
2. A DNA template is required. No primer is needed in *m*-RNA synthesis.
3. All the four types of ribonucleoside triphosphates (*i.e.*, ATP, CTP, GTP and UTP) are required.
4. Divalent metal ions Mg^{2+} or Mn^{2+} are used as a cofactor.
5. As in the case of DNA biosynthesis, the RNA chain grows from the 5′ to the 3′ end. The nucleotide at the 5′ end of the chain retains its triphosphate group.
6. The enzyme uses only one strand of the DNA as the termplate for RNA synthesis. This single strand is called the informational strand. The other, the supporting strand, somehow remains inactive during synthesis of *m*-RNA. If both strands of a given gene serve as templates, two different RNA products will be formed that will code for two different proteins. Therefore, only one

strand of DNA serves as a template for the synthesis of a single strand of
m-RNA.

7. The base sequence of the DNA contains signals for initiation and termination
   of RNA synthesis. The enzyme binds to the template strand and moves along
   it in the 3' to 5' direction.

8. The template is unchanged.

4.24   TRANSLATION

A. Introduction

We have seen transcription as the first step in protein synthesis. Translation is the
second step by which gene expression leads to protein synthesis. The order of
bases in *m*-RNA specifies the order of the amino acids in the growing protein. This
process is called **translation**. In this, the sequence of three bases in *m*-RNA directs
the incorporation of a particular amino acid into the growing protein chain. Thus,
translation is the process by which the genetic information encoded in *m*-RNA
directs the synthesis of a specific protein. This results in one language of nucleic
acids getting translated into another language of proteins.

During translation, the codons of an *m*-RNA base-pair with the anticodons of *t*-RNA
molecules, carrying specific amino acids. The order of the bonds determines the
order of the *t*-RNA molecules and the sequence of amino acids in a polypeptide.
The process of translation must be extremely orderly so that the amino acids of a
polypeptide are sequenced correctly.

B. Overall Process of Translating the Genetic Message

Protein biosynthesis is a complex process. It requires ribosomes, amino acids,
m-RNA, *t*-RNA, a number of protein factors including various activating enzymes
and cofactors like ATP, GTP, Mg^{2+} ions *etc*. The *m*-RNA and *t*-RNA are responsible
for the correct order of amino acids with the growing protein chain.

Before an amino acid can be incorporated into a growing chain, it must be activated.
This process involves *t*-RNA and a specific enzyme known as aminoacyl-*t*-RNA
synthetase. In this process, the amino acid is covalently bonded to *t*-RNA forming
an aminoacyl *t*-RNA.

The actual process of protein synthesis involves three steps—chain initiation, chain
elongation and chain termination. Enzymes are required for each of the three steps.

In the initiation step, the first aminoacyl-*t*-RNA is bound to the ribosome and to
m-RNA. A second aminoacyl-*t*-RNA forms a complex with the ribosome and with
m-RNA. The binding site for the second aminoacyl-*t*-RNA is close to that for the first
aminoacyl-*t*-RNA. A peptide bond is formed between the amino acids. This is the
chain elongation step. The process also involves translocation of the ribosome along

the *m*-RNA in addition to peptide bond formation. This process repeats itself until the polypeptide chain is complete. Finally, in the third step, the chain-termination takes place. We have already mentioned the three chain terminating codons. They serve as stop signals in the *m*-RNA message marking the end of the polypeptide.

4.25   GENE THERAPY

A. Introduction

In simple terms, **gene therapy** means diagnosis and cure of genetic diseases. There are several hundred genetic diseases in human beings, which are due to single recessive mutations. Some of them are fatal and for most of these diseases no definite treatment is available. Scientists anticipated that some of these diseases can be cured by the introduction of a normal gene. This is called gene therapy.

Gene therapy is the use of DNA as a pharmaceutical agent to treat a disease. It derives its name from the idea that DNA can be used to supplement or alter genes within individual's cells as a therapy to treat a disease. The most common form of gene therapy involves using DNA that encodes a functional therapeutic gene in order to replace a mutated gene. Other forms involve directly correcting a mutation or using DNA that encodes a therapeutic protein drug to provide treatment. In gene therapy, DNA that encodes a therapeutic protein is packaged within a vector, which is used to get the DNA inside cells within the body. Once inside, the DNA becomes expressed by the cell machinery, resulting in the production of therapeutic protein, which in turn treats the patient's disease.

Gene therapy was first introduced in 1972 by **Friedmann and Roblin**. But the authors urged caution before commencing gene therapy studies on human beings.

Although early clinical failures led many to dismiss gene therapy as over-hyped, clinical successes in 2009-11 have given optimism in the promise of gene therapy. Efforts are now on to provide gene therapy for *sickle-cell anaemia, haemophilia, diabetes, AIDS, HIV* infection, *asthma heart diseases*, various types of *cancer, muscular dystrophy*, cystic fibrosis *etc*. The normal gene for most of these diseases has been successfully cloned.

B. Types of Gene Therapy

Gene therapy may be classified into the following two types:

i. Somatic Gene Therapy

In this, the therapeutic genes are transferred into the somatic cells or body of a patient. Any modifications and effects will be restricted to the individual patient only and will not be inherited by the patient's offspring or later generations. Somatic gene therapy represents the mainstream line of current basic and clinical research, where *m*-RNA is used to treat a disease in an individual.

ii. Germ Line Gene Therapy

In this technique, germ cells *i.e.,* sperm or eggs are modified by the introduction of functional genes, which are integrated into their genomes. This would allow the therapy to be heritable and passed on to later generations. This technique may be highly effective in counteracting genetic disorders and hereditary diseases, but many juridictions prohibit this for application in human beings for a variety of technical and ethical reasons.

C. The Process

The process of gene therapy utilises the delivery of DNA into the cells which can be accomplished by a number of methods. The two major classes of methods are those that use recombinant viruses (sometimes called viral vectors) and those that use naked DNA or DNA complexes (non-viral methods).

i. Viral Vectors

All viruses bind to their hosts and introduce their genetic material into the host cells as part of their replication cycle. Therefore, this has been recognised as a plausible strategy for gene therapy, by removing the viral DNA and using the virus as a vehicle to deliver the therapeutic DNA. A number of viruses have been used for human gene therapy including restrovirus, lentivirus, herpes simplex virus, pox virus and adeno-associated viruses.

ii. Non-Viral Methods

There are several methods for non-viral gene therapy. These include the injection of naked DNA, electroporation and the use of oligonucleotides and inorganic nanoparticles etc.

Non-viral methods can present certain advantages over viral methods such as large scale production and low host immunogenicity.

Scientists have taken the logical step of trying to introduce genes directly into human cells by focusing on diseases caused by single-gene defects such as cystic fibrosis, haemophilia, muscular dystrophy and sickle-cell anaemia. However, this has proven more difficult than modifying bacteria, primarily because of the problems involved in carrying large sections of DNA and delivering them to the correct site on the gene. Today more gene therapy studies are aimed at cancer and hereditary diseases linked to a genetic defect.

D. Problems and Risks Involved in Gene Therapy

Some of the problems of gene therapy are:

i. Short-lived Nature of Gene Therapy

The therapeutic DNA introduced into target cells must remain functional and the cell containing therapeutic DNA must be long lived and stable. When therapeutic

DNA integrates into the genome many cells may divide rapidly. This prevents gene therapy from achieving any long-term benefits. Patients will have to undergo multiple rounds of gene therapy.

ii. Immune Response

Whenever a foreign object is introduced into human tissues, the immune system has evolved to attack the invader. The risk of stimulating the immune system, in a way that reduces gene therapy effectiveness, is always a possibility.

iii. Problems with Viral Vectors

Viruses used in most gene therapy studies may present a variety of potential problems to the patient. They may result in toxicity, immune and inflammatory responses and gene control. In addition, there is always the fear that the viral vector, once inside the patient, may recover its ability to cause disease.

iv. Multigene Disorders

Disorders that arise from mutations in a single gene can be treated by gene therapy. However, some of the most commonly occurring disorders such as heart disease, high blood pressure, arthritis, diabetes and Alzheimer's disease are caused by the combined effects of variations in many genes. Multigene disorders such as these would be difficult to treat effectively using gene therapy.

v. Chance of Inducing a Tumor

If the DNA is integrated in the wrong place in the genome, it could induce a tumor.

□□□

Lipids-Oils and Fats

5.1   WHAT ARE LIPIDS?

Lipids are broadly defined as any fat-soluble naturally occurring molecules. They are insoluble in water but soluble in ether, chloroform and hot alcohol. Lipids include many types of compounds containing a wide variety of functional groups such as fats, oils, waxes, cholesterol, sterols, fat-soluble vitamins (*A*, *D*, *E* and *K*), phospholipids and others. Lipids constitute a major part of stored food material in animals and human beings and a major structural component. The main biological functions of lipids include energy storage, acting as structural components of cell-membranes and participating as important signalling molecules.

Although the term lipid is sometimes used as a synonym for fats, fats are a subgroup of lipids called triglycerides and should not be confused with the term fatty acid. Lipids also include molecules such as fatty acids and their derivatives, cholesterol and all other compounds obtained from cholesterol. Terpenoids and carotenoids also come under lipids. Lipids are hydrophobic in nature. They originate entirely or in part from two distinct types of biochemical subunits or "building blocks"– ketoacyl and isoprene groups.

We consume oils and fats in our daily diet. Dietary fats supply energy, carry fat-soluble vitamins-*A*, *D*, *E*, *K* and are a source of antioxidants and bioactive compounds. Fats are also incorporated as structural components of the brain and cell membranes.

5.2   CLASSIFICATION OF LIPIDS—TYPES OF LIPIDS

Lipids may be classified into four major groups:

1. Simple Lipids
    i. Fats and oils
    ii. Waxes
2. Complex Lipids
    i. Phospholipids
    ii. Glycolipids
    iii. Sphingolipids
3. Derived Lipids
    i. Sterol lipids — Steroids and sterols
    ii. Prenol lipids — Isoprenoids or Terpenes
4. Unclassified Lipids
    Tocopherols (vitamin E) and vitamin K

A. Simple Lipids

The simple lipids are esters of monocarboxylic fatty acids and aliphatic alcohols. Simple lipids are of the following types:

i. Fats and Oils

Fats and oils are triglycerides formed of glycerol and fatty acids. Glycerol contains three hydroxyl groups. When all the three alcoholic groups form ester linkages with fatty acids, the resulting compound is a triacylglycerol (1). The older name of this type of compound is triglyceride. The three ester groups are the polar parts of the molecule, whereas the tails of the fatty acids are non-polar. Three different fatty acids are usually esterified to the alcoholic groups of the glycerol molecule.

$$
\begin{array}{c}
\quad\quad\quad \overset{\displaystyle O}{\overset{\|}{}} \\
CH_2-O-C-R_1 \\
| \quad\quad \overset{\displaystyle O}{\overset{\|}{}} \\
CH-O-C-R_2 \\
| \quad\quad \overset{\displaystyle O}{\overset{\|}{}} \\
CH_2-O-C-R_3
\end{array}
$$

(1)

Triacylglycerols do not occur as components of the membranes as do other types of lipids. They accumulate in adipose tissue and provide a means of storing fatty acids particularly in animals. They serve as concentrated stores of metabolic energy. Complete oxidation of fat yields about 9.4 K cal/g in contrast to 4 K cal/g for carbohydrates and proteins.

Triglycerides are neutral fats. They are lighter than water with a specific gravity of about 0.86. They occur both as solids and liquids.

* **Fats** are solid at room temperature. They have saturated fatty acids.
* **Oils** are fluid at room temperature. They have a large number of unsaturated fatty acids.

ii. Waxes

Waxes are a diverse class of organic compounds that are lipophilic, malleable solids at normal temperatures. They include higher alkanes and lipids typically with melting points about 40°C, melting to give low viscosity liquids. They are insoluble in water but soluble in organic non-polar solvents. All waxes are water resistant materials.

A wax is a simple lipid that is an ester of a long chain alcohol and a higher fatty acid. The alcohol may be made of 12–32 carbon atoms. These waxes can be of animal, plant or mineral origin and are found in nature as coatings on leaves and stems of plants and prevent the plants from losing an excessive amount of water.

All natural occurring waxes are typically esters of higher fatty acids and long-chain alcohols. Plant waxes are derived from long-chain hydrocarbons containing functional groups. Similarly, animal wax esters are derived from a variety of carboxylic acids and fatty alcohols. For example, beeswax is composed of mainly palmitic acid esterified with either hexacosanol or triacontanol.

$$CH_3(CH_2)_{14}COOH + HOCH_2 \text{---} (CH_2)_{24} \text{---} CH_3 \longrightarrow$$

Palmitic acid   Hexacosanol

$$CH_3 \text{---} (CH_2)_{14} \text{---} \overset{\displaystyle O}{\overset{\|}{C}} \text{---} OCH_2 \text{---} (CH_2)_{24} \text{---} CH_3$$

Waxes are chemically inert and resistant to atmospheric oxygen. They are insoluble in water. Waxes frequently serve as protective coatings for both plants and animals. In plants, they coat stems, leaves and fruits. In animals, they are found on fur, feathers, and skin. Myricyl cerotate is the principal component of carnauba wax. It is extensively used in floor wax and automobile wax. The principal component of the wax produced by whales is cetyl palmitate. This wax is an important component of cosmetics.

$$CH_3 \text{---} (CH_2)_{14} \text{---} \overset{\displaystyle O}{\overset{\|}{C}} \text{---} O \text{---} (CH_2)_{15} \text{---} CH_3$$

Cetyl palmitate

$$CH_3 \text{---} (CH_2)_{24} \text{---} \overset{\displaystyle O}{\overset{\|}{C}} \text{---} O \text{---} (CH_2)_{29} \text{---} CH_3$$

Myricyl cerotate

Waxes are neutral lipids that are of considerable physiological importance. As these are completely water insoluble and are generally solid at biological temperatures, their strongly hydrophobic nature allows them to function as water repellents on the leaves of some plants, on feathers, and on the cuticles of certain insects. Waxes also serve as energy storage substances in microscopic aquatic plants and animals.

We know the common uses of waxes in making candles, shoe polishes, crayons, coloured pencils, wood polishes and automobile polishes. There are many more uses of waxes. They are consumed industrially as components of complex formulations often for coating formulations of colourants and plastics. Sealing wax has been used for ages. Waxes are used extensively to make wax paper and cards to make them waterproof. Paraffin wax, obtained from petroleum is made of long-chain alkane hydrocarbons. It finds many uses in our lives. Millions of tons of paraffin wax are produced annually. They are used in foods, such as chewing gum and cheese wrappings, in candles and cosmetics, as non-stick and waterproof coatings and in polishes.

B. Complex Lipids or Conjugated Lipids

Complex lipids or conjugated lipids are esters of fatty acids and alcohol but also have a phosphate, carbohydrate or protein group. They are of the following types:

i. Phosphoacylglycerols (Phospholipids)

If one of the alcoholic groups of glycerol is esterified by a phosphoric acid molecule and two fatty acids are esterified to the glycerol molecule, the resulting compound is called a phosphatidic acid (2). One molecule of phosphoric acid can form ester bonds both to glycerol and some other alcohol creating a phosphatidyl ester (3). Phosphatidyl esters are classed as phosphoacylglycerols or phospholipids. Fatty acids present may be widely different.

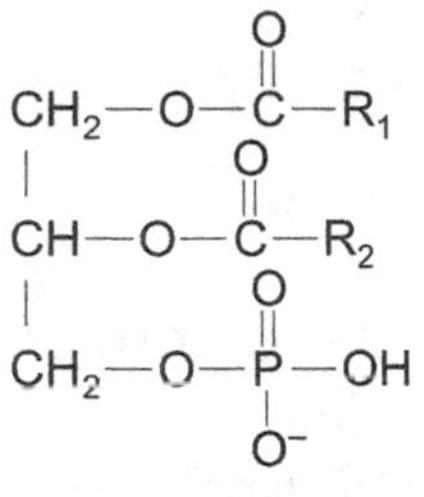

$$CH_2-O-\overset{\overset{\text{O}}{\|}}{C}-R_1$$
$$CH-O-\overset{\overset{\text{O}}{\|}}{C}-R_2$$
$$CH_2-O-\overset{\overset{\text{O}}{\|}}{P}-OH$$
$$\underset{O^-}{}$$

(2) Phosphatidic acid

$$CH_2-O-\overset{\overset{\text{O}}{\|}}{C}-(CH_2)_{16}-CH_3 \longrightarrow \text{Stearyl group}$$
$$CH-O-\overset{\overset{\text{O}}{\|}}{C}-(CH_2)_7-CH=CH-CH_2-CH=CH-(CH_2)_4-CH_3$$
$$\text{Linoleyl group}$$
$$CH_2-O-\overset{\overset{\text{O}}{\|}}{P}-OR$$
$$\underset{O^-}{}$$

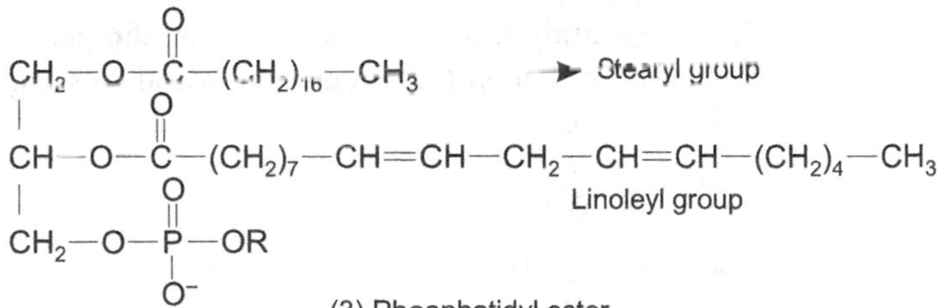

(3) Phosphatidyl ester

Phospholipids are most important lipids. They are important components of biological membranes and are essential for their proper functioning. They regulate cell permeability, control transport and metabolism of synthesised and dietary fat and participate in blood coagulation.

Phospholipids include the following types depending upon the nature of the second alcohol group esterified to the phosphoric acid.

1. Lecithins: Lecithins are monoamino-monophospholipids. An example is that of phosphatidyl choline (4). They usually occur along with cholestrol in the nervous tissue and in egg yolk. The fatty acids most commonly found in lecithins are palmitic acid, stearic acid, oleic acid, linoleic acid and linolenic acid. But only two of these acids are present at a time in one molecule of lecithin. Lecithins play an important role in permeability, osmotic tension and surface conditions of cells.

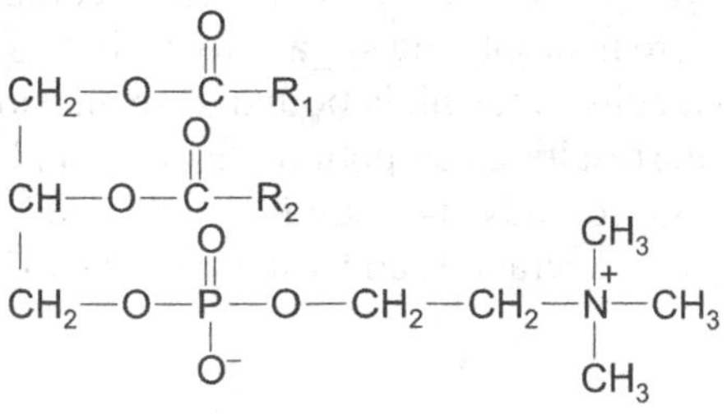

$$CH_2-O-\overset{\overset{\text{O}}{\|}}{C}-R_1$$
$$CH-O-\overset{\overset{\text{O}}{\|}}{C}-R_2$$
$$CH_2-O-\overset{\overset{\text{O}}{\|}}{P}-O-CH_2-CH_2-\overset{\overset{CH_3}{|}}{\underset{CH_3}{N^+}}-CH_3$$
$$\underset{O^-}{}$$

(4) Phosphatidylcholine, a Lecithin

2. Cephalins: The composition of cephalins is very similar to lecithins but the choline radical of lecithin is replaced by an aminoethanol group (5). The fatty acids present in cephalins are also different. Cephalins form components of certain lipoproteins and form an important factor in blood coagulation.

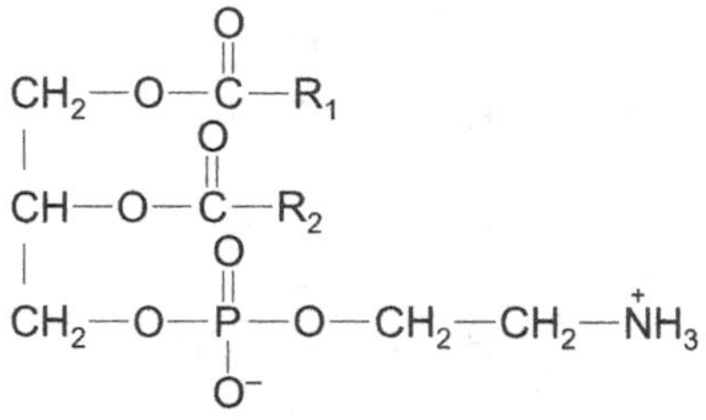

$$
\begin{array}{l}
CH_2-O-\overset{\overset{O}{\|}}{C}-R_1 \\
| \qquad \overset{O}{\|} \\
CH-O-\overset{\|}{C}-R_2 \\
| \qquad \overset{O}{\|} \\
CH_2-O-\overset{\|}{P}-O-CH_2-CH_2-\overset{+}{N}H_3 \\
\qquad \underset{O^-}{|}
\end{array}
$$

(5) Phosphatidylethanolamine

Other important phospholipids are phosphatidylserine (6) and phosphatidylinositol (7). Phosphatidylinositols and phosphatidic acids are either precursors of or are themselves membrane-derived second messengers.

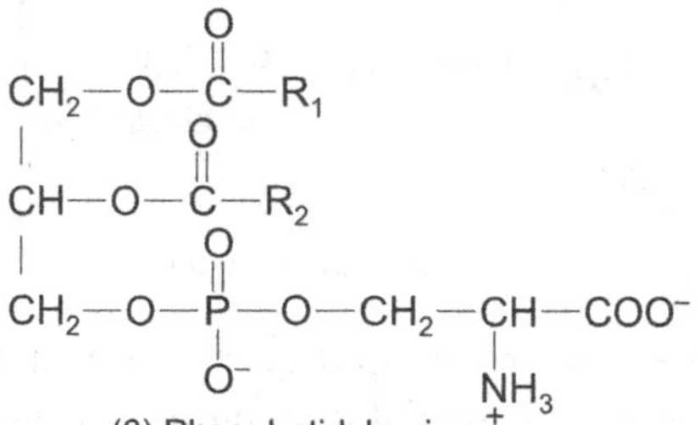

$$
\begin{array}{l}
CH_2-O-\overset{\overset{O}{\|}}{C}-R_1 \\
| \qquad \overset{O}{\|} \\
CH-O-\overset{\|}{C}-R_2 \\
| \qquad \overset{O}{\|} \\
CH_2-O-\overset{\|}{P}-O-CH_2-CH-COO^- \\
\qquad \underset{O^-}{|} \qquad\qquad \underset{\underset{+}{NH_3}}{|}
\end{array}
$$

(6) Phosphatidylserine

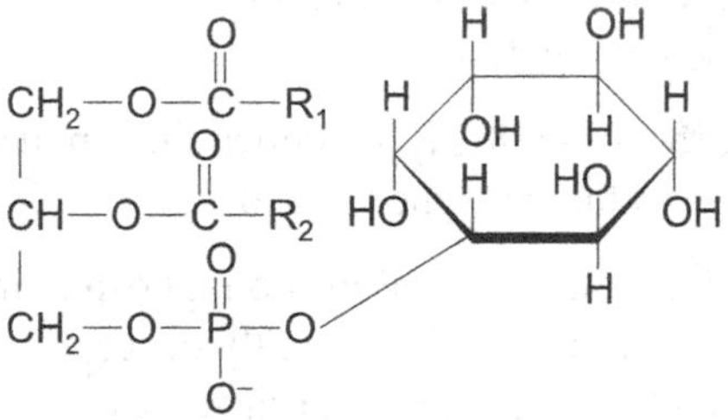

(7) Phosphatidylinositol

ii. Glycolipids

Glucolipids are the components of cellular membranes comprised of a hydrophobic lipid tail and one or more hydrophobic sugar groups linked by a glucosidic bond. In simple terms, if a carbohydrate unit is bound to an alcoholic group of a lipid by a glycosidiec linkage the resulting compound is a glycolipid. Ceramides (8) are the parent compounds for glycolipids and the glycosidic bond is formed between the primary alcohol group of the ceramide and a sugar residue. The resulting compound is called a cerebroside. In most cases, the sugar is glucose or galactose. For example, glucocerebroside (9) contains glucose.

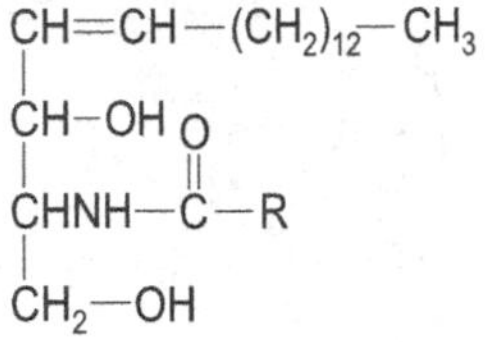

$$CH=CH-(CH_2)_{12}-CH_3$$
$$|$$
$$CH-OH$$
$$|$$
$$CHNH-\overset{\overset{\displaystyle O}{\|}}{C}-R$$
$$|$$
$$CH_2-OH$$

(8) A Ceramide

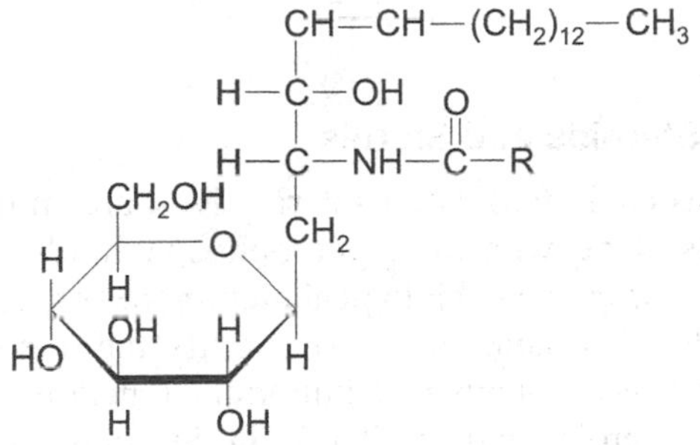

(9) A Glucocerebroside

Cerebrosides are found in nerve and brain cells, primarily in cell membranes. The basic role of glycolipids is to maintain the stability of the cell membranes. They also facilitate cellular recognition which is crucial to immune response and in the connection that allows cells to connect to one another to form tissues.

iii. Sphingolipids

Sphingolipids do not contain glycerol, but they do contain the long-chain amino alcohol sphingosine (10). Sphingolipids are found in both plants and animals; they are particularly abundant in the nervous system. As shown earlier, the simplest compounds of this class are the ceramides, which consist of one fatty acid linked to the amino group of sphingosine by an amide bond.

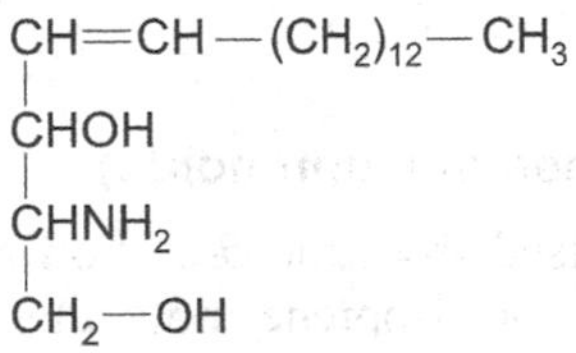

$$CH=CH-(CH_2)_{12}-CH_3$$
$$|$$
$$CHOH$$
$$|$$
$$CHNH_2$$
$$|$$
$$CH_2-OH$$

(10) Sphingosine

In sphingomyelins (11), the primary alcohol group of sphingosine is esterified to phosphoric acid, which in turn is esterified to another amino alcohol, choline. Sphingomyelins occur in cell membranes in the nervous system.

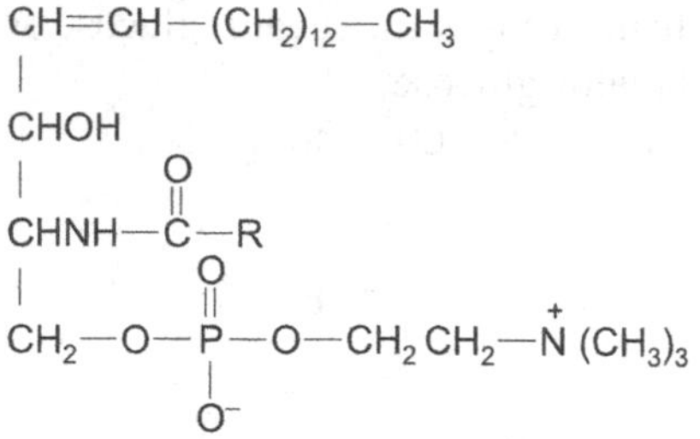

$$CH=CH-(CH_2)_{12}-CH_3$$
$$|$$
$$CHOH$$
$$|$$
$$CHNH-\overset{\overset{O}{\parallel}}{C}-R$$
$$|$$
$$CH_2-O-\overset{\overset{O}{\parallel}}{\underset{\underset{O^-}{|}}{P}}-O-CH_2CH_2-\overset{+}{N}(CH_3)_3$$

(11) A sphingomyelin

C. Derived Lipids

i. Sterol Lipids – Steroids and Sterols

Sterol lipids, such as cholesterol and its derivatives, are an important component of membrane lipids along with the glycerophospholipids and sphingomyelins. Cholesterol (12) is widespread in biological membranes especially in animals. The presence of cholesterol in membranes can modify the role of membrane-bound proteins. Cholesterol has a number of important functions including its role as a precursor of other steroids and of vitamin D. Steroids contain the same fused four-ring core structure. They contain three six-membered rings and one five-membered ring. There are many important steroids including sex hormones. All steroid hormones are derived from cholesterol. However, cholesterol is best known for its harmful effects on health when it is present in excess in the blood. Its excess in the blood develops a condition in which lipid deposits block the blood vessels and lead to heart disease.

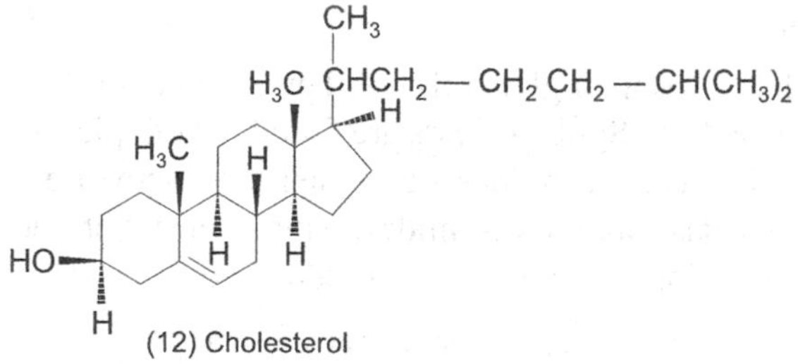

(12) Cholesterol

ii. Prenol Lipids (Terpenes or Isoprenoids)

Prenol lipids are synthesised from the 5-carbon precursors. The simplest unit is isoprene (13). Two or more isoprene molecules take part in the synthesis of isoprenoids. Examples of isoprenoids are various terpenes (discussed in Chapter 6 of this book), fat soluble vitamins, bile salts, different types of pigments like β-carotene. Terpenoids are classified according to the number of C5 units present. Structures containing greater than 40 carbons are known as polyterpenes. Carotenoids are important for their function as anti-oxidants and as precursors of vitamin A.

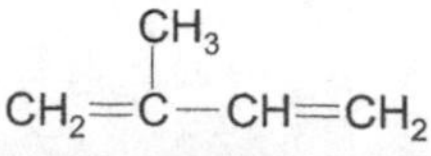

$$\overset{\overset{CH_3}{|}}{CH_2=C-CH=CH_2}$$

(13)

Sterols and steroid hormones mentioned above are also supposed to be derived from isoprenoids.

5.3 BIOLOGICAL FUNCTIONS OF LIPIDS

The main biological functions of lipids include energy storage, acting as structural components of cell-membranes. They play an important role in metabolism, nutrition and health.

1. Energy Storage and Rich Source of Energy

Triglycerides or triacylglycerols are a major form of energy storage in animals because of their high calorific value (9.4 K cal/g). Fats are stored in the adipose tissue and act as reserve food material. Animals use triglycerides for energy storage whereas plants, which do not require energy for movement, can afford to store food for energy in the form of starch.

2. Structural Constituents

Phospholipids and glycolipids are the structural components of all the membrane systems of the cell. The glycerophospholipids are the main structural component of biological membranes, such as the cellular plasma membrane and the intracellular membranes of organelles. In animal cells, the plasma membrane physically separates the intracellular components from the extracellular environment.

3. Solvents and Fat Transport

Lipids act as solvents for fat-soluble vitamins *A*, *D*, *E* and *K*. The fat-soluble vitamins, which are isoprene-based lipids are essential nutrients stored in the liver and fatty tissues. Phospholipids play an important role in the transport and metabolism of fatty acids.

4. Heat Insulators and Shock Absorbers

Fats deposited in the subcutaneous tissue act as insulators conserving body heat. The fat deposited around the visceral organs and underneath the skin acts as a cushion and absorbs mechanical shocks.

Lipids form a protective waxy covering on the aerial parts of plants to check loss of water by evaporation.

5. Metabolism

Lipids play an important role in the metabolism. Triglycerides and phospholipids are broken down into free fatty acids by the action of lipases. Beta oxidation is the process by which fatty acids in the form of acyl-CoA molecules are broken down to generate acetyl-CoA. The acetyl-CoA is then ultimately converted into ATP, CO_2 and H_2O using the citric acid cycle.

6. Nutrition and Health

Lipids play diverse and important roles in nutrition and health. Many lipids are absolutely essential for life. Adrenocorticoids, sex hormones and vitamin *D* are synthesised from cholesterol. However, there is also considerable awareness that abnormal levels of cholesterol and trans fatty acids are risk factors for heart disease amongst others.

Humans have a requirement for essential fatty acids, such as linoleic acid and alpha linolenic acid in the diet because they cannot be synthesised from simple precursors in the diet.

We have seen earlier that most vegetable oils are rich in linoleic acid (safflower, sunflower and corn oils). Alpha-linolenic acid is found in the green leaves of plants and in selected seeds, nuts and legumes. Most of the lipid found in food is in the form of triacyl glycerol, cholesterol and phospholipids.

Most of the saturated fatty acids in the diet are incorporated into adipose tissue stores because we get higher energy per carbon than is obtained from oxidation of unsaturated fatty acids. Dietary fat provides an average energy intake which is approximately twice that of carbohydrate or protein. A minimum amount of dietary fat is necessary to facilitate absorption of fat-soluble vitamins (*A*, *D*, *E* and *K*) and carotenoids. A minimal amount of body fat is also necessary to provide insulation that prevents heat loss and protects vital organs from shock due to ordinary activities.

High fat intake contributes to increased risk of obesity, diabetes and coronary and cardiovascular diseases. These diseases are due to the build up of plaque on the inside walls of arteries. Plaque is made up of cholesterol-rich low density lipoprotein (LDL), macrophages, smooth muscle cells, platelets and other substances. Saturated fats have a profound effect in increasing blood cholesterol levels and tend to increase LDL. They are found predominantly in animal products, butter, cheese and meat, but coconut oil and palm oil are common vegetable sources. Intake of monosaturated fats in oils such as olive oil is thought to be preferable to consumption of polysaturated fats in oils such as corn oil because monosaturated fats apparently do not lower high density lipoprotein (HDL) cholesterol levels. Keeping cholesterol in the normal range not only helps prevent heart attacks and strokes but may also prevent the progression of other coronary cardiovascular diseases. "Statins" are a class of drugs that lower the level of cholesterol in the blood by inhibiting the enzyme which is responsible for the biosynthesis of cholesterol in the liver.

5.4 OILS AND FATS AS SIMPLE LIPIDS

A. Introduction

We have already seen that oils and fats are simple lipids. They are triglycerides of fatty acids. Glycerol is a trihydric alcohol; *i.e.*, it contains three —OH groups that can combine with up to three fatty acids to form monoglycerides, diglycerides and triglycerides. Fatty acids may combine with any of the three hydroxyl groups to create a wide diversity of compounds. The most common glycerides are triglycerides, also known as triacylglycerols, in which all three —OH groups are esterified by fatty acids. For example, **tristearin** is a component of beef fat in which the three —OH groups of glycerol are esterified by stearic acid.

$$
\begin{array}{l}
CH_2-O-\overset{\overset{\displaystyle O}{\|}}{C}-(CH_2)_{16}-CH_3 \\[2mm]
\ \ |\qquad\quad\overset{\displaystyle O}{\|} \\[-1mm]
CH-O-\overset{}{C}-(CH_2)_{16}-CH_3 \\[2mm]
\ \ |\qquad\quad\overset{\displaystyle O}{\|} \\[-1mm]
CH_2-O-\overset{}{C}-(CH_2)_{16}-CH_3
\end{array}
$$

Tristearin

Triglycerides are commonly called **fats** if they are solid at room temperature and **oils** if they are liquid at room temperature. Most triglycerides derived from mammals are fats such as beef fat. These fats are solid at room temperature; the body temperature of the living animal keeps them somewhat fluid allowing for body movement. In plants and cold-blooded animals, triglycerides are generally oils, such as corn oil, peanut oil or fish oil. A fish requires liquid oils rather than solid fat because it would have difficulty in moving if its triglycerides solidified whenever it swam in cold water.

Oils and fats are commonly used for long-term energy storage in plants and animals. Fat is a more efficient source of energy than carbohydrates because 1 gram of fat releases over twice as much energy as 1 gram of sugar or starch.

Most of the triglycerides which occur in nature are mixed triglycerides; *i.e.*, they contain two or three different fatty acids in one molecule. Triglycerides have lower densities than water. That's why they float on water.

$$
\begin{array}{l}
CH_2-O-\overset{\overset{\displaystyle O}{\|}}{C}-R_1 \\[2mm]
\ \ |\qquad\ \ \overset{\displaystyle O}{\|} \\[-1mm]
CH-O-\overset{}{C}-R_2 \\[2mm]
\ \ |\qquad\ \ \overset{\displaystyle O}{\|} \\[-1mm]
CH_2-O-\overset{}{C}-R_3
\end{array}
$$

A Mixed Triglyceride

B. Animal and Vegetable Sources

Animal fats and oils are derived both from terrestrial and marine animals. Animal fats are located in the adipose tissue cells. We have tallow from cattle, sheep and goat. More or less of the fat occurs in the muscles, in the connective tissue, under the skin and in the bones. The amounts found in different parts of the body vary from species to species and in any given species with the age of the individual animal and its condition. The fats from the different parts and organs of a given animal differ somewhat in their properties. The fat from the interior of the animal is somewhat firmer than the fat from near the body surface *i.e.*, it melts at a higher temperature.

Marine oils include liver oils and fish oils. The different types of marine fats have been of great importance in the past and still possess considerable significance. For example, cod liver oil is used for some special purposes - some are used as food stuffs and others serve industrial uses.

Vegetable fats and oils are found in greatest abundance in fruits and seeds. They are also present in small amounts in other parts of the plants *i.e.*, roots, branches and leaves. In some seeds and fruits, however, the fat content is very high; in some cases it is as high as 35 per cent. For example, in dried coconuts the oil is present up to 65 per cent. These are the commercial sources of vegetable oils. Oils are stored in seeds to serve as nourishment for the germination of the embryo. Other important sources of vegetable oils are soya bean, groundnut, palm kernel, mustard, sesame and sunflower *etc.* All these give edible oils. However, cotton seed, linseed, castor seed and mowrah give non-edible oils.

The olive contains a large amount of fat in the pulp surrounding the kernel and only a small amount in the kernel itself, while in the oil palm both the pulp and the kernel contain large amounts. The fat from the pulp may have characteristics quite different from those of the fat in the kernel.

C. Chemical and Physical Characteristics

We know oils and fats are simple triglycerides of fatty acids. The reactions of oils and fats are the reactions of the triglycerides. Thus, they can undergo hydrolysis at all the three ester groups giving glycerol and three molecules of fatty acids. The fatty acids generated may be the same or different. Triglycerides are easily hydrolysed by enzymes called lipases, which are present in the digestive tracts of human beings and animals. Three molecules of water are used up for each molecule of fat when one molecule of glycerol and three molecules of fatty acid are obtained. Conversely, the fat may be reconstituted from glycerol and fatty acids, in which event three molecules of water are set free for each molecule of fat synthesised.

$$
\begin{array}{ccc}
\underset{\text{A Triglyceride}}{
\begin{array}{l}
CH_2-O-\overset{\displaystyle O}{\overset{\|}{C}}-R_1 \\
| \quad\quad\; \overset{\displaystyle O}{} \\
CH-O-\overset{\displaystyle O}{\overset{\|}{C}}-R_2 \\
| \quad\quad\; \overset{\displaystyle O}{} \\
CH_2-O-\overset{\displaystyle O}{\overset{\|}{C}}-R_3
\end{array}}
&
\xrightarrow{\;3H_2O\;}
&
\underset{\text{Glycerol}}{
\begin{array}{l}
CH_2OH \\
| \\
CH\,OH \\
| \\
CH_2OH
\end{array}}
\quad
\underset{\text{Fatty acid}}{
\begin{array}{l}
R_1\,COOH \\
+ \\
R_2\,COOH \\
+ \\
R_3\,COOH
\end{array}}
\end{array}
$$

The common physical properties of oils and fats are:
- They float on water but are not soluble in it.
- When pure they are colourless, odourless and tasteless.
- Triglycerides have lower densities than water.

- They are greasy to the touch and have lubricating properties.
- They are easily soluble in organic solvents like diethylether, acetone, chloroform and carbon tetrachloride.
- They are not readily volatile and may be burned without leaving any residue.

D. Saponification

When triglycerides are hydrolysed by alkalis, glycerol and salts of the fatty acids are obtained. This process is called *saponification*. Generally, sodium hydroxide or potassium hydroxide is used for hydrolysis. Sodium salts or potassium salts of these fatty acids are produced, which are called soaps. Sodium salts are hard soaps whereas potassium salts are soft soaps. The fatty acid salts of ammonium are also sometimes used for cleansing. Only a few other soaps are of practical importance; for example lead soaps are used as medicinal plasters, zinc soaps are used in ointments and aluminium soaps are used in waterproofing. Salts of lower fatty acids are soluble in water and have no cleansing action.

$$
\begin{array}{l}
CH_2{-}O{-}\overset{\displaystyle O}{\overset{\|}{C}}{-}R_1 \\
CH{-}O{-}\overset{\displaystyle O}{\overset{\|}{C}}{-}R_2 \;+\; 3NaOH \longrightarrow \\
CH_2{-}O{-}\overset{\displaystyle O}{\overset{\|}{C}}{-}R_3
\end{array}
\qquad
\begin{array}{l}
CH_2OH \\
CH\,OH \\
CH_2OH
\end{array}
\qquad
\begin{array}{l}
R_1\ COO^{-}\overset{+}{Na} \\
+ \\
R_2\ COO^{-}\overset{+}{Na} \\
+ \\
R_3\ COO^{-}\overset{+}{Na}
\end{array}
$$

Mixed Triglyceride   Mixture of
                                                                                 salts
                                                                               (soap)

5.5 FATTY ACIDS IN FATS AND OILS

A fatty acid has a carboxyl group at the polar end and a hydrocarbon chain at the non-polar end. Fatty acids are, therefore, amphipathic compounds because the carboxyl group is hydrophilic and the hydrocarbon tail is hydrophobic. Their carbon chain is long and unbranched. All fatty acids that occur in the living system contain an even number of carbon atoms because they are derived from two carbon acetyl units. If there are carbon-carbon double bonds in the chain, the fatty acid is unsaturated. If there are only single bonds, the fatty acid is saturated. The number of carbon atoms in the chain may vary from **4–30** but normally ranges from **14–18**.

Tables 5.1 and 5.2 give some examples of the saturated and unsaturated acids with their melting points. Butyric acid is one of the saturated short-chain fatty acids responsible for the characteristic flavour of butter.

In unsaturated fatty acids, the stereochemistry at the double bond is usually *cis* rather than *trans*. The difference between *cis* and *trans* acids is very important to their overall shape. A *cis* double bond puts a kink in the long chain hydrocarbon tail whereas the shape of a *trans* fatty acid is like that of a saturated fatty acid.

Table 5.1   Typical Naturally Occurring Saturated Fatty Acids

Acid	Number of Carbon Atoms	Formula	Melting Point (°C)
Lauric	12	$CH_3\text{-}(CH_2)_{10}\text{-}COOH$	44
Myristic	14	$CH_3\text{-}(CH_2)_{12}\text{-}COOH$	58
Palmitic	16	$CH_3\text{-}(CH_2)_{14}\text{-}COOH$	63
Stearic	18	$CH_3\text{-}(CH_2)_{16}\text{-}COOH$	71
Arachidic	20	$CH_3\text{-}(CH_2)_{18}\text{-}COOH$	77

TABLE 5.2   Typical Naturally Occurring Unsaturated Fatty Acids

Acid	Number of Carbon Atoms	Degree of Unsa-turation	Formula	Melting Point (°C)
Palmitoleic	16	16 : 1	$CH_3\,(CH_2)_5\,CH = CH\,(CH_2)_7\text{-}COOH$	− 0.5
Oleic	18	18 : 1	$CH_3\,(CH_2)_7\text{-}CH = CH\text{-}(CH_2)_7\text{-}COOH$	4
Linoleic	18	18 : 2	$CH_3\text{-}(CH_2)_4\text{-}CH = CH\text{-}CH_2\text{-}CH = CH\text{-}$ $(CH_2)_7\text{-}COOH$	− 5
Linolenic	18	18 : 3	$CH_3\text{-}(CH_2\text{-}CH = CH)_3\text{-}(CH_2)_7\text{-}COOH$	− 11
Arachidonic	20	20 : 4	$CH_3\text{-}(CH_2)_4\text{-}(CH = CH\text{-}CH_2)_4\text{-}(CH_2)_2\text{-}$ $COOH$	− 50

The notation used for fatty acids indicates the number of carbon atoms and the number of double bonds. Thus, 18:0 denotes an 18-carbon saturated fatty acid with no double bonds and 18:1 denotes an 18-carbon fatty acid with one double bond and so on. The double bond in oleic acid is at the ninth carbon atom, in linoleic acid there are two double bonds at the 9 and 12 carbon atoms whereas linolenic acid has three double bonds at 9, 12 and 15 carbon atoms.

Tables 5.1 and 5.2 show that saturated fatty acids have higher melting points than unsaturated fatty acids. Thus, stearic acid has a melting point of 71°C whereas oleic acid with one double bend has a melting point of 4°C. This lowering of melting point is due to the 'kink' in the *cis* configuration of oleic acid (Fig. 5.1). In such a case, the molecule cannot pack tightly.

A second double bond lowers the melting point further (lenoleic acid, m.p. − 5°C) and a third double bond lowers it still further (m.p. − 11°C).

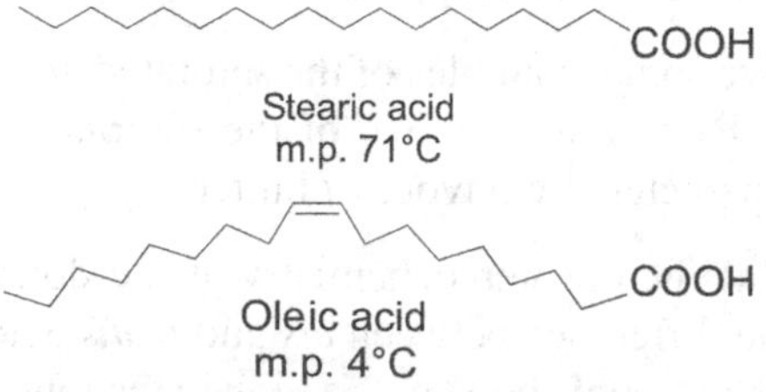

Stearic acid
m.p. 71°C

Oleic acid
m.p. 4°C

Fig. 5.1.

Fats and oils also have melting points that depend upon the degree of unsaturation in their fatty acids. Thus, tristearin has m.p. 72°C. It is a saturated fat that packs well in a solid lattice. Triolein (m.p. − 4°C) has the same number of carbon atoms but it has three double bonds whose kinked conformations prevent optimum packing in the solid.

5.6   FATTY ACID COMPOSITION IN SOME FATS AND OILS

Fatty acid compositions depend on the sources of the oils. While a large variety of fatty acids is found in natural fats and oils, only a few of them are of commercial importance. These are lauric acid, myristic acid, palmitic acid, stearic acid, oleic acid, linoleic acid and linolenic acid. The triglycerides of these seven acids make up the great bulk of natural fats and oils. Fats and oils are practically always mixtures of triglycerides in varying proportions. In some fats one triglyceride predominates, in others another and in still others several of them are present. It appears that no natural fat or oil consists solely of a single triglyceride. They are always present as mixtures. The fats of different species of animals and plants vary widely. They largely depend upon the fatty acids that are present in them.

Table 5.3 gives the composition of various fatty acids in different oils and fats.

TABLE 5.3   Per cent of Weight of Total Fatty Acids

Oil or Fat	Saturated					Unsaturated		
	Capric	Lauric acid	Myristic acid	Palmitic acid	Stearic acid	Oleic acid	Linoleic acid	Linolenic acid
Almond oil	–	–	–	7	2	69	7	–
Beef Tallow	–	–	3	24	19	43	3	1
Butterfat (cow)	3	3	11	27	12	29	2	1
Coconut oil	6	47	18	9	3	6	2	–
Corn oil	–	–	–	11	2	28	58	1
Cotton - seed oil	–	–	1	22	3	19	54	1
Olive oil	–	–	–	13	3	71	10	1
Palm oil	–	–	1	45	4	40	10	–
Palm - kernel oil	4	48	16	8	3	15	2	–
Peanut oil	–	–	–	11	2	48	32	–
Safflower oil	–	–	–	7	2	13	78	–
Soyabean oil	–	–	–	11	4	24	54	7
Sunflower oil	–	–	–	7	5	19	68	1
Walnut oil	–	–	–	11	5	28	51	5

5.7   WHAT ARE OMEGA-3 AND OMEGA-6 FATTY ACIDS?

Omega-3 (ω 3) and omega-6 (ω 6) fatty acids are unsaturated "Essential Fatty Acids" that need to be included in the diet because the human metbolism cannot create them from other fatty acids. These fatty acids use the Greek alphabet (α, β, γ, ..., ω) to identify the location of the double bonds. The "alpha" carbon is the carbon closest to the carboxyl group and the "omega" is the last carbon of the chain because omega is the last letter of the Greek alphabet. Thus, linoleic acid is omega-6 fatty acid because it has a double bond six carbons away from the omega carbon. Linoleic acid plays an important role in lowering cholesterol levels. Linolenic acid is an omega-3 fatty acid because it has a double bond three carbons away from the omega carbon. Therefore, oleic acid is an omega-9 fatty acid.

Linolenic acid (Omega-3 acid)

Linoleic acid (Omega-6 acid)

5.8   BIOLOGICAL ROLE OF FATTY ACIDS

Fatty acids are important components of the human body, having biological, structural and functional roles. A variety of fatty acids exists in the dict of humans, in blood stream and in cells and tissues. Fatty acids are energy sources and act as a main constituent of cellular membranes. The great importance of fatty acids resides in the fact that they are the main constituent of the human cell. The type of fatty acid, saturated or unsaturated, long-chained or short-chained can influence the physiology of the cell. They act to influence cell and tissue metabolism and function. Fatty acids are responsible for the regulation of membrane structure and function, regulation of intracellular signaling pathways, gene expression and regulation of the production of lipid mediators.

It is now clear that through these effects fatty acids influence the health, well-being and disease risk. Fatty acids are responsible for development of some chronic diseases including rheumatoid arthritis, coronary vascular diseases and cancer. They also influence metabolic diseases such as type-2 diabetes and inflammatory diseases.

Rheumatoid arthritis is an auto-immune disease and causes chronic inflammation of the joints and progressive joint destruction. Increased free fatty acid concentration in blood plasma that the fat metabolism is accelerated in rheumatiod arthritis. Stearic acid was found to be at higher levels in patients with established rheumatoid arthritis.

Fatty acids are highly enriched in the adipose tissues in the brain where they participate in the development and maintenance of the central nervous system.

Obesity is a metabolic disease and it increases the risk to develop diabetes, non-alcoholic fatty liver disease or other cardiovascular diseases. Studies show that

40% of the people diagnosed with diabetes will develop a kidney or cardiovasculor disease.

Fatty acid metabolism produces a huge quantity of ATP; β-oxidation of the fatty acids is mostly used by heart and the muscular tissue to obtain energy.

As phospholipids fatty acids assure the fluidity, flexibility and permeability of the membrane.

We have seen omega-3 and omega-6 are essential fatty acids. They are essential for survival of humans and other mammals as they cannot be synthesised in the body and hence, have to be obtained in our diet. Linoleic acid and α-linolenic acid are widely distributed in vegetable oils. Linoleic acid is the most abundant fatty acid in nature.

Essential fatty acids form an efficient constituent of all cell membranes and affect the properties of fluidity and thus determine the behaviour of membrane bound enzymes and receptors.

Omega-3 and omega-6 fatty acids are more important due to their multiple biological roles such as influencing the inflammatory cascade, reducing the oxidative stress giving cardiovascular protection. Omega-3 acids have high impact on health as the anti inflammatory properties of omega-3 are very important.

A high level of omega-3 and omega-6 ratio is associated with a low risk to develop a kidney disease.

The type of fatty acids found in the structure of the cell membranes can affect their fluidity, stability and functions. For example, saturation of the fatty acids influences the fluidity of the membranes. If membranes are composed of mostly saturated fatty, which have a straight rigid chain, the phospholipids will form more rigid bilayers, whereas membranes formed by many cis-unsaturated fatty acids will be more flexible.

5.9   TRANS FATS AND THEIR EFFECT

A. What are Trans Fats ?

We already know that double bonds bind carbon atoms tightly and prevent rotation of the carbon atoms along the bond axis. This gives rise to configurational isomers which we call *cis* and *trans* as shown below :

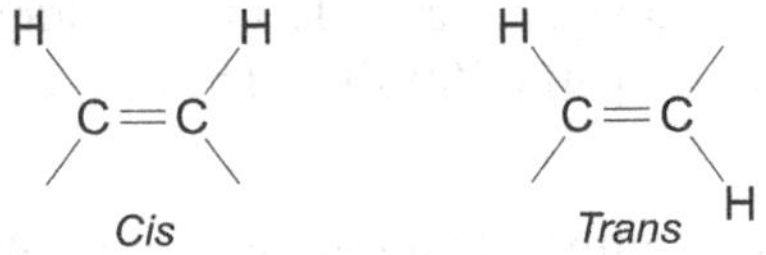

Cis   *Trans*

Naturally occurring fatty acids generally have the *cis* configuration. The natural form of 9-octadecenoic acid (oleic acid) found in olive oil has a "V" shape. The *trans* configuration looks like a straight line.

Low levels of *trans* fatty acids occur naturally in some animal food products. Polyunsaturated *trans* fatty acids cannot be used to produce useful products because the molecules have unnatural shapes that are not acted upon by enzymes. Partially hydrogenated oils contain a large proportion of diverse *trans* fatty acids. Trans fatty acids that are incorporated into the cell membranes create denser membranes that alter the normal functions of the cell.

B. Effect of Trans Fats on the Heart

Dietary *trans* fats raise the level of low density lipoproteins (LDL, or "bad cholesterol") increasing the risk of coronary heart disease. *Trans* fats also reduce high-density lipoproteins (HDL or "good cholesterol) and raise the levels of triglycerides in the blood. Both of these conditions are associated with insulin resistance which is linked to diabetes, hypertension and cardiovascular disease. Research has shown that people who take partially hydrogenated oils, which are high in *trans* fats, worsen their lipid profiles and have nearly twice the risk of heart attacks compared with those who do not consume hydrogenated oils.

C. Effect of *Trans* Fats on the Brain

Trans fats also have a detrimental effect on the brain and nervous system. Neural tissue consists mainly of lipids and fats. Studies show that trans fatty acids in the diet get incorporated into the brain cell membranes and alter the ability of neurons to communicate and may cause neural degeneration and diminished mental performance.

D. Effect of Fats on Blood Cholesterol Levels

High blood serum cholesterol levels are associated with increased risk of cardiovascular diseases. Blood cholesterol levels are lowered by reducing the sources of dietary cholesterol, increasing the amount of fibre in the diet and by consuming oils high in polysaturated fatty acids while reducing the intake of saturated fats. It has been observed that myristic acid (C14:0) and to some extent palmitic acid (C16:0) increase cholesterol levels whereas polyunsaturated acids such as linoleic (C18:2) reduce cholesterol levels. Other saturated fatty acids (*e.g.*, stearic acid) or monounsaturated fatty acids (*e.g.*, oleic acid C18:1) do not affect cholesterol levels significantly.

Animal fats and tropical oils which are high in myristic acid and low in linoleic acid increase cholesterol levels. Oils high in linoleic acid such as sunflower oil and safflower oil can play a significant role in reducing blood cholesterol levels when consumed regularly as part of the diet. The effect of olive oil on cholesterol is relatively neutral.

The best quality oils are unrefined cold pressed or expeller pressed oils packed in dark glass bottles filled with an inert gas.

5.10 EXTRACTION OF OILS AND FATS

The raw materials for the fats and oils are animal by-products from the slaughter of cattle and sheep ; fatty fish and marine mammals; a few fleshy fruits like palm and olive and various oil seeds. Most oil seeds are grown specifically for processing to oils but several important vegetable oils are obtained from by-product raw materials. For example, cotton-seed is a by-product of cotton grown for fibre and corn oil is obtained from the corn germ that accumulates from the corn-milling industry whose primary products are corn grits and starch.

Fats and oils may be obtained from oil bearing tissues by three general methods with varying degrees of mechanical simplicity :

(1) rendering, (2) pressing with mechanical presses and (3) extracting with volatile solvents

A. i. Rendering-Fruits and Seeds

This is the crudest method of rendering oil from fruits, which is still practiced in some countries. It consists of heaping fruits in piles, exposing them to the sun and collecting the oil that exudes. This process is used in the preparation of palm oil. The fresh palm fruits are boiled in water and the oil is skimmed from the surface. Such processes can be used only with seeds or fruits, such as olive and palm that contain large quantities of easily released fatty matter.

ii. Rendering-Animal Fats

The rendering process is applied on a large scale to the production of animal fats such as tallow, bone fat and whale oil. It consists of cutting the fatty pieces into small pieces that are boiled in open vats or cooked in steam digestors. The fat, which is liberated from the cells, floats on the surface of water, where it is collected by skimming. Finally, the fat is separated from the aqueous phase by pressing in hydraulic or screw presses. Additional fat is thereby obtained. The residue is used for animal feed or as fertilizer. Several centrifugal separation processes have also been developed. Compared with conventional rendering, the centrifugal methods provide a higher yield of better-quality fat and the separated protein has potential as an edible meat product.

B. Pressing

Many different mechanical devices have been used for pressing. The screw press has been used for the production of olive oil. In India, the most important source of edible oils are seeds such as mustard, groundnut, cottonseed, sesame and sunflower. The extraction from seeds is done by a variety of methods right from the village ghani to the more sophisticated expeller. The oil bearing material is screened and crushed by passing between steel rollers.

In order to get the maximum yield the crushed material is heated at 70°–100°C in a steam jacketed vessel to rupture the oil cells. The heated material is then pressed by a high pressure expeller. The modern screw press has replaced many of the hydraulic presses because it is a continuous process and has greater capacity. It also requires less labour and will generally remove more oil. The expeller consists of a perforated cylindrical tube in which a screw shaft moves. While the oil expelled from the seeds flows out of the holes in the expeller tube, the pressed oil cake is ejected from the other end of the tube.

The exhausted oil cake still contains oil. Most of the exhausted cake is used for cattle feed or subjected to solvent extraction for recovery of more oil.

C. Solvent Extraction

This is a process which involves extracting oil from oil-bearing materials by treating them with low boiling solvents. The solvent extraction method recovers almost all the oils and leaves behind only 0.5% to 0.7% residual oil in the raw material. In case of mechanical pressing the residual oil left in the oil cake may be anywhere from 6% to 14%. The solvent extraction method can be applied directly to any low oil content materials. The method can also be used to extract pre-pressed oil cakes obtained from high oil content materials. As the percentage of recovered oil is quite high, solvent extraction has become the most popular method of extraction of oils and fats. Fat is one of the essential components of the human diet; therefore, the demand for oils and fats is increasing with the increase in population and standards of living. Today large quantities of oil cakes such as peanut, cottonseed, sunflower, linseed, neem, castor etc., are extracted. Direct extraction of rice bran and soya bean is also used.

The Process

Solvent extraction is basically a process of diffusion of a solvent into oil-bearing cells of the raw material resulting in a solution of the oil in the solvent. Various solvents can be used for extraction. Most common solvents are various grades of petroleum ether and hexane. Hexane is considered to be the best and it is exclusively used for this purpose. Thus, the extraction process consists of treating the raw material with hexane and recovering the oil by distillation of the resulting solution of the oil in hexane called miscella. Evaporation and condensation from the distillation of miscella recovers the hexane absorbed in the material. The hexane thus recovered is reused for extraction. The boiling point of hexane is very low (67°C) and oils and fats are very soluble in hexane.

A typical extraction system consists of:

1. Cleaning to remove tramp iron, foreign weed seeds and stones.
2. Removing hulls in cracking, aspirating or screening operations.
3. Cracking or rough grinding the kernels or prepressed cake.
4. Steaming (tempering or cooking of the meats.)
5. Flaking the small pieces between smooth flaking rolls.

6. Extracting the oil with solvent.

7. Separating the marc from the oil-solvent solution, miscella.

8. Removing the solvent from both the miscella and the marc.

5.11 REFINING OF OILS AND FATS

The extent of processing applied to oils and fats depends on their source, quality and ultimate use. Many oils are used for edible purposes after only a single processing step, *i.e.*, clarification by settling or filtering. Sometimes filtering is done twice. Many cold-pressed oils can be used in food products without further processing. For example, cold-pressed olive, peanut and some coconut and sunflower oils are used in this way. Large quantities of butter and lard are used without special treatment after churning or rendering. However, large demands for refined oils have led to extensive processing techniques.

A. Alkali Refining

Oils and fats may contain some free fatty acids. They can be removed by treating fats at 40° to 85°C with an aqueous solution of sodium hydroxide or sodium carbonate. The refining may be done in a tank, in which case it is called batch refining. Alkali refining may also be carried out in a continuous system. In batch refining, the aqueous emulsion of soaps formed from free fatty acids, along with other impurities settles at the bottom and is drawn off. In the continuous system the emulsion is separated with centrifuges.

After the fat has been refined, it is usually washed with water to remove traces of alkali and soapstock. Oils that have been refined with sodium carbonate or ammonia generally require a light re-refining with sodium hydroxide to improve colour. After washing with water, the oil may be dried by heating in a vacuum or by filtering through a dry filter-aid material.

The refined oil may be used for industrial purposes or may be processed further to edible oils. Usually, the refined oils are neutral, free of material that separates on heating, lighter in colour, less viscous and more susceptible to rancidity.

B. Water Refining

Water refining is usually called degumming. It is generally carried out before alkali refining. Water refining consists of treating the natural oil with a small amount of water, followed by centrifugal separation. The process is applied to many oils that contain phospholipids in significant amounts. Since the separated phospholipids are rather waxy or gummy solids, the term degumming is quite naturally applied to the separation. The separated phosopholipid emulsion layer from oils such as corn and soyabean oils may be dried and used as emulsifiers in chocolate products and emulsion paints. The degummed oil may also be used directly in industrial applications such as in paints or alkyd resins or refined with alkalis for ultimate edible consumption.

C. Bleaching

Most crude vegetable oils are deeply coloured. They must be bleached. In certain cases, some colour may go after water and alkali refining. If further colour removal is desired, the oil may be treated with various bleaching agents. Heated oils are treated with fuller's earth, activated charcoal or activated clays. Fuller's earth is a natural earthy material that will decolourise oils. Many impurities, including chlorophyll and carotenoid pigments, are absorbed onto such agents and removed by filtration.

To the dry oil dry fuller's earth is added, usually a few percent, the quantity depending upon the character of the oil and the temperature. The mixture is then heated usually to less than 80°C and agitated for one-half to one hour. It is then pumped through a filter press which retains the fuller's earth and permits the oil to run out clear. An appreciable amount of the oil remains in the earth retained in the press at the end of the operation. A considerable proportion of this is sometimes recovered by forcing the dry steam through the press which carries out with it much of the residual oil.

Bleaching often reduces the resistance of the oils to rancidity, because some material antioxidants are removed together with impurities. When many oils are heated to more than 175°C, a phenomenon known as heat bleaching takes place. The heat decomposes some pigments such as carotenoids and converts them to colourless materials.

D. Deodourisation

Many oils even after they have been refined and decolourised to transparent whiteness retain diagreeable odour and flavour. Vegetable fats especially have a relatively strong taste that is foreign to that of butter. Odourless and tasteless fats first came into high demand as ingredients for the manufacture of margarine, a product designed to duplicate the flavour and texture of butter.

The deodourisation process consists of blowing steam through heated fat or oil held under vacuum. Since the substances responsible for odours and flavours are usually volatile, they are removed leaving a neutral, virtually odourless fat or oil. Deodourisation is normally carried by blowing steam through the oil after it has been heated to a high temperature say 175°–245°C. The most effective method is to carry out this treatment in a vacuum. Originally, deodourisation was a batch process, but continuous systems are being used in which hot fat flows through an evacuated column countercurrent to the upward passage of stream.

5.12 HYDROGENATION

Unsaturated fats and oils exposed to air get oxidised to form compounds that have rancid or unpleasant odours or flavours. Hydrogenation is a commercial chemical process to add more hydrogen to natural unsaturated fats to decrease the number of double bonds and retard or eliminate the potential for rancidity. Unsaturated glycerides react with hydrogen in the presence of a metal catalyst, usually nickel, to give saturated glycerides. This reaction is similar to the catalytic hydrogenation of alkenes. The hydrogenation process saturates the double bonds present in the fatty acid components of glycerides, thereby converting them to saturated acid components.

Oleic acid and linoleic acid are both converted to stearic acid when fully saturated. Thus, the liquid vegetable oil becomes a solid saturated fat. For example, glyceryltrioleate (m.p. – 5°C) upon hydrogenation yields glyceryl tristearate (m.p. 71°C)

$$CH_2-O-\overset{\overset{O}{\|}}{C}-(CH_2)_7-CH=CH-(CH_2)_7-CH_3$$
$$CH-O-\overset{\overset{O}{\|}}{C}-(CH_2)_7-CH=CH-(CH_2)_7-CH_3 \quad\xrightarrow[\Delta]{H_2/Ni}\quad$$
$$CH_2-O-\overset{\overset{O}{\|}}{C}-(CH_2)_7-CH=CH-(CH_2)_7-CH_3$$

$$CH_2-O-\overset{\overset{O}{\|}}{C}-C_{17}H_{35}$$
$$CH-O-\overset{\overset{O}{\|}}{C}-C_{17}H_{35}$$
$$CH_2-O-\overset{\overset{O}{\|}}{C}-C_{17}H_{35}$$

Glyceryl trioleate   Glyceryl tristearate

All the important unsaturated triglycerides *e.g.*, triolein, trilinolein and trilinolenin are liquids at ordinary temperatures. Hence, fats which contain them in considerable amounts are oils or if the amount is smaller, they are soft solids. On the other hand, the important saturated triglycerides, *e.g.*, tristearin and tripalmitin are solid and fats which contain them in considerable amounts are firm at ordinary temperatures.

Hydrogenation converts unsaturated triglycerides into saturated ones *i.e.*, it changes oils into solid fat. This process is of greatest commercial and economic importance. In practice, hydrogenation is widespread, but it is used far more for edible oils than for industrial products.

Fully saturated fats are too waxy and solid to use as food additives. So manufacturers use partially hydrogenated oils. In this, the process is stopped when the oil has the proper consistency for its application. The high temperature and catalysts used for this chemical reaction weaken the double bonds and cause a large percentage of the natural *cis* double bonds to change to *trans* double bonds. Trans fatty acids are present mainly in partially hydrogenated fats, but they are also present in hydrogenated fats because chemical reactions never achieve 100% efficiency.

5.13   IDENTIFICATION OF FATS AND OILS

A. Reichert-Meissel Value

The Reichert-Meissel value is a measure of the volatile water-soluble acid constituents of the fat. It is the number of millilitres of N/10 potassium hydroxide solution required to neutralise the volatile water-soluble fatty acids obtained by five grams of fat. The sample is heated with a standardised solution of potassium hydroxide until completely saponified. The mixture is then acidified with sulphuric acid and distilled. Acids up to C_{10} are volatile and appear in the distillate. The water-soluble acids in the distillate are neutralised with N/10 potassium hydroxide solution.

The Reichert-Meissel value allows us to distinguish between butter and butter substitutes made from vegetable oils. The hydrolysis of the butter fat gives C_4, C_6, C_8 acids which are volatile in steam, while practically no acids below C_{10} are obtained from vegetable oils. Butter has a Reichert-Meissel number of 17 to 34 while that of cotton seed oil is less than 1.

B. Saponification Number

We have already seen that oils and fats on hydrolysis give glycerol and fatty acids. When this hydrolysis is carried out under alkaline conditions, the process is called *saponification* since it was first observed to take place in the manufacture of soap. Oils and fats are characterised by the saponification number of an oil or a fat. It is defined as the number of milligrams of potassium hydroxide required to saponify one gram of the fat or oil *i.e.*, to convert one gram of the fat completely into glycerol and potassium soap.

Since there are three ester bonds in a molecule to hydrolyse, three equivalents of potassium hydroxide are needed to saponify one mole of any fat or oil. We can calculate the saponification number by taking the example of glycerol tripalmitate

$$CH_2-O-\overset{\overset{\displaystyle O}{\|}}{C}-(CH_2)_{14}-CH_3$$
$$CH-O-\overset{\overset{\displaystyle O}{\|}}{C}-(CH_2)_{14}-CH_3 + 3KOH \longrightarrow 3\ CH_3-(CH_2)_{14}-\overset{\overset{\displaystyle O}{\|}}{C}-\bar{O}K^+ + \begin{array}{l} CH_2OH \\ | \\ CH\ OH \\ | \\ CH_2OH \end{array}$$
$$CH_2-O-\overset{\overset{\displaystyle O}{\|}}{C}-(CH_2)_{14}-CH_3$$

Glyceryl tripalmitate

Here, molecular weight of tripalmitate = 806

Three equivalents of potassium hydroxide are required to completely saponify

i.e.,   806 g of fat = 168 g of KOH

= 168,000 milligrams of KOH

Therefore, 1 g of the fat will require $\dfrac{168,000}{806}$ milligrams of KOH

Saponification number of glycerol tripalmitate = $\dfrac{168,000}{806}$ = 208.

By following the same procedure, we can determine the saponification number of any oil or fat. The higher the saponification number of a fat, the greater is the percentage of low-molecular glycerides it contains. In that case, the soap formed will be more soluble. This information is of special importance to soap makers. Table 5.4 gives the saponification numbers of some common commercial oils and fats.

TABLE 5.4   Saponificatioin Numbers of Common Oils and Fats

Oil or Fat	Saponification Number
Rapeseed oil	170 – 179
Corn oil	188 – 193
Olive oil	185 – 196
Soya bean oil	193
Linseed oil	192 – 195
Mutton tallow	192 – 195
Cottonseed oil	193 – 195
Peanut oil	190 – 196
Palm oil	196 – 205
Butter	220 – 233
Palm kernel oil	242 – 250
Coconut oil	246 – 260

Examination of Table 5.4 shows that butter ranks with palm kernel oil and coconut oil as having a very high saponification number. This is due to the fact that its triglycerides contain appreciable quantities of myristic acid and small quantities of lauric acid. In butter, there is a high percentage of low molecular weight fatty acids; their sodium soaps are quite soluble in water. The high saponification number of coconut oil and palm kernel oil is due to large proportion of lauric acid and myristic acid that they contain. These oils, therefore, yield quite soluble soaps.

C. Iodine Number

We know oils and fats contain different saturated and unsaturated acids. When the fatty acid molecule contains the maximum of hydrogen possible, the acid is said to be a saturated fatty acid. It is saturated with respect to hydrogen. We have seen myristic, lauric, palmitic and stearic acids are such saturated acids. They are solid at ordinary temperatures. When the fatty acid molecule does not contain the maximum of hydrogen possible, the acid is said to be an unsaturated fatty acid. It is unsaturated with respect to hydrogen. Such unsaturated acids are oleic, linoleic and linolenic acids. They are liquids at ordinary temperatures.

We have already discussed that these unsaturated acids can be made to combine with hydrogen. This process is known as hydrogenation. However, unsaturated fatty acids can be made to combine with other substances instead of hydrogen. They may be made to take up iodine. This gives us a method to know the degree of unsaturation of a given fat or oil *i.e.*, by determining the amount of iodine that can be made to combine with the fat or oil. So, the degree of unsaturation in a fat or oil is expressed in terms of **Iodine Number** or **Iodine Value**. It is an index to the degree of unsaturation of the fat. Iodine number is defined as the number of grams of iodine that will add to 100 g of the fat or oil.

Table 5.5 gives the iodine number of some common oils and fats.

TABLE 5.5   Iodine Numbers of Common Oils and Fats

Oil or Fat	Iodine Number
Linseed oil	173 – 201
Tung oil	170
Whale oil	121 – 146
Soyabean oil	137 – 143
Sunflower oil	119 – 135
Corn oil	111 – 130
Cottonseed oil	108 – 110
Rapeseed oil	103 – 108
Peanut oil	83 – 100
Olive oil	79 – 88
Palm oil	51 – 57
Milk fat	26 – 50
Beef tallow	38 – 46
Palm kernel oil	13 – 17
Coconut oil	8 – 10

Examination of the above table shows that oils with the highest iodine numbers are very good drying oils *i.e.*, linseed and tung oil. As shown soya bean oil and sunflower oil are good from a nutrition point of view as they contain appreciable amounts of unsaturated fatty acids.

We consider here an example of glyceryl trioleate (triolein) to calculate its iodine number. Triolein adds with three molecules of iodine to give the addition product. The molecular weight of triolein is 884.

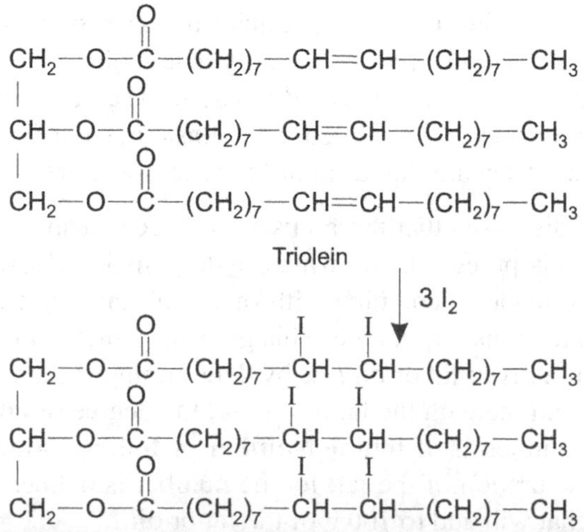

$$CH_2-O-\overset{\overset{\displaystyle O}{\|}}{C}-(CH_2)_7-CH=CH-(CH_2)_7-CH_3$$
$$CH-O-\overset{\overset{\displaystyle O}{\|}}{C}-(CH_2)_7-CH=CH-(CH_2)_7-CH_3$$
$$CH_2-O-\overset{\overset{\displaystyle O}{\|}}{C}-(CH_2)_7-CH=CH-(CH_2)_7-CH_3$$

Triolein

$\downarrow$ 3 I$_2$

$$CH_2-O-\overset{\overset{\displaystyle O}{\|}}{C}-(CH_2)_7-\overset{\overset{\displaystyle I}{|}}{CH}-\overset{\overset{\displaystyle I}{|}}{CH}-(CH_2)_7-CH_3$$
$$CH-O-\overset{\overset{\displaystyle O}{\|}}{C}-(CH_2)_7-\overset{\overset{\displaystyle I}{|}}{CH}-\overset{\overset{\displaystyle I}{|}}{CH}-(CH_2)_7-CH_3$$
$$CH_2-O-\overset{\overset{\displaystyle O}{\|}}{C}-(CH_2)_7-\overset{\overset{\displaystyle I}{|}}{CH}-\overset{\overset{\displaystyle I}{|}}{CH}-(CH_2)_7-CH_3$$

The above equation indicates that $6 \times 127 = 762$ g of iodine will add to 884 g of triolein.

The number of grams of iodine that will add to 100 g of triolein will be

$$= \frac{762 \times 100}{884}.$$

Therefore, the iodine number of triolein

$$= \frac{760 \times 100}{884} = 86.$$

It is clear that the value of iodine number depends on the number of double bonds present in the acid component of the triglycerides. A high value of iodine number indicates that the glycerides contain a large number of double bonds, while a low value gives the presence of a few double bonds. The iodine number of tripalmitate with no double bonds would be zero.

D. Acid Number

The acid number of an oil or fat tells us the amount of free fatty acids present in it. The acid number is expressed as the number of milligrams of potassium hydroxide required to neutralise the free fatty acids in one gram of the oil or fat. It is determined by dissolving a weighed amount of the fat in alcohol and titrating the solution against a standard alkali solution using phenolphthalein as indicator. The acid number of a fat can give the extent of rancidity in a stored sample.

5.14 FLAVOUR CHANGES IN OILS AND FATS

A. Rancidity

Fresh fats and oils are almost odourless and are neutral in reaction. However, they are quite unstable substances. When stored for any considerable length of time and the temperature is high and the air has free access to them, they deteriorate and spoil. The fats and oils acquire a peculiar disagreeable odour and flavour. They become rancid and are inedible. When the fat or oil develops a disagreeable odour, we say it is the onset of rancidity.

Rancidity is mainly caused by hydrolysis of the ester links and oxidation of the double bonds of the triglycerides. In this respect different oils and fats differ markedly. Some spoil more readily than others. Some oils and fats become rancid when kept for longer periods or exposed to high temperature. Onset of rancidity is also caused when they are kept in some metal containers.

We observe at our homes that when mustard oil is kept in a brass container and the air has free access to it, the rancidity develops so fast that after a few days the colour changes from brown to green and the oil gives a very offensive disagreeable odour.

The rancidity of a particular fat or oil is not necessarily the result of a long storage under unfavourable conditions. The fat may have been spoiled and become rancid from the moment of its production. This will be true when the materials from which it was produced have undergone decomposition. Thus, the fat obtained from putrifying carcasses will be rancid. Similarly, oil obtained from fermented cottonseeds will have a disagreeable odour. In other words, to obtain a good quality oil or fat, the raw material must be of good quality. It must be worked up speedily before giving it time to decompose. The fat thus obtained must be stored under favourable conditions and must be processed quickly and its consumption should not be too long delayed.

B. Different Types of Rancidity

As mentioned above there are two basic types of rancidity - hydrolytic rancidity and oxidative rancidity

i. Hydrolytic Rancidity

Fats and oils become rancid through the addition of water because water hydrolyses the bonds in the triglycerides causing them to break down into smaller compounds. The liberation of lower fatty acids by hydrolysis of ester links is responsible for this type of rancidity. Hydrolytic rancidity is particularly applicable to butter. Under moist and warm conditions the hydrolysis of the glycerides in butter liberates the odourous butyric acid, caproic acid, caprylic acid and capric acid. Microorganisms present in the air catalyse the hydrolytic process. The butter starts giving disagreeable odour even when hydrolysis may have proceeded to only a small extent. Thus, butter becomes unpalatable when even very small amounts of butyric acid are produced by the hydrolysis of triglycerides of butter.

In general, hydrolytic rancidity is caused by the lipase enzymes and heat. The rancidity has implications for deep-fat frying, because placing cold, wet food in heated frying oil introduces water and makes the oil prone to hydrolytic rancidity. On the other hand, fats that have not been heated are more prone to hydrolytic rancidity because the lipase enzymes have not yet been destroyed by heat. Room temperature is ideal for the activity of the lipase enzyme. If butter is left out at room temperature, it quickly decomposes. As mentioned above, butter's volatile fatty acids, such as butyric and caproic acids create a rancid odour. The long-chain fatty acids are also freed but they are not volatile and therefore, do not contribute to the odour of the rancid butter. Rancidity so caused can be prevented by keeping butter covered in a refrigerator.

ii. Oxidative Rancidity

Fats and oils can also become rancid when they are exposed to the oxygen in the air. This occurs in triglycerides containing unsaturated fatty acids. The higher the degree of unsaturation, the more likely it is that the fat will be subject to oxidative rancidity. In oxidative rancidity, the double bonds of an unsaturated fatty acid react chemically with oxygen resulting in two or more smaller molecules, which in turn give rise to the formation of offensive aldehydes and acids.

Unlike hydrolytic rancidity, oxidative rancidity occurs in a series of steps. The initiation period is slow and is caused by light, high temperature, food particles in the frying oil and certain metals such as iron, copper and nickel. The initial stage is followed by quicker, irreversible and self-perpetuating chain reaction. Oxygen atoms attach to the carbon next to the double bond of the fatty acid, creating very reactive and unstable free radicals. These free radicals contribute to further breakdown of fats and oils into smaller molecules, resulting in unpleasant odours and off flavours. Once this process starts, it is difficult to stop. The free radicals generated by the reaction create more free radicals, which in turn keep producing free radicals until all the double bonds have been used in the process.

The addition of antioxidants will preserve edible oils and fats for long periods of storage. Antioxidants, found naturally in the fat or commercially added, inhibit oxidative rancidity and extend shelf life. Two antioxidants occurring in natural fats are vitamin E and ascorbic acid.

5.15   CHOLESTEROL

We have already mentioned that cholesterol (12) belongs to the sterol group of lipids. The name cholesterol is derived from the Greek words, *chole*, means bile and *stereos* (solid) followed by the suffix-ol for an alcohol. It is an essential structural component of mammalian cell membranes and is required to establish proper membrane permeability and fluidity. In addition, cholesterol is an important component for the manufacture of bile acids, steroid hormones and vitamin D. Cholesterol is the principal sterol synthesised by animals. In vertebrates, it is predominantly formed in the liver.

Although cholesterol is important and necessary for human health, high levels of cholesterol in the blood have been linked to damage to arteries and cardiovascular disease.

A. Dietary Sources of Cholesterol

We have seen earlier animal fats are complex mixtures of triglycerides with lesser amounts of phospholipids and cholesterol. Therefore, all foods containing animal fat contain cholesterol in varying amounts. Major dietary sources of cholesterol are cheese, egg yolks, beef, pork, poutry and fish. Human breast milk also contains significant quantities of cholesterol.

Cholesterol is not found in appreciable amounts in plant sources. However, plant products such as flax seeds and peanuts contain cholesterol like compounds called phytosterols. Phytosterols have been proved to possess L D L cholesterol lowering efficacy. Phytosterols compete with cholesterol for absorption in the intestine.

B. Physiology and Biological Importance of Cholesterol

Since cholesterol is essential for all animal life, it is primarily synthesised from simpler substances within the body. When we take more of cholesterol from dietary sources, it increases the cholesterol levels in the blood.

Cholesterol is recycled. It is excreted by the liver via the bile into the digestive tract. About 50% of the excreted cholesterol is reabsorbed by the small intestine back into the bloodstream. Phytosterols can compete with cholesterol reabsorption in the intestinal tract, thus reducing cholesterol reabsorption.

Cholesterol is required to build and maintain membranes. It modulates membrane fluidity over the range of physiological temperatures. The hydroxyl group on cholesterol interacts with polar head groups of membrane phospholipids and sphingolipids (p. 319) while the bulky steroid and hydrocarbon chain are embedded in the membrane, alongside the nonpolar fatty acid chain of other liquids. Thus, cholesterol increases membrane packing, which reduces membrane fluidity. In this structural role, cholesterol reduces the permeability of the plasma membrane to neutral solutes, protons and sodium ions.

Within the cell membrane, cholesterol also functions in intracellular transport, cell signaling and nerve conduction.

Within cells, cholesterol is the precursor molecule in several biochemical pathways. In the liver, cholesterol is converted to bile, which is then stored in the gallbladder. Bile contains bile salts, which solubilise fats in the digestive tract and aid in the intestinal absorption of fat molecules as well as fat soluble vitamins, A, D, E and K.

Cholesterol is an important precursor molecule for the synthesis of vitamin D and steroid hormones including the adrenal gland hormones cortisol and aldosterone as well as sex hormones, progesterone, estrogen and testosterone and their derivatives.

C. High Cholesterol and Heart Disease

We know cholesterol is required in our body to perform various biological functions. But abnormal cholesterol levels lead to cardiovascular diseases. Total fat intake plays an important role in determining the blood cholesterol levels in our body. That is also affected by the changes in the quantity of cholesterol and lipoproteins that are synthesised by the body. Saturated, monosaturated and polysaturated fats have been shown in increase HDL based cholesterol levels. However, saturated fats also increase LDL based cholesterol levels. Trans fats have been shown to reduce levels of HDL and increase level of LDL. Higher concentrations of LDL and lower concentration of functional HDL are strongly associated with cardiovascular disease because these promote atheroma development in arteries (atheroselerosis). This disease process leads to myocardial infarction (heart attack), stroke and peripheral vascular disease. Since higher blood LDL contributes to this process more than the cholesterol content of the HDL particles, LDL particles are often termed "*bad cholesterol*" because they have been linked to atheroma formation. On the other hand high concentrations of functional HDL, which can remove cholesterol from cells and atheroma, offer protection and are referred to as "*good cholesterol*". These balances can be changed by body build, medications, food choices and other factors.

According to world-wide accepted norms the total blood cholesterol should be: < 200 mg/dL—as normal blood cholesterol, 200–239 mg/dL – borderline high risk and > 240 mg/dL – cholesterol. However, as today's testing methods determine LDL ("bad") and HDL ("good") cholesterol separately, this simplistic view has become outdated. Based on the fact that low HDL and high LDL levels are associated with cardiovascular disease, many health authorities advocate reducing LDL cholesterol through changes in diet in addition to other life style. The desirable LDL level is considered to be less than 100 mg/dL, although a newer upper limit of 70 mg/dL can be considered as good in higher risk individuals. However, the accepted range for LDL is normal < 130 mg/dL, borderline 130–159 mg/dL and high > = 160 mg/dL. The HDL–cholesterol level should be in the range of 35–60 mg/dL.

Elevated levels of the lipoprotein, LDL and VLDL are regarded as atherogenic *i.e.,* they are prone to cause atherosclerosis. Levels of these fractions, rather than the total cholesterol, are responsible for the progress of atherosclerosis. Even if the total cholesterol is within normal limits, having high levels of LDL results in atheroma growth rates. However, if LDL is low and there is a high percentage of HDL, then atheroma grow rates are low.

The actual cholesterol level itself is not the most important risk factor for determining cardiovascular disease potential. It is actually the ratio between the HDL "good" cholesterol and total cholesterol, which is a very important risk factor. In adults, the HDL "good" cholesterol/total cholesterol ratio should be higher than 0.24. The accepted HDL/total cholesterol ratio of 0.24 or higher is considered ideal. Under 0.24 is considered low-risk and anything less than 0.10 is considered very dangerous. It is much better to have a higher ratio which means you have a lower risk of cardiovascular disease.

Elevated cholesterol levels are treated with a diet consisting of low saturated fat, and *trans* fat-free low cholesterol foods, followed by one of the various hypolipidemic agents such as statins, fibrates or cholesterol absorption inhibitors.

Studies have also found that statins reduce atheroma progression. As a result, people with a history of cardiovascular disease may derive benefit from statins irrespective of their cholesterol levels.

5.16   IMPORTANCE OF TRIGLYCERIDES

We consume oils and fats daily in our diet. They are triglycerides of fatty acids. In the earlier chapters, we have seen carbohydrates act as a ready source of energy *i.e.,* energy can be released by the catabolic breakdown of carbohydrates in aerobic and anaerobic processes. In animals, energy is stored in the form of glycogen. However, lipids represent a more efficient way of storing chemical energy.

The potential energy of lipids resides in the fatty-acid chains of triacylglycerols. That means triacylglycerols are the main storage form of the chemical energy of lipids.

The bond between the fatty acid and the rest of the molecule can be hydrolysed by a suitable group of enzymes called lipases. Also when there are excess carlories, fatty acids are synthesised and stored in the fat cells. When energy demands are great, fatty acids are catabolised to liberate energy. Long chain fatty acyl groups are converted into acetyl Co A, which is subsequently oxidised in the citric acid cycle to provide energy that is temporarily stored as ATP.

We observe triglycerides are essential for good health because our tissues rely on triglycerides. They are formed in the liver from the fats you eat or from the body's own synthesis of fat. When triglycerides are in excess, they circulate in the blood. Like high cholesterol, elevated triglycerides have also been implicated in arterial disease.

Recent studies and evidences have strengthened the connection between high triglycerides and heart disease. Triglyceride levels can range from 150–200 mg/dL. However, most doctors consider 100 or less as ideal.

Researchers have stated that HDL "good" cholesterol is closely related to the triglycerides. Generally, people with high triglycerides have a tendency to have low HDL "good" cholesterol levels. The triglyceride/HDL "good" cholesterol ratio should be below 2. In a nut shell the triglyceride/HDL ratio, which is considered ideal is 2 or less; 4 is high and 6 or greater is considered too high.

The lower your triglycerides or the higher your HDL, it is better for your health.

Many doctors and researchers are finding the triglyceride/HDL ratio to be one of the better predictors of heart disease, much more so than just high cholesterol and LDL/HDL ratios. Research has shown that people with the highest ratio of triglycerides to HDL have 16 times the risk of heart attack as those with the lowest ratio of triglycerides to HDL.

Elevated triglycerides can be a strong indicator of biliary function, fat metabolism, the function of the liver and hereditary. There is a sugar-handling issue with elevated triglycerides or adult onset diabetes.

5.17   LIPID MEMBRANES

A. Introduction

We know that every cell has membrane, also called a plasma membrane. Membranes are usually made of phospholipids, which are molecules that have a head group and a tail group. The head is attached to water and the tail, which is made of a long hydrocarbon chain is repelled by water.

In nature, phospholipids are found in stable membranes composed of two layers, called a bilayer. In the presence of water, the heads are attracted to water and line up to form a surface facing the water. The tails are repelled by water and line up to form a surface away from the water. The hydrocarbon tails of one layer face the hydrocarbon tails of the other layer and the combined structure forms a **bilayer.**

The molecular basis of membrane's structure lies in its lipid and protein components. The most important difference between lipid bilayers and the cell membranes is that the latter contain proteins as well as lipids. In the lipid-bilayer part of the membrane, the polar head groups are in contact with water and the non polar tails lie in the interior of the membrane. The whole bilayer arrangement is held together by non-polar interactions such as *van der Waals* and hydrophobic interactions. The surface of the bilayer is polar and contains charged groups. The non-hydrocarbon interior of the bilayer consists of saturated and unsaturated chains of fatty acids and the fused ring system of cholesterol. Both the inner and outer layers of the bilayer contain mixtures of lipids, but their compositions differ and can be used to distinguish the inner and outer layers from each other. Bulkier groups tend to occur in the outer layer and smaller molecules tend to occur in the inner layer.

B. Properties

The lipid bilayer is a thin membrane made of two layers of lipid molecules. These membranes are flat sheets that form a continuous barrier around cells. The cell membrane of almost all living organisms and many viruses are made of a lipid bilayer. The lipid bilayer is the barrier that keeps ions, proteins and other molecules where they are needed and prevents them from diffusing into areas where they should not be.

Lipid bilayers are ideally suited to the role of controlling the transport of small molecules and ions. However, they are impermeable to most water soluble (hydrophilic) molecules. Bilayers are particularly impermeable to ions, which allow cells to regulate salt concentrations and pH by pumping ions across their membranes.

The arrangement of the hydrocarbon interior of the bilayer can be ordered and rigid or disordered and fluid. The bilayer's fluidity depends on its composition. When saturated fatty acids are present, a linear arrangement of the hydrocarbon chains leads to close packing of the molecules in the bilayer, and thus to rigidity. In unsaturated fatty acids there is a kink in the hydrocarbon. The kinks cause disorder in the packing of the chain which causes greater fluidity in the bilayer. The lipid components of a bilayer are always in motion to a greater extent in more fluid bilayers and to a lesser extent in more rigid bilayers.

The presence of cholesterol may also enhance order and rigidity. The fused-ring structure of cholesterol (12) is itself quite rigid and presence of cholesterol stabilises the extended straight chain arrangement of saturated fatty acids by *van der Waals* interactions. Further, the presence of cholesterol is characteristic of animal rather than plant membranes. Therefore, animal membranes are less fluid than plant membranes. Thus, cholesterol helps strengthen the bilayer, and decrease its permeability.

We have seen that natural bilayers are usually made of phospholipids, which have a hydrophilic head and two hydrophobic tails, When phospholipids are exposed to water, they arrange themselves into a two layered sheet (bilayer) with all of their tails pointing toward the centre of the sheet. The centre of this bilayer contains almost no water and excludes molecules like sugars or salts that dissolve in water but not in oil.

Lipid bilayers are quite fragile and are so thin that they are invisible in a traditional microscope.

Phospholipids with certain head groups can alter the surface chemistry of a bilayer and can mark a cell for destruction by the immune system.

The bilayer can adopt a solid gel state at lower temperatures but undergoes phase transition to a fluid state at higher temperatures.

The packing of lipids within the bilayer also affects its mechanical properties, including its resistance to streching and bending.

C. Membrane Proteins

We know biological membranes include several types of lipids other than phospholipids. For example, cholesterol is present in animals. Proteins in a biological membrane can be associated with the lipid bilayer in either of the two ways as **peripheral proteins** on the surface of the membrane or as **integral proteins** within the lipid bilayer.

Peripheral proteins are usually bound to the charged head groups of the lipid bilayer by polar interactions, electrostatic interactions or both.

Proteins can be attached to the membrane in a variety of ways. When a protein completely spans the membrane, it is often in the form of an α-helix or β-sheet. These structures minimise the contact of the polar parts of the peptide backbone with the nonpolar lipids in the interior of the bilayer. Proteins can also be anchored to the lipids *via* covalent bonds.

Membrane proteins have a variety of functions. Integral membrane proteins function when incorporated into a lipid bilayer. Cholesterol also helps regulate the activity of certain integral membrane proteins. As bilayers define the boundaries of the cell and its compartments, these membrane proteins are involved in many inter- and intra-cellular signaling processes. Certain kinds of membrane proteins are involved in the process of fusing two bilayers together. They allow cells to identify each other and interact.

Most of the important functions of the membrane proteins as a whole are those of the protein component. Transport proteins help to move substances in and out of the cell and receptor proteins are important in the transfer of extracellular signals such as those carried by hormones or neurotransmitters into the cell. In addition some enzymes are lightly bound to membranes. For example many of the enzymes responsible for aerobic oxidation reactions are found in specific parts of mitrochondrial membranes.

D. Functions of Membranes

We know that the structural role in membranes is to act as the boundaries and containers of all the cells and of the organelles within cells. In addition to this, three important functions take place in or on membranes. The first of these functions is **transport**. Membranes are semipermeable barriers to the flow of substances into and out of cells and organelles. Transport through the membrane can involve the lipid bilayer as well as the membrane proteins. The two important functions primarily involve the membrane proteins. One of these is catalysis; the third significant function is the receptor property, in which proteins bind specific biological responses in the cell. We have already studied enzymatic catalysis in the 4th chapter of this book. Here we consider the other two important functions of membranes.

E. Membrane Transport

The transport of substances across biological membranes takes place by two processes.

In **passive transport**, a substance moves from a region of higher concentration to one of lower concentration. In other words, the movement of the substance is in the same direction as a concentration gradient and the cell does not expend energy.

In **active transport**, a substance moves from a region of lower concentration to one of higher concentration and this process requires the cell to expend energy. Active transport requires moving substances against a concentration gradient. It is identified by the presence of a carrier protein and the need for the energy source to move solutes against a gradient. An example of active transport is the movement of potassium ions into a cell and simultaneously moving sodium ions out of the cell. This is referred to as a Na^+/K^+ pump.

5.18   LIPOSOMES

A. Introduction

When membrane phospholipids are disrupted, they can reassemble themselves into tiny spheres, smaller than a normal cell, either as bilayers or monolayers. The bilayer structures are **liposomes.** The monolayer structures are called **micelles**. Thus, a liposome is a tiny vesicle made out of the same material as a cell membrane. They are stable structures based on a lipid bilayer that form a spherical vesicle. These vesicles can be filled with therapeutic agents on the inside and then used to deliver the agent to a target tissue.

We have already seen that phospholipids are molecules that have polar head groups and long non-polar tails of hydrocarbon chains. Under suitable conditions, a double-layer arrangement is formed so that the polar head groups of many molecules face the aqueous environment, while the non-polar tails are in contact with each other and are kept away from the aqueous environment. These bilayers form three-dimensional structures, called liposomes. Liposomes can be filled with drugs and used to deliver therapies for cancer and other diseases.

B. Properties of Liposomes

The name liposome is derived from two Greek words, '*Lipos*' meaning fat and '*Soma*' meaning body. A liposome can be formed in a variety of sizes as unilameller or multi-lamellar construction and its name relates to its structural building blocks, phospholipids, and not to its size. Liposomes, usually contain a core of aqueous solution; lipid spheres that contain no aqueous material are called micelles.

Liposomes were first described by British haematologist **Dr. Alec D. Bangham** in 1961 at the Babraham Institute in Cambridge and the microscope pictures served as the first real evidence for the cell membrane being a bilayer lipid structure.

Liposomes are spherical, self-closed vesicles of colloidal dimensions. Due to their structure chemical composition and colloidal size, all of which can be well controlled by preparation methods, liposomes exhibit several properties which may be useful in various applications. The most important properties are colloidal size and special membrane and surface characteristics. They include bilayer phase behaviour, its mechanical properties and permeability, charge density etc.

Due to their amphiphilic character liposomes act as powerful solubilisers for a wide range of compounds.

Liposomes also exhibit many special biological characteristics including interactions with biological membranes and various cells.

These properties point to several possible applications with liposomes acting as solubilisers for difficult-to-dissolve substances, dispersants, sustained release systems, delivery systems for many substances, stabilisers and protective agents.

Liposomes can be made entirely from naturally occurring substances and are therefore, nontoxic, biodegradable and non immunogenic.

C. Applications of Liposomes

i. Liposomes in Medicine

Liposomes act as drug delivery vehicles in medicine, adjuvants in vaccination, signal enhancers/carriers in medical diagnostics and analytical biochemistry, solubilisers for various ingredients and penetration enhancers in cosmetics.

The therapeutic concentrations of many drugs is not much lower than the toxic one. In such cases, the toxicity of the drug can be reduced or the efficacy enhanced by the use of liposomes, which act as drug carriers. They change the temporal and spatial distribution of the drug . The benefits and limitations of liposome drug carriers depend on the interaction of liposomes with cells and their fate in *vivo* after administration.

Liposomes are used for drug delivery due to their unique properties. A liposome encapsulates a region on aqueous solution inside a hydrophobic membrane; dissolved hydrophobic solutes cannot readily pass through the lipids. Hydrophobic chemicals can be dissolved into the membrane and in this way, liposomes can carry both hydrophobic molecules and hydrophilic molecules. To deliver the molecules to sites of action, the lipid bilayer can fuse with other bilayers such as cell membrane and thus deliver the liposome contents. By making liposomes in a solution of DNA or drugs (which would normally be unable to diffuse through the membrane) they can be delivered past the lipid bilayer.

The use of liposomes for transformation or transfection of DNA into a host cell is known as lipofection.

ii. Liposomes in Cosmetics

The use of liposomes in nano cosmetology, also has many benefits. These are improved penetration and diffusion of active ingredients, selective transport of active ingredients, longer release time, greater stability of active substances, reduction of unwanted side effects and high biocompactibility.

iii. Use in Textiles

In addition to gene and drug delivery applications, liposomes can be used as carriers for the delivery of dyes to textiles, pesticides to plants, enzymes and nutritional supplements to food and cosmetics to the skin.

iv. Diagnostic Applications

The diagnostic applications include their use as a model tool or reagent in the basic studies of cell interactions, recognition processes and the mode of action of certain substances.

v. Other Uses

In addition to these applications which have significant impact in several industries, the properties of liposomes offer a very useful model system in many fundamental studies from topology, membrane biophysics, photophysics and photochemistry, colloid interactions and many others.

Terpenoids

In this chapter we will study

- Terpenoids—Occurrence in Nature, Classification and Isolation from Natural Sources.
- Citral— Isolation, Structure and Synthesis.
- Dipentene—Isolation, Structure and Synthesis.
- Structures and Synthesis of Geraniol and a-Terpineol.
- Menthol, α-Pinene and Camphor—Structures and Synthesis.
- Diterpenoids –Phytol
- Triterpenoids–Squalene
- Carotenoids –β-Carotene, α-Carotene, γ-Carotene and Lycopene

6.1 INTRODUCTION – THE ESSENTIAL OILS

We are quite familiar with the characteristic odour of flowers, fruits, leaves and even wood of several plants. The flowers like rose and jasmine have pleasant odours. Lemon oil and clove oil have been known to us for many years. Sandalwood oil has been in use as a perfume since ancient times. The pleasant smell of these is actually due to the presence of certain volatile oils known as **essential oils**. The essential oils obtained by steam distillation or solvent extraction of these plants or their parts have been widely used in perfumery, food flavourings and medicine from the earliest times. The simple mono- and sequiterpenes are the chief constituents of these essential oils.

The term **'terpene'** is usually restricted to hydrocarbons derived from $C_{10}H_{16}$ while the term terpenoids represents the hydrocarbons as well as the oxygenated derivatives. This is due to the fact that since the suffix *'ene' signifies unsaturated* hydrocarbons, the name terpene is inappropriate to include compounds such as alcohols, aldehydes, ketones *etc.*

6.2   OCCURRENCE IN NATURE

Mono-and sesquiterpenoids are the chief constituents of essential oils. The essential oils are complex mixtures; however, some may consist mainly of other constituents. For example, oil of bitter almonds contains benzaldehyde and oil of winter green has methylsalicylate.

Some of the important essential oils with their terpenoid constituents are given below:

Essential Oil	Terpenoids
Turpentine	Pinene
Caraway	Carvone, limonene
Citronella	Geraniol, citronellal
Eucalyptus	Cineole
Jasmine	Linalool
Lavender	Linalool
Lemon	Limonene, citral
Sweet orange	Limonene, linalool
Rose	Geraniol
Cardamom	Terpineol
Camphor	Camphor
Ginger	Zinziberene

Mono- and sesquiterpenoids are of considerable commercial importance. Due to their pleasant smelling nature they are particularly used in the perfumery industry. Various other terpenoids show biological activity and are useful in pharmaceuticals.

The di - and triterpenoids are not steam volatile. They are obtained from plant and tree gums by solvent extraction. The tetraterpenoids form a group of compounds known as **carotenoids.** Rubber is the most important polyterpenoid.

6.3   ISOPRENE RULE

On thermal decomposition almost all terpenoids yield isoprene as one of the products.

$$Terpenoid \longrightarrow CH_2{=}\underset{\underset{CH_3}{|}}{C}{-}CH{=}CH_2$$

Isoprene
(2-Methyl-1,3 butadiene)

This observation led to the suggestion that all naturally occuring terpenoids can be built up of 'isoprene units'. This is known as the **'isoprene rule'** and was first pointed out by **Wallach**. This means that all terpenoids can be divided into isoprene units linked together in certain fashion.

It was pointed out by **Ingold** that the isoprene units in natural terpenoids were linked with one another through C_1 to C_4 positions *i.e.*, head to tail union. Head is the C_1 position or the branched end of isoprene. This divisibility into isoprene units and their 'head to tail' union is referred to as the **'special isoprene rule'**. For example, in a terpenoid of the formula $C_{10} H_{16}$ head to tail linkage can be shown in the following manner.

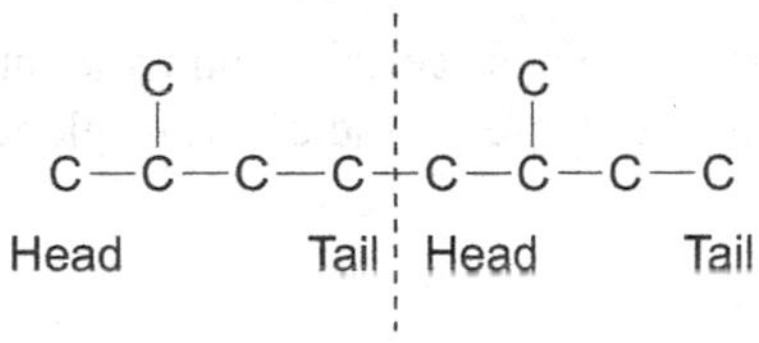

However, the above rule cannot be taken as a fixed rule as several exceptions occur. There are also some terpenoids whose carbon content is not a multiple of five and some whose C content is a multiple of five but cannot be divided into isoprene units.

The isoprene units making up some naturally occurring terpenoids are shown below:

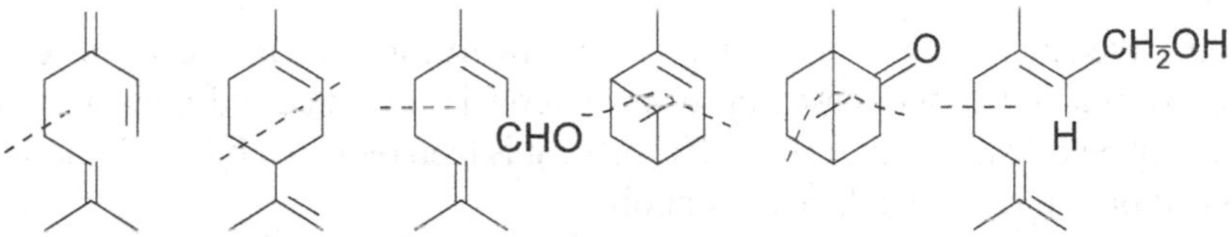

Myrcene   Limonene   Citral   Pinene   Camphor   Geraniol

6.4   CLASSIFICATION OF TERPENOIDS

Terpenoids are classified on the basis of the number of isoprene units present in them. Natural terpenoid hydrocarbons have the molecular formula as $(C_5H_8)_n$ where the value of *n* is used as a basis for classification. For monoterpenoids *n* = 2 and *n* = 4 for diterpenoids and so on. Thus, we have

1. **Monoterpenoids** with 2 isoprene units, Mol. formula is $C_{10} H_{16}$.

2. **Sesquiterpenoids** ($C_{15} H_{24}$) with 3 isoprene units.

3. **Diterpenoids** ($C_{20} H_{32}$) with four isoprene units.

4. **Sesteterpenoids** ($C_{25} H_{40}$) with five isoprene units.

5. **Triterpenoids** ($C_{30} H_{48}$) with six isoprene units.

6. **Tetraterpenoids** ($C_{40} H_{64}$) with eight isoprene units.

7. **Polyterpenoids** ($C_5 H_8)_n$ where $n > 40$ with several isoprene units.

The sesquiterpenoids have been discovered recently and so far only very few are known. In addition to the hydrocarbons there are oxygenated molecules of each class which also occur naturally. These are mainly alcohols, aldehydes and ketones. The terpenoids are further classified as acyclic and cyclic terpenoids depending upon whether they possess an open chain or a cyclic structure.

6.5   ISOLATION FROM NATURAL SOURCES

The essential oils obtained from certain plants contain mostly mono- and sesquiterpenoids. Essential oils are extracted from plants by several methods as outlined below:

1. Steam distillation

This is the most widely used method. The plant material is macerated and then steam distilled. The essential oils go into the distillate. The distillate is extracted by pure volatile solvents like petroleum ether. The solvent extracts are dried and the solvent evaporated or distilled under reduced pressure. However, the method should be used with great care since some essential oils decompose under these conditions.

2. Extraction by solvents

As described above some essential oils are sensitive to heat and hence decompose during distillation. In such cases, the plant material is powdered and extracted directly with light petroleum ether at 50°C. The solvent is then removed by distillation under reduced pressure to give the essential oil.

3. Adsorption in purified fats (Enfleurage method)

The fat taken in glass plates is warmed to about 50°C. The flower petals are spread on the surface of the fat for several days until the fat is saturated with the essential oils. The petals are then removed and the fat is digested with ethyl alcohol when the essential oils present in fat are dissolved in ethyl alcohol. If some fat is also dissolved during digestion, it is removed by cooling to about 20°C. The solvent is distilled off to get the essential oils.

The essential oils so obtained contain a number of terpenoids and these are separated by fractional distillation. The terpenoid hydrocarbons distill first and these are followed by oxygenated derivatives. Distillation of the residue under reduced pressure gives the sesquiterpenoids and these are separated by fractional distillation.

More recently, various forms of chromatography have been used for the isolation and separation of terpenoids. Adsorption chromatography, using alumina or silica gel as adsorbents, has provided the most important method for the isolation and separation of terpenoids particularly triterpenoids. Gas chromatography is also being used to a great extent for this purpose.

6.6 CITRAL

A. Introduction

Citral is the most important member of acyclic monoterpenoids and finds many industrial applications. Citral is widely distributed in nature and occurs in lemon grass oil to the extent of 60–80 percent. It also occurs in citrus fruits–in oils of lime, lemon, citronella *etc.* Citral is a yellow liquid, b.p. 225°–228°C and has the smell of lemons. It is optically inactive.

Two geometrical isomers of citral are possible. Both these exist in natural citral. The *trans* form is called **citral-a,** also named **geranial.** It is more stable and present up to 90 percent in natural citral. The *cis*-isomer is known as **citral-b** or **neral,** which is present to the extent of 10 percent.

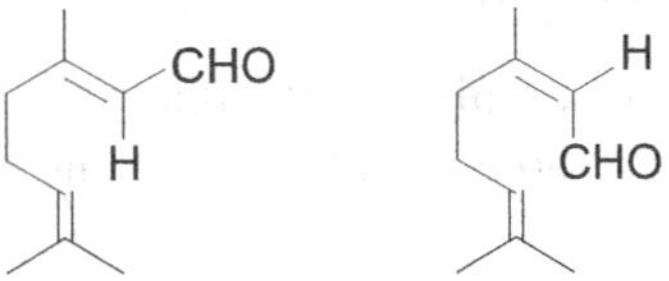

Citral-a (*trans* form)   Citral-b (*cis* form)

Both the forms have been obtained. Citral is obtained as its crystalline bisulphite product which on hydrolysis gives back citral.

B. Isolation of Citral

Citral can be isolated from lemons. Lemons are crushed and subjected to steam distillation. Lemon oil is obtained which contains citral. The lemon oil is subjected to fractional distillation. The fraction collected between 225°–230°C contains citral, which is treated with sodium bisulphite. The bisulphite addition product of citral is separated and hydrolysed with sodium carbonate solution to give citral.

Citral is also obtained from lemon grass oil by fractional distillation in vacuum and purified *via* the crystalline bisulphite addition product.

Citral is used for the preparation of β-ionone which is a special perfume.

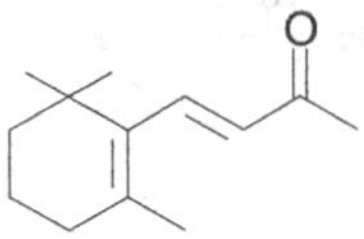

β-Ionone

C. Structure of Citral

The structure of citral has been derived on the basis of the following observations:

1. The molecular formula of citral is $C_{10}H_{16}O$.

2. **Presence of double bonds:** Citral is found to have two double bonds, as two molecules of bromine can add to citral to give tetrabromo derivative.

3. **Presence of aldehyde group:** Citral forms bisulphite addition product and oxime and 2,4-dinitrophenylhydrazone indicating the presence of a carbonyl group (oxo group). It reduces Fehling solution and Tollen's reagent. Citral is reduced by sodium - amalgam to 1° alcohol, geraniol, $C_{10}H_{18}O$.

    On oxidation with silver oxide it forms geranic acid, $C_{10}H_{16}O_2$ containing the same number of carbon atoms.

    These reactions indicate that citral has got an aldehyde group.

4. The **U. V. absorption spectrum** of citral has λ_{max} = 238nm (ε_{max} 13500), showing the presence of α, β – unsaturated system in the citral molecule.

5. **Oxidative degradation:**

    (a) Oxidation of citral with alkaline potassium permanganate followed by chromic acid gives acetone, oxalic acid and laevulic acid.

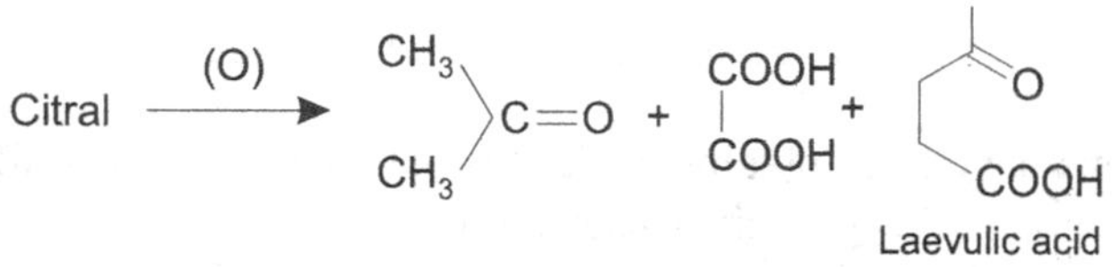

    Laevulic acid

    (b) Ozonolysis of citral gives, acetone, glyoxal and laevulic aldehyde.

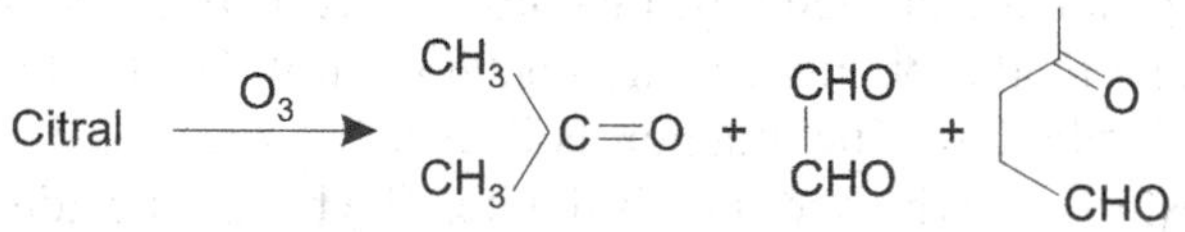

6. The carbon-skeleton of citral is shown by cyclisation of citral. On heating with $KHSO_4$ citral forms p-cymene. This reaction was used by **Semmler** to determine the position of methyl and isopropyl groups in citral.

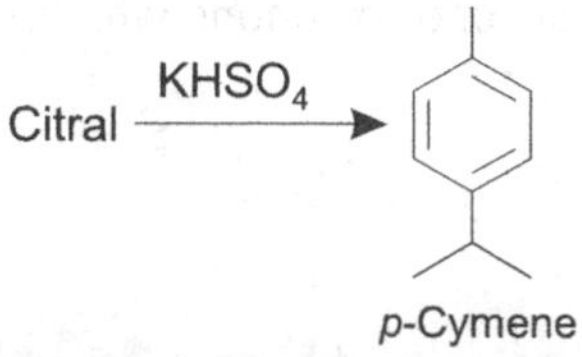

    p-Cymene

Therefore, citral has the following skeleton structure (1) in which two isoprene units are joined in head and tail fashion.

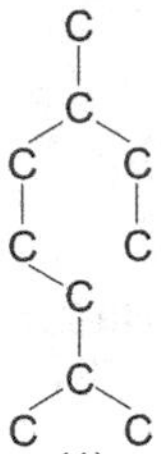

(1)

The formation of the above oxidation products can be explained by assuming structure (2) for citral.

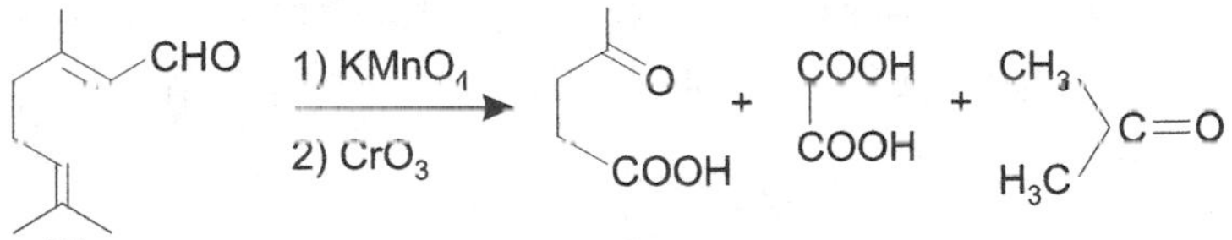

(2)

7. Hydrolysis of citral to methylheptenone:

The above structure of citral is supported by the hydrolysis of citral to methylheptenone (3). Citral on treatment with aqueous potassium carbonate is converted into 6-methylhept-5-en-2-one and acetaldehyde. Thus, we have

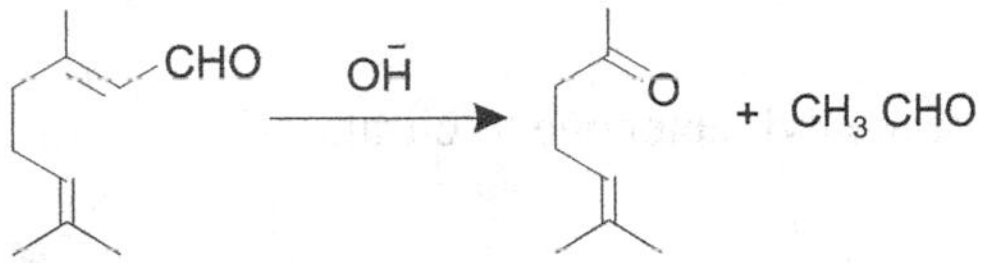

(2)   (3)

Further, heptenone is itself oxidised to acetone and laevulic acid and on ozonolysis it gives acetone and laevulic aldehyde.

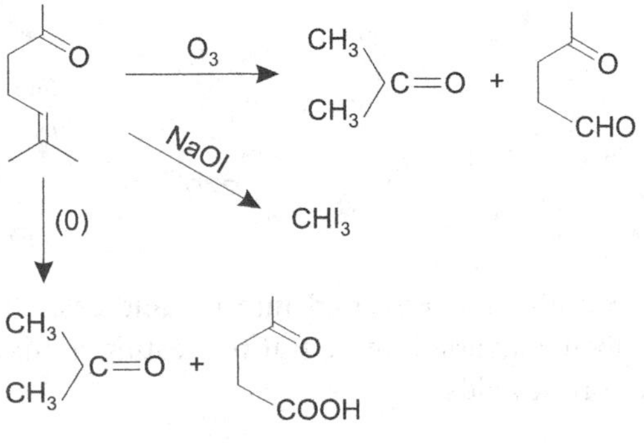

The structure of citral is finally confirmed by the synthesis of heptenone which is then converted into citral.

D. Synthesis of Citral

1. Synthesis of methylheptenone:

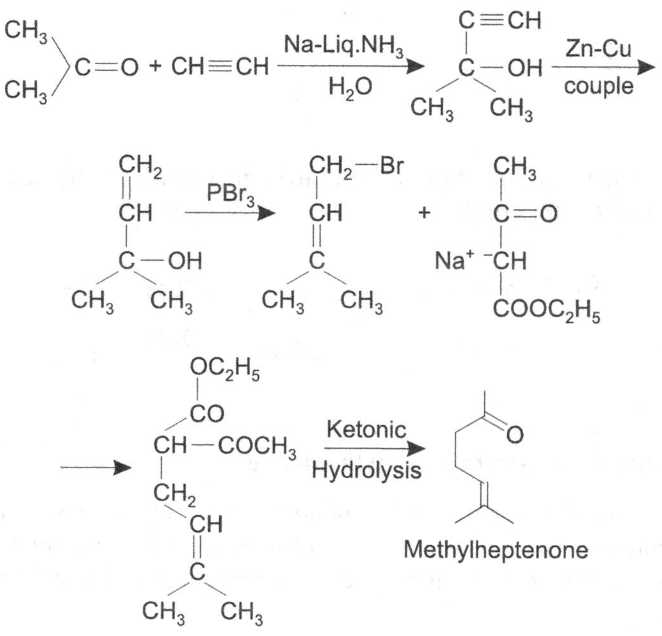

2. Conversion of methylheptenone to citral:

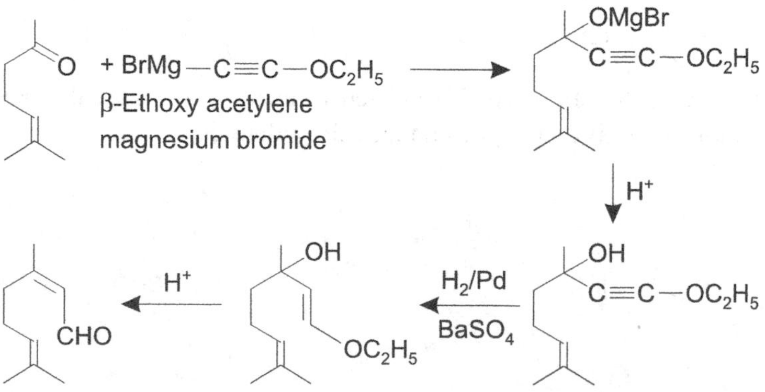

Methylheptenone can also be converted into geranic ester by the Reformatsky reaction, which is then converted into citral by heating a mixture of the calcium salt of geranic and formic acids.

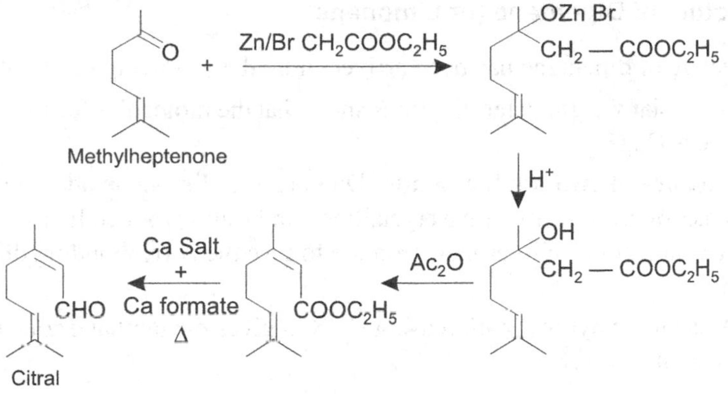

Methylheptenone

Citral

6.7 DIPENTENE

A. Introduction

The racemic modification of limonene is known as dipentene. It is a cyclic monoterpenoid. Limonene is optically active. The (+) form of limonene occurs in lemon and orange oils.

The (–) limonene occurs in peppermint oil and along with (+) pinene in pine needle oil. The (±) limonene, known as dipentene is present in turpentine oil along with α–pinene and terpinolene.

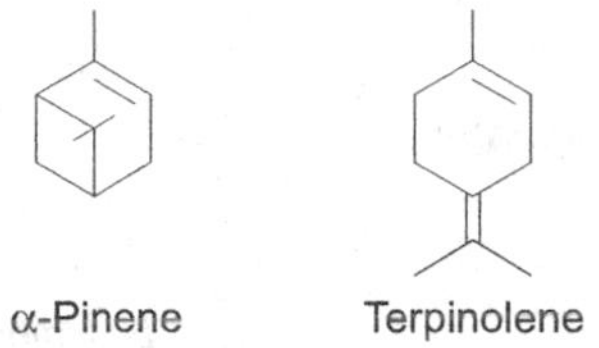

α-Pinene   Terpinolene

The racemic modification is also produced by racemisation of the optically active forms at about 250° C. Of the three optical isomers dipentene is the most important. Dipentene is used as a flavour in foods, beverages and pharmaceutical preparations. It is also used in the manufacture of resins, p–cymene and p–menthane.

B. Isolation of Dipentene

1. From turpentine oil: The turpentine oil is subjected to fractionation. The steam volatile fractions contain monoterpenoids. On fractional distillation the (±) form is separated.

2. From bitter orange oil: The oil obtained from bitter oranges is subjected to fractionation. (+) Limonene is obtained at 117°C. When this is heated to 250°– 300°C, we get dipentene.

C. Structure of Dipentene (or Limonene)

The structure of dipentene has been arrived at by the following observations:

1. Molecular weight determinations show that the molecular formula of dipentene is $C_{10}H_{16}$.

2. **Presence of two double bonds:** Dipentene or limonene adds on two molecules of bromine to form a crystalline tetrabromo product. It also reacts with two molecules of hydrogen bromide to give the corresponding dihydrobromide.

   On catalytic hydrogenation using H_2/Ni it gives p–menthane consuming two molecules of H_2,

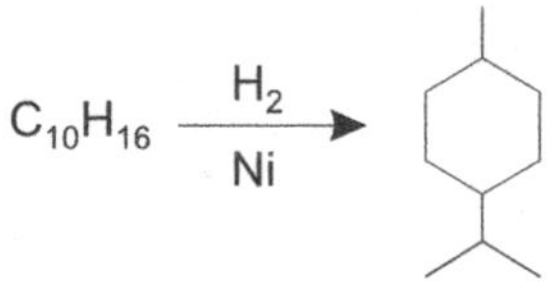

$$C_{10}H_{16} \xrightarrow[Ni]{H_2}$$

p-Menthane

   p-Menthane is a completely saturated compound and its corresponding alkane is $C_{10} H_{22}$. p-Menthane has got cyclic structure; therefore, dipentene has a cyclic structure.

3. **Carbon skeleton:** The carbon skeleton of dipentene is further shown by the conversion of limonene to p-Cymene when it is heated with sulphur or selenium.

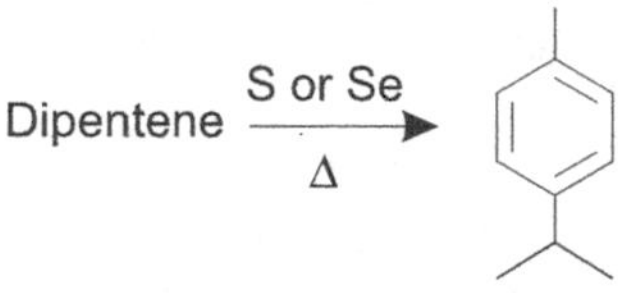

$$Dipentene \xrightarrow[\Delta]{S \text{ or } Se}$$

p-Cymene

These reactions show that the carbon skeleton of dipentene is

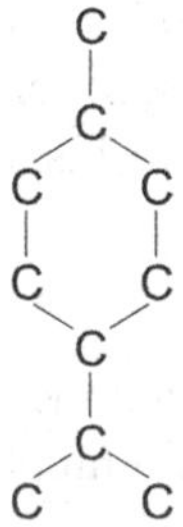

4. Position of the double bonds:

(a) Dipentene when treated with NOCl consumes only one mole of NOCl. The nitrosochloride derivative obtained is colourless. This shows that one double bond is present in the ring. As the product is colourless the C-atoms of the double bond are not tertiary.

(b) Dipentene does not undergo Diels-Alder reaction. That means double bonds are not conjugated.

(c) Dipentene nitrosochloride on treatment with alcoholic KOH gives carvoxime

$$\text{Dipentene} \xrightarrow{\text{NOCl}} \underset{\text{nitrosochloride}}{\text{Dipentene}} \xrightarrow[\text{EtOH}]{\text{KOH}} \text{Carvoxime}$$

The structure of carvoxime is known. It, therefore, follows that dipentene must have a double bond in position 8 of the molecule.

(d) When dipentene is treated with 5% H_2SO_4, we get α-terpineol (4). α-Terpineol has a double bond at 1, 2-position.

(e) (+)-α-Terpineol on dehydration with $KHSO_4$ gives limonene or dipentene.

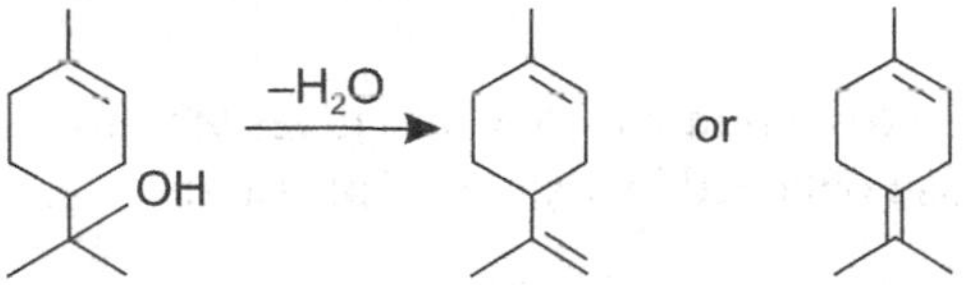

(4)	(5)	(6)
α-Terpineol	Limonene	

The position of one double bond in the ring is decided *i.e.*, it is at 1,2-position. The position of the second double bond comes from the reaction with NOCl.

Moreover, structure (5) contains a chiral centre C-4 and hence can exhibit optical activity.

The reactions with NOCl can be shown as below:

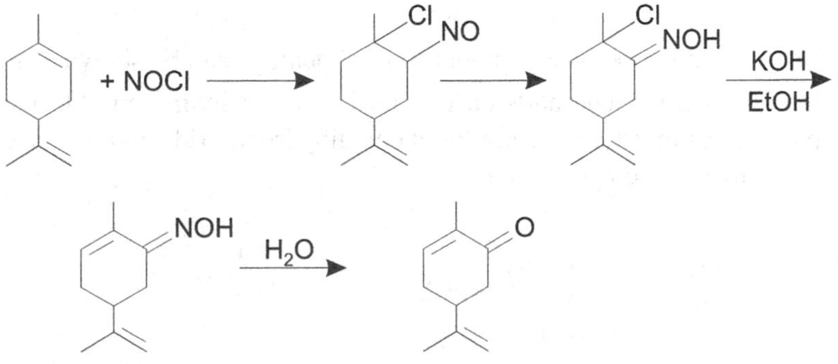

Carvoxime	Carvone

(f) Presence of a double bond at 8 (9) position can also be shown by the following reactions:

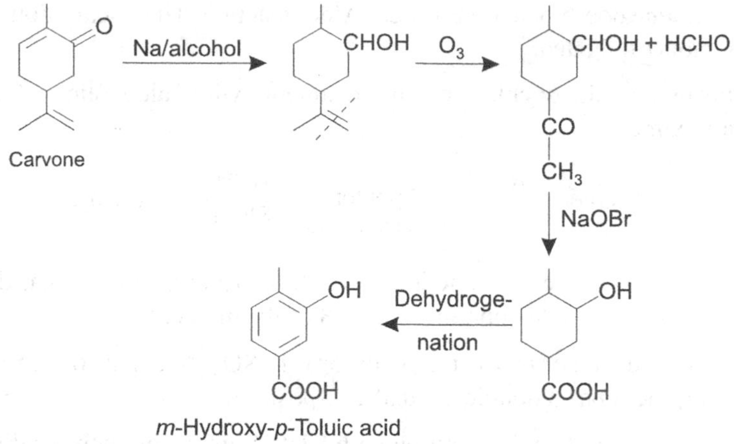

Carvone

m-Hydroxy-p-Toluic acid

Formation of formaldehyde shows the presence of a double bond at position 8 (9).

(g) Dipentene on oxidation with $KMnO_4$ gives terpenylic acid and terebic acid which are also obtained by oxidation of α-terpineol.

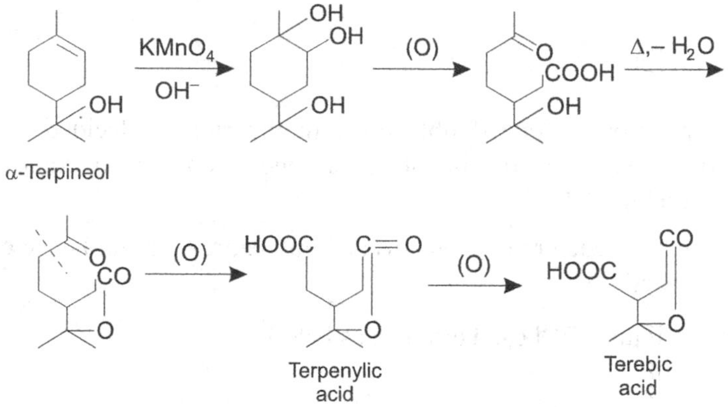

α-Terpineol

Terpenylic
acid

Terebic
acid

(h) The correlation between limonene and dipentene can be shown as follows:

(+) or (−) Limonene adds on two molecules of hydrogen chloride in the presence of moisture to form limonene dihydrochloride and this is identical with dipentene dihydrochloride

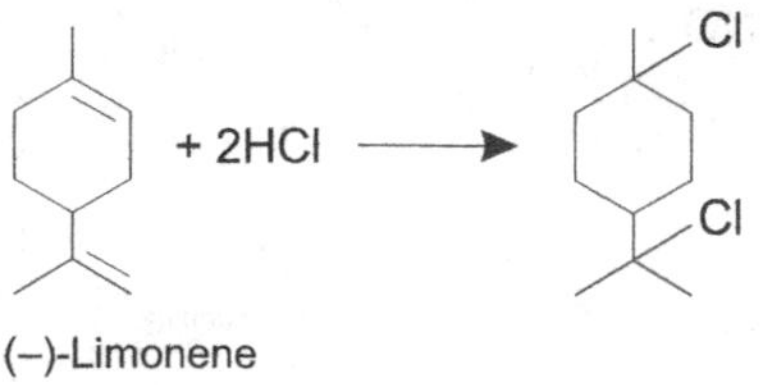

(+) or (−)-Limonene

Limonene dihydrochloride no longer contains a chiral centre. It is, therefore, optically inactive. Dipentene can be regenerated by heating the dihydrochloride with sodium acetate in acetic acid or boiling with aniline.

D. Synthesis of Dipentene

1. **From Isoprene :** By dimerisation of isoprene at 280° C, we get dipentene.

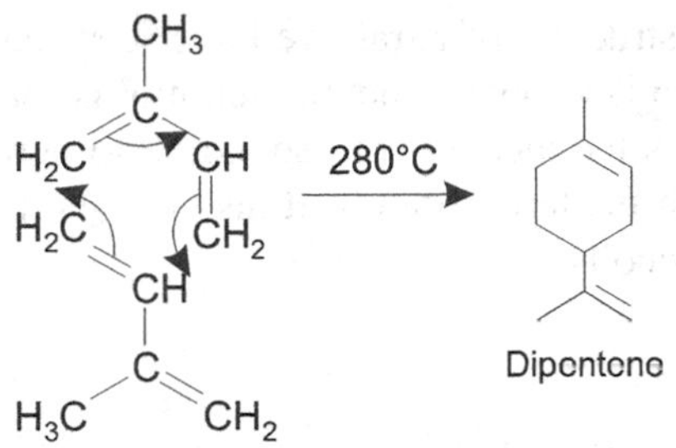

Dipentene

2. **From α-terpineol by dehydration with KHSO₄:**

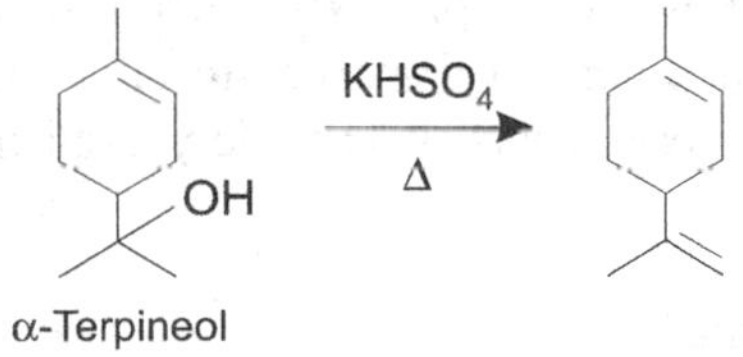

α-Terpineol

3. **Synthesis of α-Terpineol can be carried out by the following reactions:**

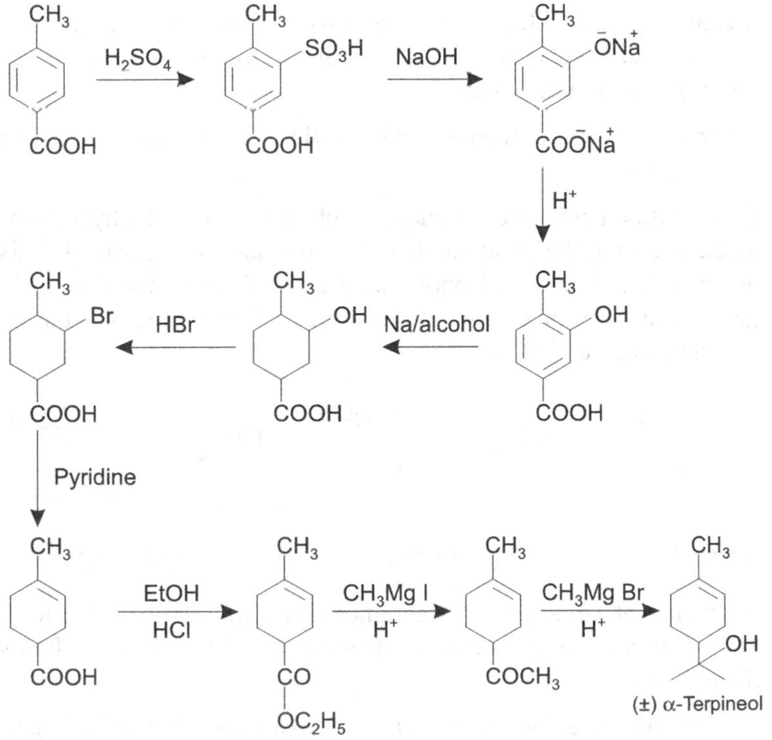

(±) α-Terpineol

6.8 GERANIOL

A. Introduction

While discussing the structure of citral, we have seen that citral-a is also known as geranial and citral-b is known as neral. Geraniol is the alcohol corresponding to geranial and nerol is the alcohol corresponding to neral. Geraniol is found in many essential oils, particularly rose oil. It also occurs in lemon-grass, lavender, citronella and palmrosa oils.

B. Isolation

Geraniol is isolated from the oil of palmrosa by treating it with anhydrous calcium chloride and decomposing the crystalline derivative with water. Geraniol is obtained as a colourless liquid (b. p. 229°–230°C). It has a pleasant rose – like odour. It is insoluble in water but dissolves in ethanol. Geraniol is used in the preparation of artificial rose scents. It is widely used in perfume, cosmetic and flavour industries.

C. Structure of Geraniol

1. Molecular formula formula of geraniol has been found to be $C_{10}H_{18}O$.

2. It adds on 2 molecules of bromine to form a tetrabromide and on catalytic hydrogenation it consumes two molecules of hydrogen. Geraniol, therefore, contains two double bonds.

3. Geraniol does not form an oxime indicating the absence of an aldehyde or a ketone group.

4. On oxidation geraniol gives successively citral-a (an aldehyde) and geranic acid containing the same number of carbon atoms as geraniol. This means geraniol is a *primary* alcohol and the positions of the double bonds in geraniol are the same as in citral-a. Geraniol, therefore, has the following structure as given below:

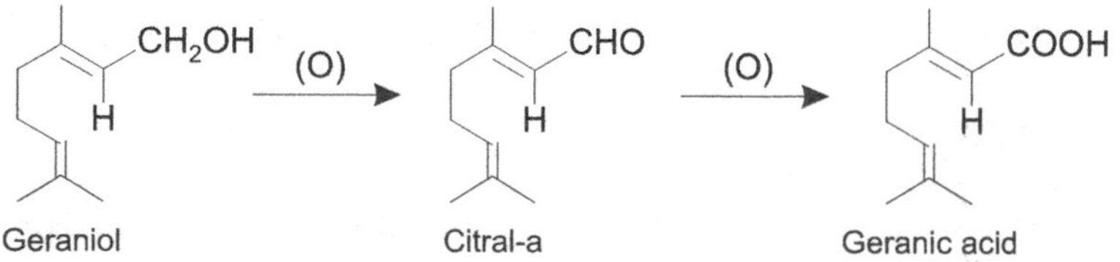

Geraniol   Citral-a   Geranic acid

5. Reduction of citral produces geraniol and some nerol is also formed. We know that geraniol and nerol are structurally related. This can be shown by the following facts:

   * Both, geraniol and nerol add on two molecules of hydrogen during catalytic hydrogenation and give the same saturated alcohol, $C_{10}H_{22}O$.

* On oxidation geraniol and nerol give the same oxidation products. Hence, geraniol and nerol are geometrical isomers.

6. Geraniol and nerol both can be cyclised to give α-terpineol by means of dilute sulphuric acid. This cyclisation is 9 times as fast with nerol as it is with geraniol. The faster rate of cyclisation indicates that the alcoholic group is very near to the carbon which is involved in the ring formation. This means the configuration of nerol is *cis* and geraniol has been assigned the *trans* configuration.

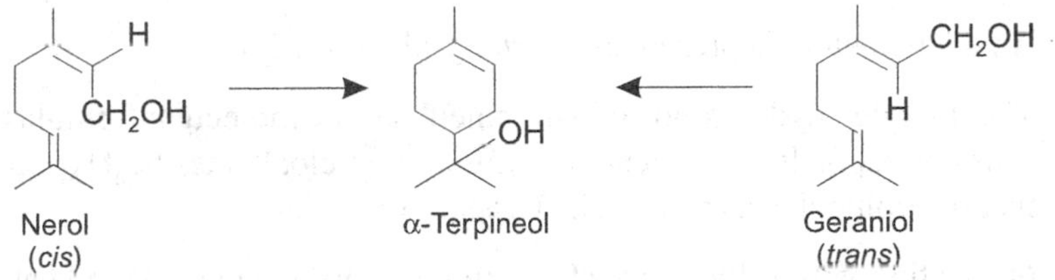

Nerol
(*cis*)   α-Terpineol   Geraniol
                                                                     (*trans*)

7. The hydration of geraniol and nerol to α-terpineol involves the formation of the same intermediate as given below:

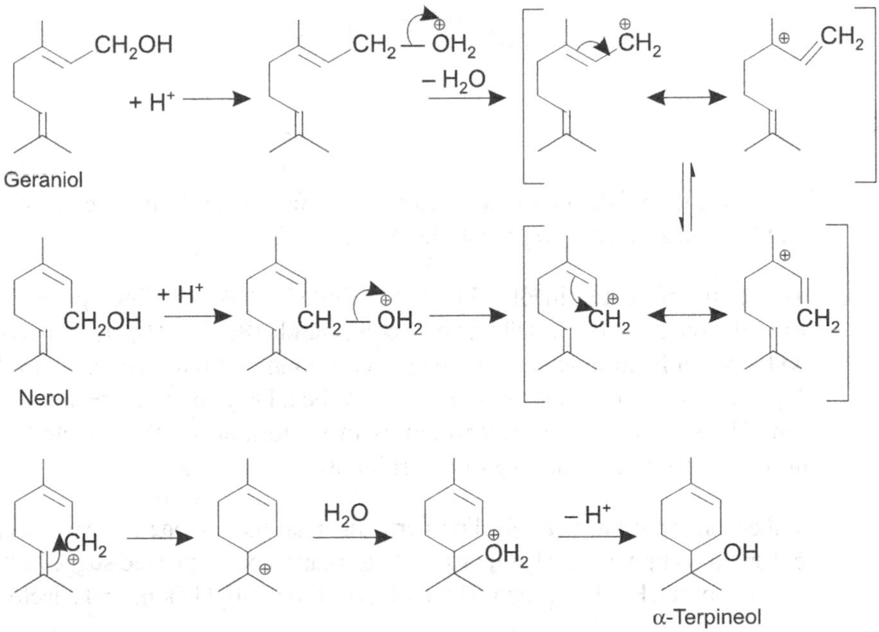

Geraniol

Nerol

α-Terpineol

6.9 α-TERPINEOL

A. Introduction

α-Terpineol is the most important monocyclic terpenoid. Its structure confirms the structures of several other monoterpenoids because α-terpineol can be converted into many other terpenoids and *vice versa*.

α-Terpineol is an optically active monoterpenoid. It occurs naturally in the (+), (–) and (±) forms. The racemic modification is a crystalline solid, m.p. 35°C. α-Terpineol is extensively used in the preparation of certain perfumes.

B. Structure of α-Terpineol

1. The molecular formula of α-*terpineol* is $C_{10}H_{18}O$.

2. It adds on only one molecule of bromine to form a dibromide. It, therefore, contains one double bond.

3. An oxygen atom is present as a *tertiary* alcoholic group.

4. The saturated hydrocarbon of α-terpineol has the molecular formula $C_{10}H_{20}$. This corresponds to the general formula of cycloalkanes (C_nH_{2n}). It means that α-terpineol is a monocyclic terpenoid.

5. On heating with sulphuric acid α-terpineol forms p-cymene indicating that α-terpineol contains a p-cymene skeleton

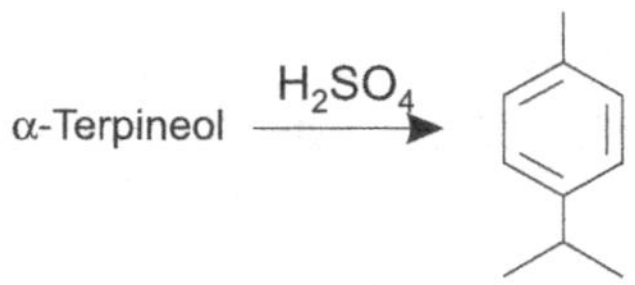

$$\alpha\text{-Terpineol} \xrightarrow{\text{H}_2\text{SO}_4} $$

p-Cymene

It follows, therefore, that α-terpineol is a p-menthane derivative with one double bond and a *tertiary* alcoholic group.

6. Oxidation of α-terpineol (7) with dilute (1%) alkaline potassium permanganate gives a trihydroxy compound (8), $C_{10}H_{20}O_3$. This on oxidation with chromic acid produces a compound with molecular formula, $C_{10}H_{16}O_3$ (9). Compound (9) was shown to be a lactone of monocarboxylic acid. There is no loss of carbon atoms in its formation; the double bond must, therefore be in the ring of α-terpineol.

Further, on warming with alkaline permanganate the lactone (9) gave acetic acid and a compound $C_8H_{12}O_4$ (10). The formation of acetic acid suggests that (9) contains a CH_3CO- group. Therefore, (9) is a methyl ketone and a lactone.

The acid (10) was named as terpenylic acid. Its reactions showed that it was the lactone of a monohydroxy dicarboxylic acid. Oxidation of terpenylic acid gives terebic acid, $C_7H_{10}O_4$ (11), which is the lactone of a monohydroxy dicarboxylic acid. The reactions are shown on next page:

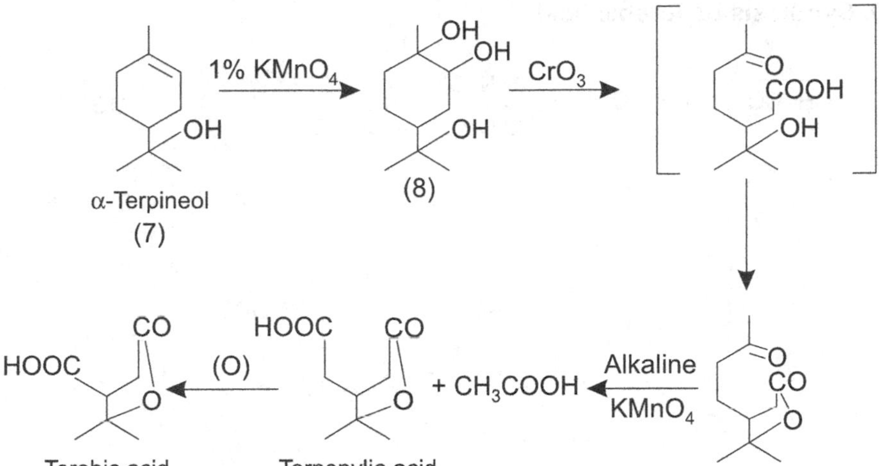

α-Terpineol
(7)

(8)

Terebic acid
(11)

Terpenylic acid
(10)

(9)

The above reactions confirm structure (7) for α-terpineol.

The structures of terpenylic acid (10) and terebic acid (11) are confirmed by synthesis.

C. Synthesis of Terpenylic Acid

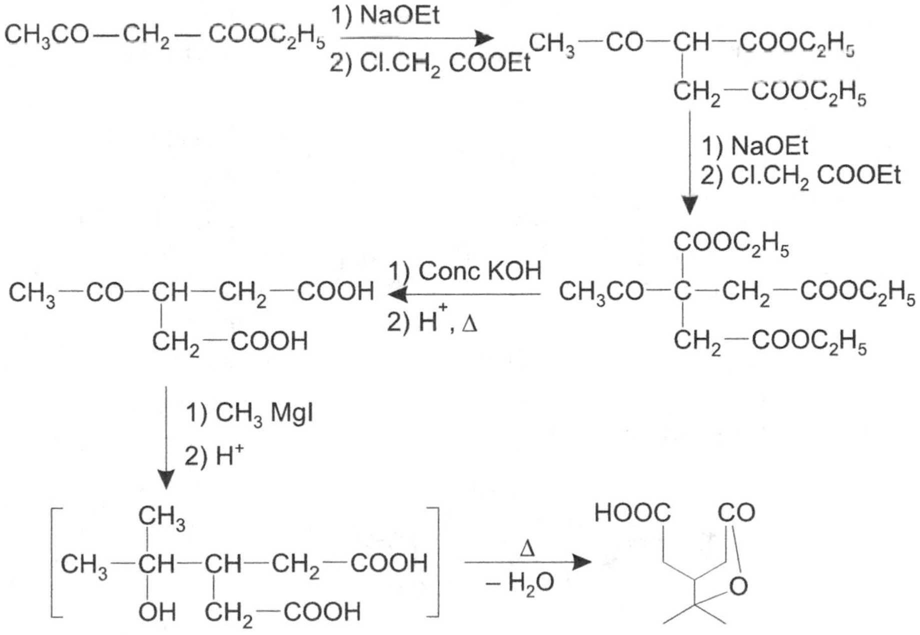

$CH_3CO-CH_2-COOC_2H_5$ $\xrightarrow{\text{1) NaOEt}}_{\text{2) Cl.CH}_2 \text{ COOEt}}$

$CH_3-CO-CH-COOC_2H_5$
$CH_2-COOC_2H_5$

1) NaOEt
2) Cl.CH$_2$ COOEt

$CH_3-CO-CH-CH_2-COOH$ $\xleftarrow{\text{1) Conc KOH}}_{\text{2) H}^+, \Delta}$ $CH_3CO-\overset{COOC_2H_5}{\underset{CH_2-COOC_2H_5}{C}}-CH_2-COOC_2H_5$
CH_2-COOH

1) CH$_3$ MgI
2) H$^+$

$\left[CH_3-\overset{CH_3}{\underset{OH}{CH}}-\overset{}{\underset{CH_2-COOH}{CH}}-CH_2-COOH \right]$ $\xrightarrow[-H_2O]{\Delta}$

Terpenylic acid

D. Synthesis of Terebic acid

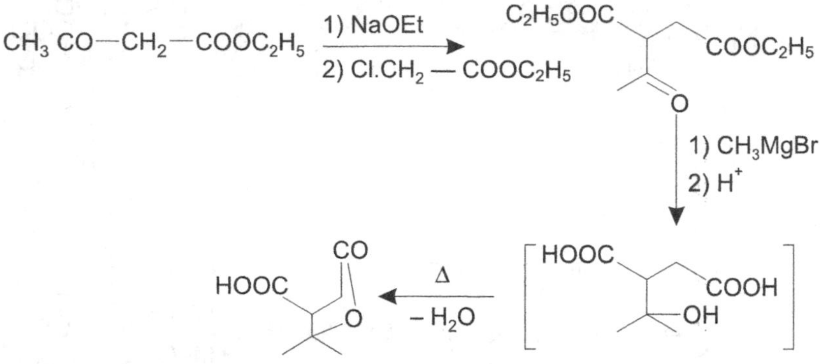

Terebic acid

E. Synthesis of α-Terpineol

1. The structure of α-terpineol is finally confirmed by its synthesis which starts with *p*-toluic acid.

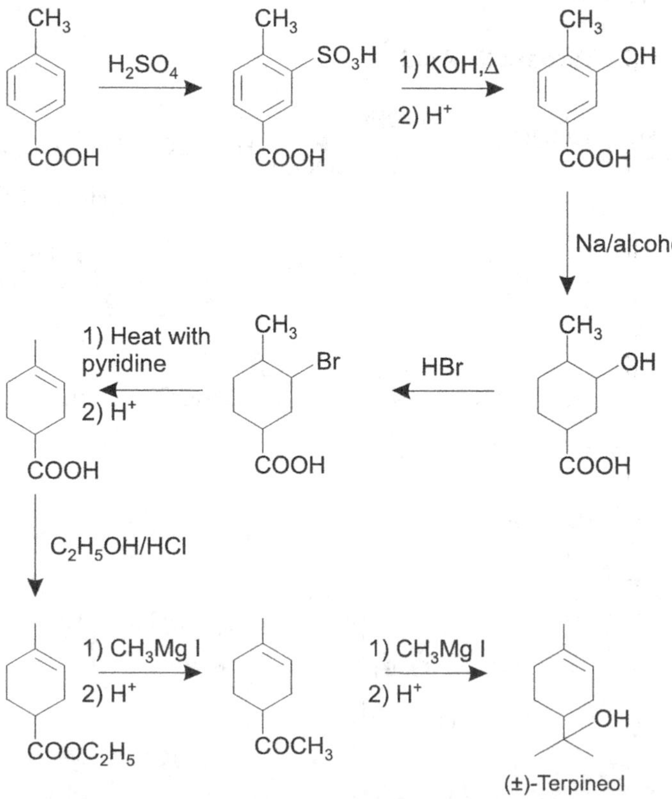

(±)-Terpineol

2. A much simpler synthesis of α-terpineol has been carried out by **Alder** and Vogt. In this isoprene and methyl vinyl acetate condense by Diels-Alder reaction to give the ketone which is converted to α-terpineol.

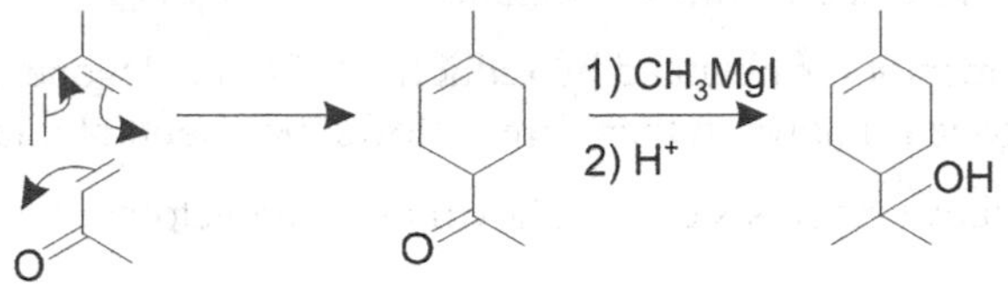

Two other terpineols β-terpineol and γ-terpineol also occur naturally.

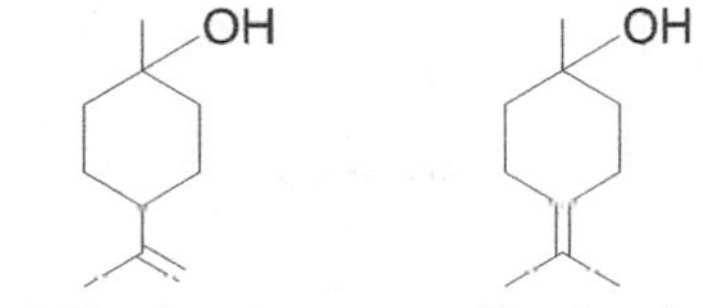

β-Terpineol   γ-Terpineol

6.10   MENTHOL

A. Introduction

Menthol occurs in the oil of peppermint to the extent of 80 percent. When this oil is cooled, crystals of menthol are obtained. Menthone obtained by the oxidation of menthol also occurs naturally in the oils of peppermint and geranium.

Menthol is an optically active compound. Only the (–) form of menthol occurs naturally. (–)-Menthol is a colourless solid, m.p. 43° C. Menthol has anaesthetic and antiseptic action. It is widely used in many pharmaceutical preparations, in face creams, shaving creams, tooth pastes and in ointments because it imparts cooling sensation to the skin. It is used in **Vicks**™ inhalers because its vapours have a cooling and mildly anaesthetic action on the inflamed nasal tissues. It is also used in chewing gums and certain brands of cigarettes.

B. Structure of Menthol

1. Menthol has the molecular formula $C_{10}H_{20}O$.

2. Menthol is a saturated compound. No addition of bromine and hydrogen is there.

3. Menthol forms esters with acids. Therefore, it must possess an alcoholic group.

4. On oxidation with chromic acid menthol is converted into menthone, which is a ketone. The alcoholic group in menthol is, therefore, *secondary*.

5. Reduction of menthol with hydrogen iodide gives *p*-menthane. Menthol has, therefore, a *p*-menthane skeleton.

6. Finally, menthol is obtained by reduction of (+) pulegone. As the structure of pulegone is known, it therefore follows that menthol must be (12).

All the above reactions can be shown as given below:

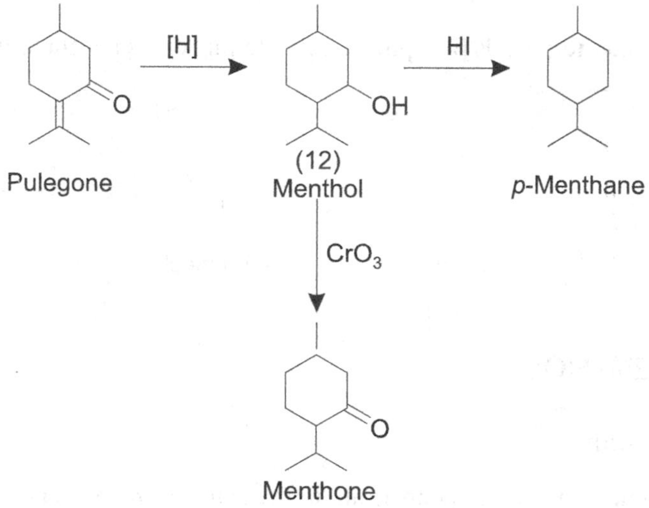

| Pulegone | (12) | *p*-Menthane |
| | Menthol | |

Menthone

7. The structure of menthol is confirmed by the oxidation products of menthone. Menthone on oxidation with $KMnO_4$ gives a keto acid, $C_{10}H_{18}O_3$ (13), which on further oxidation forms acetone and β-methyladipic acid. The reactions can be explained on the basis of the structure given above for menthone.

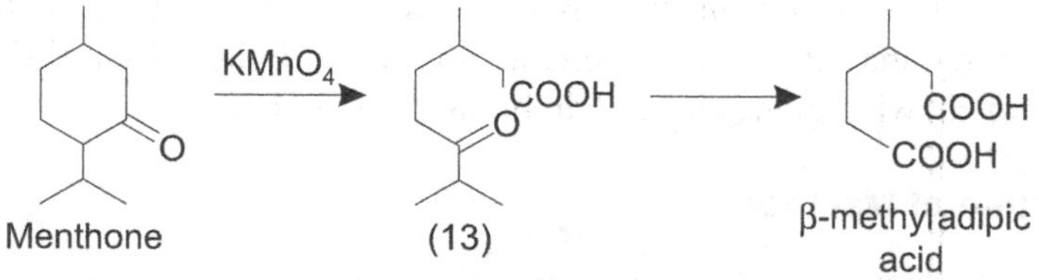

| Menthone | (13) | β-methyladipic acid |

C. Synthesis of Menthol

The structure of menthone and menthol have been confirmed by their synthesis as given below:

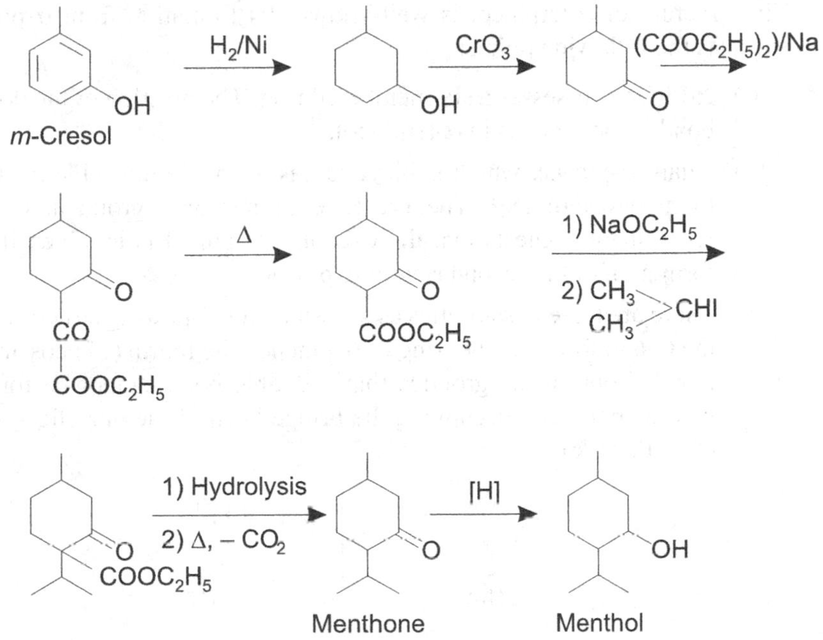

Menthone   Menthol

6.11 α-PINENE

α-Pinene is the most widely distributed terpenoid. It is present in many essential oils. The most common source of α-pinene is turpentine oil, in which it occurs in both (+) and (−) forms. α-Pinene is a liquid, b.p. 156°C. Turpentine oil, which contains a large amount of pinene, is used as a solvent for resins and as a thinner for paints and varnishes.

A. Structure of α-Pinene

1. The molecular formula of α-pinene is $C_{10}H_{16}$.

2. It adds on two bromine atoms indicating the presence of one double bond. Thus the molecular formula of the parent hydrocarbon is $C_{10}H_{18}$. This corresponds to the general formula of a bicyclic compound. α-Pinene is, therefore, bicyclic.

3. α-Pinene on treatment with alcoholic sulphuric acid gives α-terpineol.

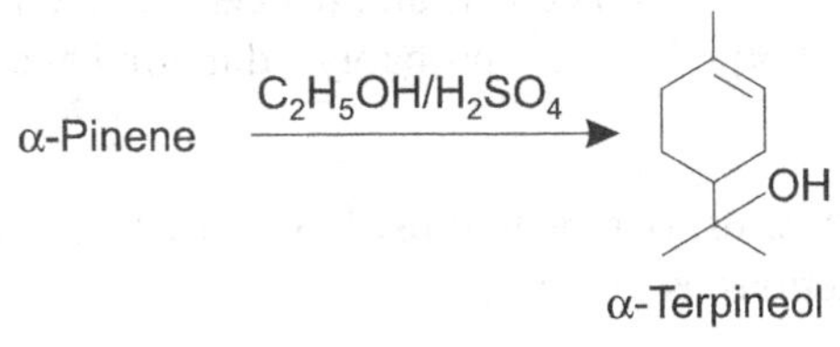

α-Terpineol

The structure of α-terpineol is well-known. Its formation from α-pinene suggests the following points:

(i) α-Pinene possesses a six-membered ring. The position of the double bond is the same as in α-terpineol.

(ii) Again α-pinene, which is bicyclic, takes up one molecule of water to give α-terpineol. The presence of hydroxyl group at C-6 in α-terpineol suggests that the C-6 of α-terpineol is involved in the formation of the second ring in α-pinene.

There are three possibilities as shown by the following structures for the C-6 to form an other ring of α-pinene. The fourth (17) possibility is ruled out on the grounds that a double bond cannot be formed by a carbon atom occupying the bridge-head of the bicyclic system (**Brett's rule**).

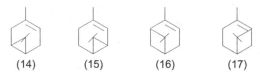

(14)   (15)   (16)   (17)

(iii) Moreover, the gem-dimethyl group of α-terpineol is not present in the six-membered ring of α-pinene and hence, it must be in the other ring.

4. The second ring was shown to be four-membered by **Baeyer** by carrying out the following oxidative degradation of α-pinene.

$$\text{α-Pinene} \xrightarrow[\text{KMnO}_4]{\text{1\% alkaline}} \text{Pinene glycol (18)} \xrightarrow[\text{KMnO}_4, \Delta]{\text{Alkaline}} \text{Pinonic acid (19)}$$

$$\text{(19)} \xrightarrow{\text{NaOBr}} \text{Pinic acid} + \text{(20)} \; \text{CHBr}_3$$

$$\text{Pinic acid} \xrightarrow[\substack{2) \text{ Ba (OH)}_2 \\ 3) \text{CrO}_3}]{1) \text{ Br}_2} \text{Cis-Norpinic acid (21)}$$

The formation of bromoform suggests that the pinonic acid has an acetyl group. The final product of oxidation, cis-norpinic acid, $C_8H_{12}O_4$, was shown to be a saturated dicarboxylic acid. Therefore, the gem – dimethyl group must be present in cis-norpinic acid. Thus, the formula for this acid may be written as $(CH_3)_2–C_4H_4–(COOH)_2$, which indicates that the saturated parent hydrocarbon of this acid is C_4H_8 and hence, the acid must be a cyclobutane derivative. So, norpinic acid is probably a dimethyl cyclobutane dicarboxylic acid.

Working backwards, α-pinene must have structure (16) which explains all the above reactions as given below:

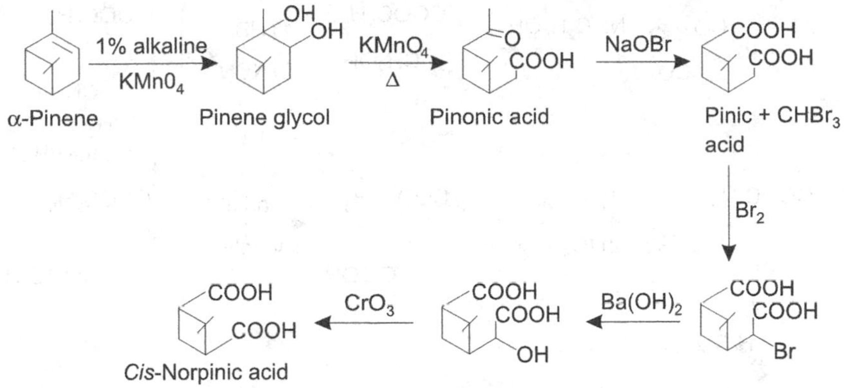

α-Pinene → (1% alkaline KMnO₄) → Pinene glycol → (KMnO₄, Δ) → Pinonic acid → (NaOBr) → Pinic acid + CHBr₃

5. **Synthesis of nor–pinic acid:** The synthesis of nor–pinic acid confirms its structure.

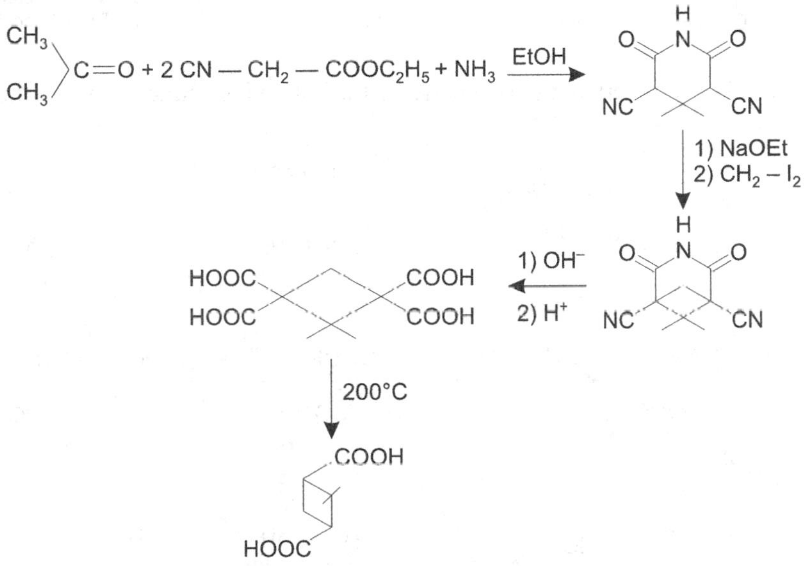

The above method gives the *trans*-isomer of nor–pinic acid. This is readily converted into the *cis*-isomer by heating the *trans*-isomer with acetic anhydride when *cis*-anhydride is formed. On hydrolysis, this gives the *cis* acid

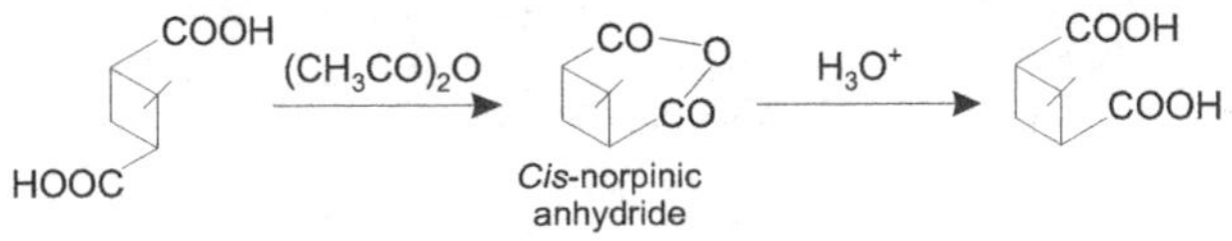

B. Synthesis of α-Pinene

Finally, the structure of α-pinene is confirmed by its synthesis from pinonic acid, which is synthesised from nor–pinic acid as given below:

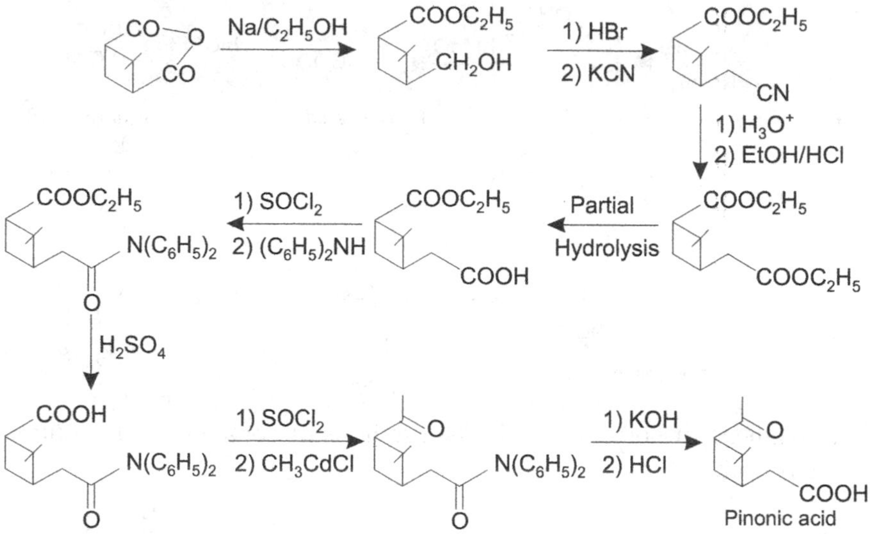

Pinonic acid is converted into α-pinene by the following sequence of reactions.

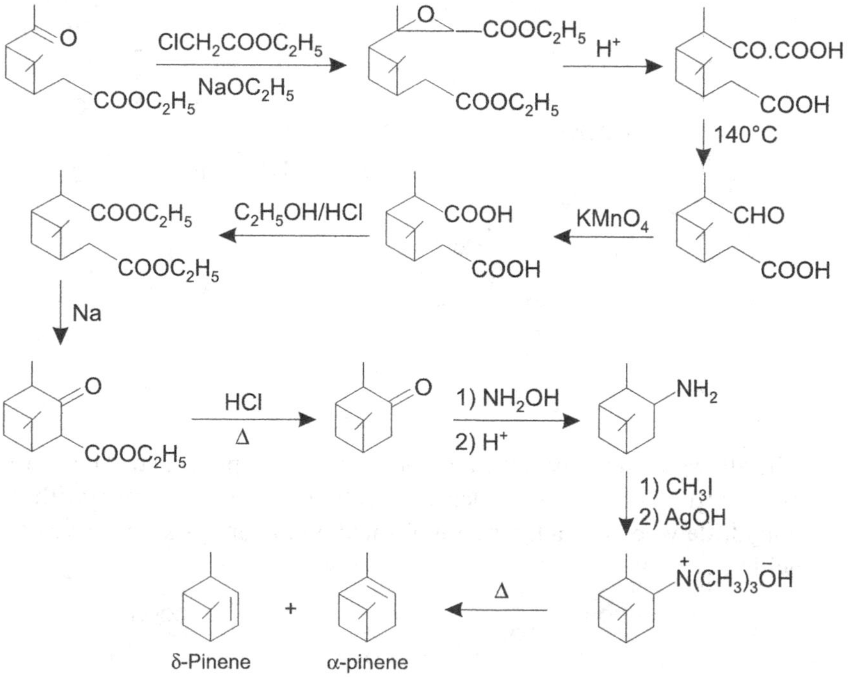

The mixture of α-pinene and δ-pinene is separated and identified.

6.12   CAMPHOR

A. Introduction

Camphor is the most important constituent of the oil of camphor. It occurs naturally in the camphor tree of **Formosa** and **Japan**. In this tree, although camphor is present in all the parts, the highest proportion is in the trunk.

Camphor is a solid, m.p. 180°C and is optically active. Both the (+) and (–) forms occur naturally. The racemic form is usually a synthetic product. Camphor is mainly used for smokeless powders and as an explosive. It is also used for medicinal purposes as a mild disinfectant and stimulant for the heart muscles.

B. Structure of Camphor

1. The molecular formula of camphor is $C_{10}H_{16}O$.

2. The general reactions of camphor show that it is a saturated molecule.

3. The oxygen atom present is in the form of a keto group. This is shown by the fact that camphor forms an oxime. However, oxidation of camphor gives a dicarboxylic acid containing 10 carbon atoms. If camphor contained an aldehyde group, oxidation would have given a mono carboxylic acid containing 10 carbon atoms.

   This means that the parent hydrocarbon of camphor has the molecular formula, $C_{10}H_{18}$ corresponding to the general formula, C_nH_{2n-2}. Camphor is, therefore, bicylic.

4. On treatment with nitrous acid, camphor forms an oxime indicating the presence of a $-CH_2CO-$ group.

5. Camphor has a p-cymene skeleton as is shown by the fact that on distillation with zinc chloride or phosphorus pentoxide it gives p-cymene.

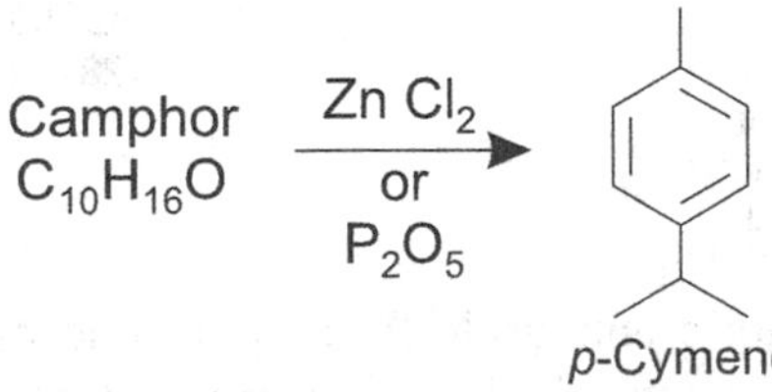

$$\text{Camphor} \quad \xrightarrow[\substack{\text{or} \\ P_2O_5}]{Zn\ Cl_2}$$

Camphor
$C_{10}H_{16}O$

p-Cymene

6. Oxidation of camphor with nitric acid gives camphoric acid ($C_{10}H_{16}O_4$) (22). Further oxidation of camphoric acid gives camphoronic acid, $C_9H_{14}O_6$ (23).

7. Structure of camphoric acid and camphoronic acid:

   (i) Various reactions of camphoric acid show that it is a dicarboxylic acid and it is saturated. As camphoric acid is formed from camphor, which means the ring containing the keto group is opened and therefore, camphoric acid must be a monocyclic compound.

(ii) Camphoronic acid (23) was shown to be a saturated tricarboxylic acid. On distillation at atmospheric pressure it gives isobutyric acid, trimethyl succinic acid, carbon dioxide and carbon. The formation of these products can only be explained from the following structure (23) for camphoronic acid.

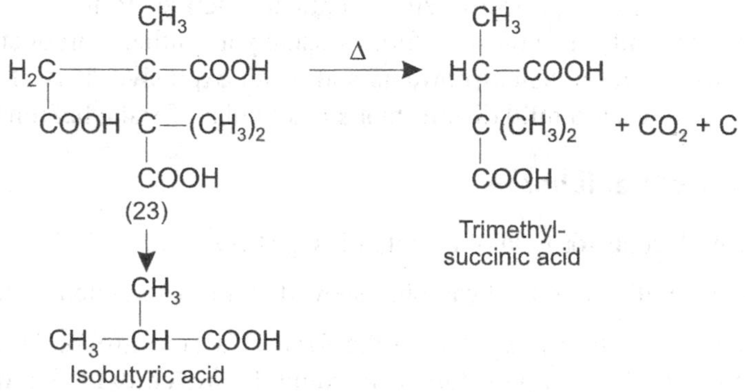

(iii) If camphoronic acid has structure (23), then camphoric acid must contain three methyl groups *i.e.*, it has the molecular formula $(CH_3)_3$–C_6H_5–$(COOH)$, which corresponds to C_5H_{10} as the parent hydrocarbon. Therefore, camphoric acid is a cyclopentane derivative. We have already seen that camphoric acid is monocyclic. Thus, oxidation of camphoric acid to camphoronic acid may be written as

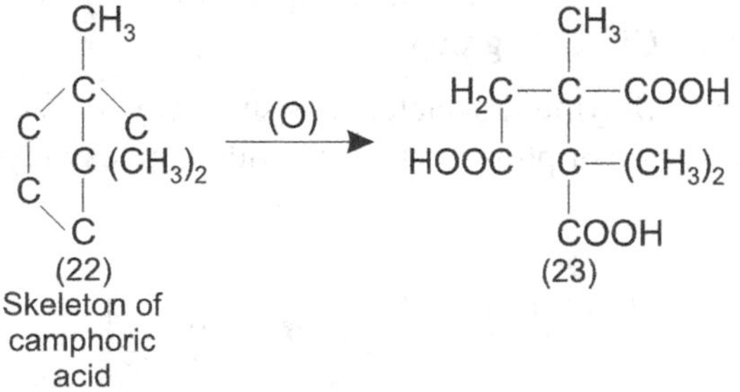

(iv) As camphoric anhydride forms only one mono bromo derivative, there is only one α-hydrogen atom in camphoric acid. Therefore, the carbon of one carboxyl group must be C1 because this is the only C atom attached to a *tertiary* carbon atom.

(v) Two –COOH groups are produced as a result of oxidation of camphor. One –COOH group comes from –CO, so the other must come from –CH_2 group. Therefore, C1 carbon must be the carbon of the keto or methylene group in camphor.

Thus, two structures (22) and (24) are possible for camphoric acid. Structure (24) suggests that camphor will contain a six-membered ring with a gem dimethyl group. The conversion of camphor into *p*-cymene cannot be explained by this structure. If we assume structure (22) for camphoric acid, then camphor would have structure (25). We can explain the above reactions *i.e.*, conversion of camphor to camphoric acid and camphoronic acid.

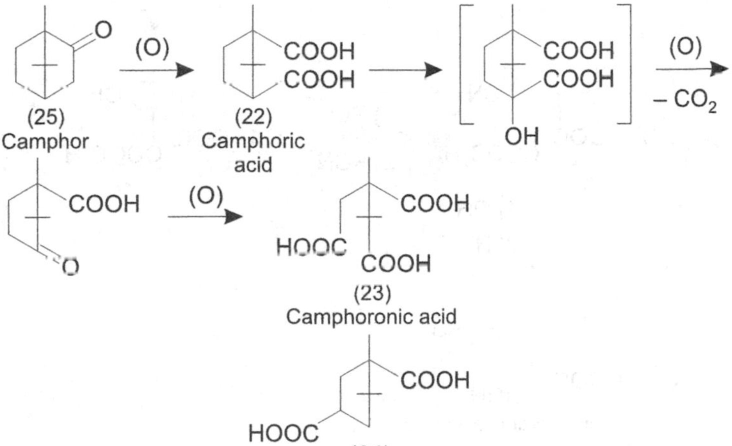

(25)
Camphor

(22)
Camphoric
acid

OH

(O)
− CO₂

(23)
Camphoronic acid

(24)

On the basis of above observations **Brett** suggested two structures for camphor

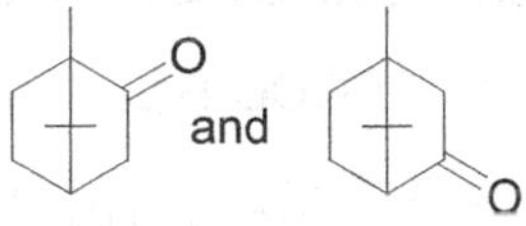

(25) and (26)

He rejected structure (26) in favour of (25) because camphor gives carvacrol (27) when distilled with iodine. The formation of this compound can be expected from (25) but not from (26)

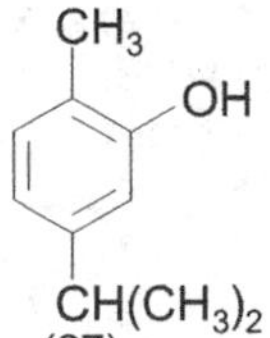

(27)

The above structure (25) for camphor was finally confirmed by the synthesis of camphoronic acid, camphoric acid and camphor.

C. Synthesis of (±) Camphoronic Acid

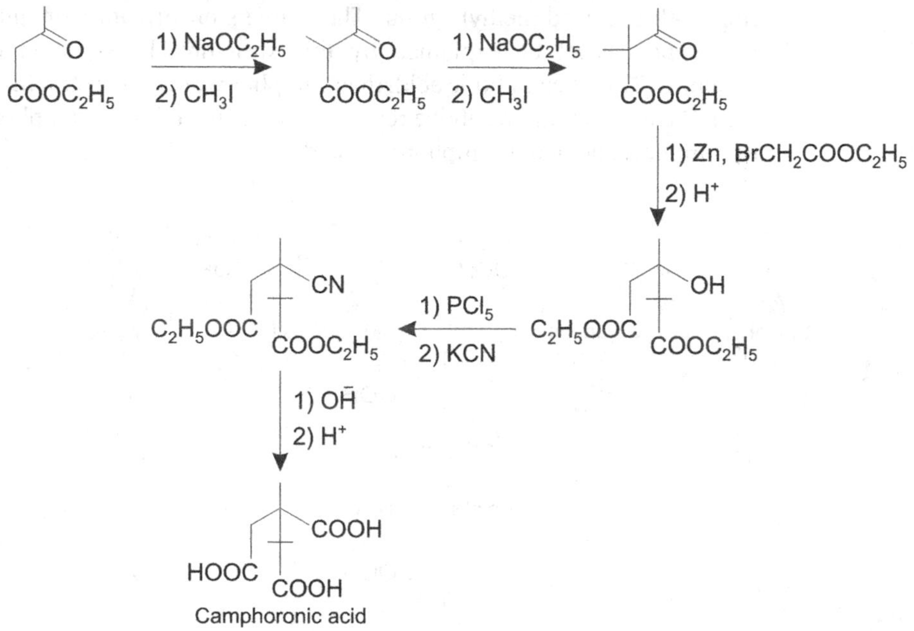

Camphoronic acid

D. Synthesis of (±) Camphoric Acid

3,3 – Dimethyl glutaric ester is synthesised first starting with mesityl oxide and diethylmalonate. The complete reactions are as follows:

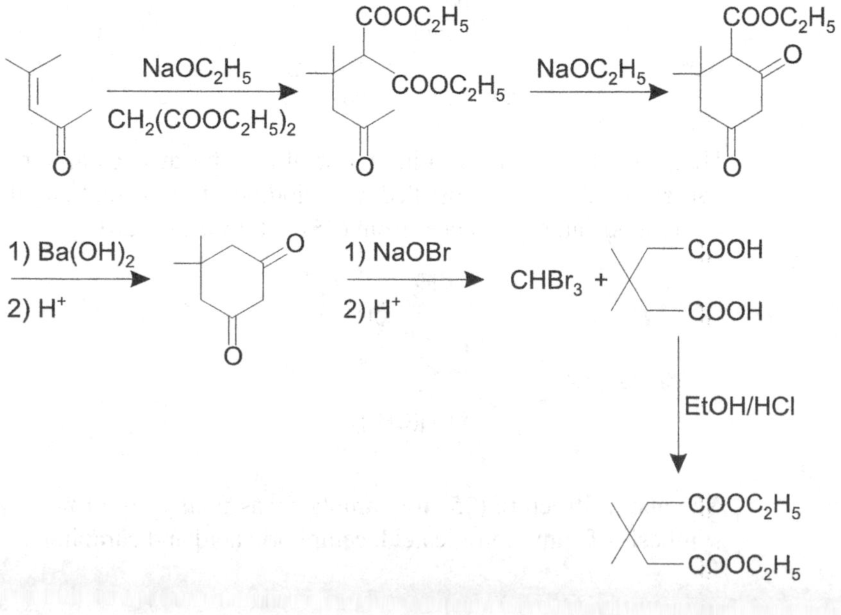

The conversion of dimethyl glutaric ester to camphoric acid involves the following reactions:

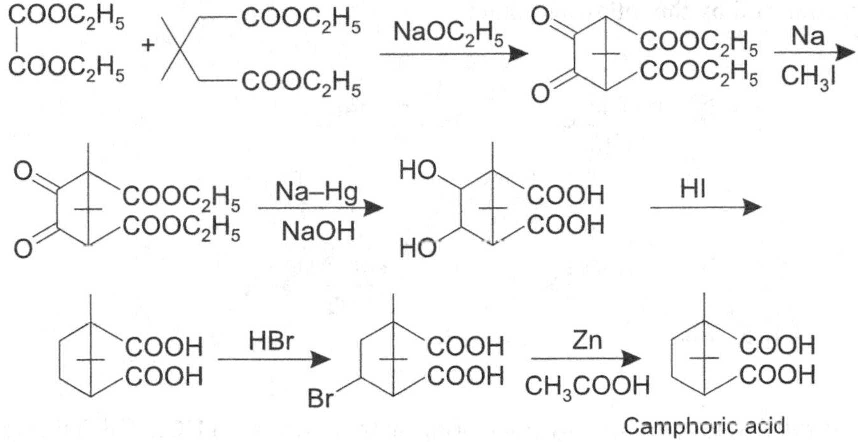

E. Synthesis of Camphor

Camphoric acid obtained above can be converted to camphor by the following reactions:

Camphoric acid

Camphoric anhydride

Homocamphoric acid

Camphor

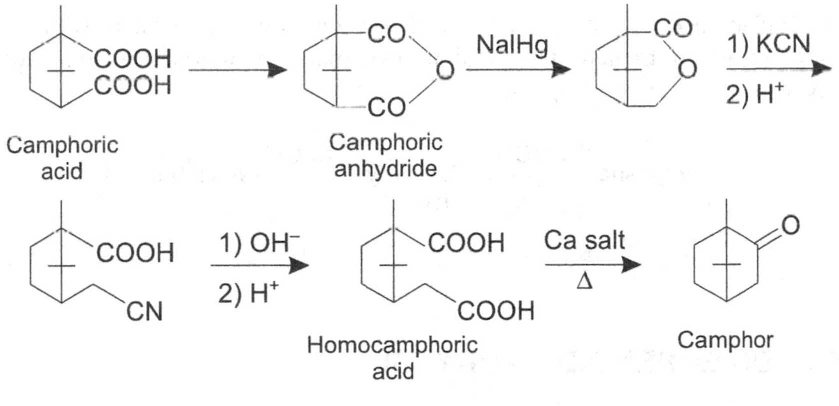

Money *et al.* synthesised (±) camphor from dihydrocarvone in just two steps.

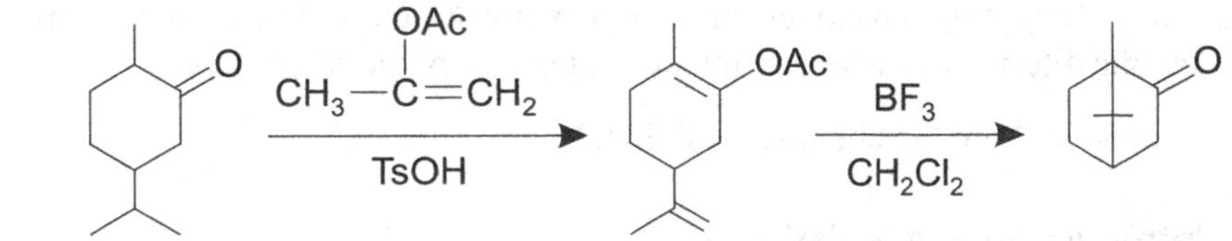

Dihydrocarvone

F. Commercial Preparation of Camphor

Camphor is prepared industrially from α-pinene which in turn is obtained from turpentine oil by the following route:

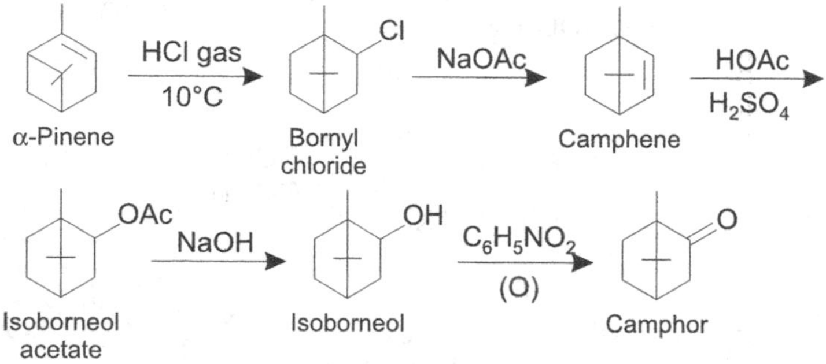

α-Pinene first gives pinene hydrochloride on treatment with HCl, which then gets converted to bornyl chloride

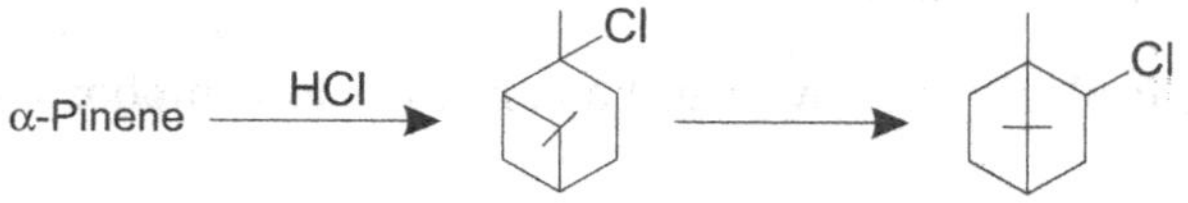

If we replace acetic acid by formic acid, we get isobornyl formate which gives isoborneol when treated with sodium hydroxide. Isoborneol is oxidised by atmospheric oxygen to camphor.

$$\text{Camphene} \xrightarrow{\text{HCOOH}} \text{Isobornyl formate} \xrightarrow{\text{NaOH}} \text{Isoborneol}$$

$$\xrightarrow{\text{O}_2} \text{Camphor}$$

6.13   DITERPENOIDS – PHYTOL

We have seen in Section 6.4 that diterpenoids have the carbon skeleton $C_{20}H_{40}O$ consisting of four isoprene units. There are many diterpenoids known to occur naturally. They may have open chain or a cyclic structure. For example, phytol is an acyclic diterpenoid whereas abietic acid is a tricyclic diterpenoid.

We will study here the chemistry of phytol.

A. Introduction and Isolation

Phytol occurs in *Lactuca sativa*, which is a well-known vegetable as well as a medicinal plant. Various classes of natural products have been isolated from *Lactuca sativa*, including phytol. The plant is famous as a folk remedy for inflammation, pain, stomach problems including indigestion and for lack of appetite. Considerable studies have been conducted to evaluate the significance of a crude extract of *Lactuca sativa*. It showed anticonvulsant, sedative, hypnotic, antioxidant, analgesic and anti-inflammatory activity.

Phytol was isolated from the ethyl acetate fraction of *L. sativa* through repeated column chromatography. However, it is produced from the hydrolysis of chlorophyll. It also forms part of the molecules of vitamins E and K.

Chemical investigation of the methanolic extract of *L. sativa* has now led to the isolation of three new sesquiterpenes together with eight known ones including phytol.

Various therapeutic activities of phytol have been reported including its activity against mycobacteria, anticonvulsant, antispasmodic and anti-cancer activities. In addition to the antimicrobial potential, phytol is also known to be one of the compounds with a highly potent anti-inflammatory property.

Phytol is a novel therapeutic agent for rheumatoid arthritis and other related chronic inflammatory diseases.

Phytol also evidently reduces oxidative stress and is widely used in perfumes, skin products, cosmetics, shampoos, soaps, food and detergents.

Phytol is commonly used as a precursor for the manufacture of synthetic forms of vitamin E and vitamin K_1.

B. Structure of Phytol

1. The molecular formula of phytol is $C_{20}H_{40}O$.
2. The reactions of phytol showed that it is a *primary* alcohol.
3. On catalytic reduction phytol forms dihydrophytol, $C_{20}H_{42}O$. Therefore, it follows that phytol contains one double bond.

   Thus, the parent hydrocarbon is $C_{20}H_{42}$ ($C_n H_{2n+2}$), so phytol is acrylic.
4. Ozonolysis of phytol gives a saturated ketone, $C_{18}H_{36}O$ (28).

   Thus, the reaction may be written as.:

$$C_{18}H_{35}CH{=}CH{-}CH_2OH \longrightarrow \underset{(28)}{C_{18}H_{36}O} + \underset{\text{Glycolaldehyde}}{\overset{\displaystyle CHO}{\underset{\displaystyle CH_2{-}OH}{|}}}$$

The formula of phytol suggested that it is composed of four reduced units of isoprene and if the units are joined head to tail, the structure of the ketone (28) would be:

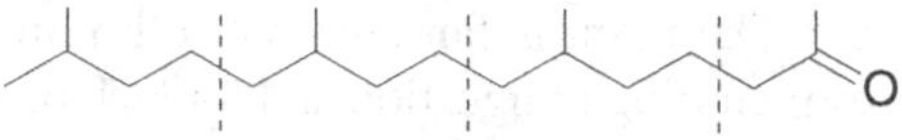
O

(28)

5. The above structure for the saturated ketone was proved to be correct by the synthesis of the ketone from farnesol

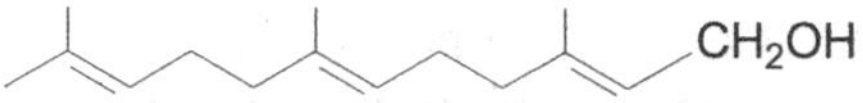
CH_2OH

Farnesol

Catalytic hydrogenation of farnesol produces hexahydrofarnesol which on treatment with phosphorus tribromide gives hexahydrofarnesyl bromide. This on treatment with sodio-acetoacetic ester followed by ketonic hydrolysis forms the saturated ketone (28)

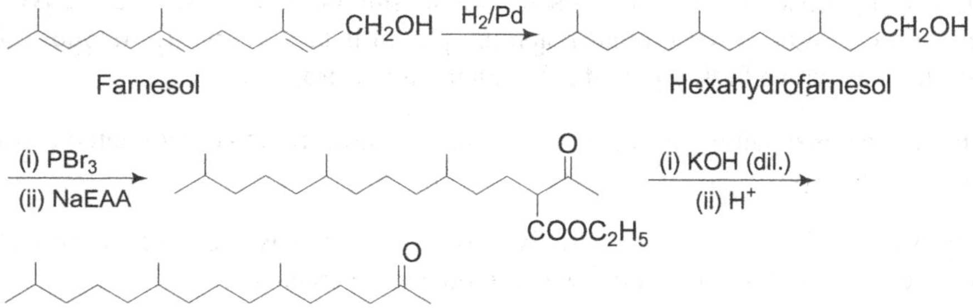
CH_2OH
H_2/Pd
CH_2OH
Farnesol
Hexahydrofarnesol
(i) PBr_3
(ii) NaEAA
O
$COOC_2H_5$
(i) KOH (dil.)
(ii) H^+
O

(28)

C. Synthesis of Phytol

The above saturated ketone (28) was converted into phytol by the following method (**F. Fischer** *et al.*) (Fig 6.1).

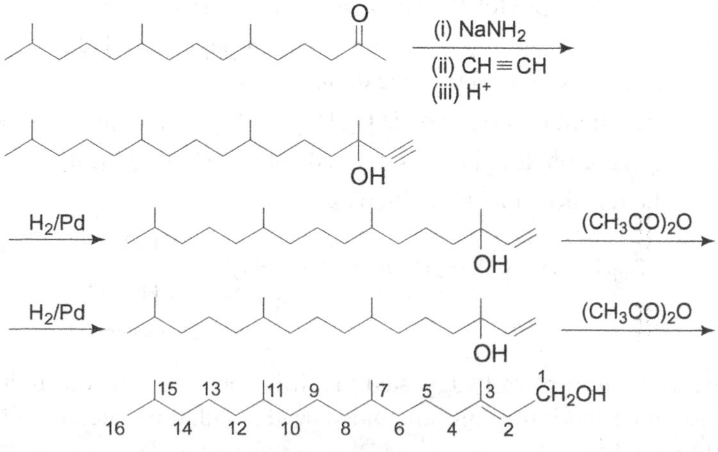
O
(i) $NaNH_2$
(ii) $CH \equiv CH$
(iii) H^+
OH
H_2/Pd
OH
$(CH_3CO)_2O$
H_2/Pd
OH
$(CH_3CO)_2O$
15 13 11 9 7 5 3 1 CH_2OH
16 14 12 10 8 6 4 2

Fig 6.1 Synthesis of phytol

The phytol molecule has got two chiral centres at C 7 and C 11. Natural phytol is weakly dextrorotatory, **Weedon** *et al.* (1959) have now synthesised the naturally occurring stereoisomer and have assigned (R)-configuration to the two chiral centres. (R)-Configuration to C7 was also given by **Djerassi** *et al.* from ORD studies. The configuration at the double bond was shown to be *trans* by NMR spectroscopy (1959).

6.14   TRITERPENOIDS – SQUALENE

A. Introduction and Importance of Squalene

Squalene is a linear triterpene synthesised in plants, animals, bacteria and fungi as a precursor for the synthesis of *secondary* metabolites such as sterols, hormones and vitamins. It is a natural organic compound originally obtained for commercial purposes primarily from shark liver oil. Now plant sources are also available for its extraction. However, olive is only used for extracting commercial squalene despite the highest content reported for amaranth. Other sources of squalene are wheat-germ oil, grapeseed oil, peanut, corn and amaranth.

Squalene is found in the human body and secreted by glands for skin protection and forms part of 10-15% of lipids on the skin surface. It is synthesised in the liver and circulates in the human blood stream.

Squalene is a precursor of various hormones in animals and sterols in plants. It has pharmaceutical, cosmetic and nutritional potential.

- It reduces skin damage by UV light.
- It reduces LDL levels and cholesterol in the body.
- It prevents the suffering of cardiovascular diseases and has antitumor and anticancer effects against ovarian, breast, lung and colon cancer.

B. Structure of Squalene

1. Squalene is a triterpenoid. It has molecular formula, $C_{30}H_{50}O$.
2. Catalytic hydrogenation of squalene using H_2/Ni gives perhydrosqualene having molecular formula $C_{30}H_{62}$.

   This suggests that squalene is acyclic and has six double bonds.

   Since squalene cannot be reduced by sodium and amyl alcohol, there are no conjugated double bonds present in the molecule.
3. Ozonolysis of squalene gives among other products, laevulic acid.

   The formation of laevulic acid indicates that the group (29) is present in squalene

(29)

4. Perhydrosqualene was found to be identical to the product obtained by subjecting hexahyderofarnesyl bromide to Wurtz reaction as given below:

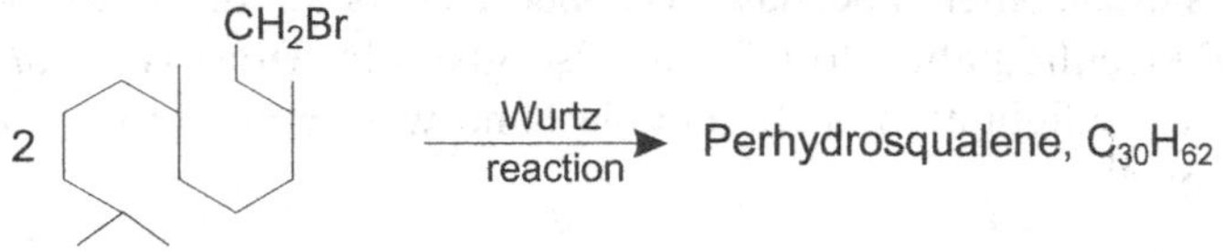

Hexahydrofarnesyl bromide

5. The structure (30) of squalene was assigned by its synthesis by **Karrer** *et al.* from farnesyl bromide by a Wurtz reaction

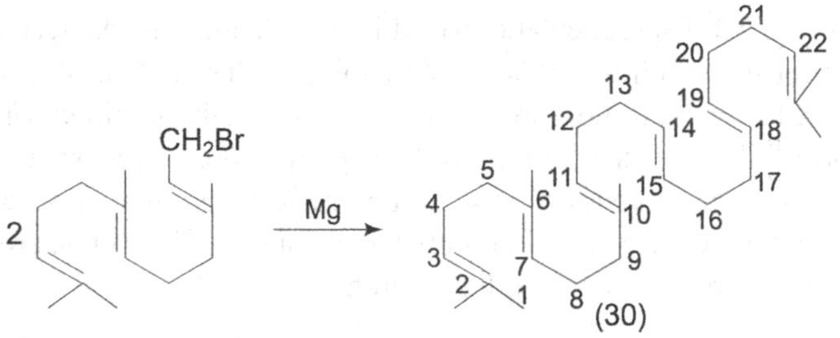

Farnesyl bromide   Squalene

From the above structure of squalene it is clear that the central portion of the squalene molecule has two isoprene units joined tail to tail.

6. Squalene forms a thiourea inclusion complex. Therefore, it was inferred that squalene is all *trans* steroisomer (**Schiesler** *et al.* 1952)

This is also supported by X-ray crystallographic studies of the thiourea inclusion complex (**Nicolaides** *et al.* 1954)

C. Synthesis of Squalene

1. Squalene has been synthesised by **Whiting** *et al.* (1958) by using Wittig reaction. They used pure *trans*-geranylacetone in the Wittig reaction and obtained a mixture of geometrical isomers of squalene. From this all *trans*-squalene was isolated *via* the thiourea complex.

This isomer was found to be identical with the natural squalene (Fig. 6.2)

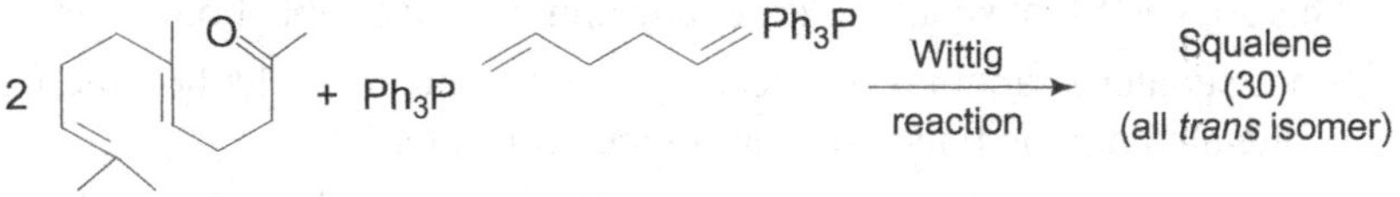

Geranylacetone

Fig. 6.2. Synthesis of squalene by Wittig-reaction

2. A highly stereoselective synthesis of all *trans* squalene was carried out by **Cornforth** *et al.* (1959) (Fig. 6.3)

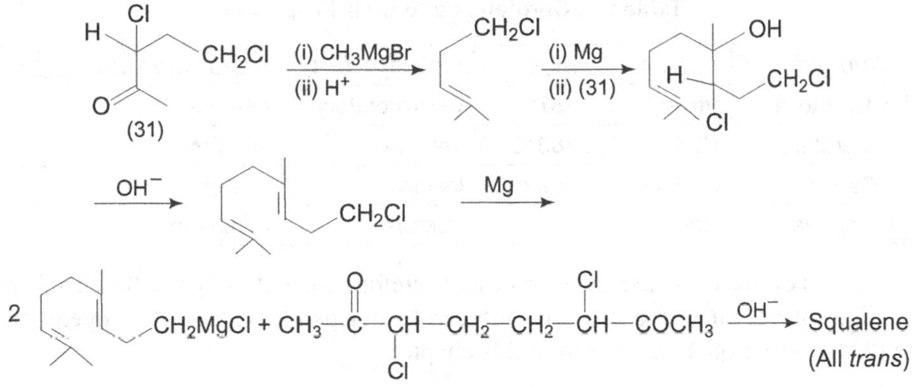

Fig. 6.3. Synthesis of squalene

Thus squalene is a linear triterpene having six double bonds at 2, 6, 10, 14, 18, 22 positions with all E-configuration.

6.15 TETRATERPENOIDS–CAROTENOIDS

The tetraterpenoids are commonly referred to as carotenoids. All the carotenoid hydrocarbons have the molecular formula as $C_{40}H_{56}$.

The carotenoids are yellow or red plant pigments which are widely distributed in plants and animals.

Chemically all carotenoids are polyenes and contain eight isoprene units. The central portion of the carotenoid molecule is composed of a long conjugated chain, which has four isoprene units, two of which are joined head to tail. The ends of the chain may be two open-chain structures or one open-chain and one ring or two rings.

As the chain of the carotenoids has extended conjugation, carotenoids, in general, are highly coloured compounds.

Generally, the carotenoids show a characteristic colour reaction. They give a deep blue colour with concentrated sulphuric acid and chloroform solution of antimony trichloride.

The carotenoids are isolated from plant sources by extraction with petroleum ether. All the carotenoid hydrocarbons are obtained in the petroleum ether fraction. They are separated by repeated column chromatography. On further extensive chromatographic adsorption, three isomeric hydrocarbons, α, β and γ-carotene were obtained.

All the three carotenes occur together in nature in varying amounts. Their relative proportion depends on the source. For example, carrots contain 15, 85 and 0.1 % of α–, β– and γ-carotene respectively.

Table 6.1 below gives the comparison of these three carotenes along with lycopene, which is also a carotenoid.

Table 6.1. Carotenes and their Properties

Carotene	Colour	m.p.	Optical activity	Petroleum ether, λ_{max}
α-Carotene	Violet	187°C	dextrorotatory	444 nm
β-Carotene	Red	183°C	inactive	460 nm
γ-Carotene	Dark red	152.4°C	inactive	462 nm
Lycopene	Red		inactive	475.5 nm

β-Carotene is the most abundant carotene. Therefore, we will study first the chemistry of β-carotene. Since the structure of γ-carotene depends on that of lycopene, we will study γ-carotene at the end of this chapter.

6.16   β-CAROTENE

A. Introduction and Importance

Plant carotenoids are the primary source of provitamin A world wide with β-carotene as the best known provitamin A carotenoid. One molecule of β-carotene can be cleaved by intestinal enzyme into two molecules of vitamin A.

The human body converts β-carotene into vitamin A (retinol) *i.e.*, β-carotene is a precursor of vitamin A. We need vitamin A for healthy skin and mucus membranes, for our immune system, good eye health and vision.

β-Carotene is found in many foods and is sold as a dietary supplement also. β-Carotene contributes to the orange colour of many different fruits and vegetables. Common sources of β-carotene are mangoes, pumpkin, papaya, carrots, sweet potatoes, apricots and dark green leafy vegetables like spinach, broccoli and peas.

β-Carotene is an antioxidant *i.e.*, it inhibits the oxidation of other molecules – it protects the body from the attack of free-radicals. Free-radicals damage cells through oxidation; β-carotene protects against free radicals and lowers the risk of developing cancer. β-Carotene is used to decrease asthma symptoms caused by exercise, heart disease, cataract and age related muscular regeneration, headache, heart burn, high blood pressure, infertility and skin disorders.

Eating more β-carotene in the diet seems to help prevent bronchitis and difficult breathing in smokers. It reduces asthma attacks, and may prevent osteoarthritis from getting worse. It may decrease sunburn in people sensitive to sun.

Isolation of β-carotene from fruits abundant in carotenoids is commonly done using column chromatography. It can be extracted from the β-carotene rich algae *Dunaliella salina*. The separation of β-carotene from the mixture of other carotenoids is based on the polarity of β-carotene. β-Carotene is a non-polar compound; it is separated with a non-polar solvent such as hexane.

Being highly conjugated, it is deeply coloured and as a hydroxyl lacking functional group it is very lipophilic.

β-Carotene is not an essential nutrient but vitamin A is. An advantage of dietary β-carotene is that the body only converts as much as it needs.

B. Structure of β-Carotene

1. The molecular formula of β-carotene is $C_{40}H_{56}$.

2. When catalytically hydrogenated β-carotene forms perhydro-β-carotene $C_{40}H_{78}$.

   Thus, β-carotene contains eleven double bonds. Since the formula of per-hydro- β-carotene corresponds to the general formula C_nH_{2n-2}, it follows that the compound contains two rings.

3. β-Carotene on aerial oxidation develops characteristic odour similar to that of β-ionone.

   It was concluded that β-ionone residue is present in β-carotene.

4. Oxidation of a benzene solution of β-carotene with cold aqueous potassium permanganate gives β-ionone. The structure of β-ionone is known as given below:

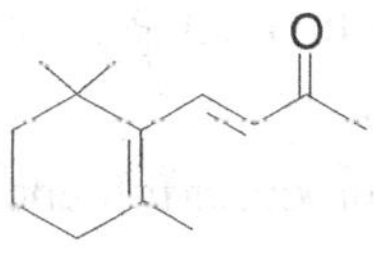

β-Ionone

5. Now β-ionone on ozonolysis gives, among other things, geronic acid

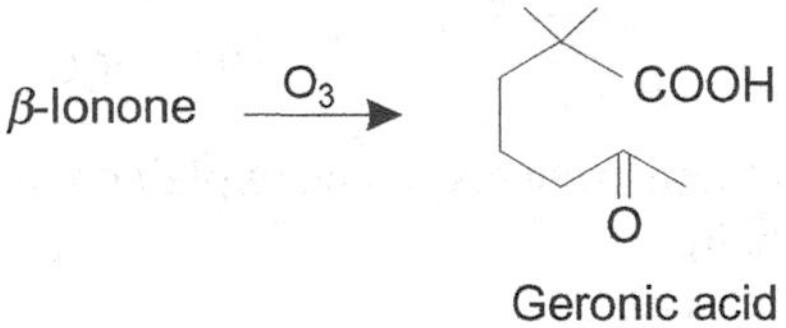

Geronic acid

6. β-Carotene also on ozonolysis gives geronic acid in an amount that corresponds to the presence of two β-ionone residues.

   Therefore, a tentative structure for β-carotene is as given below:

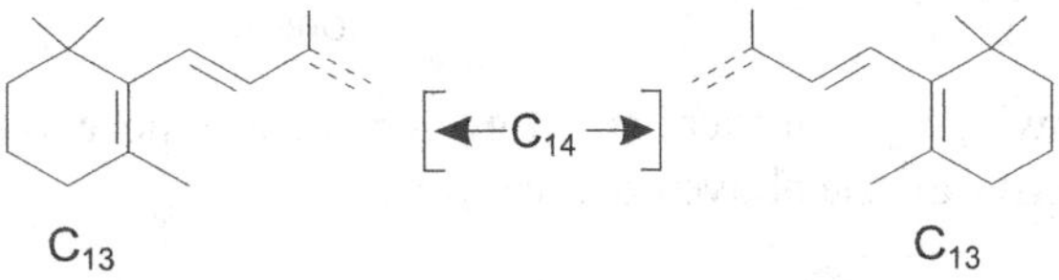

As the colour of β-carotene is due to extended conjugation, the C_{14} segment of the molecule should be conjugated. This is confirmed by the fact that β-carotene forms an adduct with five molecules of maleic anhydride.

7. Geronic acid, on oxidation with cold aqueous $KMnO_4$ gives a mixture of acetic acid, dimethylglutaric and 2, 2-dimethyl succinic acid and dimethylmalonic acid.

| | Dimethyl-
glutaric acid | Dimethyl-
succinic acid | Dimethyl-
malonic acid |

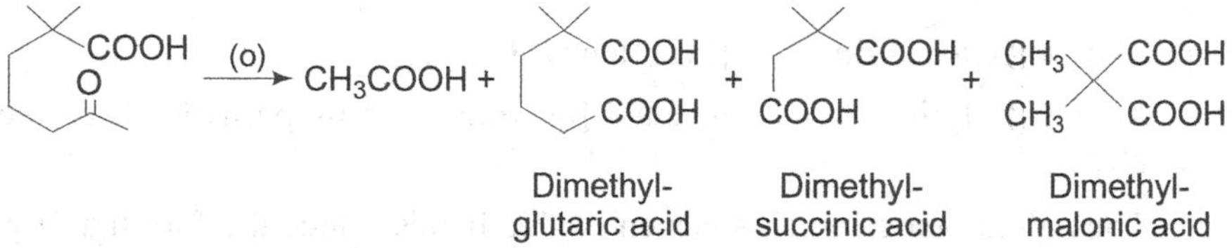

8. Oxidation of β-carotene in benzene solution under similar conditions gives β-ionone and all the above products. The amount of acetic acid was more than that expected. This suggested the presence of two ionone residues.

Thus, there must be some methyl side chains in the central C_{14} segment of the molecule.

9. Further, oxidative studies of β-carotene suggested that there are four $-C(CH_3)=$ groups in the chain. Two are already known in the two end β-ionone residues.

The problem still remains to find the positions of the remaining two.

This was done as follows:

10. Distillation of β-carotene produces toluene, m-xylene and 2,6-dimethylnaph-thalene.

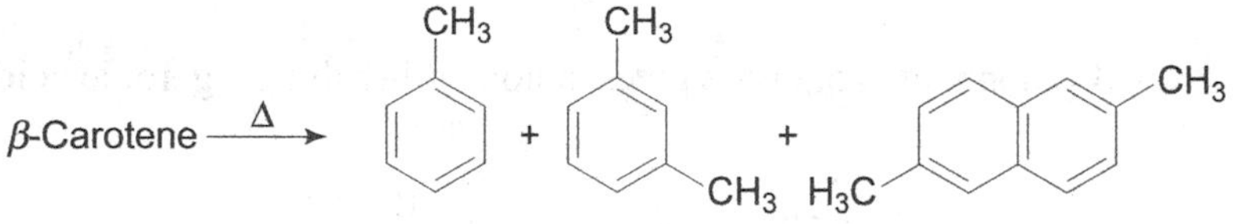

The formation of these products can be explained by cyclisation of fragments of the polyene chain.

11. Heating β-carotene in *vacuo* gave ionene.

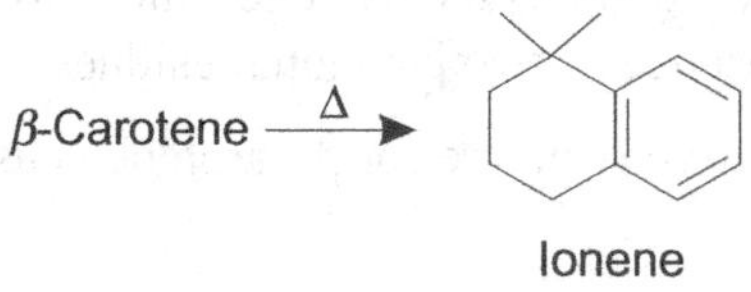

Ionene

The following symmetrical structure for β-carotene was proposed, which would satisfy all the above observations.

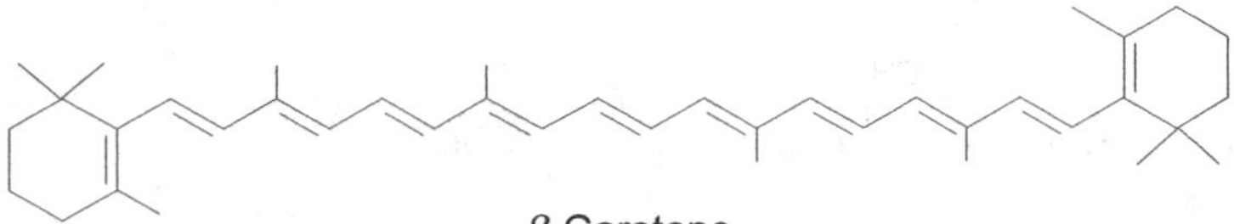

β-Carotene

The structure of β-carotene was finally confirmed by synthesis.

C. Synthesis of β-Carotene

1. The first total synthesis of β-carotene was carried out by **Karrer** *et al.* (1950). It involves the following three steps. (Fig. 6.4)

(i) Synthesis of acetylenic carbinol (31).

It was prepared by treating β-ionone with propargyl bromide in the presence of zinc (Reformastky reaction).

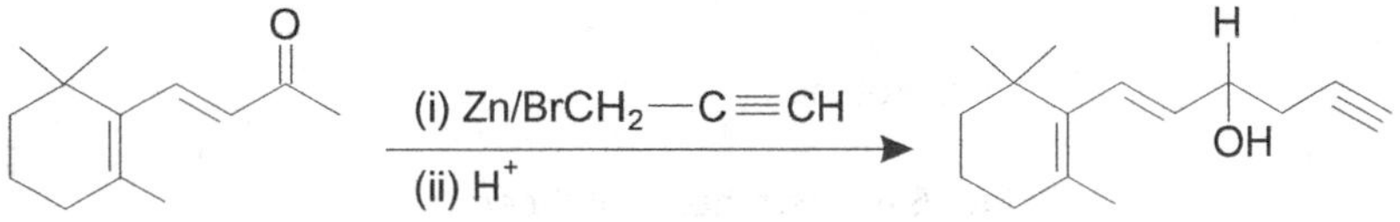

β-Ionone

(31)
Acetylenic carbinol

(ii) Synthesis of oct-4-en-2,7-dione (32) from acetoacetic acid and glyoxal.

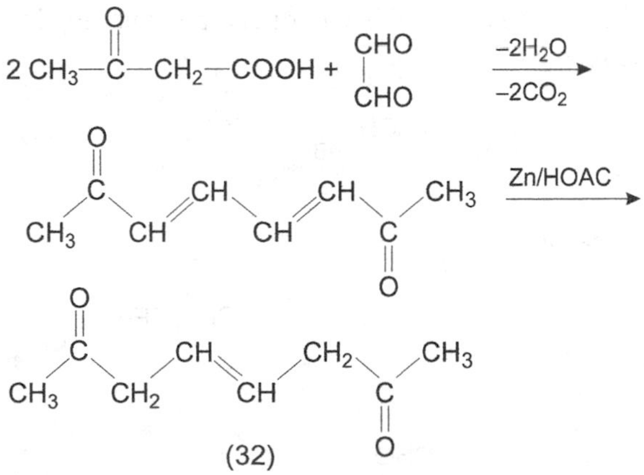

(32)

(iii) Condensation of acetylenic carbinol *via* its Grignard reagent with oct-4-en-2, 7-dione.

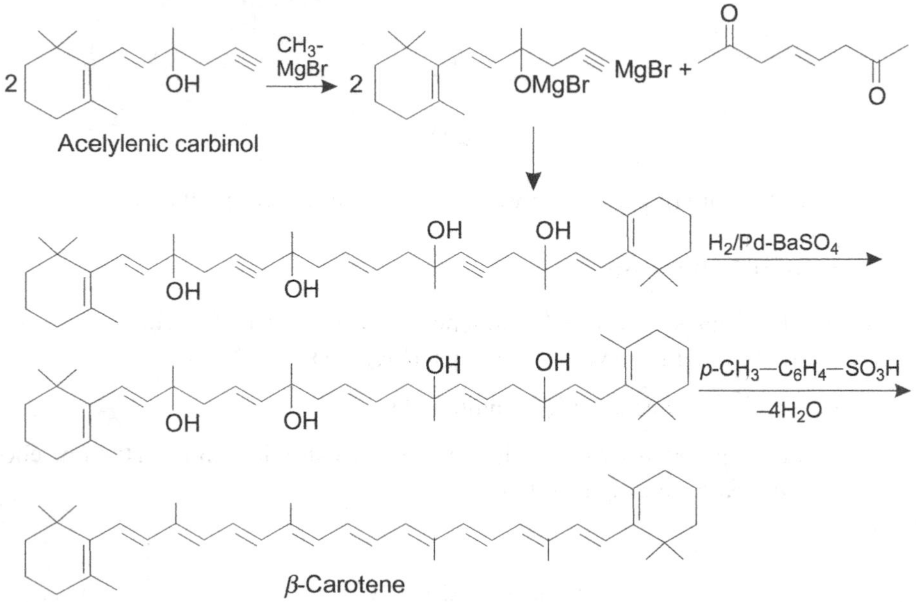

β-Carotene

Fig. 6.4. Synthesis of β-carotene.

β-Carotene obtained by the above synthesis was found to be identical with the natural β-carotene.

2. A very interesting synthesis of β-carotene was carried out by **Isler** *et al.* (1962). The starting material is vitamin A_1 and one step involves the **Wittig** reaction (Fig. 6.5)

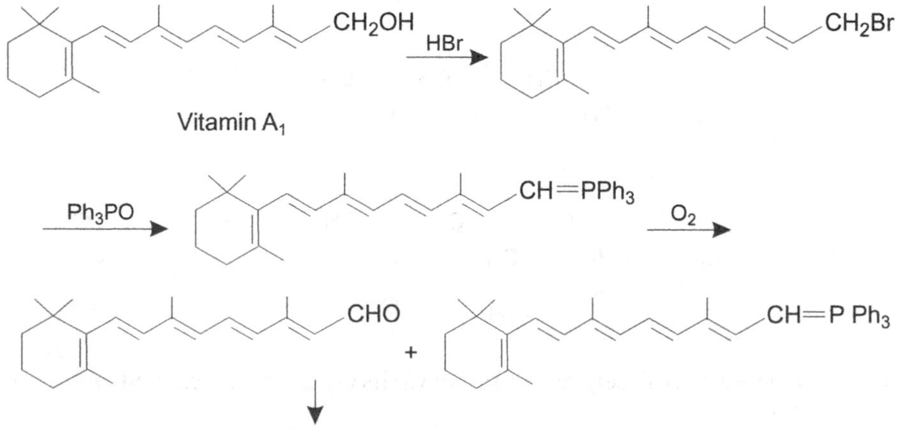

β-Carolene

Fig. 6.5. One step synthesis of β-carotene.

6.17 α-CAROTENE

A. Introduction and Importance of α-Carotene

α-Carotene is a form of carotene with a β-ionone ring at one end and an α-ionone ring at the opposite end. It is a second common form of carotene.

Leafy carotenoids are highly hydrophobic and most are yellow or orange in colour. The common sources are orange, yellow and green coloured vegetables and fruits.

Both α-and β-carotene are primarily regarded as precursors of vitamin A. Our body can convert α-and β-carotene into vitamin A for maintenance of healthy skin and bones, good vision and a robust immune system. They act as antioxidants and stop free radicals from causing cells to break. They remove destructive free radicals from the body before they cause any damage to the tissue. However, α-carotene is more effective than β-carotene as an antioxidant. α-Carotene may help to prevent cancer by stimulating cell to cell communication. High blood levels of α-carotene may reduce the risk of dying from cardiovascular disease, cancer and all other causes up to 39 percent.

α-Carotene is considered 10 times more effective than β-carotene in inhibiting the development of cancer.

B. Structure of α-Carotene

1. α-Carotene is isomeric with β-carotene having the same molecular formula as $C_{40}H_{56}$.
2. Oxidation experiments on α-carotene led to results similar to those obtained for β-carotene.

   However, oxidation of α-carotene with cold aqueous $KMnO_4$ gives isogeronic acid along with geronic acid.

   We know isogeronic acid is the oxidation product of α-ionone; therefore, it is concluded that α-carotene contains one β-ionone ring and one α-ionone ring.

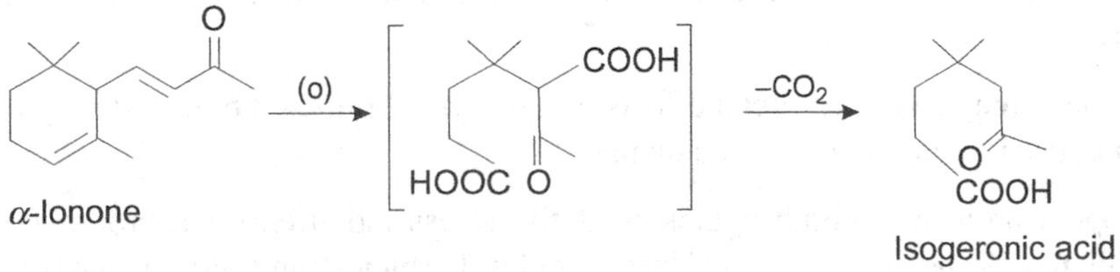

α-Ionone   Isogeronic acid

Thus, the structure of α-carotene is

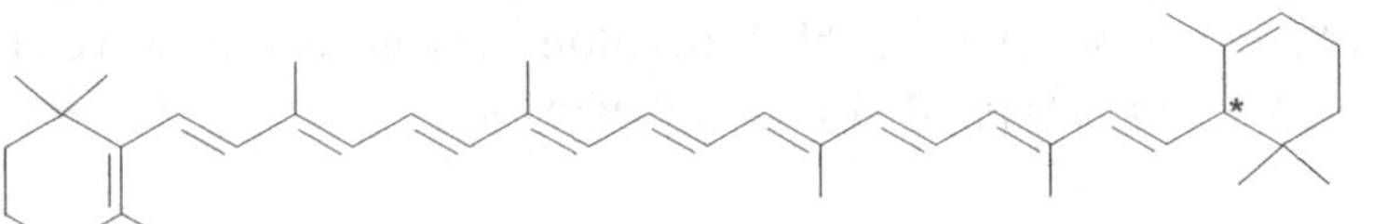

α-Carotene

As we know α-carotene is optically active and this is due to the presence of a chiral centre as marked in its structure. The optical activity is due to a chiral carbon marked in the α-ionone ring. Therefore, α-and β-carotene are geometrical isomers.

The above structure for α-carotene has been confirmed by synthesis.

C. Synthesis of α-Carotene

The synthesis of α-carotene was carried by **Karrer** *et al*. The method is the same as that described for β-carotene except that one molecule of the acetylenic alcohol (31) is used together with one molecule of the corresponding α-ionone derivative (32)

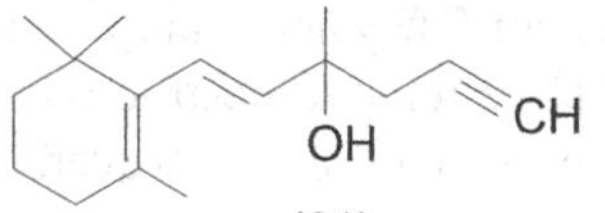

(31) 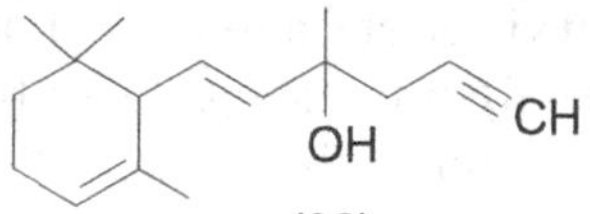   (32)

6.18 LYCOPENE

A. Introduction and Importance of Lycopene

Lycopene is a naturally occurring chemical that gives fruits and vegetables a red colour. It is found in tomatoes, watermelons, red carrots, papaya, apricots and guava.

Lycopene is a powerful antioxidant in the carotenoid family. Antioxidants protect our body from damage caused by free-radicals. When free radical levels outnumber the antioxidant levels, they can create oxidative stress in our body. This stress is linked to certain chronic diseases such as cancer, diabetes, heart disease and Alzheimer's disease.

Lycopene lowers the risk of developing pancreatic cancer, helps reduce exercise related asthma attacks and helps in protecting the skin from sunburn.

Lycopene is known to prevent cell damage and helps in the production of healthy cells. It lowers the levels of 'bad cholesterol', LDL and increases good cholesterol, HDL.

Lycopene may have beneficial effects on our eyes, brain and bones. It may reduce pain caused by nerve and tissue damage.

Lycopene protects our body against oxidative stress and offers some protection from certain toxins and damage caused by pesticides, herbicides and certain types of fungi.

People take lycopene for preventing heart disease, the hardening of arteries and cancer of prostate, breast, lungs, bladder, colon and pancreas. However, there is no scientific evidence to support many of these uses.

B. Structure of Lycopene

1. Lycopene also has the same molecular formula $C_{40}H_{56}$, as that of α- or β-carotene.

2. On catalytic hydrogenation using H_2/Pt, lycopene is converted into perhydro lycopene, $C_{40}H_{82}$.

   Therefore, lycopene has thirteen double bonds and is an acyclic compound.

$$C_{40}H_{56} \xrightarrow{\quad H_2/Pt \quad} C_{40}H_{82}$$

$$(C_nH_{2n+2})$$

3. Ozonolysis of lycopene gives acetone and laevulic acid.

   This suggests that lycopene has the following terminal group (33)

(33)

The presence of the above group is supported by the fact that controlled oxidation of lycopene with CrO_3 gives methyl heptenone *i.e.*, 6-methylhept-5-en-2-one (34)

Lycopene + CrO_3 $\longrightarrow$

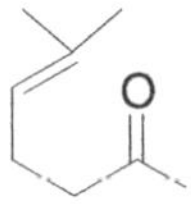

(34)
Methylheptenone

Further, estimation of the products from ozonolysis suggested that the above grouping is present at each end of molecule.

4. Quantitative oxidation of lycopene with CrO_3 gives eight molecules of acetic acid per molecule of lycopene.

   This suggests that there are six $-C(CH_3) =$ groups present in the chain.

5. Controlled oxidation of lycopene with CrO_3 gives one molecule of methyl-heptenone (34) and one molecule of lycopenal ($C_{32}H_{42}O$).

   The latter is further oxidised to give another molecule of methylheptenone and one molecule of the dialdehyde ($C_{24}H_{28}O_2$).

   It means that this dialdehyde is present in the central part of the chain and the two molecules of methylheptenone must have come by the oxidation of each end of the chain in lycopene.

6. The dialdehyde is converted into the corresponding dioxime and this on dehydration to the dicyanide followed by hydrolysis forms the dicarboxylic acid, $C_{24}H_{28}O_4$, which is identical with norbixin (36). Hence the dialdehyde is bixindial (35).

All the above observations suggest that the structure of lycopene is symmetrical as shown below and all the above facts can be explained (Fig. 6.6)

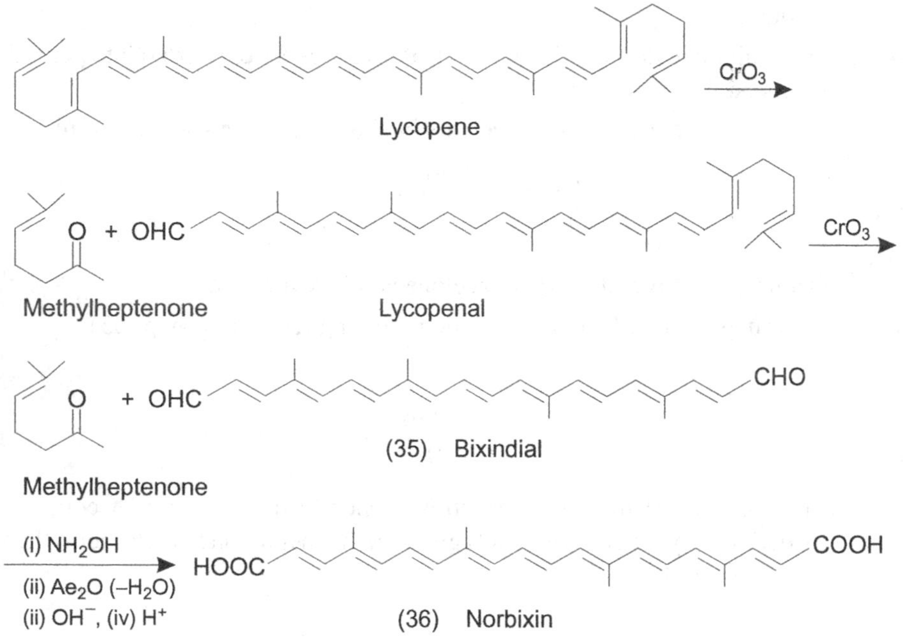

Lycopene $\xrightarrow{CrO_3}$

Methylheptenone + OHC — Lycopenal $\xrightarrow{CrO_3}$

Methylheptenone + OHC — CHO

(35)   Bixindial

$\xrightarrow[\substack{\text{(ii) Ae}_2\text{O (–H}_2\text{O)} \\ \text{(iii) OH}^-, \text{(iv) H}^+}]{\text{(i) NH}_2\text{OH}}$ HOOC — COOH

(36)   Norbixin

Fig. 6.6.

The structure of lycopene has been confirmed by its synthesis.

C. Synthesis of Lycopene

Weedon *et al.* have synthesised lycopene by using the Wittig reaction (Fig. 6.7)

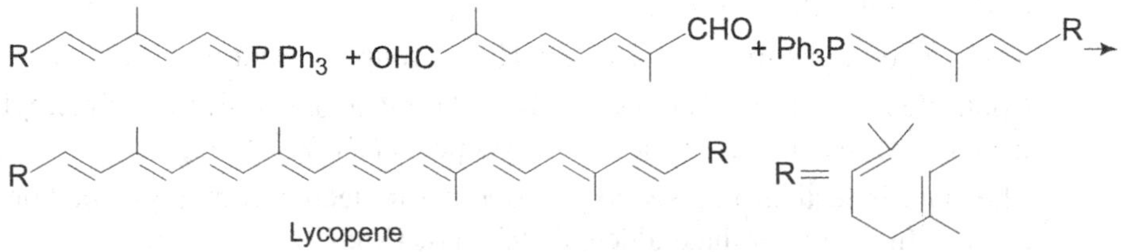

R — P Ph$_3$ + OHC — CHO + Ph$_3$P — R →

R — R   R=

Lycopene

Fig. 6.7.   Synthesis of Lycopene

6.19   γ-CAROTENE

We have seen that γ-carotene occurs in nature to a very small extent with α- and β-carotene. It is isolated from the mixture by column chromatography.

A. Structure of γ-Carotene

The structure of γ-carotene has been established on the basis of degradation studies.

1. The molecular formula of γ-carotene is $C_{40}H_{56}$. *i.e.*, it is isomeric with α- and β-carotene.

2. Catalytic hydrogenation converts γ-carotene into perhydro-γ-carotene $C_{40}H_{80}$.

   Thus, there are twelve double bonds present and the compound contains one ring.

3. Ozonolysis of γ-carotene gives among other products, acetone, laevulic acid and geronic acid

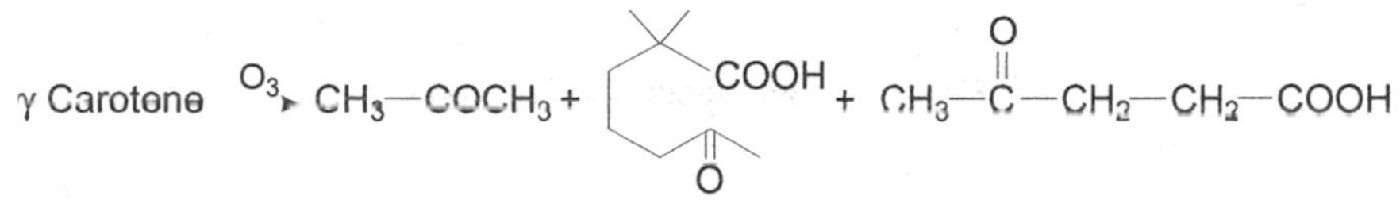

$$\gamma\ \text{Carotene} \xrightarrow{O_3} CH_3-COCH_3 + \quad + \quad CH_3-\overset{O}{\overset{\|}{C}}-CH_2-CH_2-COOH$$

Geronic acid   Laevulic acid

The formation of acetone and laevulic acid indicates the structural relationship of γ-carotene of lycopene and the formation of geronic acid indicates the presence of a β-ionone ring.

4. A growth promoting response in rats of γ-carotene was found to be half that of β-carotene. Therefore, it was suggested by **Kuhn** that γ-carotene consists of half a molecule of β-carotene joined to half a molecule of lycopene. Therefore, the structure of γ-carotene was suggested as follows:

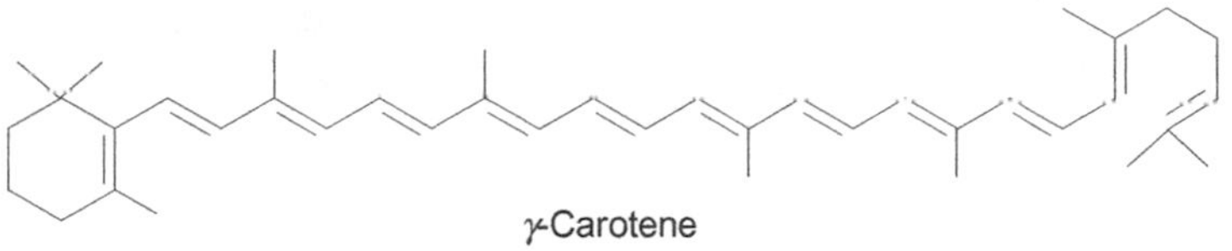

γ-Carotene

The above structure for γ-carotene is supported by the fact that the absorption maximum was found to be in between that of β-carotene and that of lycopene.

The structure was finally confirmed by the synthesis of γ-carotene.

B. Synthesis of γ-Carotene

Karrer *et al.* (1953) carried out the synthesis of γ-carotene using the combination $(C_{16} + C_8 + C_{15})$. The acetylene carbinol was treated with ethyl-magnesium bromide and the product was then treated as shown below. (Fig. 6.8)

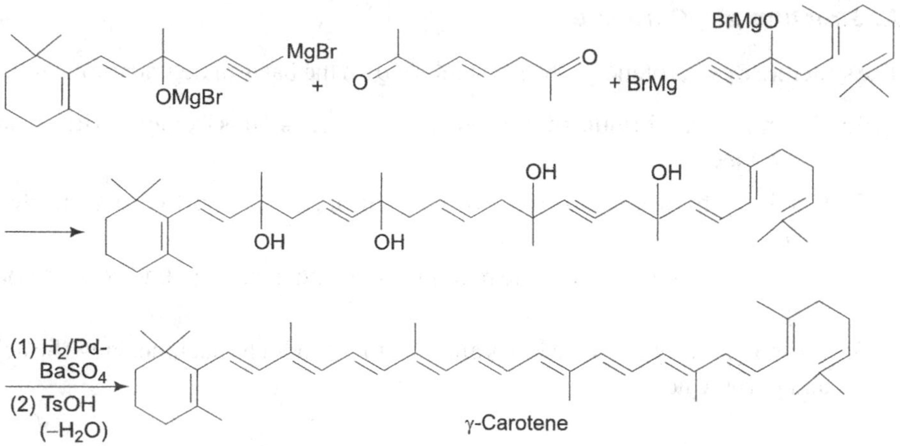

Fig. 6.8.   Synthesis of γ-Carotene.

Alkaloids

In this chapter we will study

- Occurrence of Alkaloids in Nature—Extraction from plants and their general properties.
- Hofmann's Exhaustive Methylation—Emde modification.
- Classification, Importance and Physiological Action of Alkaloids.
- Coniine—Isolation, Structure and Synthesis.
- Nicotine—Properties, Structure, Synthesis and Medicinal Importance.
- Hygrine—Structure and Medicinal Importance.
- Atropine—Isolation, Structure and Synthesis.
- Cocaine—Structure and Medicinal Importance.
- Quinine – Structure, Synthesis and Medicinal Importance.
- Morphine, Structure, Medicinal Importance.
- Medicinal Importance of Reserpine.

7.1 INTRODUCTION

In the earlier chapters, we discussed various types of organic compounds. Alkaloids are a special type of compounds having varied structures and possessing physiological activity. However, they are all basic in nature. The term alkaloid, meaning alkali like, was originally given to all basic nitrogen containing compounds of plant origin. '**Ladenberg**' defined alkaloids as all naturally occurring compounds of plant origin having a basic character and containing at least one nitrogen atom in a heterocyclic ring system. But we know that all alkaloids have strong physiological activity.

Therefore, alkaloids can be defined as compounds of plant origin, which have one or more nitrogen atoms in heterocyclic ring systems possessing physiological action.

However, the above definition is incomplete because

1. It does not contain alkaloids of animal origin and synthetic analogues. For example, ephedrine (4) and papaverine (16) are synthesised commercially.

2. Piperine (6), the alkaloid of pepper has little physiological activity and it is not basic in nature.

3. Some alkaloids, which have strong physiological action, do not contain a nitrogen atom in the ring, *e.g.,* ephedrine is a straight chain alkaloid. Its action is similar to that of adrenaline (5), which also does not have an N atom in the ring.

7.2   OCCURRENCE IN NATURE

There are more than two thousand alkaloids known to mankind. They mostly occur in the plants of dicotyledon families. Alkaloids are rarely found in lower plants. They are usually found in the seeds, root, leaves or bark of the plant.

Alkaloids generally occur as salts of various common plant acids such as acetic, oxalic, citric, malic, tartaric acids *etc.* Closely related alkaloids generally occur in the same plant; *e.g.,* nearly twenty alkaloids have been isolated from opium. However, simple alkaloids occur in various different unrelated plants while complex alkaloids such as quinine, nicotine, colchicine *etc.* generally occur in only one species or genus of a family.

7.3   EXTRACTION OF ALKALOIDS FROM PLANTS

The plant part (leaves, bark *etc.*) is dried and powdered. It is first extracted with petroleum ether to remove soluble fats. The plant material is then extracted with methanol. The solvent is distilled off and the residue is treated with dilute inorganic acids (HCl or H_2SO_4). The free bases form soluble salts. The acid solution of the alkaloids is then treated with sodium carbonate to liberate free bases and extracted with organic solvents like ether or chloroform. The solvent is removed by distillation to get a crude mixture of alkaloids. The mixture is fractionated into individual pure alkaloids by using usual techniques of separation. Most recent methods of separation involve the use of column chromatography using different adsorbents and ion-exchange methods. By using ion-exchange methods alkaloids are obtained in a high state of purity.

The basic scheme of extraction of alkaloids is given in Fig. 7.1.

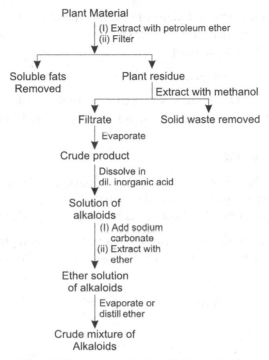

Fig. 7.1 Extraction of alkaloids from plants.

7.4 GENERAL PROPERTIES OF ALKALOIDS

Alkaloids are generally colourless, crystalline, non-volatile solids. Only a few are liquids such as coniine (8) and nicotine (11). A few alkaloids are coloured *e.g.*, berberine is yellow. Most alkaloids are insoluble in water except coniine and nicotine, which are soluble in water. But alkaloids are generally soluble in most organic solvents.

Most of the alkaloids have a bitter taste and are optically active. They have, generally, a marked physiological activity. Only a few are optically inactive *e.g.*, papaverine (16). Coniine is dextrorotatory. However, all other well-known alkaloids are laevorotatory.

All the alkaloids are basic in nature and contain one or two nitrogen atoms present in the *tertiary* state in a ring system. Most of the alkaloids also contain oxygen in the form of common functional groups.

The alkaloids are precipitated out with common alkaloidal reagents such as phosphotungstic acid, phosphomolybdic acid, picric acid, potassium mercuri-iodide *etc.* Some of these precipitates have definite crystalline shapes and may be used in the identification of an alkaloid. These reagents are also used as spray reagents in paper and thin layer chromatography for detecting alkaloids.

7.5 HOFMANN'S EXHAUSTIVE METHYLATION

This is the most important technique of studying the ring size of an alkaloid. The heterocyclic ring is opened. Nitrogen gets removed to give an aliphatic compound, which can be easily characterised.

The general procedure adopted is as follows:

1. Hydrogenate the heterocyclic ring (if it is unsaturated).
2. Convert this into quaternary methyl ammonium hydroxide by exhaustive methylation.
3. Heat the hydroxide.

We already know that when a quaternary ammonium hydroxide is heated, a molecule of water is eliminated with the loss of $\overline{OH}$ with the hydrogen atom from the β-carbon atom and *cleavage* of the C–N bond takes place to give an alkene.

The quaternary ammonium hydroxide is obtained by the complete methylation of the amine followed by hydrolysis with moist silver oxide. For example,

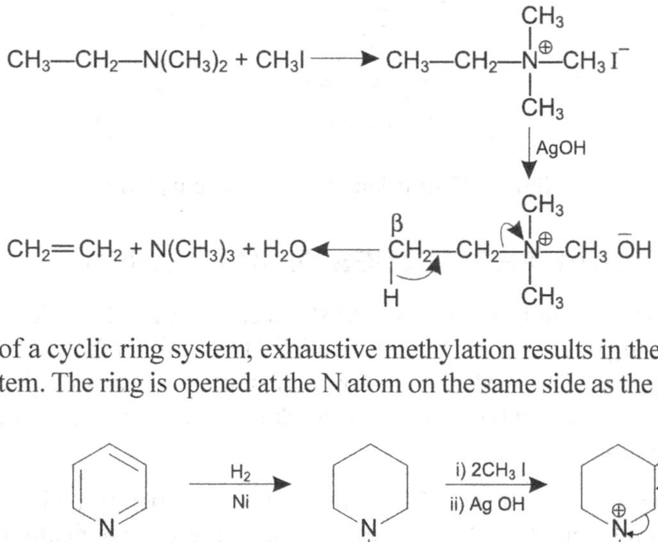

$$CH_3—CH_2—N(CH_3)_2 + CH_3I \longrightarrow CH_3—CH_2—\overset{\overset{\displaystyle CH_3}{|}}{\underset{\underset{\displaystyle CH_3}{|}}{\overset{\oplus}{N}}}—CH_3\ I^-$$

$$\downarrow AgOH$$

$$CH_2{=\!=}CH_2 + N(CH_3)_3 + H_2O \longleftarrow \underset{\underset{\displaystyle H}{|}}{\overset{\displaystyle \beta}{CH_2}}—CH_2—\overset{\overset{\displaystyle CH_3}{|}}{\underset{\underset{\displaystyle CH_3}{|}}{\overset{\oplus}{N}}}—CH_3\ \overline{O}H$$

In the case of a cyclic ring system, exhaustive methylation results in the opening of the ring system. The ring is opened at the N atom on the same side as the β-hydrogen atom.

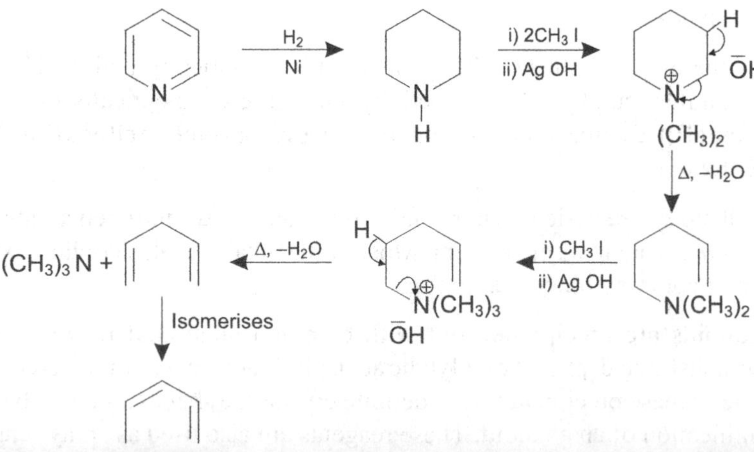

Piperylene

The process of exhaustive methylation is repeated on the product. The result is complete removal of the N atom and an unsaturated compound is obtained; *e.g.*, pyridine on exhaustive methylation gives piperylene as given above:

In this procedure, the quaternary hydroxide is heated at 200°C. The reaction may also be carried out by refluxing an aqueous or alcoholic solution of potassium hydroxide containing the methiodide.

In the above scheme of reactions, elimination of a water molecule involves the removal of a hydrogen atom from the β-position. If β-hydrogen is absent, the exhaustive methylation method fails; *e.g.*, isoquinoline is degraded only one time. It is not possible to remove the nitrogen atom from the product.

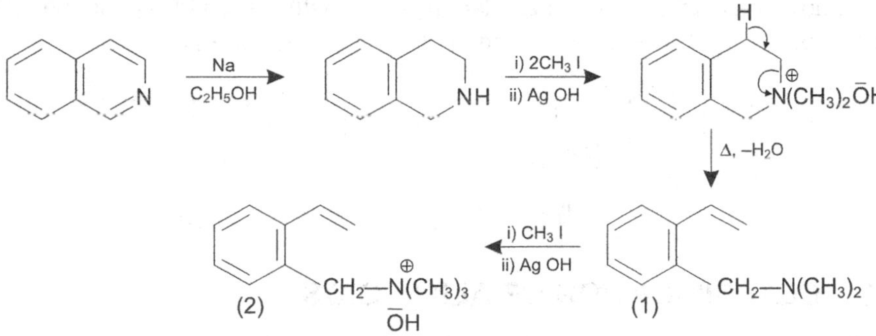

In this compound (1), there is no β-hydrogen atom, so it does not undergo Hofmann's degradation reaction.

In certain cases, even if the compound contains a β-hydrogen atom Hofmann's methylation may fail, *e.g.*, tetrahydroquinoline gives *N*-methyl-tetrahydroquinoline.

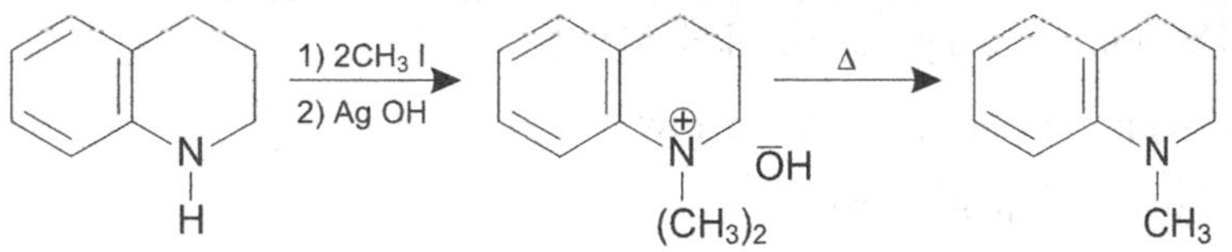

7.6   EMDE MODIFICATION

In this method, the quaternary ammonium salt is reduced with sodium-amalgam in aqueous ethanol or with sodium in liquid ammonia or is catalytically hydrogenated. For example, the product (2) obtained by exhaustive methylation of isoquinoline gives (3) by following the above procedure.

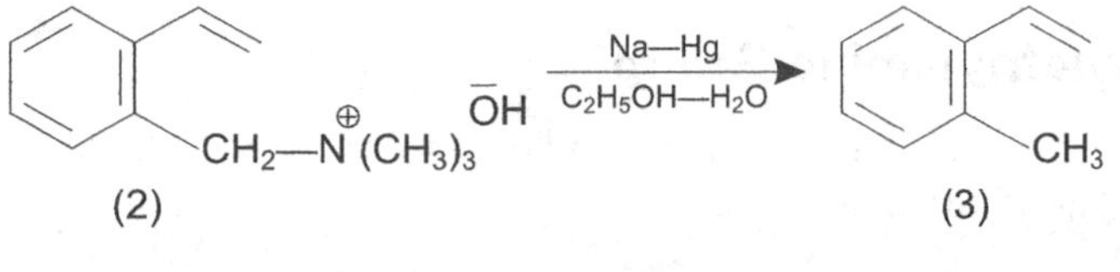

The Emde degradation can also be carried out directly.

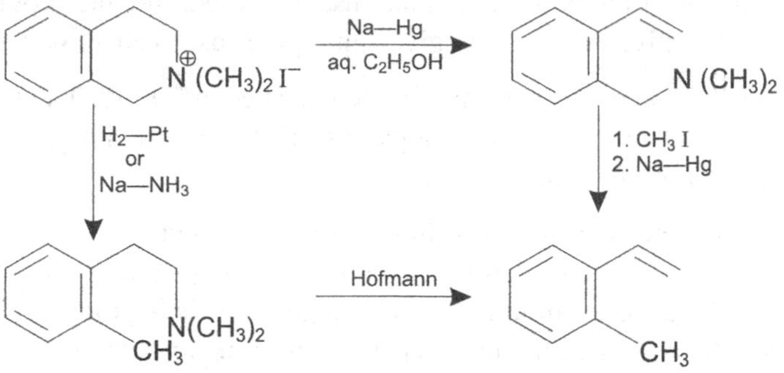

We know that exhaustive methylation fails with tetrahydroquinoline. The heterocyclic ring, however, is opened by the Emde degradation.

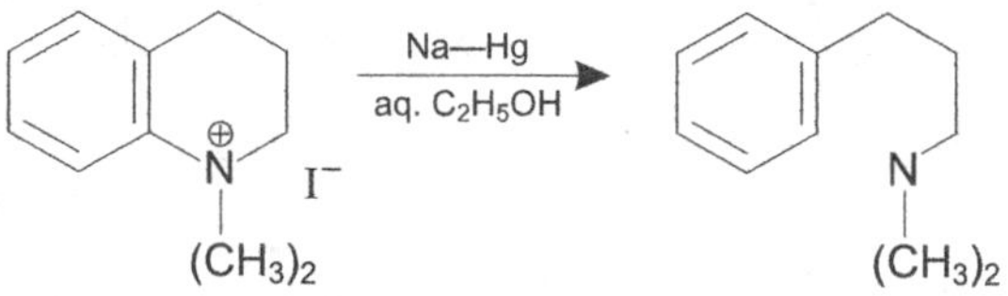

7.7   CLASSIFICATION OF ALKALOIDS

Earlier when the structures of most of alkaloids were unknown, the alkaloids were classified according to plant genera in which they occur *e.g.*, hemlock alkaloids, cinchona alkaloids, opium alkaloids, tobacco alkaloids *etc.*

Even today, it is difficult to classify them into distinct groups because a large number of alkaloids with different structural variations are known. However, the most satisfactory method of classifying alkaloids is according to the main ring system, which is common to a group of alkaloids.

Some common groups are listed below:

1. Phenylethylamine group
2. Pyrrolidine group
3. Pyridine-piperidine group
4. Pyridine-pyrrolidine group
5. Quinoline group
6. Isoquinoline group
7. Phenanthrene group
8. Indole group

A. Phenylethylamine Group

D(−)-Ephedrine (4) is an important member of this group of alkaloids. Its action is similar to that of adrenaline (5).

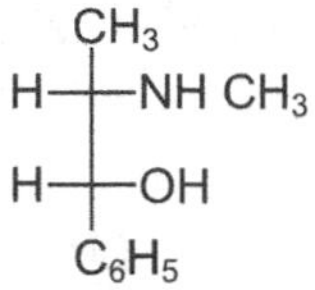

$$CH_3$$
$$H-\!\!-NH\ CH_3$$
$$H-\!\!-OH$$
$$C_6H_5$$
(4)

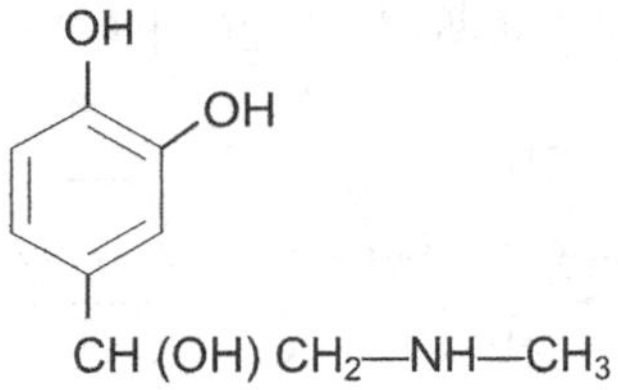

OH

OH

CH (OH) CH$_2$—NH—CH$_3$

Adrenaline
(5)

B. Pyrrolidine Group

Hygrine (7) is the important alkaloid belonging to this group. It is one of the coca alkaloids.

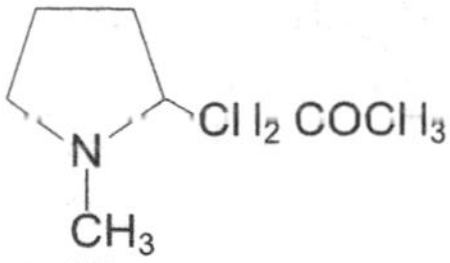

CI I$_2$ COCI I$_3$

N
|
CH$_3$
(7)

C. Pyridine-piperidine Group

Some important alkaloids of this group are listed below:

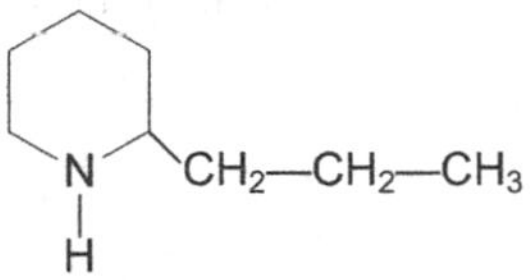

N   CH$_2$—CH$_2$—CH$_3$
|
H

Coniine (8)
(Hemlock alkaloid)

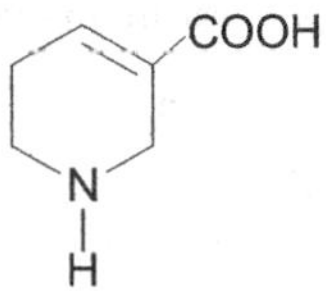

COOH

N
|
H

Guvacine (9)
(Areca nut alkaloid)

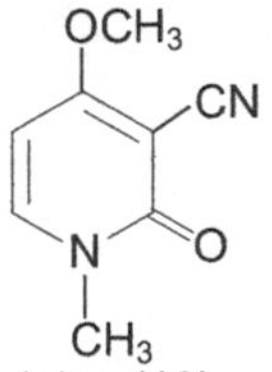

OCH$_3$

CN

N   O
|
CH$_3$

Ricinine (10)

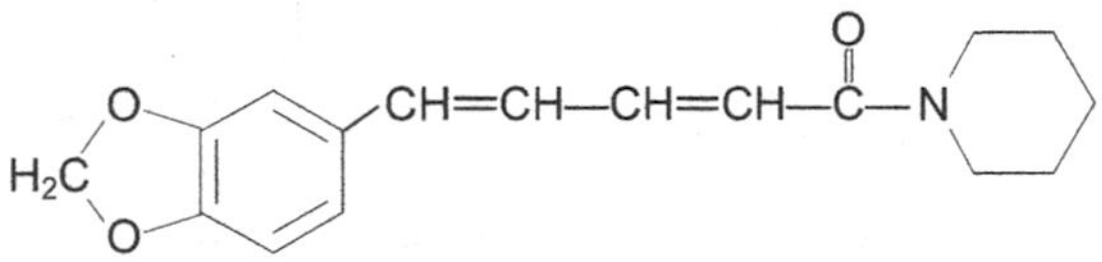

H$_2$C

O

O

CH=CH—CH=CH—C—N

O

Piperine (6)

D. Pyridine-pyrrolidine Group

Nicotine is an important alkaloid of this group. It is present in tobacco leaves.

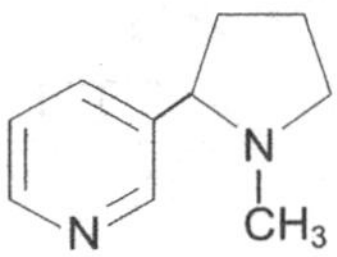

N

N
|
CH$_3$

Nicotine
(11)

Other important alkaloids of this group are atropine and cocaine.

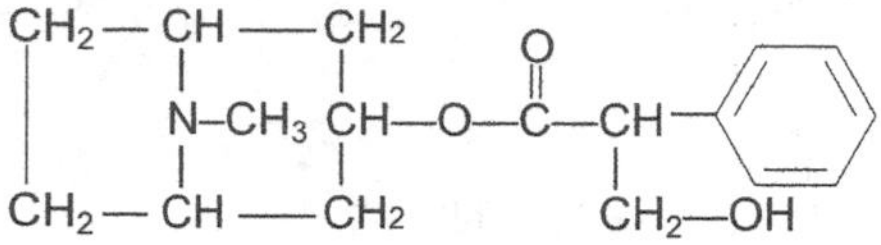

Atropine (12) 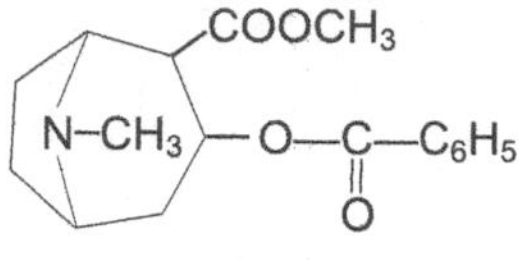   (−)-Cocaine (13)

E. Quinoline Group

Quinine and cinchonine, present in the bark of cinchona tree, belong to this group.

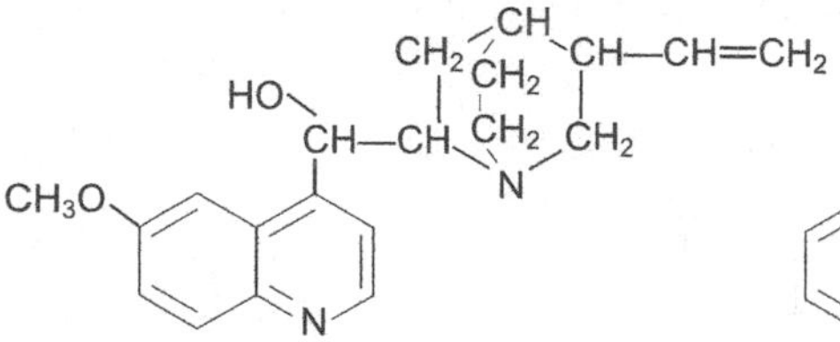

Quinine (14) 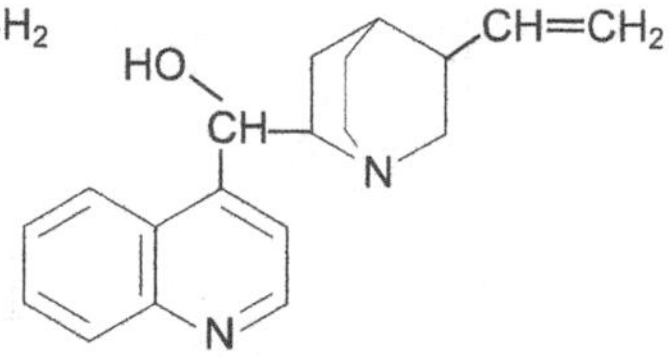   Cinchonine (15)

Primaquine is a synthetic antimalarial which contains quinoline nucleus.

F. Isoquinoline Group

Papaverine (isolated from opium) is an important alkaloid belonging to this group.

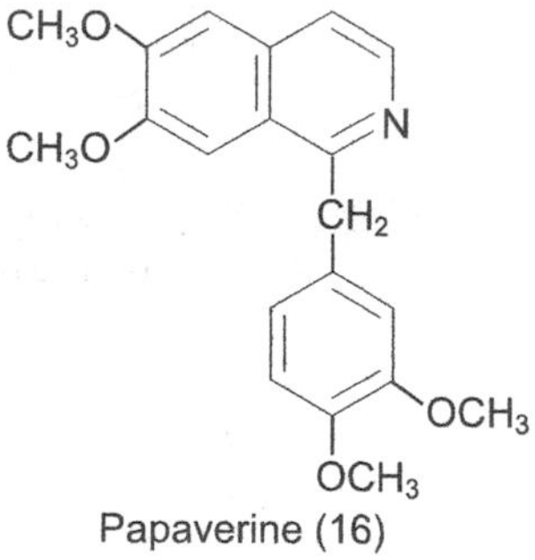

Papaverine (16)

G. Phenanthrene Group

Morphine (17) is the chief alkaloid of opium, which has a phenanthrene nucleus. Its diacetate is known as *heroin*.

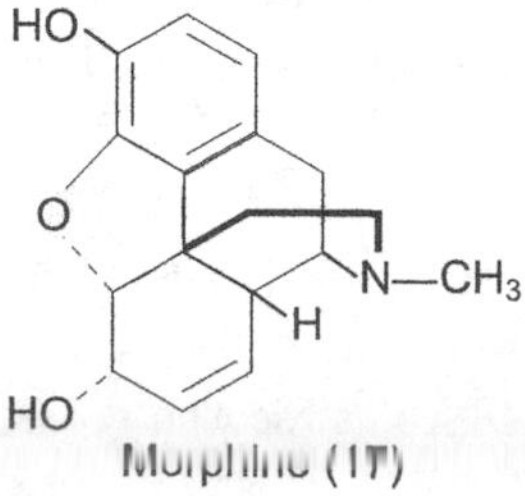

Morphine (17)

H. Indole Group

Lysergic acid diethylamide (LSD) is considered an alkaloid containing an indole nucleus. It is a potent hallucinogen.

Lysergic acid diethylamide
(LSD)

7.8   IMPORTANCE AND PHYSIOLOGICAL ACTION OF ALKALOIDS

Alkaloids are an important group of biological active compounds. They have marked physiological activity on animals and human beings. They are mostly synthesised by plants to protect them from being eaten by insects and other animals.

Alkaloids can occur in all parts of the plant but frequently they accumulate only in particular organs (*e.g.*, barks, roots, leaves and fruits) whereas at the same time other organs are alkaloid free. For example, in potato plants, the edible tubers are alkaloid-free whereas the green parts contain the poisonous *solanine.* The organ in which alkaloids accumulate is not always the site of their synthesis. For example, in tobacco nicotine is produced in the roots and translocated to the leaves where it accumulates.

The functions of alkaloids in plants are mostly unknown. A single plant species may contain over one hundred different alkaloids and the concentration can vary from a small fraction to as much as 10 percent of the dry weight. Most alkaloids are very toxic and therefore, have the potential to function as defence agents against attack by herbivores and microorganisms. For example, nicotine present in tobacco leaves inhibits the growth of tobacco hornworm larvae. The purified compound is also applied as an effective insecticide in green-houses. However, alkaloids have been suggested to serve as a storage form of nitrogen or as protectants against damage by UV light.

Alkaloids have traditionally been of great interest to humans because of their pronounced physiological and medicinal properties. When administered to animals most alkaloids produce striking physiological effects, which vary from alkaloid to alkaloid. Some alkaloids stimulate the central nervous system, others cause paralysis. Some alkaloids increase blood pressure, others lower it. Certain alkaloids act as analgesics, some others are used as tranquilisers. Some alkaloids act against infectious microorganisms.

We find that most alkaloids are toxic when their dosage is very large and some are toxic even when the dosage is very small. In spite of this a large number of alkaloids find use in medicine. Alkaloids with hallucinogenic, narcotic or analgesic properties have found medical application as pure compounds or served as model compounds for modern synthetic drugs.

The first alkaloid isolated in a pure state was *morphine* (17). It occurs in poppy plant and is responsible for the physiological effect of opium. It has marked analgesic action. Other members of the morphine family are *o*-methyl derivative *codeine* (18) and the diacetyl derivative, *heroin* (19). Morphine is a 50 times more effective pain reliever than aspirin (20).

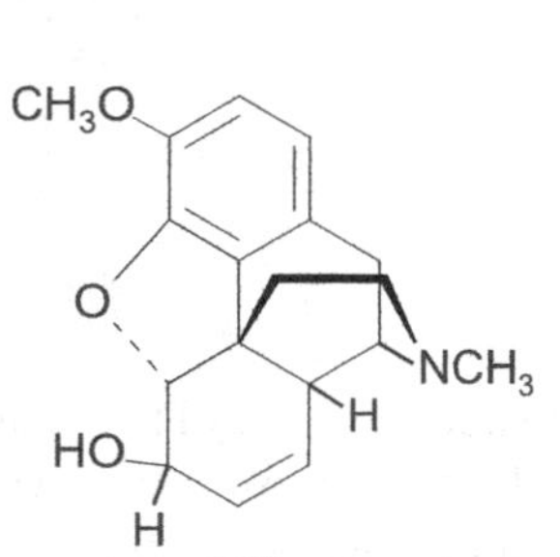

(18) 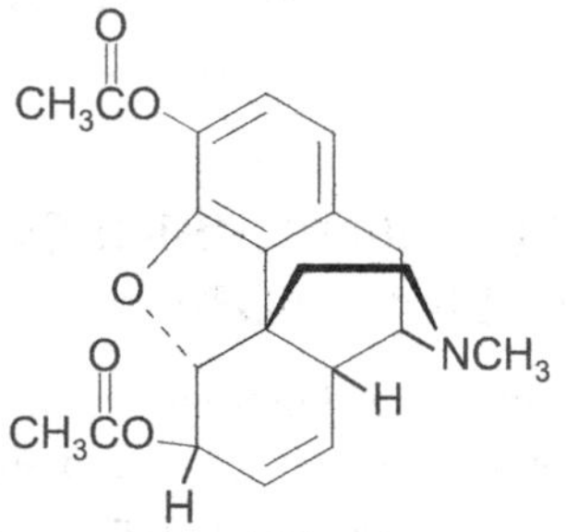 (19) 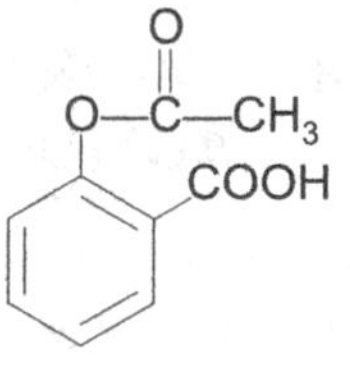 Aspirin (20)

Cocaine (13) is the tropane alkaloid, which has anaesthetic properties but is abused as an illegal drug because it is most addictive and very harmful. *Quinine* (14), an alkaloid from cinchona bark, is an important antimalarial agent. *Nicotine* (11) is the chief alkaloid of tobacco leaves, which has mood altering effects.

Strychinine (21) is a powerful stimulant. *Strychnine* and *brucine* (22) are used for resolving racemic mixtures of carboxylic acids. They have been used as rodent poisons.

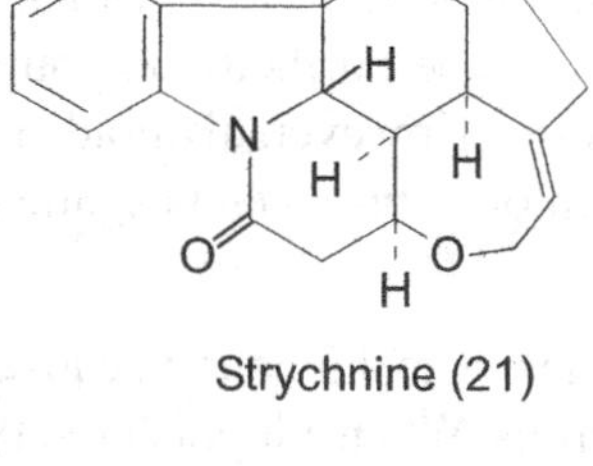

Strychnine (21) 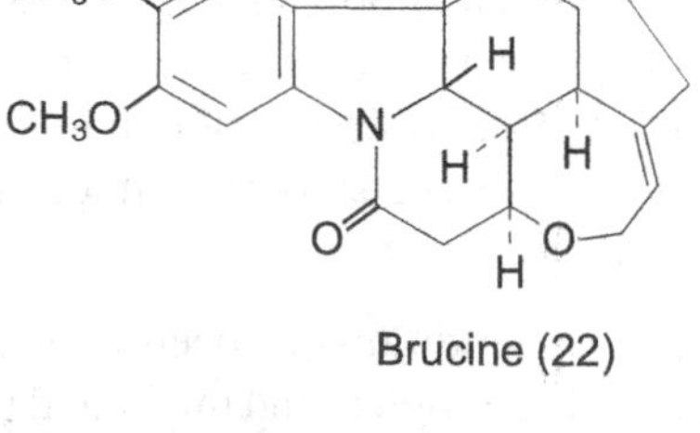 Brucine (22)

Colchinine (23) has been used in the treatment of gout. *Emetine* (24) is used as an emetic and expectorant agent. Its recent use has been in cases of amoebic dysentery.

Other alkaloids are too toxic for any therapeutic use. From the beginning of civilisation, alkaloid containing plant extracts have been used in all cultures as medicines and poisons. The Greeks chose the alkaloid coniine (8) present in oil of **hemlock** to kill Socrates, the Greek philosopher although morphine, nicotine or cocaine would have served equally well. They used the oil of hemlock as the official poison for execution of criminals. The Egyptian queen, **Cleopatra** used atropine containing plant extracts such as belladonna to dilate her pupils. Atropine (12) has a marked dilating action on the pupils of eyes. It is still used by ophthalmologists before testing the eyes. In modern times, the stimulants caffeine in coffee and tea and nicotine in cigarettes are consumed worldwide.

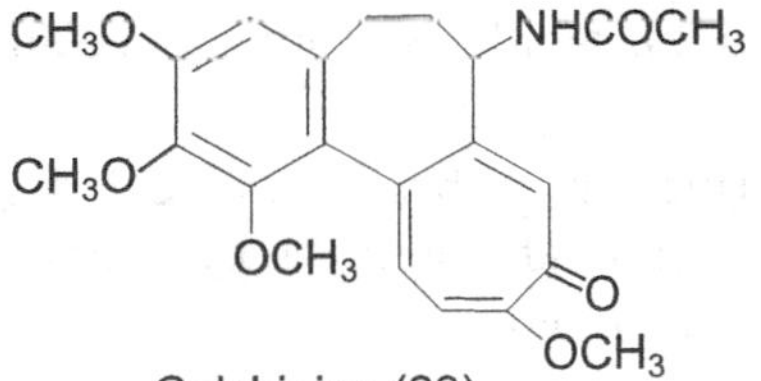

Colchinine (23)

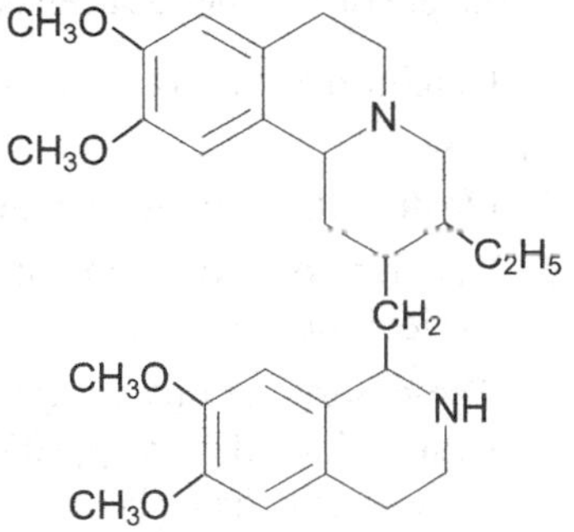

Emetine (24)

Mild doses of alkaloids can produce psychological effects that resemble peacefulness, euphoria or hallucinations. Persons seeking these effects often become addicted to alkaloids. Alkaloid addiction often turns out to be fatal.

7.9 CONIINE

Coniine belongs to the hemlock group of alkaloids. The oil of hemlock (*conium maculatum*) contains five alkaloids namely coniine, conhydrine, pseudoconhydrine *etc.*

But coniine is of special importance. It is the major constituent of oil of hemlock and the poisonous property of hemlock is due to coniine. It paralyses the central nervous system leading to death. The Great philosopher Socrates was condemned to death by drinking oil of hemlock in 399 B.C.

Coniine was the first alkaloid to be synthesised because of its simple structure.

A. Isolation of Coniine

The seeds of hemlock are powdered and distilled with sodium hydroxide solution. The alkaloids are extracted with ether. Ether is removed by distillation. It leaves behind a mixture of alkaloids, which are fractionated by the usual methods of separation.

Coniine boils at 167°C and is obtained as a colourless liquid.

B. Properties of Coniine

- Coniine has an unpleasant smell and burning taste.
- It is dextrorotatory, $[\alpha]_D = +15.7°$.
- It is sparingly soluble in water but readily dissolves in alcohol.
- Coniine is extremely poisonous to human beings.

C. Structure of Coniine

The structure of coniine has been deduced on the basis of the following observations:

1. Coniine has the molecular formula $C_8H_{17}N$ obtained on the basis of analytical data.
2. On distillation with zinc dust coniine gives a new base, conyrine, $C_8H_{11}N$.
3. Conyrine on oxidation with $KMnO_4$ yields pyridine-2-carboxylic acid *i.e.*, α-picolinic acid.

   These observations indicate that conyrine is probably a pyridine derivative with a side chain at position 2, which on oxidation gives pyridine-2-carboxylic acid. The above reactions can be written as

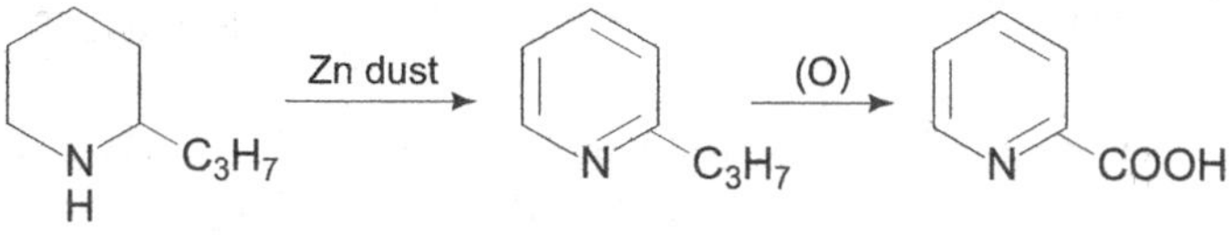

| Coniine | Conyrine | α-Picolinic acid |

The side chain in coniine may be either (25) or (26)

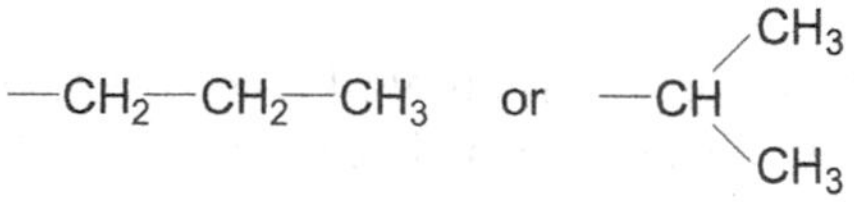

$$-CH_2-CH_2-CH_3 \quad \text{or} \quad -CH\begin{smallmatrix}CH_3\\\\CH_3\end{smallmatrix}$$

(25)   (26)

4. On heating with HI/P-coniine gives *n*-octane. That shows that the side chain in coniine is *n*-propyl and not isopropyl.

   If the side chain had been isopropyl we would have obtained isooctane in the above reaction.

   Therefore, the structure of coniine is

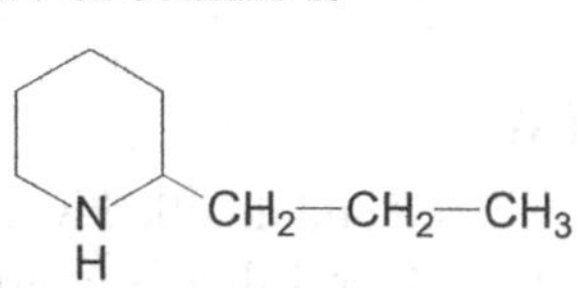

Coniine

5. Exhaustive methylation of coniine gives an alkene, conylene C_8H_{14} which is reduced to give *n*-octane.

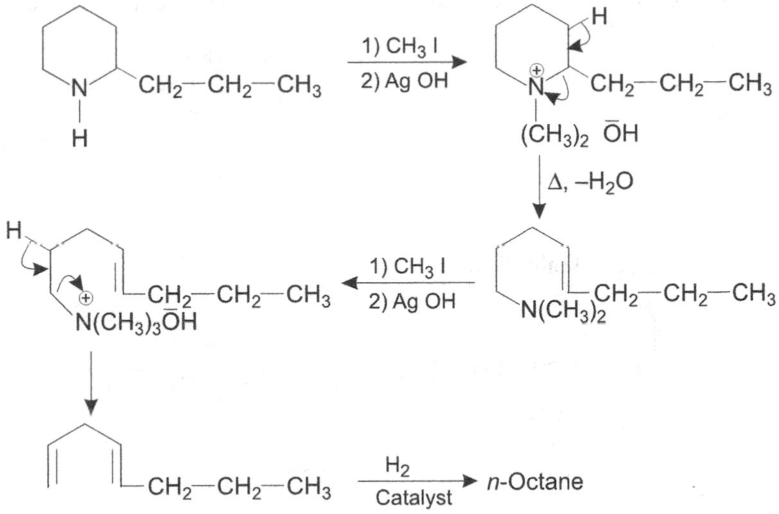

D. Synthesis of Coniine

The structure of coniine is finally confirmed by its synthesis.

1. In the **Landenberg** synthesis, α-picoline is condensed with acetaldehyde in the presence of anhydrous zinc chloride. At 250°C, it forms 2-propenyl pyridine, which on reduction with sodium and ethanol yields (±)-coniine.

   Coniine is finally resolved by using (+)-tartaric acid. Salt of (+)-coniine is less soluble and crystallises out first. This on treatment with alkali liberates (+)-coniine.

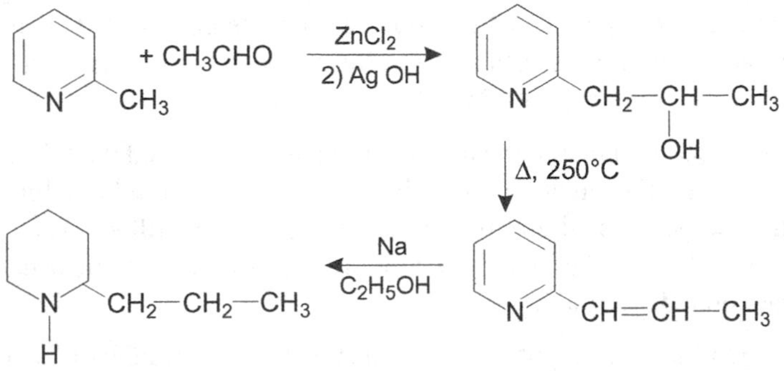

(±)-Coniine

2. In the **Bergmann** synthesis, α-picoline is treated with phenyl lithium. The resulting lithium salt is alkylated with ethyl bromide to give conyrine, which is reduced to (±)-coniine.

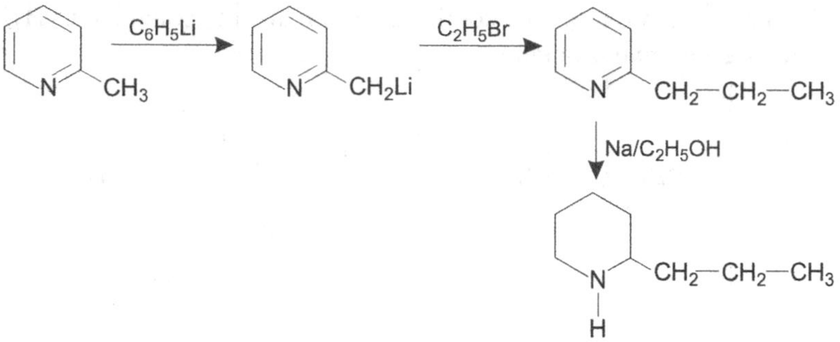

Other hemlock alkaloids are

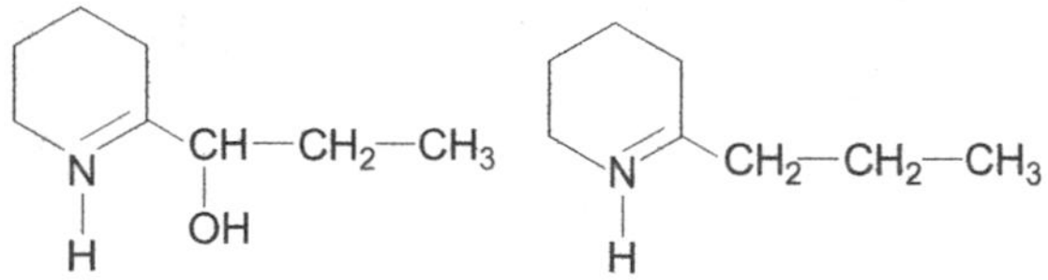

Conhydrine   γ-Coniceine

7.10 NICOTINE

Nicotine belongs to the pyrrolidine-pyridine group of alkaloids. It is the chief alkaloid of tobacco plant, **Nicotana tobacum**.

Nicotine is present in tobacco leaves (4 to 5%) as salt of malic and citric acids. Other tobacco alkaloids are nicotimine, nornicotine *etc.*

A. Isolation of Nicotine

1. Tobacco leaves are dried, powdered and extracted with water (nicotine is soluble in water). The water extract is treated with sodium hydroxide to liberate free nicotine. The product is then steam distilled. The distillate is extracted with ether. Ether is removed and the liquid is fractionated. The fraction boiling at 240 – 247°C contains nicotine.

2. In the second method, the powdered leaves are treated with dilute sulphuric acid when the alkaloids are obtained in solution. The solution is made alkaline with sodium hydroxide and steam distilled. The distillate is extracted with ether and fractionation is carried out by the usual method of fractional distillation.

The free nicotine can be purified through the formation of its oxalate. Nicotine oxalate is treated with alkali to liberate free nicotine which is extracted with ether. Ether is removed and pure nicotine is obtained.

B. Properties of Nicotine

• Nicotine is a colourless liquid, b.p. 246 – 247°C. It darkens in air.
• It has a tobacco like odour and a burning alkaline taste.

- Nicotine is soluble in water and most organic solvents. It is steam volatile.
- Nicotine is laevorotatory, $[\alpha]_D = -169°$.
- Nicotine is poisonous in nature. Small quantities (30 – 50 mg) can stimulate the central nervous system.
- It causes depression and increases blood pressure in human beings.
- Tobacco smoking can cause diseases like asthma and lung cancer. It is mainly due to presence of nicotine in tobacco.

C. Structure of Nicotine

The structure of nicotine has been arrived at on the basis of the following observations:

1. The molecular formula of nicotine is found to be $C_{10}H_{14}N_2$.

2. Nicotine does not form any acetyl or benzoyl derivative. This means that both the nitrogen atoms in nicotine exist as *tertiary*. Further, nicotine reacts with two molecules of CH_3I suggesting again the *tertiary* nature of the N atoms.

3. On heating with HI nicotine liberates one molecule of CH_3I, which is confirmed by the formation of AgI when treated with $AgNO_3$. This indicates that there is one $-NCH_3$ group present in nicotine.

4. On oxidation with acid-dichromatic or permanganate nicotine forms nicotinic acid.

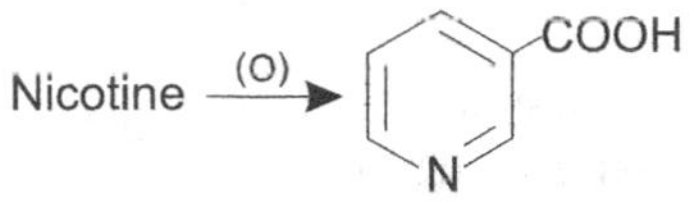

Nicotinic acid

This shows that nicotine has a pyridine nucleus with a side chain in the 3-position. The N–CH$_3$ group is present in the side chain. Therefore, nicotine may be written as

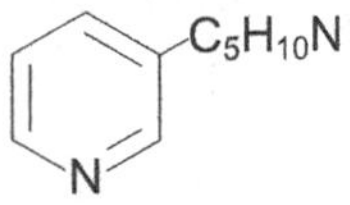

5. Nicotine forms an addition compound when treated with zinc chloride. This on distillation gives pyridine, pyrrole and methylamine. This suggests that the side chain in nicotine is a pyrrole derivative.

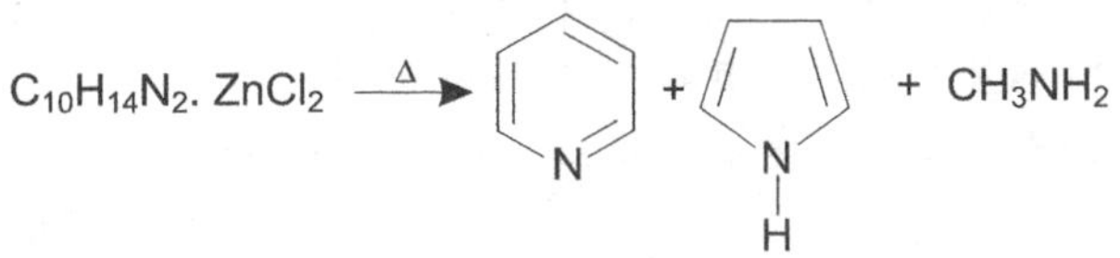

6. Catalytic hydrogenation of nicotine uses only three moles of hydrogen to give a complete saturated system. This indicates that the side chain is fully saturated. Thus the structure of nicotine may be either (3) or (4).

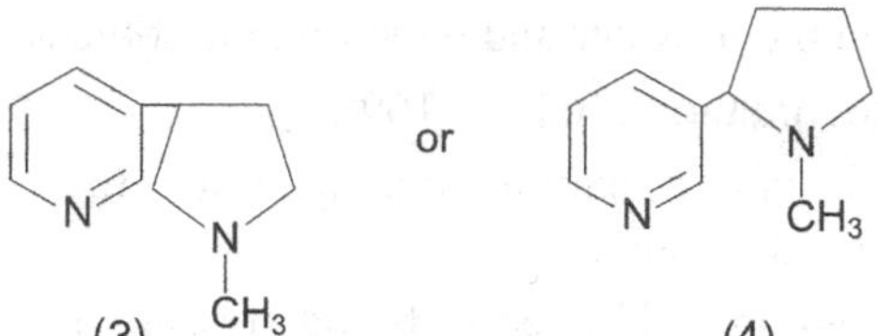

or

(3)   (4)

7. Nicotine on treatment with Br_2 in acetic acid gives hydrobromide perbromide, $C_{10}H_{10}Br_2N_2O.HBr.Br_2$, which on reaction with H_2SO_3 is converted to dibromocotinine, $C_{10}H_{10}Br_2N_2O$.

Dibromocotinine on heating with a mixture of sulphurous and sulphuric acids forms 3-acetyl pyridine, oxalic acid and methylamine.

Dibromocotinine $\xrightarrow[H_2SO_4]{H_2SO_3}$ [3-acetyl pyridine] + $\begin{matrix} COOH \\ | \\ COOH \end{matrix}$ + CH_3NH_2

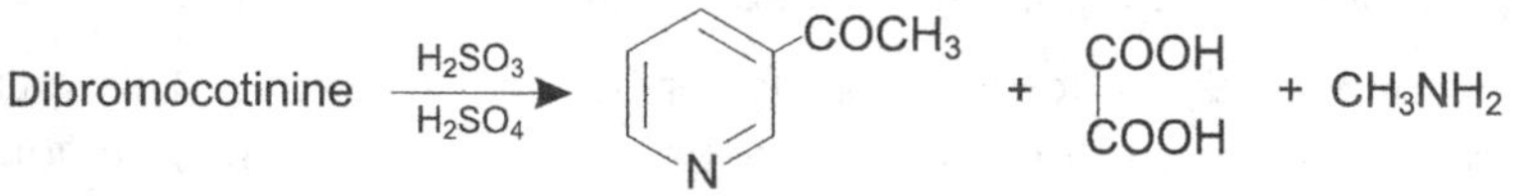

8. Nicotine gives dibromoticonine, $C_{10}H_8Br_2N_2O_2$ on treatment with Br_2 in the presence of HBr.

This on heating with $Ba(OH)_2$ solution at 100°C forms nicotinic acid, malonic acid and methylamine.

$C_{10}H_8Br_2N_2O_2 \xrightarrow[100°C]{Ba(OH)_2}$ [nicotinic acid] + $\begin{matrix} COOH \\ CH_2 \\ COOH \end{matrix}$ + CH_3NH_2

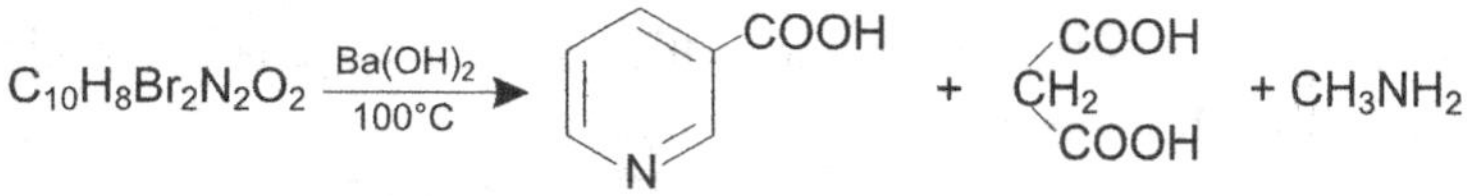

All the above reactions are possible only from a given structure for nicotine, as it must account for the following skeletons.

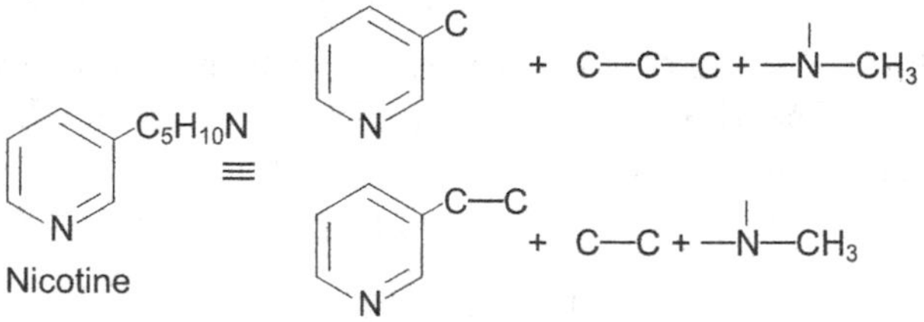

Nicotine

$\equiv$

+ C—C—C + —N—CH₃

+ C—C + —N—CH₃

Therefore, the side chain must be saturated and cyclic.

We have seen that nicotine behaves as a ditertiary base and forms two isomeric methyl iodide addition products. Moreover, nicotine is reduced to hexahydronicotine and not beyond that; *i.e.*, the only pyridine portion of nicotine is reduced to piperidine. Hence, the side chain must be saturated.

All the above observations are satisfied by the following structure of nicotine.

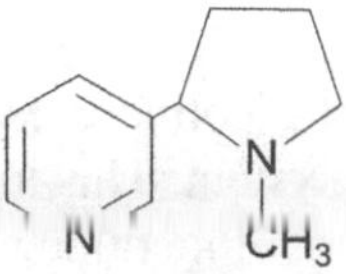

Nicotine

The formation of dibromocotinine and dibromoticonine and their reactions (given above) can be explained as follows:

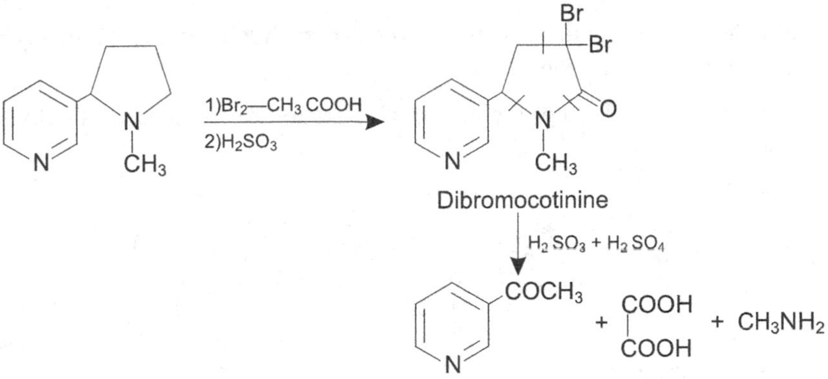

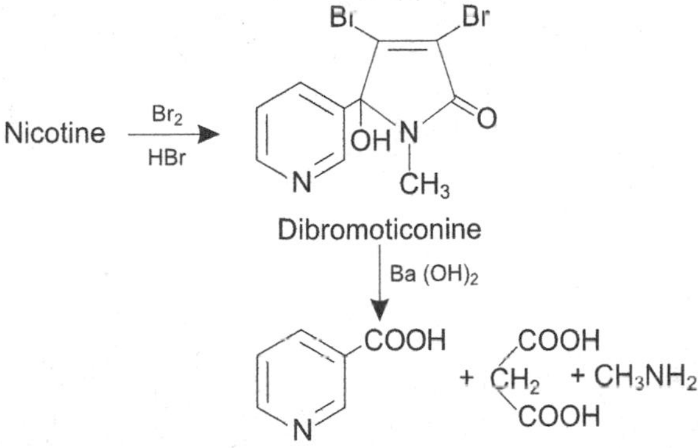

Dibromocotinine

Dibromoticonine

9. That a pyrrolidine group is present in the side chain in nicotine is proved by the following observation.

Nicotine when warmed with methyl iodide gives nicotine isomethiodide. This, on oxidation with potassium ferricyanide is converted into nicotone, which on oxidation with chromium trioxide gives hygrinic acid.

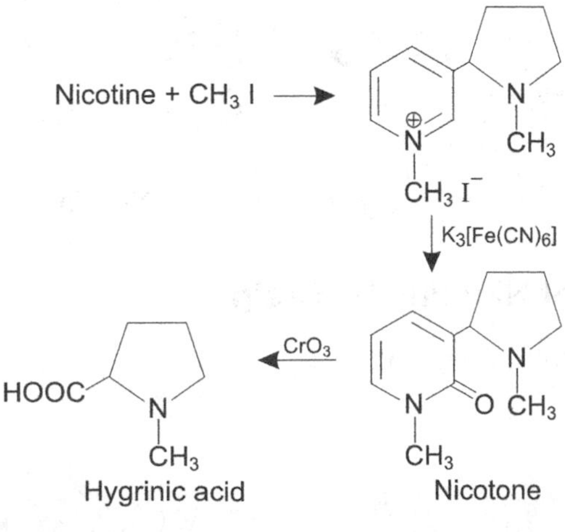

Hygrinic acid   Nicotone

D. Synthesis of Nicotine

i. Synthesis by Spath et al.

Succinimide on electrolytic reduction gives 2-pyrrolidone, which on methylation gives *N*-methyl-2-pyrrolidone.

This is condensed with ethyl nicotinate. Cyclisation of the product finally gives (±)-nicotine.

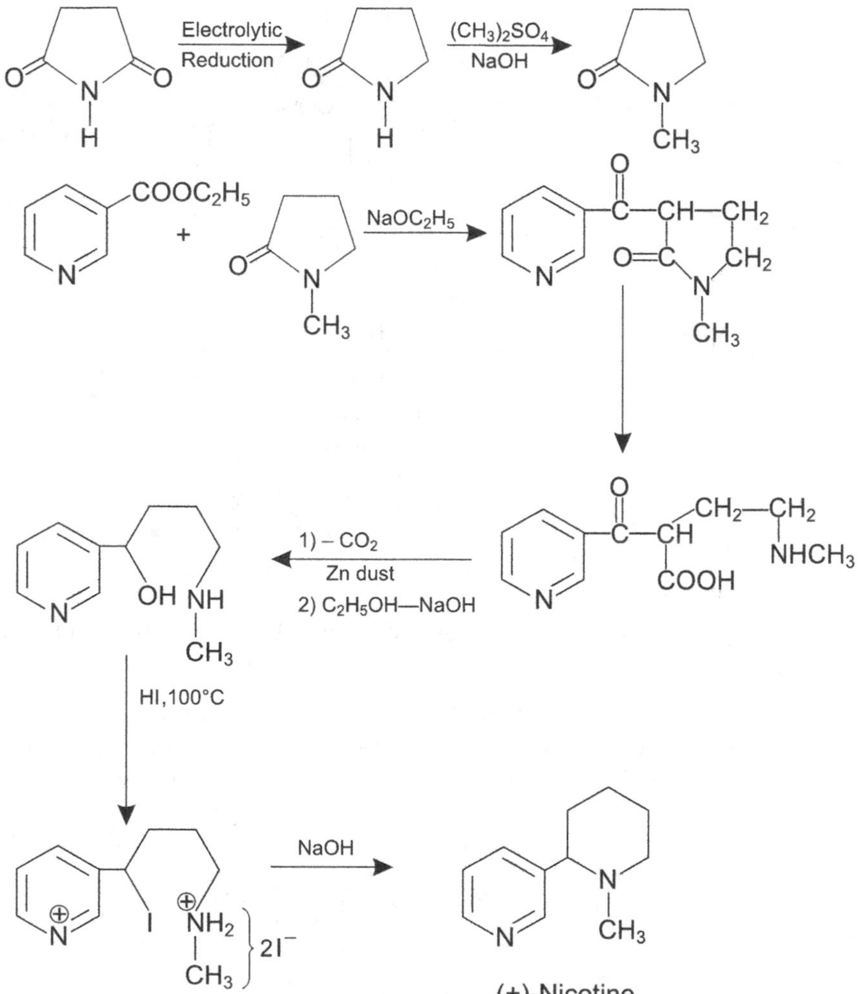

ii. Synthesis of Nicotine by Craig

This involves the following steps:

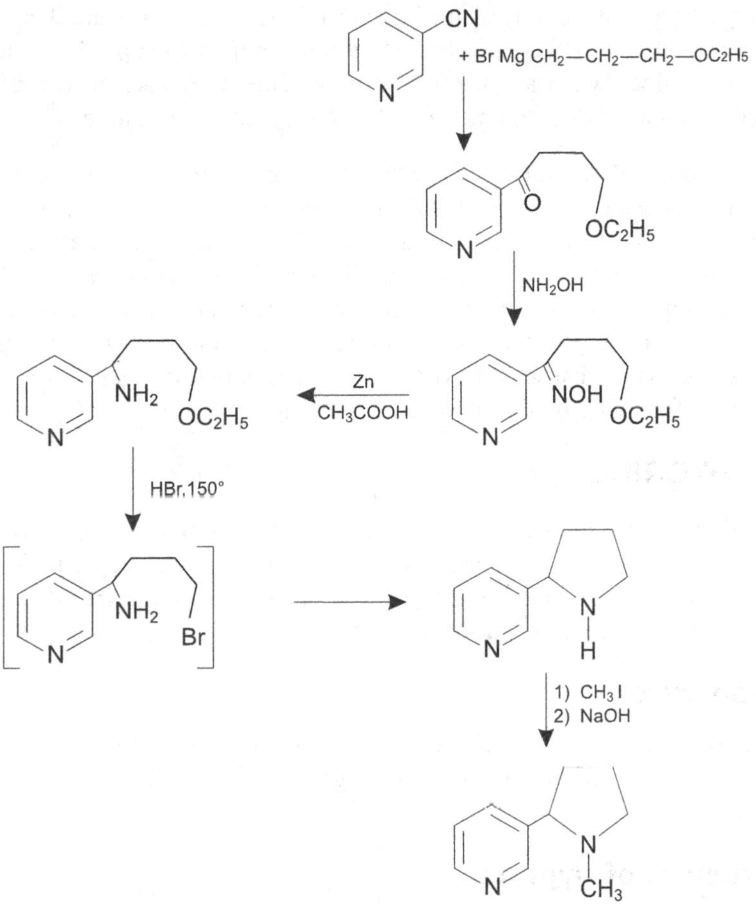

(±)-Nicotine

E. Medicinal Importance of Nicotine

We know nicotine is the prominant alkaloid of the tobacco plant. It possesses various psychoactive effects. Nicotine is both a stimulant and a relaxant. By release of glucose from the liver and adrenaline from the adrenal medulla, it causes stimulation. Users report feelings of relaxation, sharpness, calmness and alertness. As it reduces the appetite and raises the metabolism, some smokers may lose weight.

When a cigarette is smoked, nicotine-rich blood passes from the lungs to the brain within seven seconds and immediately stimulates the release of many chemical messengers. This release of neurotransmitters and hormones is responsible for most of nicotine's effects. Nicotine appears to enhance concentration and memory due to the increase of acetylcholine. Nicotine is believed to reduce pain and anxiety due to increase of acetylcholine and β-endorphin.

Most cigarettes contain 1 to 3 milligrams of nicotine. When smokers wish to achieve a stimulating effect, they take short quick puffs which produce a low level of blood nicotine. When a smoker wishes to relax, they take deep puffs which produce a high level of blood nicotine. It produces a mild sedative effect.

One of nicotine's effects is to increase the concentration of a chemical in the brain's reward system. Release of this chemical makes smokers feel good and reinforces the need to smoke. But nicotine as such is not significantly addictive as nicotine administered alone does not produce significant reinforcing properties. However, after coadministration with other materials nicotine produces significant addiction characteristics. In very small doses nicotine acts as a stimulant but in large doses it causes depression, nausea and vomiting. In still larger doses it is very toxic to humans; the fatal dose by ingestion is only 40 mg.

7.11 HYGRINE

Hygrine is one of the coca alkaloids, present up to 0.2% in coca leaves (*Erythroxylon coca*). Cocaine is another important alkaloid present in coca. Hygrine belongs to the pyrrolidine group of alkaloids. Hygroline and cuscohygrine are other alkaloids found in the leaves of the coca shrub.

A. Properties

Hygrine is isolated as thick yellow oil (b.p. 83–84°C). It is pungent in taste and odour and forms solid salts with acids. Hygrine hydrochloride has $[\alpha]_D = +34.5$ in water.

B. Structure of Hygrine

1. The molecular formula of hygrine is found to be $C_8H_{15}NO$.

2. It gives typical reactions of a keto group *e.g.*, formation of oxime and oxidation to an acid having a lesser number of carbon atoms.

3. Hygrine does not form an acetyl or benzoyl derivative. This shows nitrogen present in hygrine is *tertiary*.

4. On oxidation with chromic acid hygrine forms hygrinic or hygric acid.

$$C_8H_{15}NO \xrightarrow[\text{[O]}]{CrO_3} C_6H_{11}NO_2$$

Hygrine   Hygrinic acid

5. Dry distillation of hygrinic acid gives *N*-methylpyrrolidine. Hence, hygrinic acid is an *N*-methylpyrrolidine carboxylic acid.

As decarboxylation of hygrinic acid occurs very readily, the carboxyl group should be in the 2-position of the pyrrolidine ring, the same thing we observe in the case of α-amino acids. Therefore, hygrinic acid should be N-methylpyrrolidine-2-carboxylic acid.

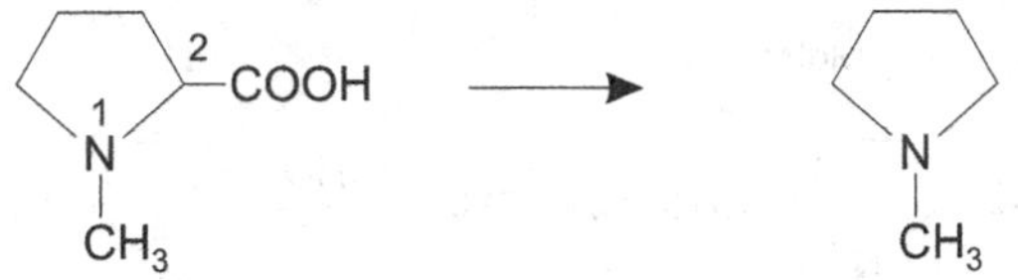

Hygrinic acid   N-Methylpyrrolidine

6. The above structure, N-methyl-2-carboxylic acid for hygrinic acid was confirmed by synthesis as given below:

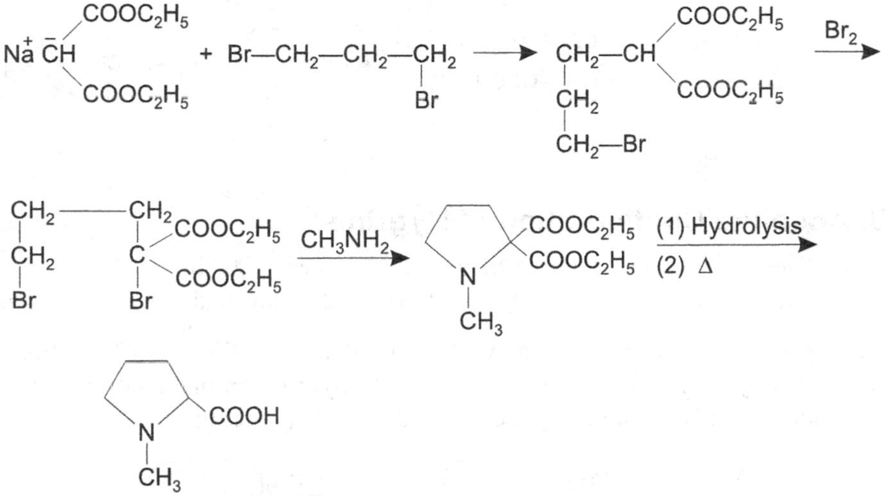

(±) Hygrinic acid

We get hygrinic acid by oxidation of hygrine. Therefore, hygrine may have two possible structures, (I) or (II)

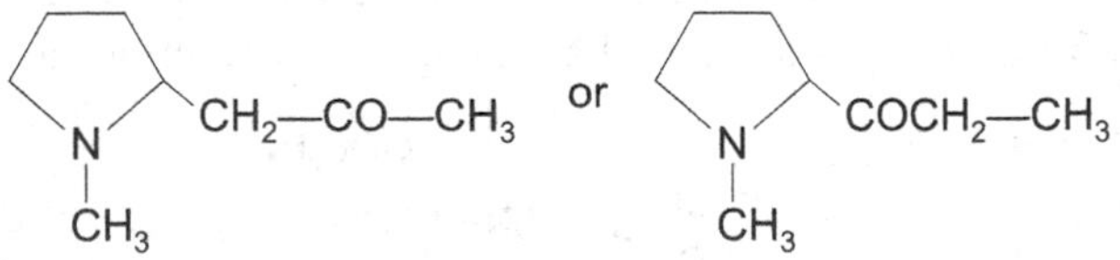

(I)   (II)

Both the compounds (I) and (II) were synthesised by **Hess**. Comparison of these with natural hygrine showed that (I) was hygrine.

C. Synthesis of Hygrine

1. Hygrine was synthesised starting from pyrrole as given below:

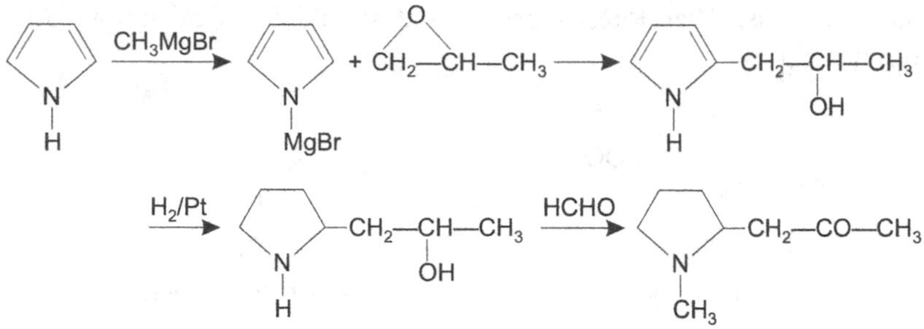

(±)-Hygrine

2. **Anet** *et al.* have synthesised (±)-hygrine by condensing γ-methylaminobutyraldehyde with ethyl acetoacetate at a pH of 7.

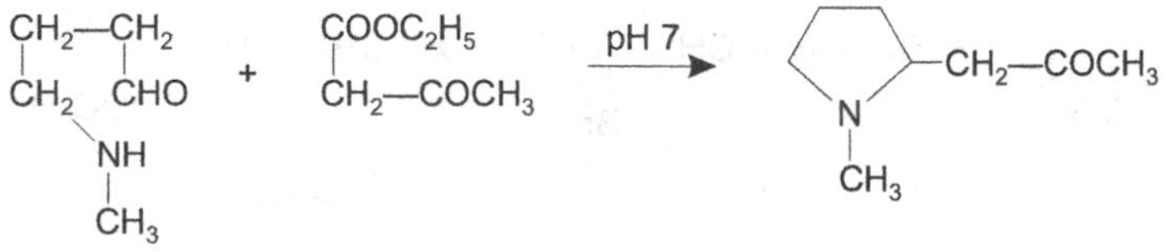

D. Absolute Configuration of Hygrine

1. The absolute configuration of hygrine has been established. (–)-Hygrinic acid was configurationally related to L(+)-glutamic acid and L(–)-proline.

(–)-Hygrinic acid, which was shown to have the same direction of rotation as that of (–)-hygrine, was converted into (–)-stachydrine on methylation and this compound was also obtained by the methylation of L(–)-proline.

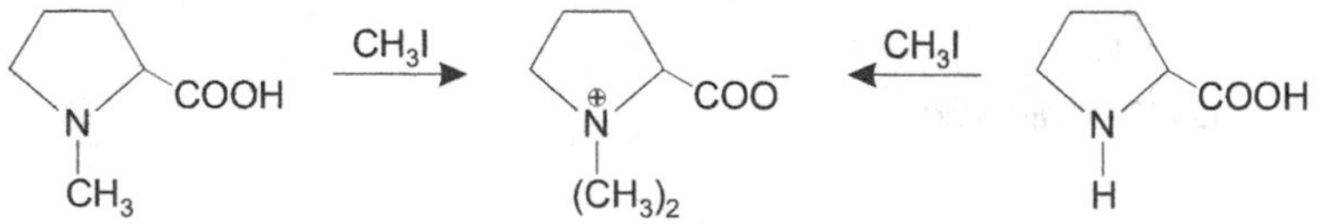

(–)-Hygrinic acid   (–)-Stachydrine   L(–)-Proline

L(–)-Proline on reduction with $LiAlH_4$ gives the corresponding alcohol, which was obtained from L(+)-glutamic acid by the series of reactions as shown below:

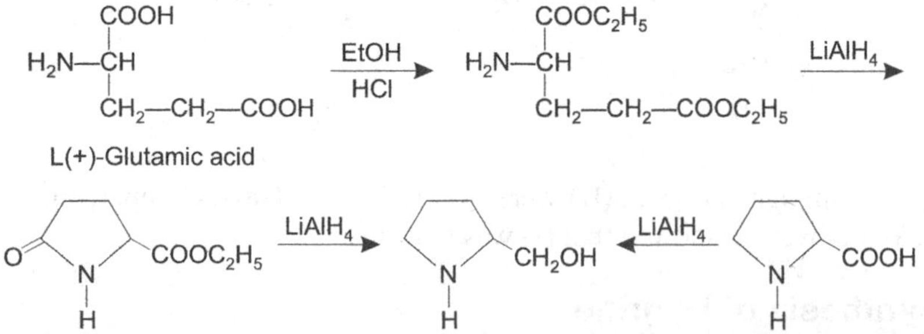

L(+)-Glutamic acid

L(–)-Proline (as ester)

2. More recently (–)-hygrine has been synthesised from a readily available *N*-methyl derivative of L(–)-proline. This establishes the configuration of (–) hygrine as given below:

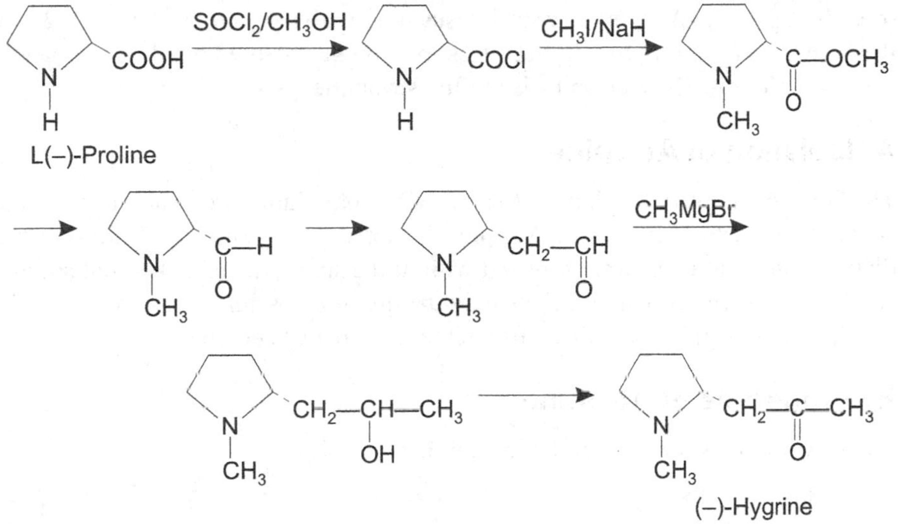

L(–)-Proline

(–)-Hygrine

Hygrine was readily transformed into its hydrochloride salt for comparison with the literature data of the same compound.

E. Medicinal Importance of Hygrine

Since it is known that hygrine is the precursor of *hyoscyamine* and *scopolamine,* both the compounds are used in the preparation of a vast number of pharmaceutical products. Therefore, hygrine has been the subject of several biological and pharmacological studies.

Hyoscyamine and scopolamine are the solanaceous alkaloids. This group includes atropine *(Atropa belladonna)*. Hyoscyamine is optically active but readily racemises to atropine (page 420).

Seopalamine, also known as *Hyoscine* is obtained from various sources, *e.g.*, Datura metel. Hyoscine is a constituent of travel sickness tablets. When administered with morphine it produces 'twilight sleep'.

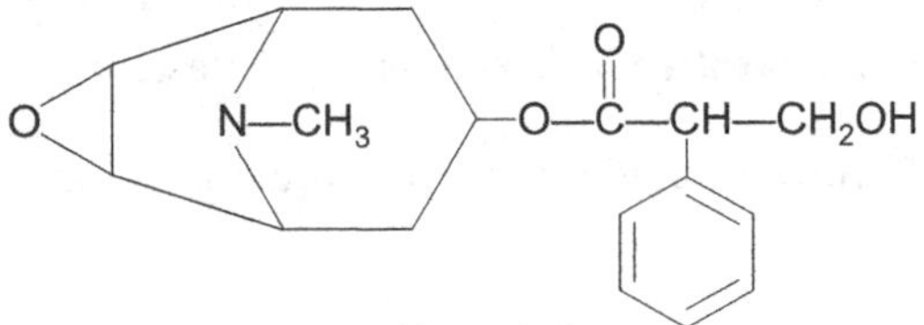

Hyoscine

7.12 ATROPINE

Atropine belongs to the *Solanaceous* group of alkaloids. This group has atropine, hyoscyamine and scopolamine alkaloids. Atropine occurs in *Atropa belladonna* (deadly nightshade) together with hyoscyamine. Hyoscyamine, $[\alpha]_D = -20°$, is the active form of atropine. It racemises to atropine when warmed with ethanolic alkaline solution. Thus, atropine is (±)-hyoscyamine.

A. Isolation of Atropine

The juice of nightshade plant is treated with potassium carbonate. The mixture is extracted with chloroform. Atropine is obtained in solution. Chloroform is distilled off and the residue is treated with dilute sulphuric acid. The salt solution so obtained is treated with a calculated quantity of potassium carbonate. Atropine precipitates out. It is crystallised from ethanol, ether or acetone.

B. Properties of Atropine

- Atropine is a white crystalline solid, m.p. 118°C.
- It is optically inactive.
- It is insoluble in water, moderately soluble in alcohol, ether and acetone, completely soluble in chloroform.
- Atropine is basic in character. It has a bitter taste and it is poisonous in nature.
- Atropine is extensively used for the dilation of the pupils of the eye.

C. Structure of Atropine

1. The molecular formula of atropine has been found to be $C_{17}H_{23}NO_3$.
2. It does not form any acetyl derivative or benzoyl derivative. That means atropine does not have any primary or secondary amino group and there is also no –OH group in atropine.
3. On methylation it uses one molecule of methyl iodide indicating the presence of a *tertiary* N-atom.
4. Atropine on treatment with hydriodic acid liberates one mole of methyl iodide indicating the presence of a $–N–CH_3$ group in atropine.
5. Atropine gives a positive hydroxamic test for esters.

When atropine is hydrolysed with barium hydroxide we get (±)-tropic acid and tropine (an alcohol).

Therefore, atropine is the ester of tropic acid.

However hyoscyamine is hydrolysed with cold water to give tropine and (–) tropic acid.

D. Structure of (±)-Tropic Acid

1. Tropic acid has the molecular formula $C_9H_{10}O_3$, m.p. 117°C.

2. No addition with Br_2 water indicates that it is a saturated molecule.

3. The presence of a –COOH group and an alcoholic group is indicated by the usual tests.

4. Vigorous oxidation of tropic acid with acidified permanganate gives benzoic acid. That indicates the presence of a benzene nucleus in tropic acid.

5. When heated strongly, tropic acid loses a molecule of water to give atropic acid, which is an unsaturated acid, $C_9H_8O_2$. This could be either (27) or (28).

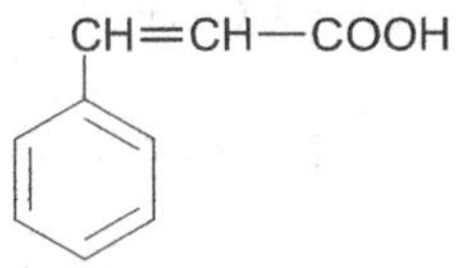

CH=CH—COOH

(27)

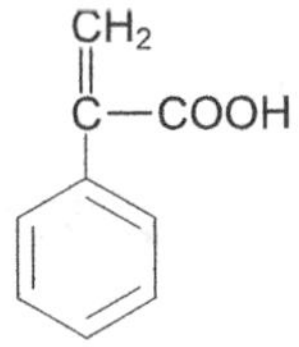

CH_2
‖
C—COOH

(28)

Structure (27) is not possible because it is a well-known cinnamic acid. Therefore, atropic acid has structure (28).

6. Further oxidation of atropic acid with permanganate gives phenyl glyoxylic acid,

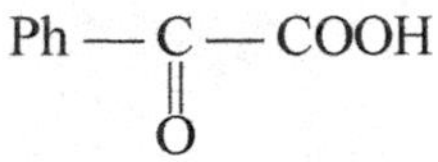

Ph — C — COOH
‖
O

Addition of water to structure (28) gives us two possible structures for tropic acid as (29) and (30).

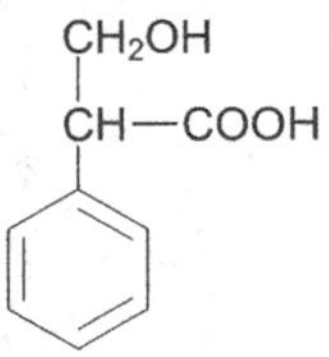

CH_2OH
|
CH—COOH

(29)

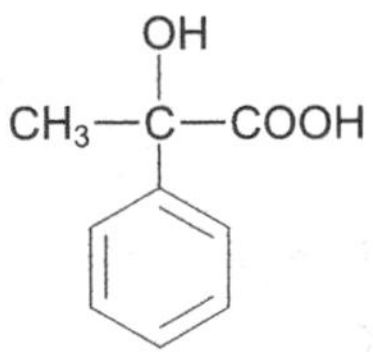

OH
|
CH_3—C—COOH

(30)

Structure (30) is rejected because it contains a *tertiary* –OH group, which is not possible. Therefore, tropic acid has structure (29).

The above structure of tropic acid is confirmed by synthesis.

E. Synthesis of Tropic Acid

1. **Mackenzie** and **Wood** synthesis starts from acetophenone. Atropic acid is obtained as an intermediate product, which adds up to a molecule of HCl and the product is hydrolysed to tropic acid.

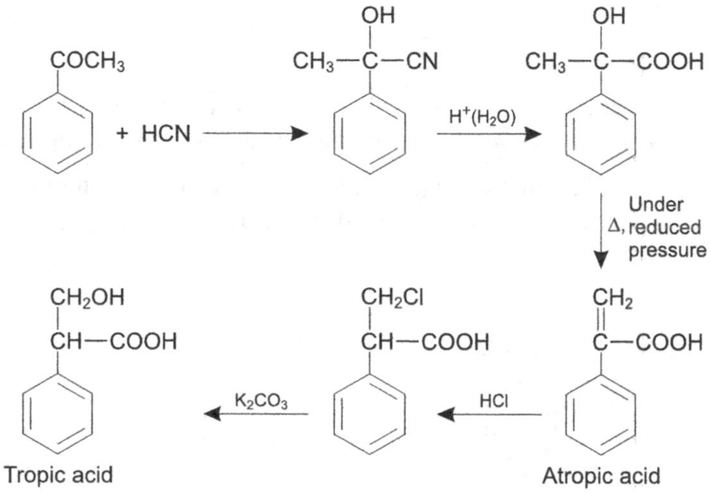

(±)-Tropic acid is resolved by using (–)-quinine.

2. **Synthesis by Blicks** *et al.*

In this synthesis, phenyl acetic acid is condensed with isopropyl magnesium chloride. The Grignard reagent so obtained is treated with formaldehyde and hydrolysed to generate a primary alcoholic group as given below:

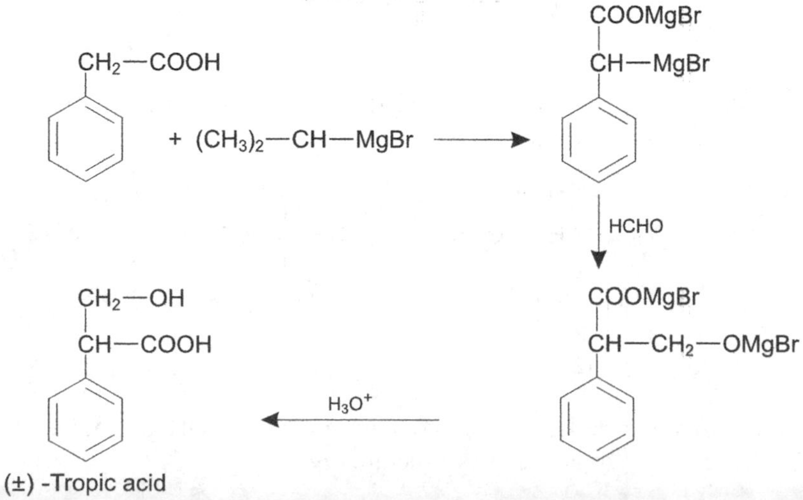

F. Structure of Tropine (Tropanol)

1. The molecular formula of tropine has been found to be $C_8H_{15}NO$, m.p. 63°C.

2. It contains one alcoholic group as it forms a monoacetate.

3. Nature of the N-atom

   Tropine does not undergo acetylation or benzoylation. However, it takes one molecule of methyl iodide. This indicates that tropine is a *tertiary* base.

4. Landenberg performed the following experiments to show that tropine contains a reduced pyridine nucleus.

   Tropine on treatment with HI gives tropine iodide, which is reduced to dihydrotropidine. This is heated with HCl and zinc distillation of the product gives 2-ethyl pyridine.

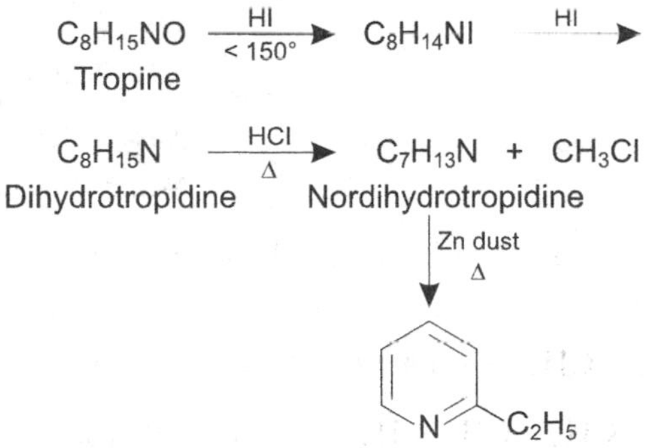

$$C_8H_{15}NO \xrightarrow[< 150°]{HI} C_8H_{14}NI \xrightarrow{HI}$$
Tropine

$$C_8H_{15}N \xrightarrow[\Delta]{HCl} C_7H_{13}N + CH_3Cl$$
Dihydrotropidine   Nordihydrotropidine

$$\Big\downarrow \begin{array}{l} Zn\ dust \\ \Delta \end{array}$$

There is an alcoholic group in tropine, which is replaced by an iodine atom to give tropine iodide. This on reduction gives dihydrotropidine. The formation of methyl chloride indicates the presence of an *N*-methyl group and the isolation of 2-ethyl pyridine shows the presence of pyridine nucleus in the reduced form.

On the basis of the above observations, Landenberg suggested two structures (31) or (32) for tropine.

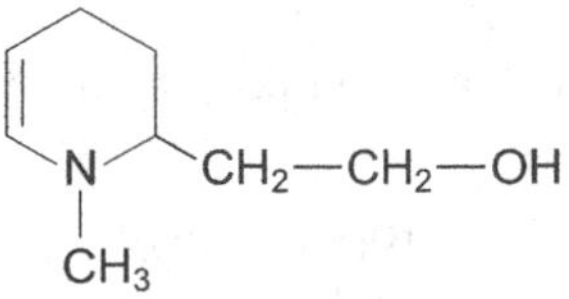

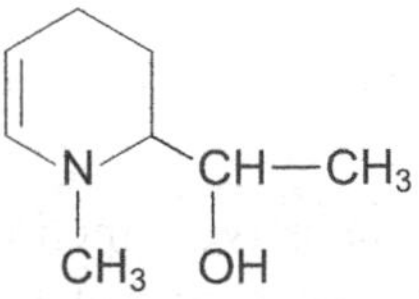

5. Both the above structures were rejected because tropine on oxidation with chromium trioxide gives (±)-tropinic acid (33), $C_8H_{13}NO_4$. There is no loss of C-atom. Therefore, –OH group is not in the side chain. It must be in the ring.

The following two structures were proposed by **Merling.**

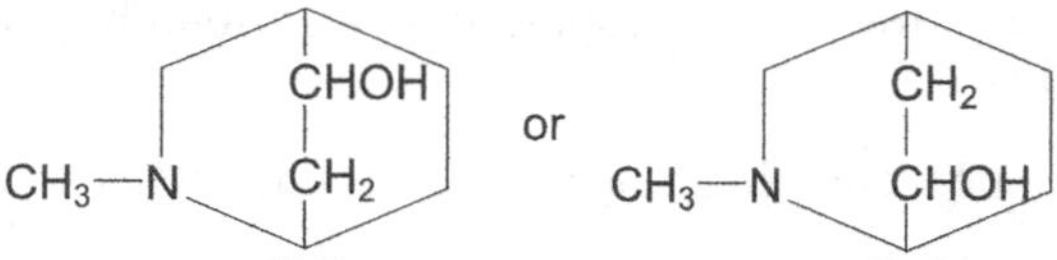

$$CH_3-N\begin{array}{c} CHOH \\ CH_2 \end{array} \quad \text{or} \quad CH_3-N\begin{array}{c} CH_2 \\ CHOH \end{array}$$

(34)   (35)

6. The oxidation products were again examined by **Willstätter.** Tropine on oxidation with chromium trioxide gives first tropinone (36), $C_8H_{13}NO$, which on further oxidation gives tropinic acid (33) and finally N-methylsuccinimide.

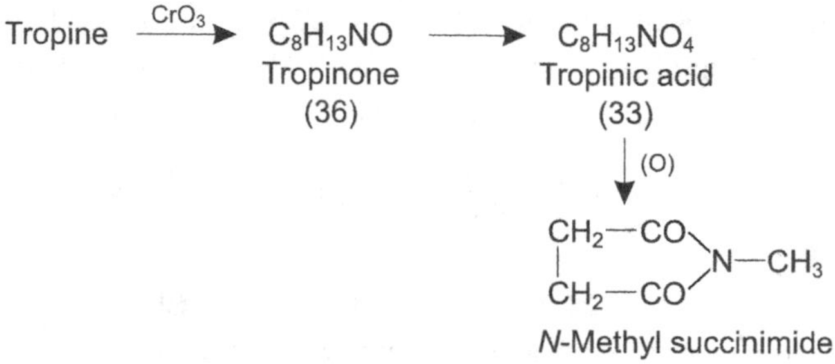

$$\text{Tropine} \xrightarrow{CrO_3} \underset{\substack{\text{Tropinone} \\ (36)}}{C_8H_{13}NO} \longrightarrow \underset{\substack{\text{Tropinic acid} \\ (33)}}{C_8H_{13}NO_4}$$

$$\downarrow (O)$$

$$\begin{array}{c} CH_2-CO \\ | \qquad\quad \ \ \rangle N-CH_3 \\ CH_2-CO \end{array}$$

N-Methyl succinimide

Tropinone is a ketone because it forms dibenzylidene derivative. That means it

contains $-CH_2-\overset{\overset{\textstyle O}{\|}}{C}-CH_2-$ grouping. Therefore, tropine is a *secondary* alcohol;

i.e., it must have $-CH_2-\underset{\substack{| \\ OH}}{CH}-CH_2-$ group, which is a part of the ring. The

following structure (37) was suggested for tropine by Willstätter, which contains a pyrrole and pyridine nucleus with an N atom common to both.

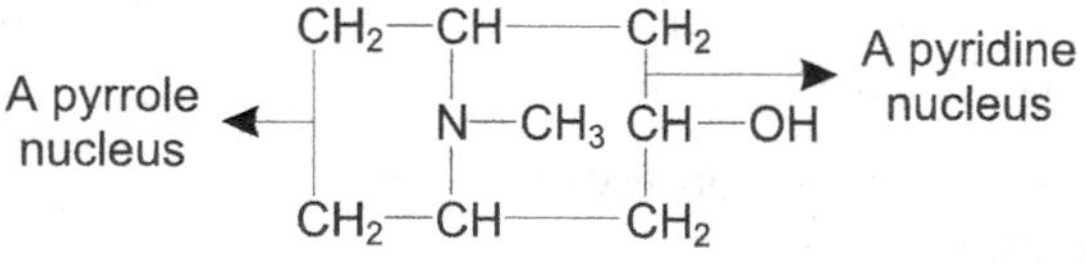

$$\begin{array}{ccc} & CH_2-CH---CH_2 & \\ \text{A pyrrole} & | \qquad\qquad\qquad | & \text{A pyridine} \\ \text{nucleus} \longleftarrow & N-CH_3 \ CH-OH & \longrightarrow \text{nucleus} \\ & | \qquad\qquad\qquad | & \\ & CH_2-CH---CH_2 & \end{array}$$

Tropine (37)

7. Finally, the following experiments clearly support the above structure (37) for tropine.

   (a) Exhaustive methylation of tropine gives tropilidene *i.e.*, cycloheptatriene, C_7H_8.

   (b) Exhaustive methylation of tropinic acid gives an unsaturated dicarboxylic acid, which on reduction forms pimelic acid, a seven carbon acid.

   (c) Tropinone on oxidation with CrO_3–H_2SO_4 gives N-methyl succinimide. This indicates the presence of a pyrrolidine nucleus in tropinone.

All the above reactions can be explained by the structure (37) for tropine.

(i) We can explain the formulation of 2-ethyl pyridine starting from tropine.

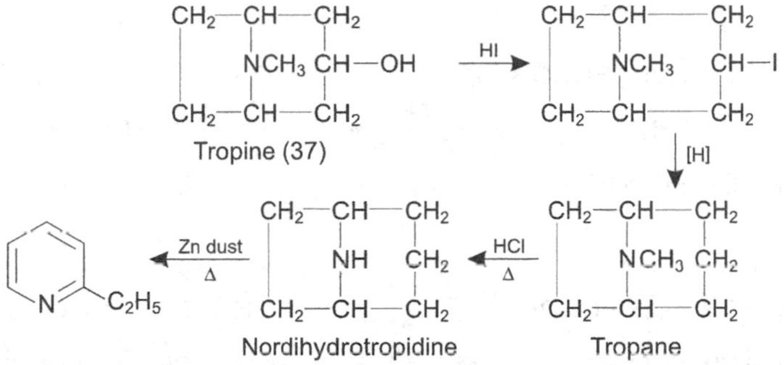

$$
\begin{array}{c}
CH_2-CH-CH_2 \\
| \quad\quad NCH_3 \;\; CH-OH \\
CH_2-CH-CH_2 \\
\text{Tropine (37)}
\end{array}
\xrightarrow{\;HI\;}
\begin{array}{c}
CH_2-CH-CH_2 \\
| \quad\quad NCH_3 \;\; CH-I \\
CH_2-CH-CH_2
\end{array}
\Big\downarrow [H]
$$

$$
\underset{\text{Nordihydrotropidine}}{
\begin{array}{c}
CH_2-CH-CH_2 \\
NH \;\; CH_2 \\
CH_2-CH-CH_2
\end{array}}
\xleftarrow[\Delta]{\text{Zn dust}}
\quad\quad
\xleftarrow[\Delta]{\text{HCl}}
\underset{\text{Tropane}}{
\begin{array}{c}
CH_2-CH-CH_2 \\
NCH_3 \;\; CH_2 \\
CH_2-CH-CH_2
\end{array}}
$$

(2-ethyl pyridine structure: pyridine ring with N and C_2H_5)

(ii) Formation of *Tropinone* and *Tropinic* acid from tropine can be shown in the following manner.

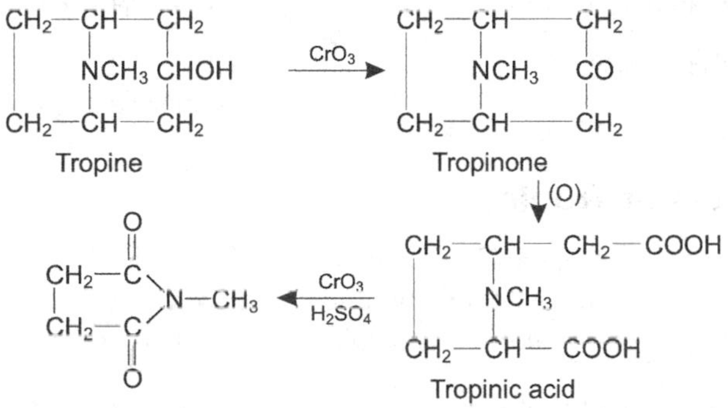

$$
\underset{\text{Tropine}}{
\begin{array}{c}
CH_2-CH-CH_2 \\
NCH_3 \;\; CHOH \\
CH_2-CH-CH_2
\end{array}}
\xrightarrow{CrO_3}
\underset{\text{Tropinone}}{
\begin{array}{c}
CH_2-CH-CH_2 \\
NCH_3 \;\; CO \\
CH_2-CH-CH_2
\end{array}}
\Big\downarrow (O)
$$

$$
\begin{array}{c}
O \\
\| \\
CH_2-C \\
| \quad\quad\quad N-CH_3 \\
CH_2-C \\
\| \\
O
\end{array}
\xleftarrow[H_2SO_4]{CrO_3}
\underset{\text{Tropinic acid}}{
\begin{array}{c}
CH_2-CH-CH_2-COOH \\
NCH_3 \\
CH_2-CH-COOH
\end{array}}
$$

(iii) As given above, exhaustive methylation of tropine forms tropilidene (cycloheptatriene).

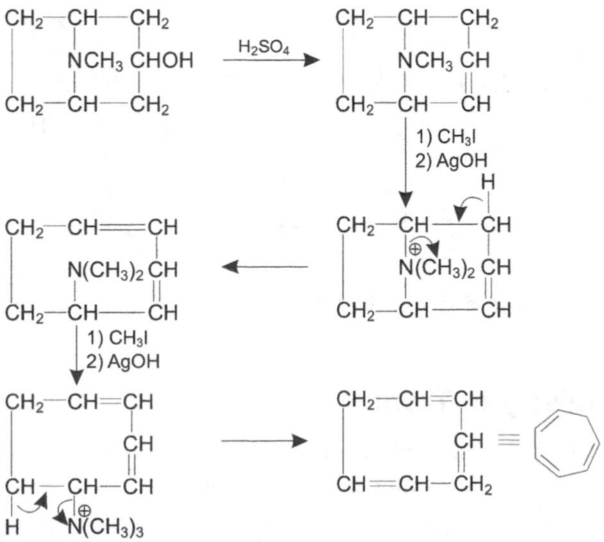

$$
\begin{array}{c}
CH_2-CH-CH_2 \\
NCH_3 \;\; CHOH \\
CH_2-CH-CH_2
\end{array}
\xrightarrow{H_2SO_4}
\begin{array}{c}
CH_2-CH-CH_2 \\
NCH_3 \;\; CH \\
\quad\quad\quad\quad \| \\
CH_2-CH-CH
\end{array}
$$

$$
\Big\downarrow
\begin{array}{l}
1)\ CH_3I \\
2)\ AgOH
\end{array}
$$

$$
\begin{array}{c}
CH_2-CH=CH \\
N(CH_3)_2\ CH \\
\quad\quad\quad\quad \| \\
CH_2-CH-CH
\end{array}
\longleftarrow
\begin{array}{c}
CH_2-CH-CH \\
N(CH_3)_2\ CH \\
\quad\quad\quad\quad \| \\
CH_2-CH-CH
\end{array}
$$

$$
\Big\downarrow
\begin{array}{l}
1)\ CH_3I \\
2)\ AgOH
\end{array}
$$

$$
\begin{array}{c}
CH_2-CH=CH \\
CH \\
\| \\
CH-CH-CH \\
H \quad N(CH_3)_3
\end{array}
\longrightarrow
\begin{array}{c}
CH_2-CH=CH \\
CH \\
\| \\
CH=CH-CH_2
\end{array}
=
\text{(cycloheptatriene ring)}
$$

(iv) Formation of pimelic acid from tropinic acid.

We get pimelic acid (38) from tropinic acid in the following way:

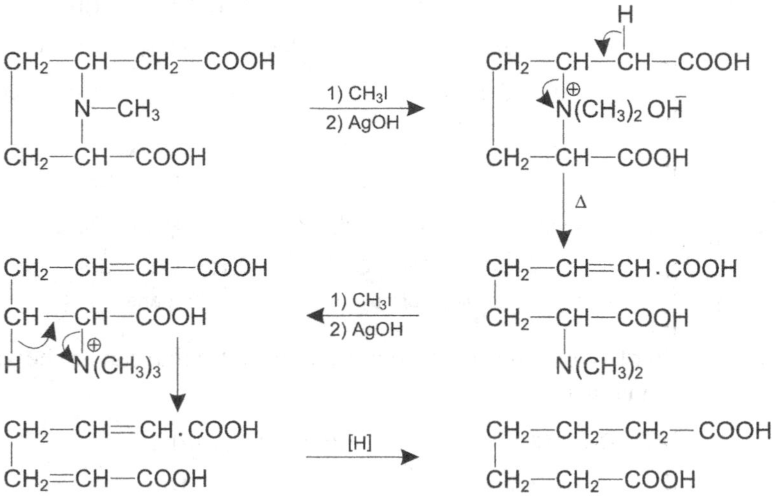

$$
\begin{array}{c}
\text{CH}_2\text{—CH—CH}_2\text{—COOH} \\
| \\
\text{N—CH}_3 \\
| \\
\text{CH}_2\text{—CH—COOH}
\end{array}
\quad
\xrightarrow[\text{2) AgOH}]{\text{1) CH}_3\text{I}}
\quad
\begin{array}{c}
\overset{\displaystyle H}{\text{CH}_2\text{—CH—CH—COOH}} \\
| \\
\overset{\oplus}{\text{N}}(\text{CH}_3)_2\ \text{OH}^- \\
| \\
\text{CH}_2\text{—CH—COOH}
\end{array}
$$

$$\downarrow \Delta$$

$$
\begin{array}{c}
\text{CH}_2\text{—CH}=\text{CH—COOH} \\
| \\
\text{CH—CH—COOH} \\
|\qquad\ \ | \\
\text{H}\quad\ \overset{\oplus}{\text{N}}(\text{CH}_3)_3
\end{array}
\quad
\xleftarrow[\text{2) AgOH}]{\text{1) CH}_3\text{I}}
\quad
\begin{array}{c}
\text{CH}_2\text{—CH}=\text{CH}\cdot\text{COOH} \\
| \\
\text{CH}_2\text{—CH—COOH} \\
| \\
\text{N}(\text{CH}_3)_2
\end{array}
$$

$$\downarrow$$

$$
\begin{array}{c}
\text{CH}_2\text{—CH}=\text{CH}\cdot\text{COOH} \\
| \\
\text{CH}_2=\text{CH—COOH}
\end{array}
\quad
\xrightarrow{\text{[H]}}
\quad
\begin{array}{c}
\text{CH}_2\text{—CH}_2\text{—CH}_2\text{—COOH} \\
| \\
\text{CH}_2\text{—CH}_2\text{—COOH} \\
\text{Pimelic acid (38)}
\end{array}
$$

G. Synthesis of Tropine

The above structure of tropine (37) was confirmed by synthesis given by **Robinson**. A mixture of succindialdehyde, methylamine and acetone was allowed to stand in water for 30 minutes; tropinone was obtained in small amounts.

However, when calcium acetone dicarboxylate is used in place of acetone, a better yield of the product is obtained. The calcium salt so produced is converted into tropinone by warming with hydrochloric acid.

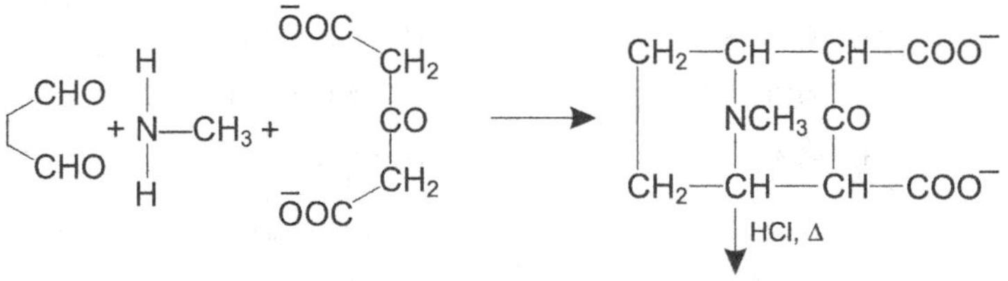

$$
\begin{array}{c}
\text{CHO} \\
\text{CHO}
\end{array}
+
\begin{array}{c}
H \\
| \\
\text{N—CH}_3 \\
| \\
H
\end{array}
+
\begin{array}{c}
\overset{-}{\text{OOC}}\diagdown \\
\qquad\text{CH}_2 \\
\qquad\text{CO} \\
\qquad\text{CH}_2 \\
\overset{-}{\text{OOC}}\diagup
\end{array}
\longrightarrow
\begin{array}{c}
\text{CH}_2\text{—CH—CH—COO}^- \\
|\qquad\ \ | \\
\text{NCH}_3\ \ \text{CO} \\
|\qquad\ \ | \\
\text{CH}_2\text{—CH—CH—COO}^-
\end{array}
$$

$$\downarrow \text{HCl, } \Delta$$

$$\text{Tropine} \xleftarrow{\text{Zn/HCl}} \text{Tropinone}$$
$$\quad(37)\qquad\qquad\qquad\quad(36)$$

Improved yields of the final product are obtained by carrying out the Robinson synthesis at pH7.

Stereochemistry of Tropine

Tropinone can be reduced to give two alcohols tropine and ψ-tropine. They are epimers. Tropine is the *anti* compound whereas ψ-tropine is the *syn* compound.

They have been shown to have the following orientation.

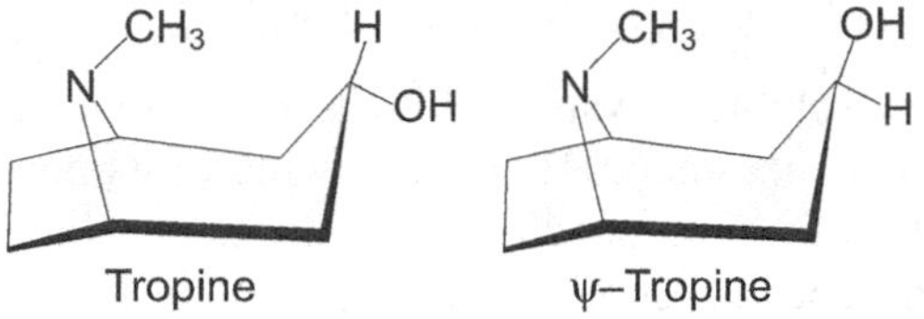

Tropine   ψ–Tropine

H. Final Structure of Atropine

Finally, tropine and tropic acid are combined to give atropine by heating the two in the presence of hydrogen chloride.

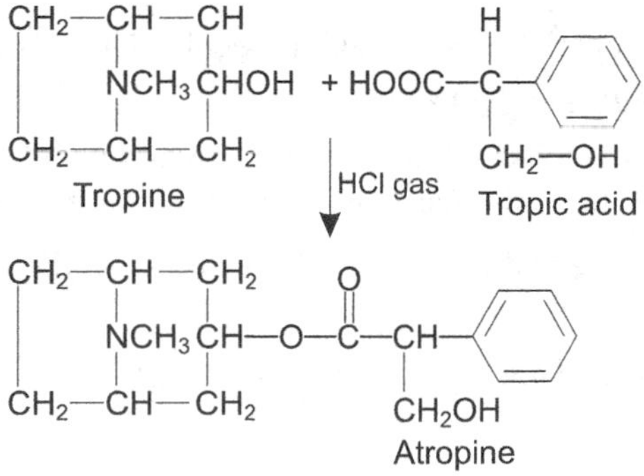

Tropine   HCl gas   Tropic acid

Atropine

If (+) or (–)-tropic acid is used, the product is (+) or (–)-hyoscyamine respectively.

7.13   COCAINE

Cocaine belongs to the coca group of alkaloids. It occurs in the coca leaves (*Erythroxylin coca*) with a number of other alkaloids namely hygrine, benzoylecgonine, tropacocaine *etc.* Of these cocaine is the most important.

The plant is mainly found in South America. Now it is grown in Java and Srilanka also.

A. Isolation of Cocaine

The coca leaves are powdered and extracted with water at 80°C. The solution contains proteins and tannins along with a mixture of alkaloids. The proteinous material is precipitated out with lead acetate. To the solution after filtration dilute sulphuric acid is added. Lead sulphate is filtered off and the alkaloid is precipitated out by adding sodium carbonate. The solid product is again treated with dilute sulphuric acid. The acid solution is evaporated when a large amount of cocaine crystallises out, which can be recrystallised again to obtain pure cocaine.

B. Properties of Cocaine

- Cocaine is a colourless crystalline solid, mp. 98°C.

- It is sparingly soluble in water, but its hydrochloride is quite soluble and is used as a local anaesthetic. However, owing to its poisonous properties, it cannot be used for prolonged anaesthesia. As cocaine is the habit forming drug, its synthetic substitutes are used.

C. Structure of Cocaine

The structure of cocaine has been established as (13) on the basis of the following observations.

1. The molecular formula of cocaine is $C_{17}H_{21}NO_4$.

2. Cocaine is a completely saturated molecule. There is no reaction with Br_2 water or alkaline permanganate.

3. It does not undergo acetylation and benzoylation and uses one molecule of methyl iodide. It shows that N atom is present as a *tertiary* base.

4. It gives a positive hydroxamic test for esters; *i.e.,* the ester group is present in cocaine.

5. Hydrolysis of cocaine with boiling water affords benzoyl ecgonine and methanol.

$$C_{17}H_{21}NO_4 \xrightarrow{H_2O} C_{16}H_{19}NO_4 + CH_3OH$$
$$\text{Cocaine} \qquad\qquad \text{Benzoyl ecgonine}$$

Thus, cocaine contains a carbomethoxyl group and benzoylecgonine has a carboxyl group.

6. When benzoyl ecgonine is heated with barium hydroxide, further hydrolysis takes place and we get benzoic acid and ecgonine (39).

$$C_{16}H_{19}NO_4 + Ba(OH)_2 \longrightarrow C_9H_{15}NO_3 + C_6H_5COOH$$
$$\text{Ecgonine (39)}$$

D. Structure of Ecgonine

1. Its molecular formula has been established as $C_9H_{15}NO_3$.

2. It uses one molecule of methyl iodide indicating that it is a *tertiary* base.

3. Usual tests show that ecgonine (39) contains one –COOH group and one –OH group.

4. Oxidation of ecgonine with chromium trioxide gives tropinone, which on further oxidation gives tropinic acid and ecgoninic acid.

$$C_9H_{15}NO_3 \xrightarrow{CrO_3} C_8H_{13}NO \xrightarrow{CrO_3} C_8H_{13}NO_4 + C_7H_{11}NO_3$$

Ecgonine   Tropinone   Tropinic acid   Ecgoninic acid

From the above reactions, it is clear that ecgonine is a tropane derivative. The alcoholic group is in the same position as in tropine (37).

During oxidation of ecgonine to tropinone (36), a –COOH group is lost and a ketonic group is formed. This means ecgonine is a β-hydroxy acid, which is oxidised to β-keto acid (40). This β-keto acid gets converted to tropinone.

Ecgonine $\xrightarrow{(O)}$ β–Keto acid $\xrightarrow{-CO_2}$ Tropinone

(Unstable)   (36)

5. The similarity in structures of ecgonine with tropine (37) is further proved by the following reactions.

   Ecgonine on dehydration gives anhydroecgonine (41) which decarboxylates to give tropidine. Tropidine (42) is also obtained by the dehydration of tropine. Therefore, ecgonine must have structure (39).

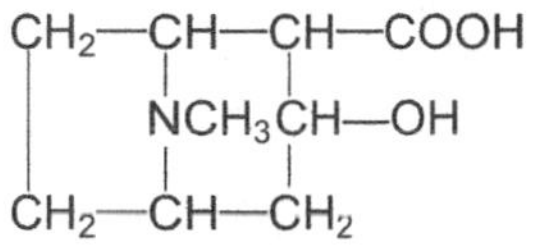

Ecgonine (39)

All the above reactions of ecgonine can thus be shown as follows:

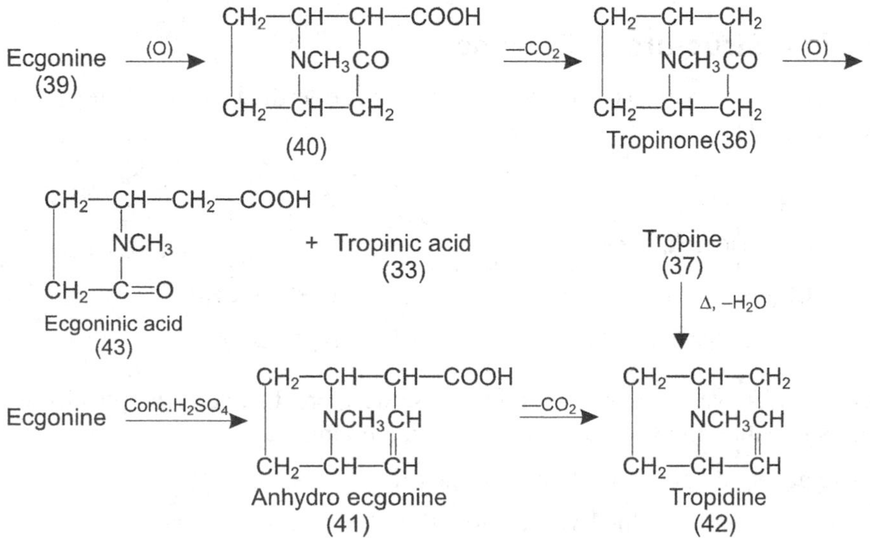

Ecgonine (39) $\xrightarrow{(O)}$ (40) $\xrightarrow{-CO_2}$ Tropinone(36) $\xrightarrow{(O)}$

Ecgoninic acid (43)   + Tropinic acid (33)   Tropine (37) $\Big\downarrow$ Δ, –H₂O

Ecgonine $\xrightarrow{Conc.H_2SO_4}$ Anhydro ecgonine (41) $\xrightarrow{-CO_2}$ Tropidine (42)

E. Synthesis of Ecgonine

The above structure of ecgonine has been confirmed by its synthesis.

1. The starting material for the synthesis of ecgonine is tropinone.

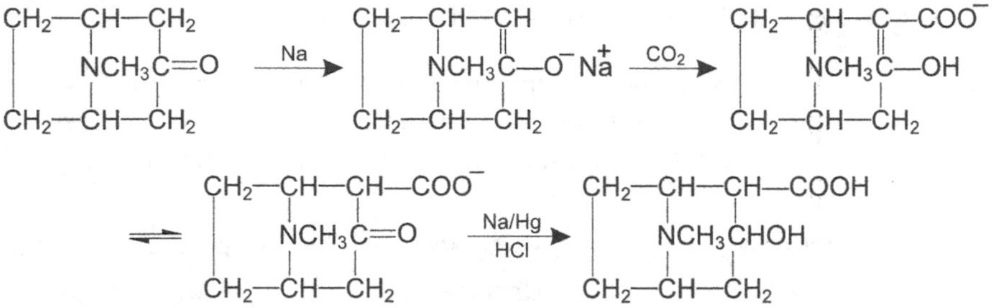

(±)-Ecgonine

2. **Willstatter** *et al.* synthesised it by using Robinson's method as given below

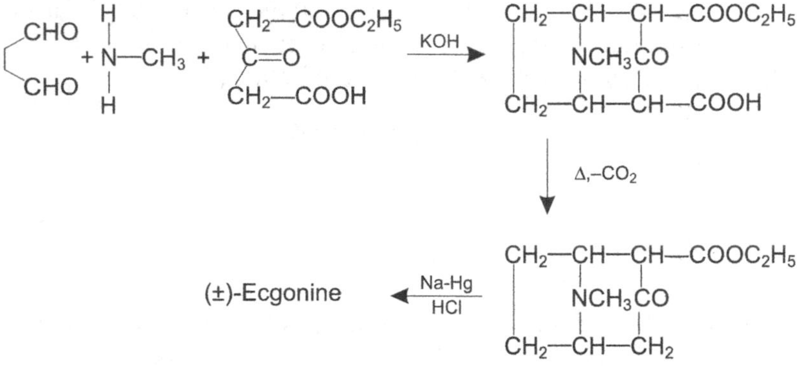

$\Delta, -CO_2$

(±)-Ecgonine $\xleftarrow[\text{HCl}]{\text{Na-Hg}}$

F. Final Structure of Cocaine

The racemic ecgonine is resolved and (–) form is esterified with methanol and then benzoylated to obtain (–)-cocaine.

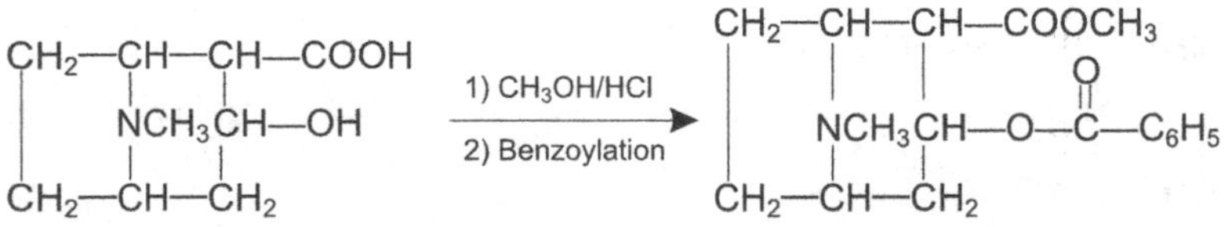

1) CH_3OH/HCl
2) Benzoylation

(–)-Ecgonine

(–)-Cocaine

As shown above, there are 4 chiral centres; therefore, there are 16 optically active forms possible. But C_1 and C_5 have the same configuration.

There are eight optically active forms possible.

Conformation of cocaine has been established by **Fodor** *et al.* as follows:

CH₃ structure diagram

(−)-Cocaine

G. Medicinal Importance of Cocaine

We know cocaine is one of the important coca alkaloids. Cocaine is a powerful nervous system stimulant. Its effects can last from 15–30 minutes to an hour depending upon the method of ingestion.

Cocaine increases alertness, feelings of well-being and euphoria, energy and motor activity. Feelings of competence and sexuality are common. In sports, where sustained attention and endurance are required, athletic performance may be enhanced with its use.

However, cocaine is ranked both the second most addictive and second most harmful of twenty popular recreational drugs. With its use anxiety, paranoia and restlessness are frequent. With excessive dosage tremors, convulsions and increased body temperature are observed. With excessive or prolonged use the drug can cause itching, tachycardia, hallucinations and paranoid delusions. Overdoses can cause a marked elevation of blood pressure which can be life threatening. The symptoms of insatiable hunger, aches, insomnia/oversleeping, lethargy and persistent runny nose are associated with its usage. Depression with suicidal tendencies may develop in very heavy users. All these effects contribute to a rise in tolerance thus requiring a larger dosage to achieve the same effect.

Physical side effects from chronic smoking of cocaine include hemoptysis, bronchospasm, fever, chest pain, lung trauma, sore throat, asthma, hoarse voice, shortness of breath, itching and a flu like syndrome. Cocaine often causes involuntary tooth grinding, which can deteriorate tooth enamel and lead to gingivitis. Smoking of cocaine may cause dehydration and dry mouth. Since saliva is an important mechanism in maintaining one's oral pH level, heavy abusers of cocaine may experience demineralisation of their teeth due to the pH of the tooth surface dropping too low.

Cocaine may also greatly increase the risk of developing rare autoimmune or connective tissue diseases. It can also cause kidney diseases and renal failure. It increases the risk of myocardial infarction.

Occasional cocaine use does not typically lead to severe or even minor physical or social problems. Physical withdrawal is not dangerous and is in fact restorative.

Physiological changes caused by cocaine withdrawal include vivid and unpleasant dreams, insomnia, increased appetite and psychomotor retardation or agitation. In small doses cocaine decreases fatigue, increases mental activity and gives a general feeling of well-being.

Besides being an extremely addictive and dangerous narcotic, cocaine has some medicinal use. Cocaine has been used as a local anaesthetic for eye and nasal surgery.

Besides being an anaesthetic it also constricts the blood vessels around the area where injected. This vasoconstriction helps reduce bleeding and the systemic circulation of cocaine into the heart.

Most recently, the use of cocaine as an anaesthetic for eye and nasal surgery has decreased. It has been replaced by other local anaesthetics. However, these newer local anaesthetics have to be combined with a vasoconstrictor in order to have the same effect as cocaine does.

7.14   QUININE

A. Introduction

Quinine belongs to the quinoline group of alkaloids, which are commonly known as cinchona alkaloids. The cinchona was known to possess some antifebrile properties. This was known to mankind in the 16th century. The infusions of cinchona bark were effective in the treatment of some tropical diseases. It was observed later on that the alkaloids isolated from cinchona were quite effective against malaria.

Quinine is one of the alkaloids present in the cinchona bark. The other important alkaloid present in the crude mixture is cinchonine. The amount of quinine present in the cinchona bark is about 5 percent. Quinine was isolated from cinchona bark and its use as an effective antimalarial started long ago.

Quinine is very slightly soluble in water. It has an intensely bitter taste and is laevorotatory.

Quinine is a derivative of 6-methoxyquinoline with a basic side chain at the 4-position. Its structure has been established on the basis of degradation products and confirmed by its synthesis.

B. Structure of Quinine

1. The molecular formula of quinine is $C_{20}H_{24}O_2N_2$.
2. It adds on two molecules of methyl iodide to form a diquaternary salt. It is, therefore, a ditertiary base.

3. When heated with hydrochloric acid, quinine eliminates one mole of CH_3I. This means that there is one $-OCH_3$ group present in quinine.

4. Oxidation of quinine with CrO_3 gives quininone.

$$C_{20}H_{24}N_2O_2 \xrightarrow{\text{CrO}_3} C_{20}H_{22}N_2O_2$$

$$\text{Quinine} \qquad\qquad \text{Quininone}$$

Quinine forms a monoacetate and a monobenzoate. That indicates the presence of one hydroxyl group.

5. Quinine contains one double bond because it adds on 1 molecule of bromine.

6. That the hydroxyl group is a *secondary* alcoholic group is shown by the fact that oxidation of quinine with CrO_3 gives a quinone.

Controlled oxidation of quinine with chromic acid produces among other products, quininic acid.

$$C_{20}H_{24}N_2O_2 \xrightarrow[\text{H}_2\text{SO}_4]{\text{CrO}_3} C_{11}H_9NO_3$$

$$\text{Quinine} \qquad\qquad \text{Quininic acid}$$

7. When heated with sodalime quininic acid is decarboxylated to a methoxy-quinoline.

Further on oxidation with chromic acid quininic acid forms pyridine –2, 3, 4–tricarboxylic acid. These two reactions suggest that the methoxyl group must be present in the benzene ring and the carboxyl group at position 4.

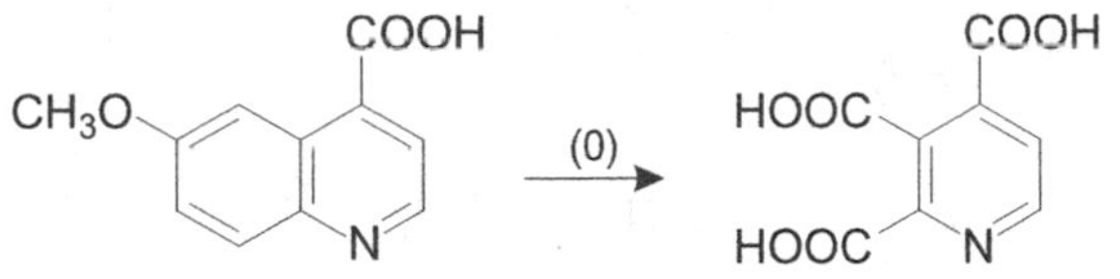

Quininic acid   Pyridine-2,3,4,-
                         tricarboxylic acid

8. The position of the methoxyl group was ascertained by heating quininic acid with HCl and decarboxylating the product. 6-Hydroxy quinoline was obtained, which is a known product. Thus, quininic acid is 6-methoxycinchoninic acid.

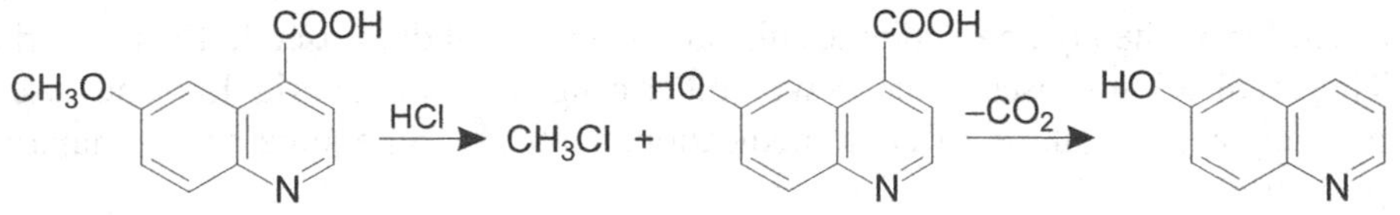

Quininic acid   6-Hydroxy-
                                                 quinoline

9. The above structure of quininic acid has been confirmed by synthesis.

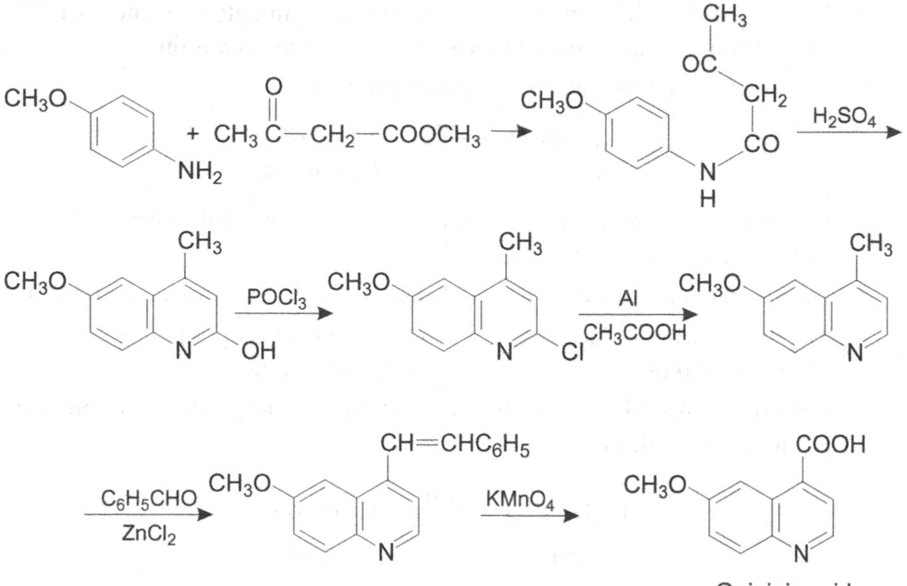

On the basis of the above evidence, the quininic acid structure of quinine is given below which was finally confirmed by its synthesis.

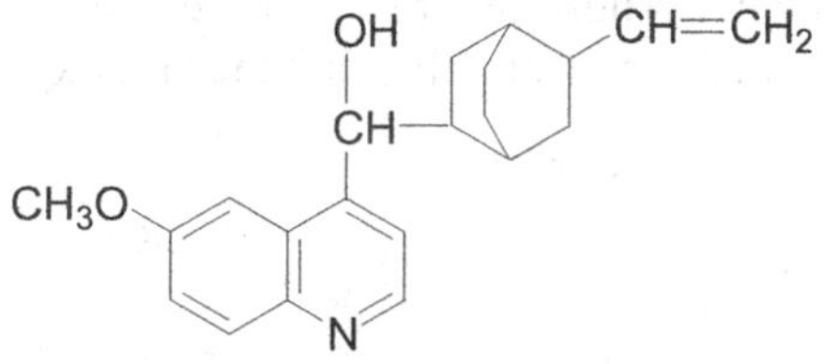

Quinine

C. Medicinal Uses of Quinine

Quinine has been in use for centuries in the prevention and treatment of malaria. Quinine is also a mild antipyretic and analgesic and has been used in common cold preparations for that purpose.

Quinine sulphate is a very common salt of quinine and is ordinarily the 'quinine' asked for by the layman. Quinine sulphate is the actual drug used to treat malaria. This medication is used alone or with other drugs to treat malaria. In some cases, one may need to take a different medication, such as 'primaquine' to kill malarial parasites.

Quinine sulphate can be used in the form of tablets or preferably as capsules for masking the bitter taste of quinine. When quinine cannot be taken by mouth its dihydrochloride may be given by slow intravenous injection because it is very soluble in water.

Quinine is mainly used in the treatment of *Plasmodium falciparum* malaria and *Plasmodium vivax* but is not effective against *P. malariae*.

As of 2006, it is no longer recommended by WHO as a first time treatment for malaria and it should be used only when alternate medicines are not available. The use of quinine has declined considerably and nowadays it is no more the drug of choice as many toxic reactions have been found to be associated with it. Frequent reactions are allergic skin reactions, tinnitus, slight deafness, vertigo, nausea, vomiting, diarrhoea, abdominal pain, hypertension and respiratory depression. Quinine can cause unpredictable serious and life threatening blood and cardiovascular reactions including low platelet count and haemolytic uremic syndrome.

7.15 MORPHINE

A. Introduction and Isolation

Morphine is the main alkaloid of opium. It belongs to the phenanthrene group of alkaloids. It is the principal alkaloid of opium, present as 8–19% by dry weight. Opium also contains another important alkaloid, papaverine (16) which belongs to the isoquinoline group. The primary source of morphine is the poppy plant.

In 2013 about 523 tons of morphine were produced. 70 percent of the product is used to make other opioids such as hydromorphone, oxymorphone and heroin.

As these compounds produce narcosis along with analgesia, they are generally called **narcotic analgesics**.

It is on the WHO list of essential drugs. Morphine is indicated for the relief of severe acute and severe chronic pain.

Morphine is isolated from the poppy plant. The crushed plant is extracted with dilute sulphuric acid 6–10 times. From the solution alkaloids are precipitated by either NH_4OH or Na_2CO_3.

Morphine is purified from other opium alkaloids using the usual physical methods.

B. Structure of Morphine

The structure of morphine as (17) was established on the basis of extensive degradative and oxidative studies.

1. Morphine has the molecular formula $C_{17}H_{19}NO_3$.
2. Nitrogen present was found to be in the *tertiary* state.
3. Morphine forms a diacetate and a dibenzoate. It means two hydroxyl groups are present in the molecule.
4. Morphine dissolves in aqueous sodium hydroxide to form a monosodium salt and it is reconverted into morphine by the action of carbon dioxide. Moreover, morphine gives a ferric chloride test for phenols.

Thus one of the hydroxyl groups is phenolic. The second hydroxyl group is *secondary* alcoholic as is shown by the following reactions.

5. Halogen acids convert morphine into a monohalogeno derivative–one hydroxyl group is replaced by a halogen atom. When heated with methyl iodide in the presence of aqueous potassium hydroxide, morphine is methylated to give (–) codeine, $C_{18}H_{21}NO_3$.

$$C_{17}H_{19}NO_3 \xrightarrow{\text{HCl}} C_{17}H_{18}NO_2Cl \xrightarrow[\text{aq.KOH}]{\text{CH}_3\text{I}} C_{18}H_{21}NO_3$$
$$\text{Morphine} \qquad\qquad\qquad\qquad\qquad\qquad \text{Codeine}$$

Since codeine is not soluble in alkalis, therefore, it follows that it is only the phenolic hydroxyl group in morphine that has been methylated.

6. Codeine can be oxidised by chromic acid to codeinone, a ketone.

So the hydroxyl group in codeine is a *secondary* alcoholic group. Thus, codeine is the monomethyl ether of morphine. Also codeine absorbs one molecule of hydrogen on catalytic reduction. Therefore, both codeine and morphine contain one double bond.

The above observations suggest the following formulae for morphine and codeine

$$C_6H_{16}NO \begin{cases} -\text{OH} \\ -\text{CHOH} \end{cases} \qquad\qquad C_{16}H_{16}NO \begin{cases} -\text{OCH}_3 \\ -\text{CHOH} \end{cases}$$

Morphine   Codeine

7. The unreactivity of the third oxygen atom suggests that it is probably of the ether type.

When heated with HCl at 140°C, both morphine and codeine give apomorphine ($C_{17}H_{17}NO_2$) with the loss of one molecule of water.

The presence of a cyclic *tertiary* base is supported by the fact that codeine when subjected to exhaustive methylation, produces α-codeimethine which contains one more CH_2 than codeine itself and nitrogen is not lost.

This observation can be readily explained only if codeine contains a *tertiary* cyclic base system

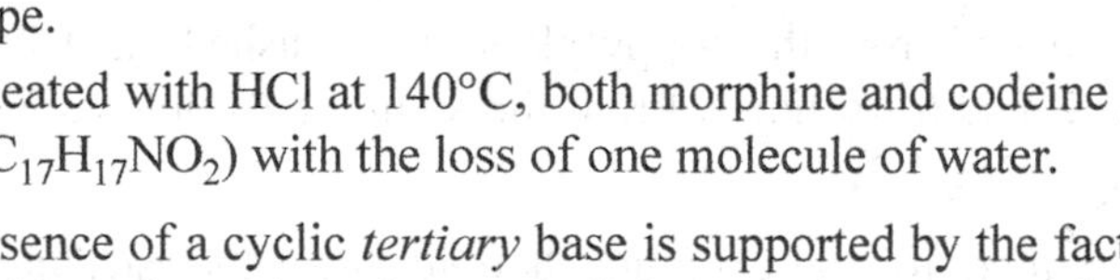

8. When morphine is distilled with zinc dust, phenanthrene and a number of bases are produced. This suggests that a phenanthrene nucleus is probably present.

This has been confirmed by the following observations:

When codeine methiodide is boiled with sodium hydroxide solution, α-methylmorphimethine is obtained. This on heating with acetic anhydride forms methylmorphol and ethanoldimethyl amine.

$$C_{18}H_{21}NO_3 \xrightarrow{\text{NaOH}} C_{19}H_{23}NO_3 \xrightarrow{(CH_3)CO/_2O} C_{15}H_{12}O_2$$

Codeine methiodide	α-Methylmorphi-methine	Methylmorphol

+

$(CH_3)_2NCH_2 \quad CH_2OH$

Ethanoldimethylamine

9. Structure of Methylmorphol

The structure of methylmorphol was ascertained by heating it with HCl at 180°C under pressure; methyl chloride and dihydroxy phenanthrene were obtained.

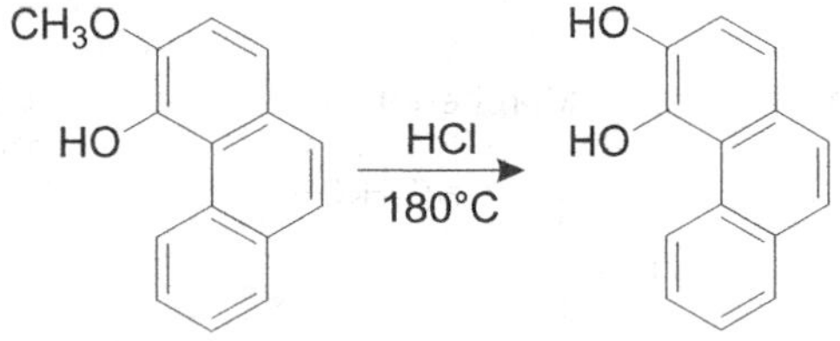

Methylmorphol   Morphol
(3,4-Dihydroxyphenanthrene)

Oxidation of diacetylmorphol gives a diacetyl phenanthraquinone; thus positions 9 and 10 are free. On further oxidation, the quinone is converted into phthalic acid. Therefore, the two hydroxyl groups are in the same ring and at *ortho* positions.

Finally **Pschorr** *et al.*, showed by synthesis that dimethylmorphol is 3,4-dimethoxyphenanthrene.

All the above reactions can be shown in the following way:

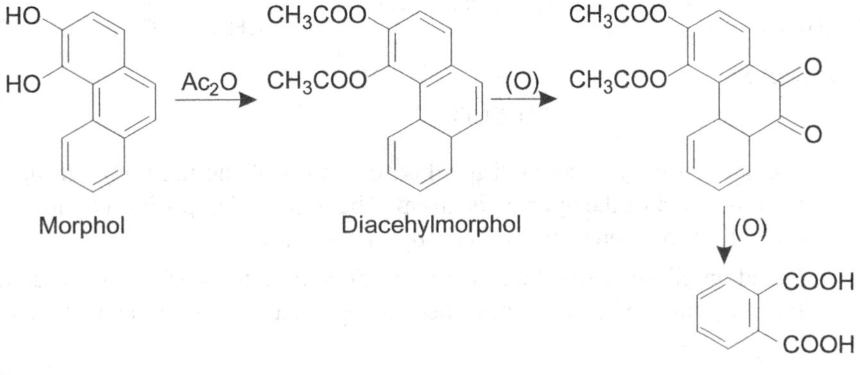

Morphol   Diacehylmorphol

Phthalic acid

10. When β-methylmorphimethine is heated with water the products are trimethylamine, ethylene and methylmorphenol. Demethylation of this with hydrochloric acid produces morphenol.

On fusion with KOH morphenol gives 3,4,5-trihydroxyphenanthrene. The synthesis of 3,4,5-trimethoxy phenanthrene obtained from morphenol by methylating confirms its structure. Reduction of morphenol with sodium and ethanol gives morphol.

The above results can be explained by assuming that morphenol has an ether linkage at positions 4, 5.

We outline below all the above reactions.

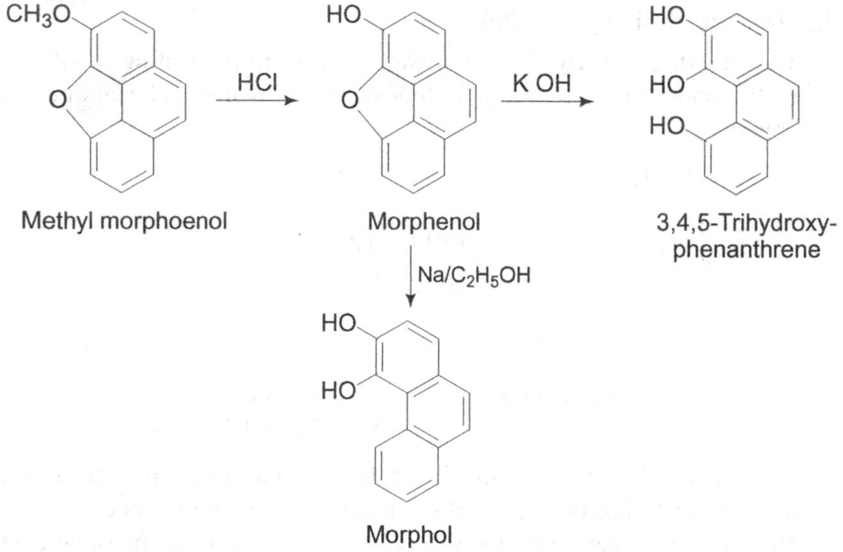

| Methyl morphoenol | Morphenol | 3,4,5-Trihydroxy-phenanthrene |

Morphol

Codeinone on heating with acetic anhydride gives ethanolmethylamine and the diacetyl derivative of 4,6,-dihydroxy-3-methoxy phenanthrene.

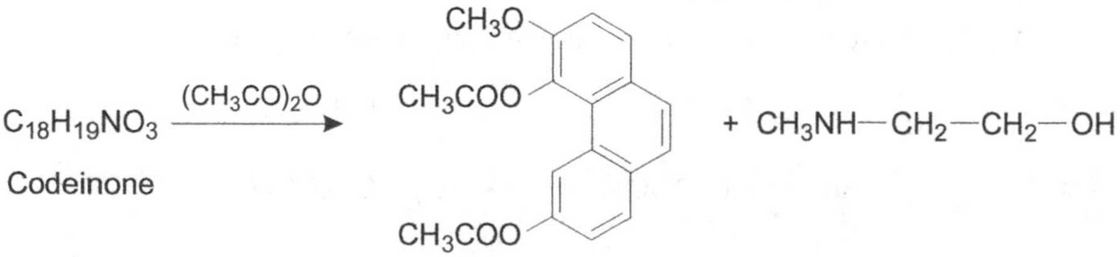

$$C_{18}H_{19}NO_3 \xrightarrow{(CH_3CO)_2O}$$

Codeinone

$+ \; CH_3NH-CH_2-CH_2-OH$

We have already accounted for the position 3 of the methoxyl group and the position 4 of the hydroxyls group. The hydroxyl at position 6 must have come from oxygen of the keto group in codeinone.

Based on all the above facts and work done by later workers the structures for morphine and codeine have been assigned as given on the next page.

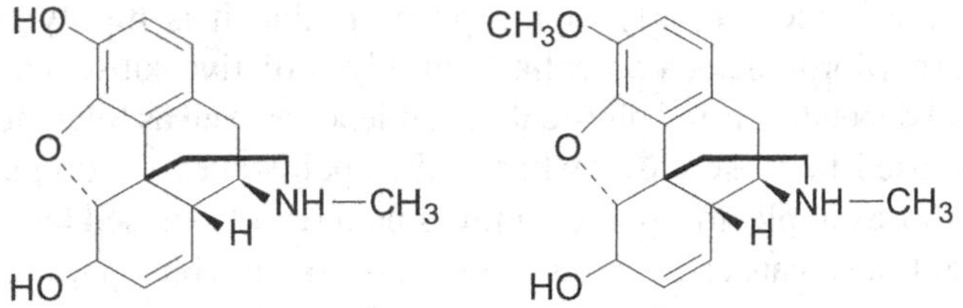

Morphine   Codeine

The structures of other products, apomorphine and codeinone are shown below:

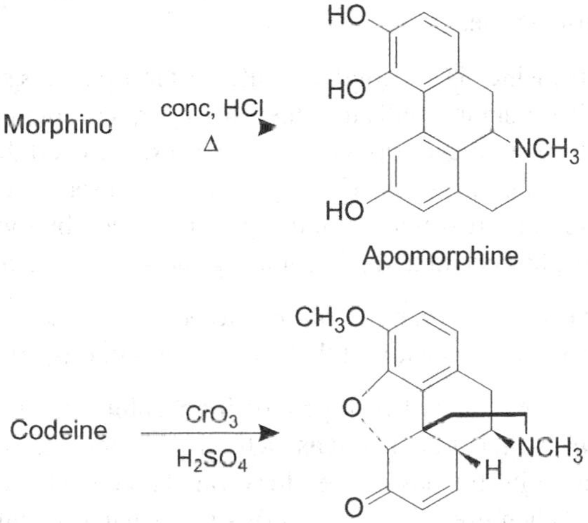

Morphine $\xrightarrow[\Delta]{\text{conc, HCl}}$

Apomorphine

Codeine $\xrightarrow[H_2SO_4]{CrO_3}$

Codeinone

As mentioned earlier morphine when heated with concentrated HCl undergoes rearrangement to form apomorphine. The details of the mechanism are uncertain.

C. Medicinal Uses and Side Effects

Morphine is used as morphine hydrochloride or as morphine sulphate. Both the salts are white crystalline solids with bitter taste and soluble in water. Morphine is a powerful analgesic and narcotic in nature but it is still used extensively to treat both acute and chronic severe pain. It is also used for pain due to myocardial infarction and labour pain. It depresses the central nervous system so that there is increased tolerance to pain. It is also used to check peristalsis and for suppression of cough and relief of anxiety.

Morphine is also beneficial in reducing the symptoms of shortness of breath due to both cancer and non-cancer causes.

It is more effective when administered by injection than when given by mouth. Its duration of analgesia is about 20 minutes when given intravenously and 60 minutes when given by mouth.

The chief undesirable property of morphine is that it is highly addictive and prone to abuse. Morphine is a potentially highly addictive substance. It can cause psychological dependence and physical dependence as well as tolerance. Therefore, it should be used in those cases where other pain relieving drugs prove to be inadequate. For example, morphine controls pain readily caused by serious injury, migraine *etc*. It also causes respiratory depression and constipation. Serious side effects include decreased respiratory effort and low blood pressure. Common side effects are drowsiness, vomiting, nausea, dizziness and sweating. Caution is advised when used during pregnancy or breast feeding as morphine may affect the baby.

Morphine may cause withdrawal reactions if it has been used for a long time. Common withdrawal symptoms are restlessness, watering eyes, runny nose, nausea, sweating and muscle ache.

The use of codeine, which is a methyl derivative of morphine is still there. Codeine occurs naturally in opium in small amounts. However, it is obtained from morphine by methylating the phenolic hydroxyl group. It is a white crystalline powder, bitter in taste, slightly soluble in water. Usually codeine is used as codeine phosphate which is fairly soluble in water. Codeine phosphate is obtained by neutralising codeine with phosphoric acid and precipitating the salt with alcohol.

Codeine is much less potent than morphine. However, it is the only natural product which is used in combination with synthetic analgesics like aspirin and paracetamol.

The acetylation of both the hydroxyl groups in morphine gives diamorphine, also known as "*heroin*", which is also used as its hydrochloride. Heroin is a more potent analgesic than morphine but has shorter duration of action. However, heroin is also highly addictive. When used illicitly, a very serious narcotic habit can develop in a matter of weeks.

7.16   MEDICINAL IMPORTANCE OF RESERPINE

Reserpine is the most important alkaloid of the **Rauwolfia** species. It is present in *rauwolfia serpentina* which is known as snake root. Its structure has been established as given below:

Reserpine

Reserpine is used to treat high blood pressure. It is also used to treat severe agitation in patients with mental disorders and insomnia. It works by slowing the activity of the nervous system, causing the heartbeat to slow and the blood vessels to relax. Thus, reserpine lowers the blood pressure and also acts as a tranquilizer.

Reserpine controls high blood pressure or symptoms of agitation, but does not cure them. It should be continued even if one feels well. Reserpine should not be stopped without consulting the doctor. If one suddenly stops taking it, one may develop high blood pressure again and experience unwanted side effects.

Reserpine comes as a tablet to be taken by mouth. It is usually taken once daily and it should be taken at around the same time. Alcohol should not be consumed while taking reserpine. It can make the side effects from reserpine worse. Now it is rarely used in the management of hypertension. However, in some countries reserpine is still available as part of a combination of drugs for the treatment of hypertension; in most cases they contain a diuretic. These combinations are currently regarded as second choice drugs.

Reserpine is also used to treat symptoms of dyskinesia in patients suffering from **Huntington's** disease. Reserpine may be used as a sedative for horses. Older adults should not usually take high doses of reserpine because it is not as safe as other medications that can be used to treat the same condition.

Index

9781032337951